Advances in Intelligent Systems and Computing

Volume 1227

The series "Advances in Intelligent Systems and Computing" contains publications on theory, applications, and design methods of Intelligent Systems and Intelligent Computing. Virtually all disciplines such as engineering, natural sciences, computer and information science, ICT, economics, business, e-commerce, environment, healthcare, life science are covered. The list of topics spans all the areas of modern intelligent systems and computing such as: computational intelligence, soft computing including neural networks, fuzzy systems, evolutionary computing and the fusion of these paradigms, social intelligence, ambient intelligence, computational neuroscience, artificial life, virtual worlds and society, cognitive science and systems, Perception and Vision, DNA and immune based systems, self-organizing and adaptive systems, e-Learning and teaching, human-centered and human-centric computing, recommender systems, intelligent control, robotics and mechatronics including human-machine teaming, knowledge-based paradigms, learning paradigms, machine ethics, intelligent data analysis, knowledge management, intelligent agents, intelligent decision making and support, intelligent network security, trust management, interactive entertainment, Web intelligence and multimedia.

The publications within "Advances in Intelligent Systems and Computing" are primarily proceedings of important conferences, symposia and congresses. They cover significant recent developments in the field, both of a foundational and applicable character. An important characteristic feature of the series is the short publication time and world-wide distribution. This permits a rapid and broad dissemination of research results.

**** Indexing: The books of this series are submitted to ISI Proceedings, EI-Compendex, DBLP, SCOPUS, Google Scholar and Springerlink ****

More information about this series at http://www.springer.com/series/11156

Vijendra Singh · Vijayan K. Asari ·
Sanjay Kumar · R. B. Patel
Editors

Computational Methods and Data Engineering

Proceedings of ICMDE 2020, Volume 1

Springer

Editors
Vijendra Singh
School of Computer Science
University of Petroleum and Energy Studies
Dehradun, Uttarakhand, India

Sanjay Kumar
Department of Computer Science
and Engineering
SRM University Delhi-NCR
Sonepat, Haryana, India

Vijayan K. Asari
Department of Electrical and Computer
Engineering
University of Dayton
Dayton, OH, USA

R. B. Patel
Department of Computer Science
and Engineering
Chandigarh College of Engineering
and Technology
Chandigarh, Punjab, India

ISSN 2194-5357 ISSN 2194-5365 (electronic)
Advances in Intelligent Systems and Computing
ISBN 978-981-15-6875-6 ISBN 978-981-15-6876-3 (eBook)
https://doi.org/10.1007/978-981-15-6876-3

This Springer imprint is published by the registered company Springer Nature Singapore Pte Ltd.
The registered company address is: 152 Beach Road, #21-01/04 Gateway East, Singapore 189721, Singapore

Preface

We are pleased to present Springer Book entitled *Computational Methods and Data Engineering*, which consists of the Proceedings of the International Conference on Computational Methods and Data Engineering (ICMDE 2020), Volume 1 papers.

The main aim of the International Conference on Computational Methods and Data Engineering (ICMDE 2020) was to provide a platform for researchers and academia in the area of computational methods and data engineering to exchange research ideas, results and collaborate together. The conference was held at the SRM University, Sonepat, Haryana, Delhi-NCR, India, from January 30 to 31, 2020.

All the 49 published chapters in the Computational Methods and Data Engineering book have been peer reviewed by three reviewers drawn from the scientific committee, external reviewers and editorial board depending on the subject matter of the chapter. After the rigorous peer-review process, the submitted papers were selected based on originality, significance and clarity and published as chapters.

We would like to express our gratitude to the management, faculty members and other staff of the SRM University, Sonepat, for their kind support during the organization of this event. We would like to thank all the authors, presenters and delegates for their valuable contribution in making this an extraordinary event.

We would like to acknowledge all the members of honorary advisory chairs, international/national advisory committee members, general chairs, program chairs, organization committee members, keynote speakers, the members of the technical committees and reviewers for their work.

Finally, we thank series editors, Advances in Intelligent Systems and Computing, Aninda Bose and Radhakrishnan for their high support and help.

Editors

Dehradun, India — Vijendra Singh
Dayton, USA — Vijayan K. Asari
Sonepat, India — Sanjay Kumar
Chandigarh, India — R. B. Patel

Contents

About the Editors

Dr. Vijendra Singh is working as Professor in the School of Computer Science at The University of Petroleum and Energy Studies, Dehradun, Uttarakhand, India. Prior to joining the UPES, he worked with the NCU, Delhi-NCR, India, Mody University, Lakshmangarh, India, and Asian CERC Information Technology Ltd. Dr. Singh received his Ph.D. degree in Engineering and M.Tech. degree in Computer Science and Engineering from Birla Institute of Technology, Mesra, India. He has 20 years of experience in research and teaching including IT industry. Dr. Singh major research concentration has been in the areas of data mining, pattern recognition, image processing, big data, machine learning, and soft computation. He has published more than 65 scientific papers in this domain. He has served as Editor-in-Chief, Special Issue, Procedia Computer Science, Vol 167, 2020, Elsevier; Editor-in-Chief, Special Issue, Procedia Computer Science, Vol 132, 2018, Elsevier; Associate Editor, International Journal of Healthcare Information Systems and Informatics, IGI Global, USA; Guest Editor, Intelligent Data Mining and Machine Learning, International Journal of Healthcare Information Systems and Informatics, IGI Global, USA; Editor-in-Chief, International Journal of Social Computing and Cyber-Physical Systems, Inderscience, UK; Editorial Board Member, International Journal of Multivariate Data Analysis, Inderscience, UK; Editorial Board Member, International Journal of Information and Decision Sciences, Inderscience, UK.

Dr. Vijayan K. Asari is a Professor in Electrical and Computer Engineering and Ohio Research Scholars Endowed Chair in Wide Area Surveillance at the University of Dayton, Dayton, Ohio. He is the Director of the University of Dayton Vision Lab (Center of Excellence for Computer Vision and Wide Area Surveillance Research). Dr. Asari had been a Professor in Electrical and Computer Engineering at Old Dominion University, Norfolk, Virginia, till January 2010. He was the Founding Director of the Computational Intelligence and Machine Vision Laboratory (ODU Vision Lab) at ODU. Dr. Asari received the bachelor's degree in Electronics and Communication Engineering from the University of Kerala (College of Engineering, Trivandrum), India, in 1978, the M.Tech. and Ph.D.

degrees in Electrical Engineering from the Indian Institute of Technology, Madras, in 1984 and 1994, respectively. Dr. Asari received several teachings, research, advising, and technical leadership awards. Dr. Asari received the Outstanding Teacher Award from the Department of Electrical and Computer Engineering in April 2002 and the Excellence in Teaching Award from the Frank Batten College of Engineering and Technology in April 2004. Dr. Asari has published more than 480 research papers including 80 peer-reviewed journal papers co-authoring with his graduate students and colleagues in the areas of image processing, computer vision, pattern recognition, machine learning, and high-performance digital system architecture design. Dr. Asari has been a Senior Member of the IEEE since 2001 and is a Senior Member of the Society of Photo-Optical Instrumentation Engineers (SPIE). He is a Member of the IEEE Computational Intelligence Society (CIS), IEEE CIS Intelligent Systems Applications Technical Committee, IEEE Computer Society, IEEE Circuits and Systems Society, Association for Computing Machinery (ACM), and American Society for Engineering Education (ASEE).

Dr. Sanjay Kumar is working as Professor in the Computer Science and Engineering Department, SRM University, India. He received his Ph.D. degree in Computer Science and Engineering from Deenbandhu Chhotu Ram University of Science and Technology (DCRUST), Murthal (Sonipat), in 2014. He obtained his B.Tech. and M.Tech. degrees in Computer Science and Engineering in 1999 and 2005, respectively. He has more than 16 years of academic and administrative experience. He has published more than 15 papers in the international and national journals of repute. He has also presented more than 12 papers in the international and national conferences. His current research area is wireless sensor networks, machine learning, IoT, cloud computing, mobile computing and cyber, and network security. He chaired the sessions in many international conferences like IEEE, Springer, and Taylor & Francis. He is the Life Member of Computer Society of India and Indian Society for Technical Education.

Prof. R. B. Patel is working as Professor in the Department of Computer Science and Engineering, Chandigarh College of Engineering and Technology (CCET), Chandigarh, India. Prior to joining the CCET, he worked as Professor at NIT, Uttarakhand, India, and Dean, Faculty of Information Technology and Computer Science, Deenbandhu Chhotu Ram University of Science and Technology, Murthal, India. His research areas include mobile and distributed computing, machine and deep learning, and wireless sensor networks. Prof. Patel has published more than 150 papers in international journals and conference proceedings. He has supervised 16 Ph.D. scholars and currently 02 are in progress.

Content Recommendation Based on Topic Modeling

Sachin Papneja, Kapil Sharma, and Nitesh Khilwani

Abstract With the proliferation in Internet usage and communicating devices, plenty amount of information is available at user disposal but on other side, it leads to a challenge to provide the fruitful information to end users. To overcome this problem, recommendation system plays a decisive role in providing pragmatic information to end users at appropriate time. This paper proposes a topic modeling based recommendation system to provide contents related to end users interest. Recommendation systems are based on different filtering mechanisms which are classified as content based, collaborative based, knowledge based, utility based and hybrid filtering, etc. The objective of this research is thus to proffer a recommendation system based on topic modeling. Benefit of latent Dirichlet allocation (LDA) is to uncover latent semantic structure from the text documents. By analyzing the contents using topic modeling, system can recommend the right articles to end users based on user interest.

Keywords Recommendation system · LDA · Topic modeling · Content filtering · Collaborative filtering

1 Introduction

In last few years, with the telecom revolution, Internet has become a powerful tool which has changed the way user communicate among themselves as well use it in the professional business. As per year 2018 statistics, there are now more than 4

S. Papneja (✉) · K. Sharma
Department of Computer Science & Engineering, Delhi Technological University, New Delhi, India
e-mail: p.sachin@samsung.com

K. Sharma
e-mail: kapil@ieee.org

N. Khilwani
RoundGlass, New Delhi, India
e-mail: nitesh.nifft@gmail.com

V. Singh et al. (eds.), *Computational Methods and Data Engineering*, Advances in Intelligent Systems and Computing 1227,
https://doi.org/10.1007/978-981-15-6876-3_1

billion people around the world using the Internet whereas around 7.5 billion mobile connections across the globe. As per the assessment, there are close to 1.5 billion Internet sites on the cyberspace today. Out of the total available sites, less than 200 million are operating. As the number of communicating devices increase rapidly, it results in the infinite amount of data in the embodiment of text, images, and videos. Fundamental test is to give the exact data to the user dependent on user intrigue.

Amidst the appearance of Internet network accessibility, user's propensities for understanding news or most recent data have alternated from magazine or booklet to advanced substance. Because of the immense amount of information accessible on the cyberspace, it is extremely awkward for the end user to have the data accessible according to his/her advantage. Recommender Systems aid conquers this issue and gives important data or administrations to the user. Various sorts of suggestion frameworks exist, for example, content based [17], collaborative [13], hybrid [7], utility based, multi-criteria, context-aware, risk-aware based, each having with their impediments. Analysts need to utilize distinctive suggestion frameworks dependent on their exploration territories.

Content-based frameworks attempt to prescribe things like those a given user has enjoyed before. For sure, the essential procedure performed by a content-based recommender comprises in coordinating up the characteristics of a client profile in which inclinations and interests are put away, with the properties of a substance object (thing), so as to prescribe to the client new intriguing things. Content-based recommenders exploit solely ratings. Content-based recommenders are capable of recommending items not yet rated by any user provided by the active user to build her own profile.

Numerous customary news recommender frameworks utilize collective sifting to make suggestions dependent on the conduct of clients in the framework. In this methodology, the presentation of new clients or new things can cause the cold start issue, as there will be lacking information on these new sections for the communitarian separating to draw any deductions for new clients or things. Content-based news recommender frameworks developed to address the cold start issue. In any case, many substance-based news recommender frameworks consider records as a sack of-words disregarding the shrouded subjects of the news stories.

Individuals have consistently been standing up to with a developing measure of information, which thusly requests more on their capacities to channel the substance as indicated by their inclinations. Among the undeniably overpowering measures of website pages, records, pictures, or recordings, it is never again natural to and what we truly need. Besides, copy or a few data sources are discovered covering similar themes. The clients are touchy to the recentness of data and their inclinations are additionally changing after some time alongside the substance of the Web.

During the previous two decades, the ideas of recommender frameworks have risen to cure the circumstance. The quintessence of recommender frameworks are profoundly connected with the broad work in psychological science, guess hypothesis, data recovery, determining speculations, and the board science. The content-based methodology of suggestion has its foundations in data recovery [18], and data separating [13] research. Content-based frameworks are planned for the most part to suggest content-based things; the substance in these frameworks is generally portrayed with keyword. Customized recommender frameworks intend to prescribe applicable things to users dependent on their watched conduct, e.g., search personalization [3], Google News personalization [4], and Yahoo! conduct focusing on [5] among others.

As of late, topic modeling approach, for example, latent Dirichlet allocation (LDA) and probabilistic latent semantic analysis (pLSA) helps in examining content substance includes by uncovering latent topics of each archive from the document archive. Reason for LDA is to reveal the semantic structure covered up in the documents which incorporates the word appropriations over the latent subjects and the inactive point disposal over archives [8]. The principle challenge is the way to suggest explicit articles from an immense measure of recently accessible data to end customer, where the chosen commodity should coordinate the buyer intrigue. In this research work, concocted a recommendation framework dependent on LDA topic modeling. In the recommended framework, LDA topic modeling is used to uncover topic from the document related to user hobbies. Once system know the user interest, based on the information system can recommend the articles related to interest by filtering required information. One of the significant qualities of probabilistic topic modeling is the capacity to uncover shrouded relations through the examination of co-event designs on dyadic perceptions, for example, document-term pairs.

2 Related Researches

The main purpose of Recommender System is to assist users to make accurate decisions without spending too much on searching this vast amount of information. Traditional Recommender System is designed to recommend meaningful items to their users. Those items depend on the purpose of the RS, for example, Google recommends news to people while Facebook recommends people (friends) to people. Recommender Systems are a sub-class of information retrieval systems and designed to predict users' future preferences by analyzing their past interaction with the system. Usage of Recommender System became more common in recent years.

From the last two decades, Recommender Systems have become the topic of interest for both academician and for the industry due to increase in overloaded information and to provide relevant information to end users [1] by filtering out the information. A knowledge-based filtering framework is a data framework intended for unstructured or semi-organized information [5]. Recommender System may anticipate whether a end user would be keen on purchasing a specific item or not. Social recommendation strategies gather assessment of commodity from numerous people, and use nearest neighbor procedures to make proposals to a user concerning new stock [4]. Recommendation system has been largely used in approximation theory [14], cognitive science [16], forecasting theory, management science. In addition to Recommender Systems works on the absolute values of ratings, [9] worked on preference-based filtering, i.e., anticipating the general inclinations of end user.

Xia et al. [19] Suggested content-based recommender framework for E-Commerce Platform and took a shot at streamlines the coupon picking process and customizes the suggestion to improve the active clicking factor and, eventually, the conversion rates. Deng et al. [10] proposed the amalgamation of item rating data that user has given plus consolidated features of item to propose a novel recommendation model.

Bozanta and Kutlu [6] proposed to gathered client visit chronicles, scene related data (separation, class, notoriety and cost) and relevant data (climate, season, date and time of visits) identified with singular client visits from different sources as each current scene suggestion framework calculation has its own disadvantages. Another issue is that basic data about setting is not ordinarily utilized in scene suggestion frameworks. Badriyah et al. [3] utilize proposed framework which suggest property-related data based on the user action via looking through publicizing content recently looked by the user. Topic modeling is based on the experience that document consist of topics whereas topics are congregation of words. Goal of the Topic modeling is to understand the documents by uncovering hidden latent variables which are used to describe the document semantic. Latent Semantic Analysis is based on singular value decomposition (SVD) whereas pLSA is based on probability distribution. LDA is a Bayesian version of pLSA which uses Dirichlet priors for the document-topic and word-topic distributions for better generalization.

Luostarinen and Kohonen [12] Studied and compared LDA with other standard methods such as Naïve Bayes, K-nearest neighbor, regression and regular linear regression and found out that LDA gives significant improvement in cold start simulation.

Apaza et al. [2] use LDA by inferring topics from content given in a college course syllabus for course recommendation to college students from sites such as Coursera, Udacity, Edx, etc. Pyo et al. [15] proposed unified topic model for User Grouping and TV program recommendation by employing two latent Dirichlet allocation (LDA) models. One model is applied on TV users and the other on the viewed TV programs.

3 Background

3.1 Content-Based Recommender Systems

Content-Based (CB) Recommender Systems prescribe things to a user as indicated by the substance of user's past inclinations. As such, framework produces proposals dependent on thing highlights that match with the user profile. The fundamental procedure can be clarified in two primary advances:

1. Framework makes user profile utilizing user past conduct, all the more exactly utilizing thing highlights that has been obtained or loved in the past by the user.
2. At that point, framework creates suggestion by breaking down the qualities of these things and contrasting them and the user profile.

Content-based calculation can be comprehended from its name that this strategy for the most part thinks about thing's substance. Content-based strategy can be effectively utilized in thing proposal; however, it necessitates that the applicable traits of the things can be separated, or at the end of the day it depends on the thing's substance. For instance, on the off chance that framework prescribes archives to its users, at that point the content-based calculation examines reports' words (content). Be that as it may, a few things' highlights cannot be removed effectively, for example, motion pictures and music, or they can be covered because of security issues consequently materialness of these techniques is constrained relying upon the idea of the things.

Probabilistic topic models are a suite of methods whose objective is to detect the concealed topical structure in enormous chronicles of documents.

3.2 Recommender Systems Major Challenges

There are numerous difficulties that recommender framework researchers face today and those difficulties can influence the algorithm outcome. Some of the challenges are as follows:

- **Data sparsity**: Nowadays a great many things are accessible particularly in online business sites and every day this number is expanding. Along these lines, finding comparative user (that purchased comparative things) is getting more enthusiastically. A large portion of the Recommender System calculations are utilizing user/things closeness to create recommenders. Along these lines, due to information sparsity calculations may not perform precisely.
- **Scalability**: Especially, enormous sites have a large number of user and a great many information. In this way, when planning a Recommender System it ought to likewise think about the computational expense.
- **Cold Start**: When new user or information enter the system, system cannot draw any data hence it cannot produce proposals either. One of the most guileless answers for the cold start issue is prescribing well known or stylish things to new users. For instance, in YouTube, when a user has no past video history story it will prescribe the most famous recordings to this user. In any case, when the user watches a video then system will have some clue regarding the client's inclination and afterward it will prescribe comparative recordings to the past video that the client has viewed.
- **Diversity and accuracy**: It is typically viable to prescribe famous things to users. In any case, users can likewise discover those things independent from anyone else without a recommender framework. Recommender framework ought to likewise locate the less famous things however are probably going to be favored by the users to suggest. One answer for this issue is utilizing mixture suggestion techniques.
- **Vulnerability to attacks**: Recommender Systems can be focus of a few assaults attempting to mishandle the Recommender System calculations utilized in the

web-based business sites. Those assaults attempt to trick Recommender System to wrongly propose foreordained things for benefit.

- **The value of time**: Customer needs/inclinations will in general change in time. Be that as it may, most Recommender Systems calculations don't think about time as a parameter.
- **Evaluation of recommendations**: There are a few Recommender System structured with various purposes and measurements proposed to assess the Recommender System. Notwithstanding, how to pick the one that precisely assesses the comparing framework is as yet not clear.

3.3 Probabilistic Topic Modeling

Today there are a large amount of articles, site pages, books and web journals accessible on the web. Besides, every day the measure of content reports are expanding with commitments from informal communities and mechanical improvements. In this way, finding what we are actually searching for is not a simple assignment as it used to be and it tends to be very tedious. For instance, for researchers, there are a million of articles accessible on the web, to locate the related ones is a challenging task for researchers. It is not practical to peruse every content and compose or classify them. Along these lines, it is important to utilize programming devices to sort out them.

For instance, most journals chronicle their issues, putting away every distributed article, and along these lines, they should store a lot of data. Without utilizing computational devices arranging such a major unstructured text assortment is unimaginable by just utilizing human work. In this way, researchers evolve distinctive probabilistic models for subject revelation from an enormous unstructured text corpus and they called them probabilistic topic models.

Probabilistic subject models are calculations intended to find the concealed topic of the article. At the end of the day, they are measurable techniques attempting to find the shrouded topic of each article by breaking down the recurrence of the words. The primary thought behind theme models is a presumption that articles are blends of points (ordinary dispersion) and subjects are typical circulation over words. Topic models are generative models which fundamentally imply that producing a document is considered as a probabilistic procedure. This procedure can be clarified in three fundamental points as pursues:

- Determine an article to be produced.
- Pick topic for every word of the article.
- Draft a word dependent on the topic that has been picked.

Despite the fact that theme models are initially intended to arrange or locate the shrouded subject of unstructured archives, they have been embraced in a wide range of spaces with various sorts of information. For instance, they are used in data retrieval, multimedia retrieval.

Probabilistic Topic Modeling comes under the non-supervised learning [11] in the sense that it does not require antecedent interpretation or document labeling. In probabilistic modeling, information is exuded from a generative procedure that incorporates latent variables. This generative procedure characterizes a joint probability distribution over both the noticed and concealed random variables. It does not make any earlier supposition how the words are showed up in the document yet rather what is important to the model is the occurrence of the word is referenced in the document.

3.4 *Latent Dirichlet Allocation*

Latent Dirichlet Allocation (LDA) is a three-level hierarchical Bayesian model, in which every collected item is demonstrated as a limited blend over a basic arrangement of topics and is utilized to reveal topics in a lot of documents. Every topic is, thus, demonstrated as a limitless blend over a hidden arrangement of topic probabilities. Document is only having some data about the topic while every topic is portrayed by dissemination over words. The LDA model is spoken to as a probabilistic graphical model as shown in Fig. 1.

As it tends to be seen from the diagram that there are three unique degrees of factors and parameters:

- First level is corpus level parameters and they are examined in the first place for example before start producing the corpus.
- Second level is record level factors and they are tested once for producing each archive.
- Third level factors are word-level factors and they are created for each expression of all records in the corpus.

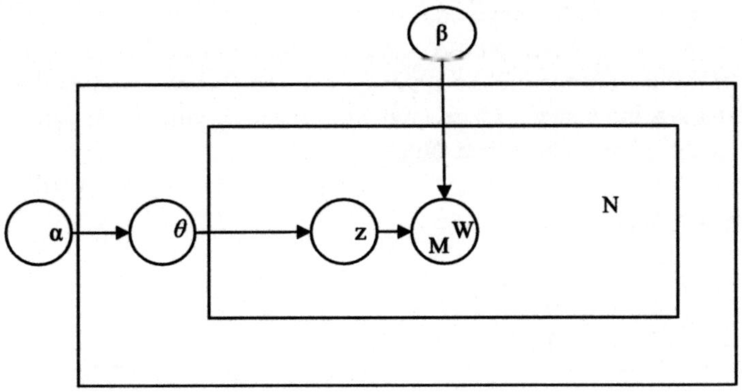

Fig. 1 LDA graphical model

In Fig. 1, document is described by M though each document is succession of N words where word is signified by w and topic variable in document is characterized by z. The parameters α and β are corpus-level parameters and are inspected once during the time spent creating a corpus. The factors θ is document level variable, examined once per document. Lastly, the factors z and ware word-level factors and are examined once for each word in each document.

4 Proposed System

To provide contents related to user interest, each article related to interest is considered as document. LDA is used to find out the semantic structure concealed in the document. LDA provided us a topic distribution for each interest area, so this learning will help to recommend the related article to end user based on the user interest. LDA consider each document as collection of topics in a certain distribution and each topic as a collection of keywords. Once number of topics is decided as input to LDA algorithm, it firstly rearranges the topic proportion within the document and keyword distribution with in a topic to have a good configuration of topic-keyword.

Accuracy of LDA algorithm depends on some key factors:

1. Quality of input text.
2. Number and variety of topics.
3. Tuning parameters.

In our experiment, we have taken three different topics (cooking, cricket and bodybuilding) as a user interest for an input to LDA algorithm. Data is gathered from different websites by writing a crawler in python. Before inputting the data to the LDA algorithm. All collected data has been cleaned by removing stop words, removing emails, remove new line characters and remove distracting single quotes. Once data is preprocessed, now all the sentences are converted into words. To have more accuracy build the bigram model and performed the lemmatization on the words followed by removing words whose count is either less than 15% or more than 50% of the words. Now corpus will be created. Now the preprocessed data is separated into training set and test set. Once model is prepared with the training set, model accuracy is checked using the test data.

In Fig. 2, all the three topics are well segregated and have a keywords weight age for all the three topics.

5 Conclusion and Future Scope

In this paper, content recommendation based on topic modeling is studied and implemented. Implementation is performed on python by considering document related to three topics and accuracy achieved is 89%. In the future, work will be extended

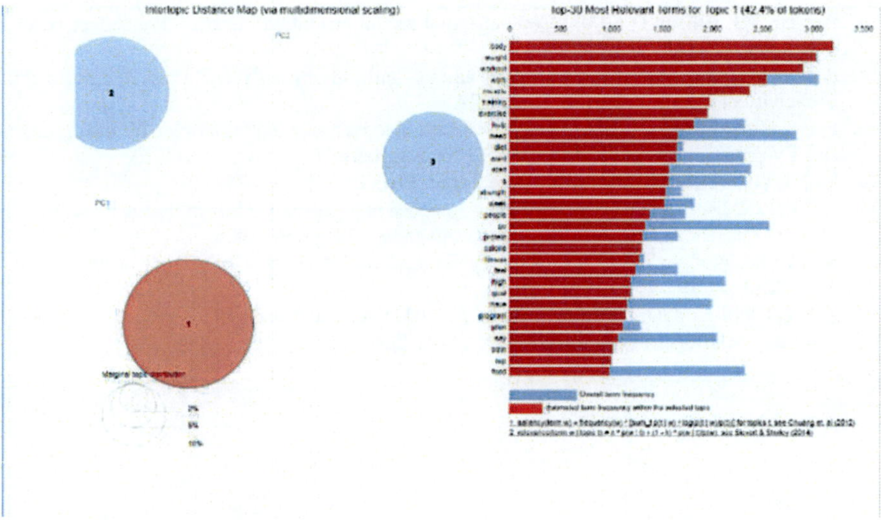

Fig. 2 Topic 1 most relevant terms

by considering more number of different topic documents and system will provide personalized content to the end users.

References

1. Adomavicius G, Tuzhilin A (2005) Toward the next generation of recommender systems: a survey of the state-of-the-art and possible extension. Ieee Trans Knowl Data Eng 17(6):734–749
2. Apaza RG, Cervantes EV, Quispe LC, Luna JO (2014) Online courses recommendation based on LDA. In: Symposium on information management and big data—SIMBig 2014. Peru, p 7
3. Badriyah T, Azvy S, Yuwono W, Syarif I (2018) Recommendation system for property search using content based filtering method. In: International Conference on Information and Communications Technology. Yogyakarta
4. Basu C, Hirsh H, Cohen W (1998) Recommendation as classification: using social and content-based information in recommendation. Am Assoc Artif Intell, p 7. (USA)
5. Belkin NJ, Croft WB (1992). Information filtering and information retrieval: two sides of the same coin? Commun ACM 35(12):29–38
6. Bozanta A, Kutlu B (2018) HybRecSys: content-based contextual hybrid venue recommender system. J Inf Sci 45(2)
7. Burke R (2007) Hybrid web recommender systems. Springer-Verlag, Berlin
8. Chang TM, Hsiao W-F (2013) LDA-based personalized document recommendation. Pacific Asia Conf Inf Sys 13
9. Cohen WW, Schapire RE, Singer Y (1999) Learning to order things. J Artif Intell Res 10:243–270
10. Deng F, Ren P, Qin Z, Huang G, Qin Z (2018) August). Leveraging image visual features in content-based recommender system, Hindawi Scientific Programming, p 8
11. Duda RO, Hart PE, Stork DG (2001) Pattern classification. Wiley
12. Luostarinen T, Kohonen O (2013) Using topic models in content-based news recommender systems. In: 19th Nordic conference of computational linguistics. Oslo, Norway, p 13

13. Pazzani MJ, Billsus D (2007) Content-based recommendation systems. The Adaptive Web. Berlin, pp 325–341
14. Powell M (1981) Approximation theory and methods. In: Press CU (ed) Press Syndicate of the University of Cambridge, New York, USA
15. Pyo S, Kim M, Kim E (2014) LDA-based unified topic modeling for similar TV user grouping and TV program recommendation. IEEE Trans Cybern 16
16. Rich E (1979) User modeling via stereotypes. Elsevier 3(4):329–354
17. Sarkhel JK, Das P (2010) Towards a new generation of reading habits in Internet Era. In: 24th national seminar of IASLIC. Gorakhpur University, U.P, pp 94–102
18. Su X, Khoshgoftaar TM (2009) A survey of collaborative filtering techniques. Adv Artif Intell 2009(421425), 19
19. Xia Y, Fabbrizio GD, Vaibhav S, Datta A (2017) A content-based recommender system for e-commerce offers and coupons. SIGIR eCom. Tokyo, p 7

Hybrid ANFIS-GA and ANFIS-PSO Based Models for Prediction of Type 2 Diabetes Mellitus

Ratna Patil, Sharvari Tamane, and Nirmal Rawandale

Abstract Type- Diabetes Mellitus (T2DM), a major threat to developing as well as developed countries, can be easily controlled to a large extent through lifestyle modifications. Diabetes increases the risk of developing various health as well as financial problems to cure these health complications. The health complications are stroke, myocardial infarction, and coronary artery disease. Nerve, muscle, kidney and retinal damage have distressing impact on the life of a diabetic patient. It is the need of the hour to halt the epidemic of T2DM in the early stage. Data science approaches have the potential to predict on medical data. Machine learning is an evolving scientific field in data science where machines learn mechanically and improve from experience without any explicit program. Our goal was to develop a system which can improve performance of a classifier for prediction of T2DM. The purpose of this work is to implement a hybrid model for prediction by integrating the advantages of artificial neural net (ANN) and fuzzy logic. Genetic algorithm (GA) and particle swarm optimization (PSO) have been applied to optimize parameters of developed predicting model. The proposed scheme used a fuzzification matrix. This matrix is used to relate the input patterns with a degree of membership to different classes. The specific class is predicted based on the value of degree of membership of a pattern. We have analyzed the proposed method and previous research in the literature. High accuracy was achieved using the ANFIS-PSO approach.

Keywords Machine learning · Fuzzy system · Diabetes mellitus · Particle swarm intelligence approach · Adaptive neuro-fuzzy inference system (ANFIS)

R. Patil (✉)
Noida Institute of Engineering and Technology, Greater Noida, Uttar Pradesh, India
e-mail: ratna.nitin.patil@gmail.com

S. Tamane
Jawaharlal Nehru Engineering College, Aurangabad, India
e-mail: sharvaree73@yahoo.com

N. Rawandale
Shri Bhausaheb Hire Government Medical College, Dhule, India
e-mail: drrawandale@gmail.com

V. Singh et al. (eds.), *Computational Methods and Data Engineering*, Advances in Intelligent Systems and Computing 1227,
https://doi.org/10.1007/978-981-15-6876-3_2

1 Introduction

Diabetes Mellitus is classified into three types. These are namely Type-I (T1DM), Type-II (T2DM), and Gestational DM (GDM). T2DM appears to be the most common form of diabetes in India where more than one crore cases are reported per year. It is developed if insulin is not produced adequately by the pancreas. The main contributing factors of T2DM include lifestyle, physical inactivity, obesity, eating habits, and genetics. In T2DM human body does not use insulin properly. We have considered T2DM for our study. Several classification algorithms are designed for classifying the patients as diabetic or healthy. ANFIS has its place in the class of hybrid structure, termed as neuro-fuzzy systems. ANFIS receives the properties of neural net as well as fuzzy systems [1]. Neural networks can learn effortlessly from the input provided but it is hard to understand the knowledge assimilated through neural net [2]. In contrast, fuzzy-based models are understood very straightforwardly. Fuzzy inference system (FIS) exploits linguistic terms instead of numeric values and generates rules in the form of if-then structure. Linguistic variables have values in the form of words in natural language having degrees of membership. Partial membership is allowed in fuzzy sets, which shows that an element exists in more than one set partially.

The usage of ANFIS makes the creation of the rule base more adaptive to the state for modeling and controlling of complex and non-linear problems. In this approach, the rule base is created by exploiting the neural network systems through the back-propagation process. To boost its performance, the properties of fuzzy logic are inherited in this model.

In the proposed method, the fusion of ANFIS with metaheuristic approach has been done. Metaheuristic algorithm follows repetitive process. Metaheuristic methods control a subordinate heuristic by exploiting and exploring the search space. These algorithms are stimulated by seeing the phenomena happening in the nature.

This paper is systematized as follows: Related work done by other researchers is discussed in Sect. 2. Section 3 includes discussion and construction of ANFIS process. Discussion on GA is represented in Sect. 4 and PSO is depicted in Sect. 5. Section 6 presents the building of proposed algorithm. Experimental results are discussed and results obtained are compared in Sect. 7. Lastly, in Sect. 8 concluding remarks are made.

2 Related Work

ANFIS has been used commonly as an effective tool for prediction due to its learning abilities and this approach facilitates rapid adaptation to deviations in systems which directed to robust groundwork for research. In this background work done by other researchers is presented here.

Author Soumadip Ghosh have has analyzed the performance of three different techniques NFS, RBFNN, and ANFIS widely used in Data Mining [3]. Performance was analyzed based on root mean square error (RMSE), Kappa statistic, F-measure, accuracy percentage on ten standard datasets from UCI. The results suggest that ANFIS has RMSE value of 0.4205.

Author Alby in his paper has developed ANFIS with GA and General Regression Neural Network (GRNN) for prediction of Type-II DM [4]. Using ANFIS with GA accuracy was 93.49% and accuracy was 85.49% with GRNN classifier.

Authors Ratna, Sharvari Tamne have done the comparison and analysis of logistic regression (LR), decision tree, K nearest neighbors (KNN), gradient boost, Gaussian Naïve Bayes, MLP, support vector machine (SVM), and random forest algorithms [5]. In this study, they have stated the strength and limitations of existing work.

Author Sinan Adnan Diwan Alalwan has carried out a detailed literature survey on different methods for predicting T2DM [6]. In his work, he has suggested random forest method and self-organizing map for improving the accuracy of prediction. Several authors have used PCA technique for dimensionality reduction of dataset. Authors Ratna et al. have used PCA for dimensionality reduction technique followed by KMeans in their study and have shown that performance was improved [7, 8]. Author Murat et al. used PCA followed by ANFIS for diagnosing diabetes [9].

Author Quan Zou has implemented three classifiers using random forest, decision tree, and neural network methods. He has analyzed and compared these classifiers on PIMA and Luzhou dataset [10]. The study shows that random forests are better than the other two. For dimensionality reduction PCA and minimum redundancy maximum relevance (mRMR) were employed. But the result shows that accuracy was 0.8084 which was better when all the features were used with random forest.

Authors Patil and Tamane have developed the genetic algorithm for feature selection with K nearest neighbor (KNN) and Naïve Bayes approach [11]. Though both the models have improved the accuracy of the prediction with reduced feature set, GA + KNN have got the better results than GA + Naïve Bayes. In GA + KNN approach, validation accuracy has been improved from 74% to 83%.

3 ANFIS

ANFIS is a fuzzy inference system introduced by Jang, 1993. It is implemented in the framework of adaptive systems. ANFIS architecture is depicted in Fig. 1. ANFIS network has two membership functions. Inputs are converted to fuzzy values using input membership function. Generally used input membership functions are Triangular, Trapezoidal, Gaussian. Fuzzy output of FIS is mapped to crisp value by output membership functions. Tuning of parameters related with the membership function is completed during the learning phase. Gradient vector is used for computation of these parameters and their tuning. For a specified set of parameters, gradient vector actually computes a measure of how fine the FIS has modeled the provided data. After getting the gradient vector one of various optimization method can be used

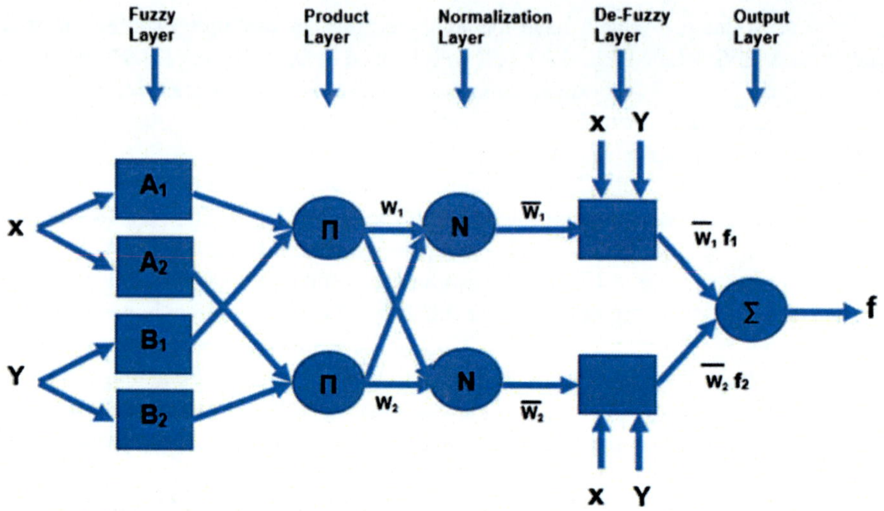

Fig. 1 5-layered architecture of ANFIS

for adjusting the parameters for minimizing error measure. This degree of error is generally calculated by the sum of the squared difference between actual and wanted outputs.

For approximation of membership function parameters, ANFIS employs either back-propagation or combination least squares estimation with back-propagation.

Fuzzy rules are created using Sugeno-type fuzzy system on a specified dataset. A typical form of Sugeno fuzzy rule is:

IF I_1 is Z_1
AND I_2 is Z_2
. . . . .
AND I_m is Z_m
THEN $y = f\ (I_1, I_2, \ldots, I_m)$
Where,
$I_1, I_2, \ldots, I_m$ are input variables;
$Z_1, Z_2, \ldots, Z_m$ are fuzzy sets.

There are five layers with different functions in ANFIS architecture. These layers are called as fuzzification, product, normalization, de-fuzzy, and output layer sequentially. Equations (1) to (6) depict function of each layer.

Layer 1: It is a fuzzy layer where the crisp signal is given as input to the *i*th node. This node is linked with a linguistic label A_i or else B_{i-2}. The function computes the membership value of the input. The input layer calculates the output from all the nodes by applying Eqs. (1) and (2).

$$O_{1,i} = \mu_{Ai}(X), \text{ where } i = 1, 2 \tag{1}$$

$$O_{1,i} = \mu_{Bi-2}(Y), \text{ where } i = 3, 4 \quad (2)$$

In Eqs. (1) and (2) the inputs to ith node are given by X, Y and A_i, B_i are representing linguistic symbols. μ_{Ai} is the membership function of A_i.

Layer 2: All the nodes in this product layer are fixed nodes characterized as Π. A rule neuron computes firing strength W_i by the product of all the incoming signals by Eq. (3). Each node output implies the firing strength of a rule.

$$O_{2,i} = W_i = \min\{\mu_{Ai}(X), \mu_{Bi}(Y)\}, \text{ where } i = 1, 2 \quad (3)$$

Layer 3: Every node in normalization layer calculates normalized firing strength of a given rule. It is proportion of the firing strength of specified rule to the summation of firing strengths of all rules. It indicates the involvement of a given rule to the ultimate result. Consequently, the output from ith neuron in layer 3 is calculated by Eq. (4).

$$O_{3,i} = \overline{W_i} = \frac{W_i}{(W_1 + W_2)}, \text{ where } i = 1, 2 \quad (4)$$

Layer 4: Each neuron in the defuzzification layer computes the weighted consequential value of a certain rule by Eq. (5).

$$O_{4,i} = \overline{W_i} f_i = \overline{W_i}(p_i x + q_i y + r_i), \text{ where } i = 1, 2 \quad (5)$$

Layer 5: The output layer has a single node. This is a fixed node having label $\sum$. It computes the overall ANFIS output by adding the outputs from all the neurons in the defuzzification layer as in Eq. (6).

$$O_{5,1} = \sum_i \overline{W_i} f_i \quad (6)$$

4 Genetic Algorithm (GA)

They are generally used to produce solutions for optimization and exploration tasks. GA simulates "survival of-the-fittest" between individuals of succeeding generation for problem-solving. Genetic algorithms use methods inspired by evolutionary biology such as selection, inheritance, alteration, and recombination. Pseudocode of GA is given below:

1. Select initial population.
2. Compute the fitness of every candidate in the populace.
3. Repetition of the next steps (a–e) until termination condition is satisfied.

a. High ranking entities are selected for reproduction.
b. Use recombination operator to yield next generation.
c. The resultant offspring is mutated.
d. Evaluate the offspring.
e. Substitution of low ranked chunk of populace with the reproduced descend ants.

5 Particle Swarm Optimization (PSO)

Kennedy and Eberhart developed PSO in 1995. It is a stochastic optimization method. The concept of PSO is analogous to flight of birds in hunt of food. It is an evolutionary optimization method built on the movement and intellect of swarms [12]. PSO is a population-based searching process where swarm of particles are the searching agents and position of particle gives solution. Each particle is considered to be a point (candidate solution) in a N-dimensional space which fine-tunes its "flying" based on its personal flying experience and the flying experience of other particles. This concept is represented in Fig. 2.

PSO has found its way in modeling of biological and sociological behavior like group of birds looking for food cooperatively. The PSO has been also extensively used in population-based hunt approach. In a search space, the position of particle is changed repetitively until it reaches to the best solution or until the computational boundaries are reached.

Pseudocode of PSO is mentioned below:

Fig. 2 PSO concept

Table 1 PSO parameters

Vel(t)	Velocity of the particle at time *t*
P(t)	Position of the particle at time *t*
W	Inertia-weight
c_1, c_2	Weight for local and global information, respectively (acceleration factors)
r_1, r_2	They are random values which are uniformly distributed between zero to one They are representing cognitive and social factor, respectively
P_{pbest}	The local best position of particle
P_{gbest}	The global best position particle

```
    For every particle
    Set particle position P_i (0) and velocity Vel_i (0)
       randomly
End
Do
   For every particle
     Evaluate fitness function
     If this fitness value is improved as compared to its pbest
          update pBest by assigning present computed value to it
     End

    Update gBest by selecting the particle with the greatest
       fitness value of all and assign this value to gBest.
   For every particle
      Evaluate velocity of particle using equation (7)
      Modify position of particle using equation (8)
   End
While terminating conditions are not reached.
```

$$\mathrm{Vel}(t+1) = w \times \mathrm{Vel}(t) + c_1 \times r_1 \times (\mathrm{Ppbest} - P(t)) + c_2 \times r_2 \times (\mathrm{Pgbest} - P(t)) \quad (7)$$

$$P(t+1) = P(t) + \mathrm{Vel}(t+1) \quad (8)$$

where description of parameters is given in Table 1.

6 Proposed Algorithm

We have presented an approach in this paper that combines ANFIS with PSO to develop ANFIS-PSO and ANFIS with GA to develop ANFIS-GA correspondingly. ANFIS approach utilizes the advantages of neural network's (NN) learning and adaptation capability, fuzzy inference system's (FIS) knowledge representation by fuzzy

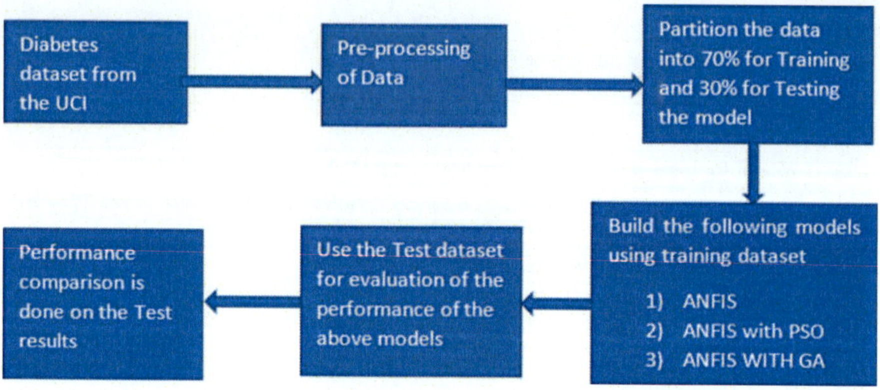

Fig. 3 Broad level phases for the proposed algorithm

if-then rules. Proposed hybrid algorithm uses systematic random search of genetic algorithms (GAs) and efficiency and probability of finding global optima of PSO with ANFIS. We have used PSO and GA for performance improvement of ANFIS by minimizing the errors by adjusting the membership functions. Broad level phases for the proposed algorithm are shown in Fig. 3

ANFIS builds FIS by extracting the set of rules using fuzzy-CMeans method. In MATLAB, FCM is provided in genfis3 function. Genfis3 creates FIS by ANFIS training used to model the data behavior. Membership function is used for writing the rules in the form of antecedents and consequents. Gaussian membership is used in this study as it is recommended in the previous study. The genfis3 function allows to specify the number of clusters which in turn confines the number of rules.

The scheme demonstrating the model establishment for ANFIS-GA and ANFIS-PSO is shown in Fig. 4. ANFIS provides the exploration space and the best solution is searched by the PSO by comparing the solution at each solution point in ANFIS-PSO. The variance among target output and the predicted output is minimized by repeating PSO. PSO does not depend on the derivative nature of the objective function and attains the optimal solution by fine-tuning the membership functions. Performance of ANFIS is raised by integrating it with GA. Error is minimized by fine-tuning the membership functions of FIS.

7 Experimental Results

The experiment is implemented in MATLAB over the PIMA—diabetes dataset which was available in the UCI repository for machine learning [13]. The existing experimental data was used for measuring the performance of approaches. MSE, RMSE, Error Mean, and Error St.D. were used to analyze the performance of ANFIS, hybrid

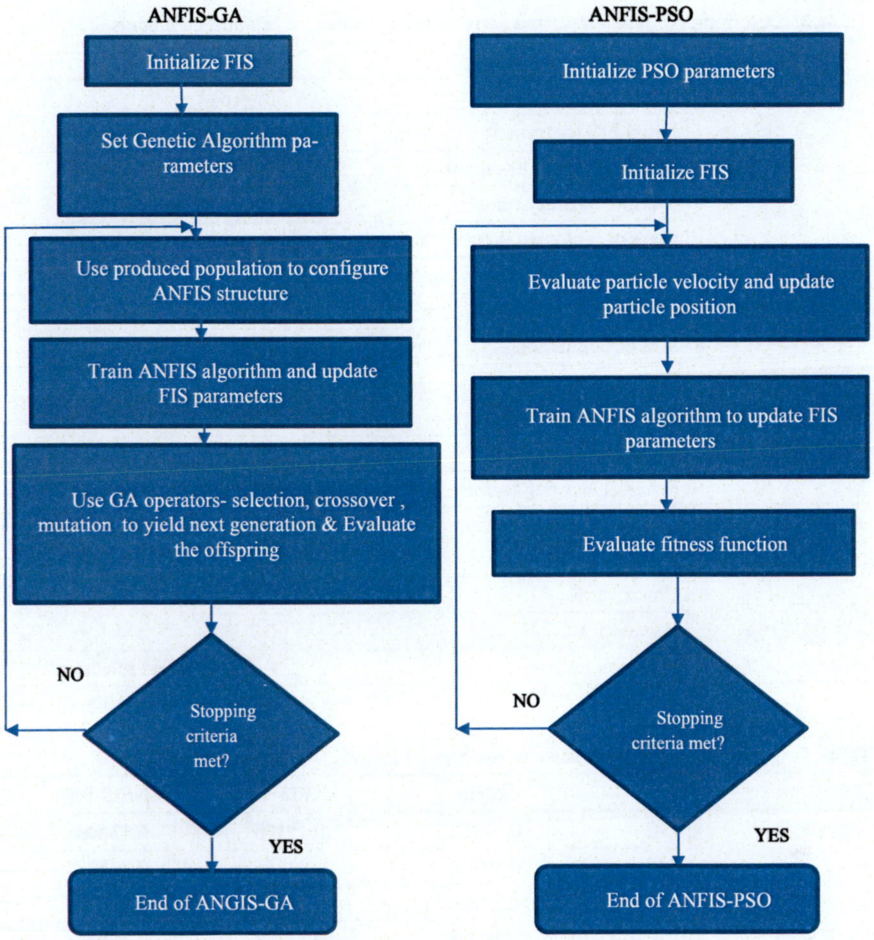

Fig. 4 Scheme of ANFIS with GA and ANFIS with PSO

approaches ANFIS with GA, and ANFIS with PSO. Table presents the summary of comparison of proposed algorithm.

We have used six PSO parameters while implementing the model which are listed in Table 2. These parameters are maximum iterations number, global and personal learning factors, inertia-weight, damping ratio, size of population. For this work by trial and error process, we have found out these parameters' optimal values. Details of ANFIS, ANFIS-GA, and ANFIS-PSO parameter values are shown in Table 2.

Comparison of results obtained during training phase and testing phase of developed hybrid models with ANFIS is provided in Table 3. MSE, RMSE, Error Mean, and Error St.D were used for the comparison of developed models by integrating ANFIS with PSO and GA. It is observed that both GA and PSO algorithm effectively improves the performance of the ANFIS model. The analysis of testing results

Table 2 Description of parameters and corresponding values for established models

Model	Parameter	Values
ANFIS	Fuzzy structure	Sugeno-type
	Initial FIS for training	Genfis3
	Maximum iterations number	500
	Number of fuzzy rules	10
	Class of input membership function	Gaussmf
	Form of output membership function	Linear
ANFIS-PSO	Maximum iterations number	1000
	Size of population	25
	Weight of inertia	1
	Damping ratio	0.99
	Global learning factor	2
	Personal learning factor	1
ANFIS-GA	Maximum number of iterations	1000
	Population-size	25
	Crossover %	0.4
	Mutation %	0.7
	Selection method	Roulette-wheel selection

Table 3 Comparison of performance of established models

		ANFIS	ANFIS-GA	ANFIS-PSO
Training set	MSE	0.15222	0.15105	0.13394
	RMSE	0.39016	0.38866	0.36598
	Error Mean	1.9914e−17	−0.0052282	0.002298
	Error St.D.	0.39052	0.38898	0.36631
Testing set	MSE	0.17627	0.16588	0.14029
	RMSE	0.41985	0.40728	0.37456
	Error Mean	0.020322	−0.023361	−0.047618
	Error St.D.	0.42027	0.4075	0.37233

of developed ANFIS, ANFIS-GA, and ANFIS-PSO models is given in Figs. 5, 6, and 7, respectively.

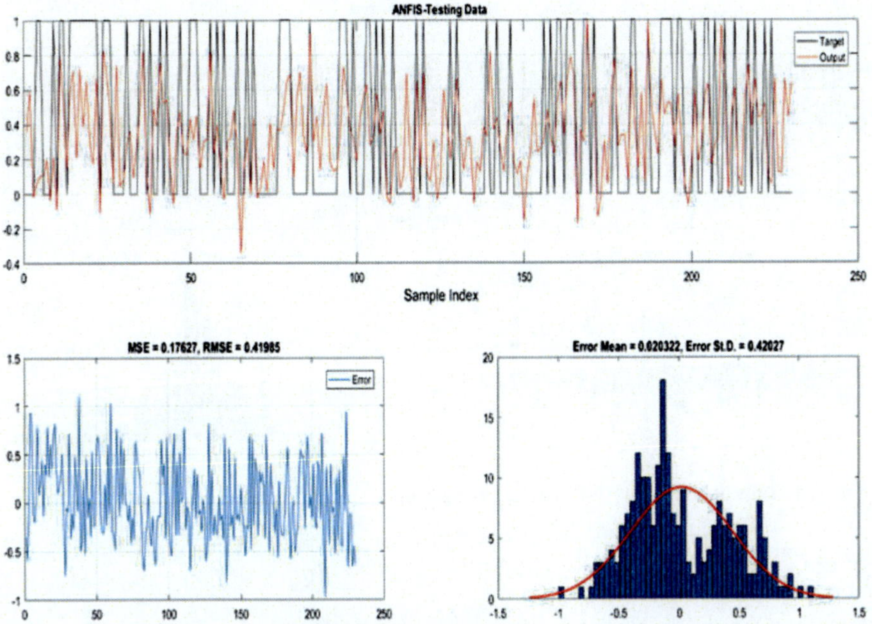

Fig. 5 Results obtained by ANFIS during testing phase

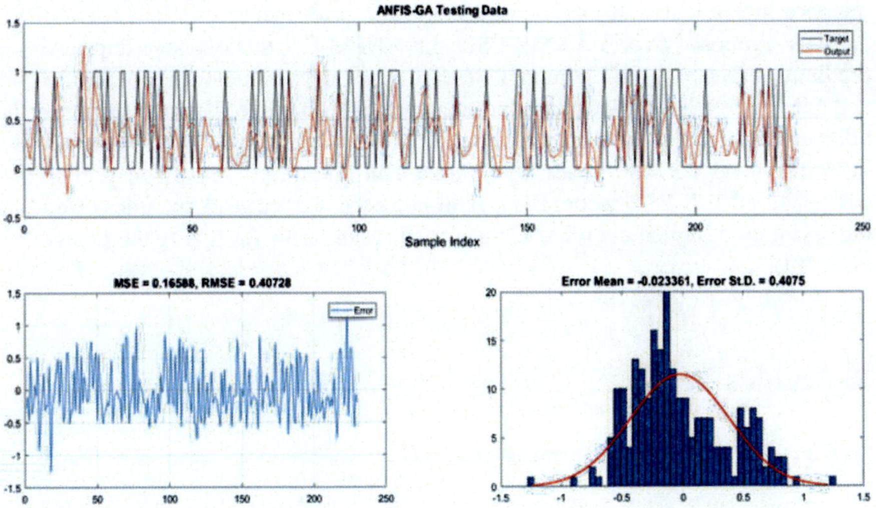

Fig. 6 Results obtained by ANFIS-GA during testing

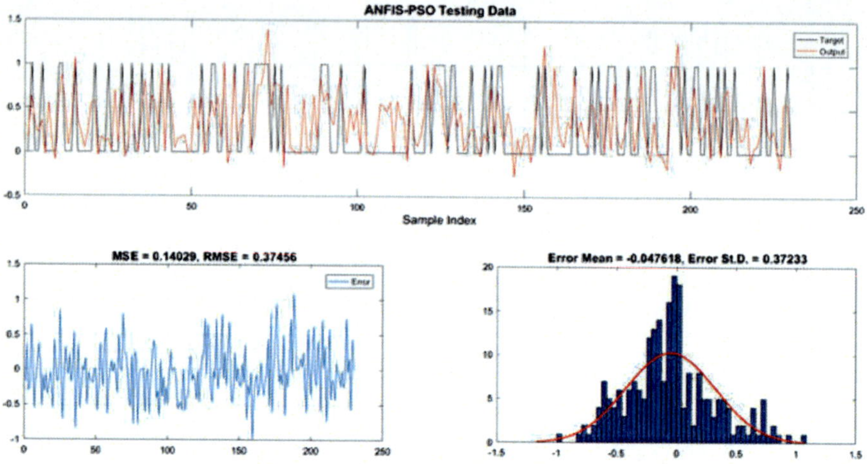

Fig. 7 Results obtained by ANFIS-PSO during testing

8 Conclusion

It is observed from the literature review that ANFIS is computationally effective. ANFIS can be integrated with optimization and adaptive techniques for tuning its membership function. It can also be combined with metaheuristic methods PSO and GA. Proposed hybrid ANFIS-PSO and ANFIS-GA models have improved the prediction efficacy of ANFIS model. Studied statistical parameters like MSE, RMSE, and Mean Error have confirmed that the ANFIS-PSO model has outperformed the other models. ANFIS-PSO beats other approaches with average RMSE value 0.36598 in training and 0.37456 in testing phases. The literature comparison proved that developed ANFIS-PSO model has a great potential. Future work includes extending the research to implement other metaheuristic algorithms for tuning the parameters of ANFIS.

References

1. Mitchell T (2007) Machine learning. Tata McGraw-Hill Education India. Genre: Computers. ISBN: 9781259096952
2. UmmugulthumNatchiar S, Baulkani S (2018) Review of Diabetes Disease Diagnosis Using Data Mining and Soft Computing Techniques. Int J Pure Appl Math 118(10):137–142
3. Ghosh S, Biswas S, Sarkar D, Sarkar P (2014) A novel Neuro-fuzzy classification technique for data mining. Egypt Inf J 129–147
4. Alby S, Shivakumar BL (2018) A prediction model for type 2 diabetes using adaptive neuro-fuzzy interface system. Biomedical Research (2018) Computational Life Sciences and Smarter Technological Advancement, 2017
5. Patil R, Tamane S (2018) A comparative analysis on the evaluation of classification algorithms in the prediction of Diabetes. Int J Electr Comput Eng (IJECE) 8(5):3966–3975

6. Alalwan SAD (2019) Diabetic analytics: proposed conceptual data mining. Indonesian J Electr Eng Comput Sci 14(1):88–95
7. Patil RN, Tamane S (2017) A novel scheme for predicting type 2 diabetes in women: using kmeans with PCA as dimensionality reduction. Int J Comput Eng Appl XI(VIII):76–87
8. Wu H, Yang S, Huang Z, He J, Wang X (2018) Type 2 diabetes mellitus prediction model based on data mining. Inf Med Unlocked 10:100–107
9. Kirisci M, Yılmaz H, Saka MU (2018) An ANFIS perspective for the diagnosis of type II diabetes. Ann Fuzzy Math Inf X, 2018
10. Zou Q, Qu K, Luo Y, Yin D, Ju Y, Tang H (2018) Predicting diabetes mellitus with machine learning techniques. Frpntiers in Genetics; Bioinf Comput Bio 9
11. Patil RN, Tamane SC (2018) Upgrading the performance of KNN and naïve bayes in diabetes detection with genetic algorithm for feature selection. Int J Sci Res Comput Sci Eng Inf Technol 3(1):2456–3307
12. Hu X [Online]. Available: http://www.swarmintelligence.org/tutorials.php
13. [Online]. Available: https://archive.ics.uci.edu/ml/datasets/Pima+Indians+Diabetes

Social Network Analysis of YouTube: A Case Study on Content Diversity and Genre Recommendation

Shubham Garg, Saurabh, and Manvi Breja

Abstract Social Network Analysis has a great potential in analyzing social networks and understand how users in communities interact with each other. It can be used to draw meaningful insights from networks, as users with similar patterns can be identified and mapped together, thereby helping provide relevant content to new users. This would not only help platforms enhance user experience but also benefit users who are new to the platform. The aim of this paper is to analyze the network of users who upload videos on YouTube. We apply social network analysis on YouTube data to analyze the diversity of video genres uploaded by a user and also find the most popular uploader in each category. A new approach is also proposed using the Apriori algorithm to recommend a category that a new user might be interested in uploading, based on what other users with similar interest are uploading.

Keywords Recommendation · Density · Betweenness · Homophily · Centrality

1 Introduction

As the number of people on social media platforms are increasing at an exponential rate, it has become more important than ever to understand the intricacies of connections between them. Social Network Analysis (SNA) utilizes the concept of networks and graph theory in order to visualize the network structures consisting of nodes and edges connecting them. Nodes represent person, group or entities and edges (ties or links) represent relationships or interactions between the nodes. SNA aims to provide visualization and mathematical analysis of relationships which are used to judge how

S. Garg (✉) · Saurabh · M. Breja
The NorthCap University, Gurugram 122017, India
e-mail: shubham16csu356@ncuindia.edu

Saurabh
e-mail: saurabh16csu332@ncuindia.edu

M. Breja
e-mail: manvibreja@ncuindia.edu

V. Singh et al. (eds.), *Computational Methods and Data Engineering*, Advances in Intelligent Systems and Computing 1227,
https://doi.org/10.1007/978-981-15-6876-3_3

people are connected together [1]. In the past, Social Network Analysis has helped researchers analyze social behaviour in animals [2], find evidence for information diffusion [3], analyze users' interactions and behaviour based on their activity on Twitter [4] and Facebook [5] and also in analyzing question answering systems [6].

Over the past few years, YouTube has become one of the most popular video-sharing platforms. The platform offers a lot of diverse content which is uploaded by its users. Till now, no significant research was done on the users who upload these videos. In this paper, using Social Network Analysis, we analyze the diversity of video genres uploaded by a user, i.e. users sharing videos of different categories or genres and also visualize the most popular uploader in a genre, based on certain metrics like views, likes, etc. In this work, we also propose a recommendation algorithm to suggest other genres to new uploaders based on various genre pairs being used by existing uploaders. This would enable new users to try different things which might seem interesting to them, making the platform richer and more diverse in terms of content and quality for its users.

2 Properties to Measure Social Networks

a. Connectivity

Important properties based on network connectivity are as follows:

Homophily. The likelihood of a node to be connected to other nodes having similar attributes than the ones showing different characteristics. For instance, two people with similar interests are more likely to be friends.

Multiplexity. Two nodes interacting and related (or connected) to each other in multiple ways. It measures the strength of their connection. For instance, two people work together and are each other's neighbour also, share a multiplex relation.

Network Closure. The likelihood of connections of a node getting connected to each other at some point in time. In other words, if the connections of a node are also connected to each other.

b. Distributions

Centrality measures help in identifying the biggest influencers, most popular and liked nodes in a network. These measures help us analyze the effect of a node in influencing other nodes within a social network.

Degree Centrality. It is the measure of nodes that are directly connected to a node in a network, i.e. it measures how many neighbours a particular node has. The degree centrality of a node, for a given graph $G = (n, i)$ with 'n' nodes and 'i' edges is defined as:

$$C_D = \frac{\sum_{i=1}^{g}[C_D(n^*) - C_D(i)]}{[(N-1)(N-2]} \tag{1}$$

where C_D is degree centrality, n^* being the node with highest degree centrality and N is the number of nodes [7].

Betweenness Centrality. Measure of occurrence of a node lying on the shortest possible path between two nodes in a network. It is useful to find the nodes that influence the flow of information in a network. The betweenness centrality of a node n is given by the following expression:

$$C_B(n) = \sum_{a<b} \frac{g_{ab}(i)}{g_{ab}} \tag{2}$$

where g_{ab} is the number of shortest paths connecting pair ab, and $g_{ab}(i)$ is the number of shortest paths passing through node i [1].

Closeness Centrality. In a connected graph, closeness centrality tells us how close a given node is from all the other nodes. It is the mean of the shortest path calculated from a node to every other node in the network. The equation of closeness centrality is given by:

$$C(a) = \frac{N}{\sum_b d(b,a)} \tag{3}$$

where $C(a)$ is closeness centrality, $d(b, a)$ is the distance between the nodes a and b and N is the number of nodes in graph [7].

Density. Network Density measures the connection between different nodes. It is the ratio of the number of actual connections a node has, to the total number of connections a node could possibly have.

$$\text{Network Density} = \frac{\text{Actual Ties}}{\text{Total Ties}} \tag{4}$$

$$\text{Total ties} = \frac{n * (n-1)}{2} \tag{5}$$

where n is the number of nodes within a network [8].

Distance. It measures the shortest route between two nodes. Distance plays an important role in helping us understand the similarities and differences in attributes among the nodes based on their position. The farther two nodes are, the longer it may take for the information to spread within the network [9].

3 Literature Review

Social Network Analysis has affected many fields and attracted researchers to study different networks and visualize their results. Since mid-1930s, SNA has been used to research social and behavioural sciences which has grown rapidly from early 2000s. Wey et al. in 2006 [2] utilized social network analysis to visualize animal social behaviour which indicates how animals live and interact with each other. With the advent of search engines and social media, there is a lot of research focusing on analyzing different patterns in social media. Vemprala in 2019 [3], analyzed the characteristics of data breach messages posted on social media, presence of evidence for breach message diffusion and users exchanging such messages. Kim et al. in 2018 [4] analyzed communications during disaster between different users such as organizations, individuals and emergency agencies on Facebook. The studied networks are used to prepare pre-and post-disaster mitigation plan. Hagen in 2017 [5] examined communications during a crisis such as a trend on Twitter at the time of Zika virus outbreak in the United States during 2015–16. The authors visualized the network structure to survey how different entities interacted with each other and probe the information flow. Rihn et al. in 2015 [10] filed a patent to identify abusive user accounts prevalent on social networks using contents of the messages being posted, images contained in the messages and time at which the message had been posted so that those accounts could be cancelled. Akrouf et al. in 2013 [11] researched Flickr and YouTube networks for information propagation. The authors utilized two networks viz. user's contacts on Flickr and users commenting on YouTube videos and visualized the structure of networks with the type of relationships. They stated that information diffusion in a network is affected by the network structure and information diffused faster through explicit connections than implicit relations. Feroz Khan et al. in 2014 [12] developed a model to find reasons for videos going viral on YouTube. The authors collected 100 most viewed videos and other related information using the YouTube API and Webometrics analyst Tool [13]. Various characteristics like age, gender, videos viewed, videos posted, postdate, type, duration (to look into video characteristics) and likes, dislikes, view count, count of comments (to judge the virality) were explored to visualize the relationships between different frames. This helped conclude network dynamics, social and non-social capital as well as view counts play a crucial role in determining the virality of the videos.

4 Evolution of YouTube

YouTube was launched in February 2005 and now operates as one of Google's subsidiaries [14]. It is a platform where content creators from all over the world can showcase their talent and the viewers can entertain themselves, learn and build their skills by watching the videos and sharing them to a wider audience.

It is a video-sharing platform where users can upload, view and share videos and also subscribe to other users. It not only offers user-generated videos but also different corporate media videos. Some of the popular media corporations include BBC, Vevo and T-Series. Majority of the content on YouTube is uploaded by the individuals who have registered on the platform. Registered users get the right to upload an unlimited number of videos, can share their views about the video by commenting on them, share the video on different platforms, 'like' the videos as well as 'dislike' them. Registered users can also create playlists and add videos to them. Unregistered users, on the other hand, cannot access these features but can only watch the videos.

Everyday millions of videos are being uploaded and billions of hours of content is being watched. YouTube has also become the main source of income for millions of content creators. YouTubers earn revenue from ads, sponsors and supporters. Most of the content on YouTube is free to watch but there are also subscription-based services and film rentals. YouTube has recently announced two services namely 'YouTube Music' and 'YouTube Premium'. Subscription to these services helps users enjoy ad-free and uninterrupted access to music streaming service and all the other content available on the platform.

5 Research Objectives

YouTube is one of the various open sources through which a network of users having common interests can be found. Thus, we carried out the research by formulating social network on a subset of YouTube data and proposed three objectives as below:

a. **Research Objective 1**

To analyze the diversity of video genres uploaded by a user and find what proportion of users actually upload videos of different categories.

b. **Research Objective 2**

To analyze and find the most influential uploader in a genre and other meaningful insights by using certain popularity metrics of uploaders.

c. **Research Objective 3**

To create a recommendation system using the Apriori algorithm to suggest other genres to new uploaders based on various genre pairs being used by existing uploaders.

6 Dataset

The dataset for carrying out our research was obtained from the website http://netsg.cs.sfu.ca/youtubedata/ [15], which was extracted from YouTube API containing information of all the metadata except three categories namely age, category and related videos. The extracted data has been stored in a tab-separated text file. The data used to carry out the analysis was extracted through 1st March 2007 crawl. We have used the relatively older dataset to gauge historical trend.

We consider all the YouTube videos in the dataset to form an undirected graph, where a set of uploaders and categories are nodes in the graph and edges show links between the uploader and video category.

a. Description

YouTube assigns each video a unique 11-digit ID composed of characters and numeric values including (0–9, a–z, A–Z). Each video contains a series of metadata: video ID, uploader, date when it was added, category, length, user rating, number of views, ratings and comments and a list of related videos.

A video also contains information about metadata. For example, Age is an integer which tells the number of days between the date (March 1, 2007) when the video was uploaded and Feb.15, 2007 (YouTube's establishment). Each video has some rating assigned to it which depends on the number of likes and dislikes on each video (Table 1).

b. Video Category

All the videos uploaded by different creators belong to a total of 13 categories. Most of the content creators are interested in categories like Entertainment, Sports, Music, People & Blogs, Comedy, Film & Animation and Gadgets (Entertainment leads the pack with most uploaded videos) while a comparatively less interest has been seen

Table 1 Information of metadata [15]

Attributes	Description
videoID	An 11-digit string, which is unique
Uploader	String of the video uploader username
Age	Integer of days between the date when video was uploaded and March 1, 2007
Category	String of video category chosen by creator
Length	An integer number of video length
Views	An integer number of the views
Rate	A float number of the video rate
Ratings	An integer number of the ratings
Comments	An integer number of the comments
Related IDs	Up to 20 strings of the related video IDs

in categories like Travel & Places, Pets & Animals, How-to & Style and Auto & Vehicles.

7 Proposed Work

We have implemented this work in Python 3.6 which is an open-source programming language. Due to its ease of use and a mature ecosystem of well-maintained open-source libraries, it is one of the most popular languages in the data science community. Python is a versatile language and can be used for a wide range of applications. It has a large number of libraries like NumPy, SciPy enabling fast mathematical computations. Libraries like Sklearn reduce the tedious task of implementing machine learning algorithms to just writing just a few lines of code. Visualization of data and results after processing are often an important part of analysis. Python also streamlines this process with libraries like Matplotlib, Seaborn, etc., which enable a user to plot interactive visualizations easily. And with libraries like TensorFlow and Pytorch, it has become much easier to implement deep learning algorithms as well.

a. **Libraries Used**

Pandas. This library provides tools which makes managing and modifying datasets really easy. The data frames created can be modified, split or merged according to the required conditions. Pandas provide an easy and fast implementation of the transformation steps applied to the data. It can also be used to read and write data from/to multiple file formats.

NumPy. This Python library provides support for homogenous data structures called arrays and also provides a number of mathematical functions that can be performed on these. NumPy enables really fast computations, even with large array size and is better suited for operations involving large datasets [16].

NetworkX. It is used for the study and analysis of the structure, dynamics and properties of complex networks. With the help of this library, we can find important attributes of a network like degree, clustering coefficient, eccentricity, degree centrality of a node within a network. NetworkX allows a user to create graph objects directly from a Pandas Dataframe, without needing to convert the data to a matrix first. Using this library also gives user freedom to select from multiple available layouts to visualize a graph.

Matplotlib. A plotting library that lets a user easily visualize various types of plots of the data. It offers a variety of graphs namely histogram, bar plot, boxplot etc. The Matplotlib package offers a number of modules, Pyplot being one of the examples. Pyplot lets users create and modify graphs by providing useful features to format plot axes, labels, font properties, colour schemes, etc.

b. **Data Pre-processing**

This dataset had initially 10 columns of video metadata. For our analysis, we've only used two columns, uploader and category. The column 'uploader' gives information about the username of uploaders who have posted the videos online and category gives information about the genre of those uploaded videos. We have also handled null values by removing the information of some uploaders for which adequate data was not available.

c. **Strategy for Research Objective 1**

With the help of NetworkX, we create a graph that maps content uploaders and video categories. In the graph, green nodes are the video uploaders and the red nodes represent different video categories. The graph is shown in Fig. 1.

d. **Strategy for Research Objective 2**

For every video uploaded on YouTube, there are a number of metrics that help us gauge the performance or popularity of a video as well as the uploader. The metrics we've used for this are the number of views, comments and ratings of a video. Rather than using these parameters individually, we combine and transform them into one parameter called average weights.

For this, we first add the values of each individual parameter of number of views, comments and ratings and then, transform the result by scaling them to a range between 0 and 1 using Sklearn'sMinMaxScaler. We then again multiply the scaled weights by a factor of 10^5 to ensure that most of the values are greater than one, with

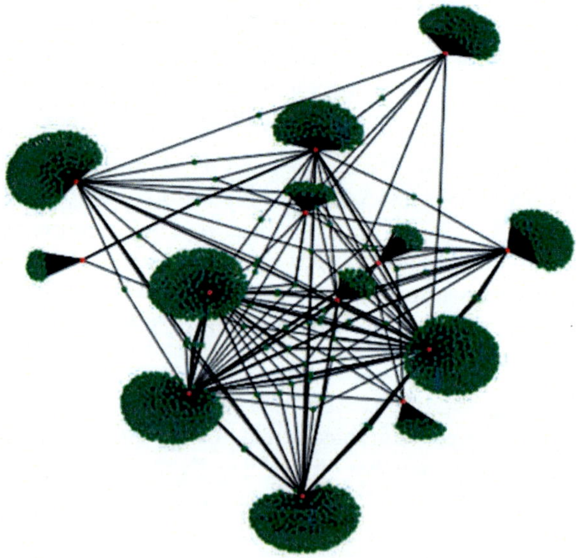

Fig. 1 Graph showing how video uploaders (green nodes) are connected to video genres (red nodes)

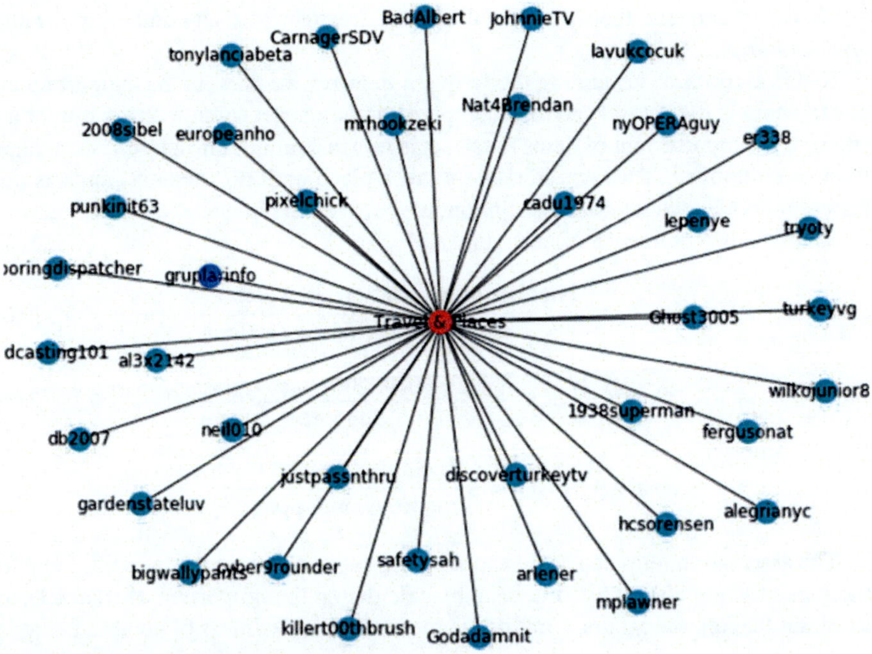

Fig. 2 Graph showing the network of uploaders with videos in 'Travel & Places' category

at least two significant digits. The final result is an average weights parameter that is used as an indicator of performance. We also visualize the network of uploaders that post videos in a particular category. Figure 2 shows the network graph of uploaders in the category 'Travel & Places'. The node representing the most popular uploader is coloured in blue and the rest of them are coloured in cyan.

e. **Strategy for Research Objective 3**

We propose a recommendation system using Apriori algorithm which recommends categories to a user based on the content frequently uploaded together by other users, who share similar interests (upload videos in the same category as the user itself). We extract the data of users who have uploaded videos of more than one category. Each video category is considered as an item set. Using the Apriori algorithm on the YouTube dataset, we try to find the video categories with strong association rules, so that it can be recommended to other uploaders. For example, If a user uploads videos of Entertainment and Comedy categories, then a new user who uploads a Comedy video, will be recommended to create videos of other categories like Entertainment.

Apriori Algorithm. Apriori algorithm is used for mining frequent items sets by learning various association rules over the databases, generally transactional databases. The Apriori principle states that 'If an item in the dataset is frequent, then all of its subsets must also be frequent' [17]. For example, If a set of elements

$\{A, B, C\}$ is frequent, then $\{A, B\}$ will also be frequent as a set containing A must also contain B.

In this algorithm, for finding the frequent itemsets we find out the support count of each item in the dataset and then compare item's support count with the minimum support. If support count of items in the set is less than minimum support, then these items are removed. This step iteratively takes place until no frequent itemsets are further found. For generating association rule between the frequent itemsets, we need to calculate the confidence of each rule.

$$\text{Support}(A) = \frac{\text{Number of transactions in which } A \text{ occurs}}{\text{Total number of transactions}} \tag{6}$$

$$\text{Confidence}(A \rightarrow B) = \frac{\text{Support}_{\text{Count}}(A \cup B)}{\text{Support}_{\text{Count}}(A)} \tag{7}$$

$$\text{Lift}(A \rightarrow B) = \frac{\text{Support}(A \cup B)}{\text{Support(A)} * \text{Support}(B)} \tag{8}$$

The association rules can be measured using support, confidence and lift. Support measures the popularity of an itemset by calculating the proportion of transactions in which the itemset occurs. Confidence measures the likeliness of an item being in an itemset if some other item is already present. While confidence is measured based on the popularity of only one item, this sometimes can lead to misinterpretation. To overcome this, lift is used. Lift also measures the likeliness of items being in an itemset while taking into account the popularity of both the items in the itemset [18].

8 Results

a. Results for Research Objective 1

As shown in Fig. 1, each uploader node is connected to at least one category node. The clusters formed around the categories show the uploaders for that category. Bigger clusters indicate a large number of uploaders for that category. For the given graph, the maximum number of uploaders belong to the Entertainment category, followed by Sports and People & Blogs. The category with the least number of uploaders is Travel & Places.

From this analysis, we find that only 4.16% of the users upload videos in more than one category and the majority of users upload videos in just one category.

b. Results for Research Objective 2

For the given dataset, we do not have much information about how different uploaders might be connected with each other. So, we try to gain some other meaningful insights about individual uploaders. We try to identify the uploader with maximum average weight among others in the network. As a higher average weight indicates the high popularity of an uploader, we try to find the most influential nodes in the network. The user node with the highest rating in this category is coloured blue and the rest of the user nodes are coloured in cyan. And the red node is the genre of the videos uploaded by the users. With the help of such visualizations, important nodes or users in the network can be identified easily, like the ones uploading the maximum number of videos in a given time period or ones with highest average ratings, and so on.

c. Results for Research Objective 3

For the given data, we are able to create significant association rules of video categories with maximum support of 30%. This means that these association rule pairs constitute 30% of the video categories for a case when users upload videos in multiple categories. We try to find frequent video categories by varying the support and confidence values. The item sets of these video categories are shown in Table 2.

The item sets show how one category is associated with the other. For example, the association rule for categories 'News & Politics' and 'People & Blogs' indicate that there is a 76% chance (or confidence) that an uploader sharing videos in 'News & Politics' category will also upload videos in the 'People & Blogs' category.

The result shows that the support threshold is inversely proportional to number of association rules. Depending on the threshold value, different frequent itemsets of video categories can be recommended to the users.

Table 2 Association rules for video categories

With 30% support	With 20% support
• **Entertainment → Comedy** Support: 0.34 Confidence: 0.58 Lift: 0.98	• **Comedy → Music** Support: 0.25 Confidence: 0.43 Lift: 0.97
• **Comedy → People and Blogs** Support: 0.34 Confidence: 0.60 Lift: 1.01	• **News and Politics → Comedy** Support: 0.22 Confidence: 0.38 Lift: 0.94
• **Entertainment → People and Blogs** Support: 0.33 Confidence: 0.55 Lift: 0.93	• **Entertainment → Music** Support: 0.25 Confidence: 0.42 Lift: 0.94
• **News and Politics → People and Blogs** Support: 0.31 Confidence: 0.76 Lift: 1.28	• **News and Politics → Entertainment** Support: 0.21 Confidence: 0.36 Lift: 0.88

9 Conclusion and Future Work

Social Network Analysis is an emerging field which has integrated various fields. Through this paper, we try to analyze YouTube data and look at the results from the perspective of social network analysis. We have mapped users based on the type of videos they upload and try to find the most popular uploader in a category based on certain metrics. The proposed approach of recommending video categories to a new user using Apriori algorithm looks promising and would perform better as more users upload diverse content.

Also, it has been perceived that there are very few uploaders who posted content in multiple genres. The most popular video category is Entertainment while Travel & Places is the least popular, based on the number of videos uploaded. However, People & Blogs category is most likely to be tried by an uploader as an alternative genre to upload videos other than the ones which they are already uploading.

This paper still has a wide scope to cover, which leaves room for future work like analyzing how the uploaders are connected to each other, connections of viewers with similar interests and use of more robust recommendation algorithms like content-based or collaborative-filtering to recommend new video categories to uploaders.

References

1. Knoke D, Yang S (2008) Social network analysis (154). Sage
2. Wey T, Blumstein DT, Shen W, Jordan F (2008) Social network analysis of animal behaviour: a promising tool for the study of sociality. Anim Behav 75(2):333–344
3. Vemprala N, Dietrich G (2019) A social network analysis (SNA) study on data breach concerns over social media. In: Proceedings of the 52nd hawaii international conference on system sciences
4. Kim J, Hastak M (2018) Social network analysis: characteristics of online social networks after a disaster. Int J Inf Manage 38(1):86–96
5. Hagen L, Keller T, Neely S, DePaula N, Robert-Cooperman C (2018) Crisis communications in the age of social media: a network analysis of Zika-related tweets. Soc Sci Comput Rev 36(5):523–541
6. Breja M (2017) Social network analysis in question answering community. In: 2017 International conference on energy, communication, data analytics and soft computing (ICECDS). IEEE, pp 314–318
7. Eaton E (2013) CS 380: network analysis spring 2013—Section 02. Bryn Mawr College. https://cs.brynmawr.edu/Courses/cs380/spring2013/section02/slides/05_Centrality.pdf
8. Saxena A, Iyengar SR, Gupta Y (2015) Understanding spreading patterns on social networks based on network topology. In: Proceedings of the 2015 IEEE/ACM international conference on advances in social networks analysis and mining 2015, ACM, pp 1616–1617
9. Hanneman RA, Riddle M (2005) Introduction to social network methods. University of California, Riverside. CA (Online book)
10. Rihn J, Oliver JJ (2015) Analyzing traffic patterns to detect infectious messages. U.S. Patent No. 8,955,136
11. Akrouf S, Meriem L, Yahia B, Eddine MN (2013) Social network analysis and information propagation: a case study using Flickr and YouTube networks. Int J Future Comput Commun 2(3):246–252

12. Feroz Khan G, Vong S (2014) Virality over YouTube: an empirical analysis. Internet Res 24(5):629–647
13. Webometrics Analyst Homepage,http://lexiurl.wlv.ac.uk/
14. Davidson J, Liebald B, Liu J, Nandy P, Van Vleet T, Gargi U, Gupta S, He Y, Lambert M, Livingston B, Sampath D (2010) The YouTube video recommendation system. In: Proceedings of the fourth ACM conference on recommender systems, pp. 293–296
15. Cheng X, Dale C, Liu J (2008) J Statistics and social network of Youtube videos. In: 2008 16th international workshop on quality of service. IEEE, pp 229–238
16. Oliphant TE (2007) Python for scientific computing. Comput Sci Eng 9(3):10–20
17. Kotsiantis S, Kanellopoulos D (2006) Association rules mining: a recent overview. GESTS Int Trans Comput Sci Eng 32(1):71–82
18. Han J, Pei J, Kamber M (2011) Data mining: concepts and techniques. Elsevier

Feature Extraction Technique for Vision-Based Indian Sign Language Recognition System: A Review

Akansha Tyagi and Sandhya Bansal

Abstract Vision-based sign language recognition system is an emerging research area in human-computer interaction (HCI). It has been proved as a powerful communication tool for deaf and mute society, irrespective of geographical differences. Automatic sign language recognition system works broadly in three phases namely image pre-processing, feature extraction and classification of gestures. The output of the feature extraction phase is crucial for various classifiers. This paper presents a comprehensive review of feature extraction techniques used in vision-based sign language recognition system. A taxonomy of currently used techniques for feature extraction has been presented. The paper concludes by presenting future direction in feature extraction technique for Indian sign language (ISL).

Keywords Indian sign language (ISL) · Human-computer interaction (HCI) · Feature extraction · Vision-based · Soft computing

1 Introduction

Sign language (SL) is the non-verbal communication mode used by the deaf and mute community. However, gestures used in sign language are not standards and universal so they varies from region to region. The roots of SL are American sign language (ASL). In India, it is commonly known as Indian sign language (ISL). Automatic SL recognition starts from data acquisition phase. This can be done in two ways, i.e., computer vision [1–6] and sensor enabled device [7–10]. Camera is used for capturing gestures in formal technique while devices (Kinect sensor, data gloves, flex sensor, LMC) with sequence of sensor are used for capturing images in later method. However trained environment and handling of devices are not always

A. Tyagi (✉) · S. Bansal
Department of Computer Science and Engineering, Maharishi Markandeshwar (Deemed to be University), Ambala, Haryana, India
e-mail: Akan7sha@gmail.com

S. Bansal
e-mail: sandhya12bansal@gmail.com

V. Singh et al. (eds.), *Computational Methods and Data Engineering*, Advances in Intelligent Systems and Computing 1227,
https://doi.org/10.1007/978-981-15-6876-3_4

Fig. 1 Processing stages of ISL recognition system

possible for signer. Hence current research is going on vision-based sign language recognition [11–14]. Further images are pre-processed to acquire accurate gesture from captured image and transform them into gray images. These images are then used to extract features that can predict the gesture. Feature extraction plays a vital role in SLRS as it further provides input to classifier phase. In classification phase trained images and test images are classified on the basis of features extracted in previous phase. Hence recognition rate of gestures depends mainly on these two phases [15–18]. It is evident that the literature [19–23] has tremendous amount of work on feature extraction techniques. However there exists a wide scope in vision-based ISL recognition system that can be helpful for solving real world problems like automatic home appliances. Various processing stages involved in recognition of automatic SL system is shown in Fig. 1. This review paper focuses on feature extraction techniques currently being used in vision-based Indian sign language.

1.1 Motivation

Around 63 million people in India are suffering from Deaf-mute loss [6]. HCI works like bridge between deaf-mute section and other. Automatic SLR is also an application of HCI which mainly depends on hand gesture recognition. On the basis of modality gestures can be of two types that is manual gestures made by hands or non-manual made by using facial gestures. We can acquire gestures by two modes, i.e., vision-based gesture and device-based gestures. However, carrying out devices and a trained environment is a vital problem for signer. Hence vision-based ISL recognition is the most widely studied topic for research in sign language recognition. After acquisition of images feature extraction is done. The output of this phase is very crucial for the performance of classifier used in recognition of gestures. Therefore, feature extraction has gained more attention in the field of ISLR. With the advancement and availability of SLR, amount of work has been done on feature extraction techniques. Most of the work on vision-based ISL recognition is present in forms of journals articles, conference proceedings, thesis, technical reports/books, etc. Pisharady and Saerbeck [24] reviews the algorithms in the field since from last sixteen years based on quantitative and qualitative comparison which provides a thorough study. In [25] feature extraction technique that has been used so far are reviewed. While [26] has only analyzed and compared the methods applied in the SLR system and classification techniques that has been used and in results suggests the best promising technique that can be used in future researches. Wiryana et al. [22] has evaluated three commonly used feature extraction technique compared their performance on the basis of invariance, speed, storage requirement and recognition rates. Further [19] has done study on feature selection and extraction used from since last 10 years and conclude how they can affect system performance. In [23] has reviewed soft computing techniques based on learning pattern that can be used for feature extraction. Indeed, of such tremendous work, the taxonomy of feature extraction techniques is yet not presented.

An effort has been made here to review the presently used techniques in this field and to compare and contrast according to various parameters.

1.2 Survey Methodology

Keeping in view removing exhaustive literature review, we have tried to overview the work available on feature extraction technique. The papers we have referred are from 2010 to 2019. This survey paper presents a review of relevant work on feature extraction technique used in ISL gesture recognition. An exhaustive search has been conducted using the Google search engine. All of the technique has been analyzed in terms of accuracy. Approximately more than 70 articles from the field of gesture recognition has been selected, where main emphasis is on vision-based ISLR. We prioritized the papers with good citations. Some other relevant literature from other journals has been also selected where we felt necessary for clarity purpose. Collected data was initially selected on the basis of relevancy to the feature extraction process. Rest of the paper is organized as follows. Section 2 presents introduction to feature extraction and various techniques used for feature extraction in ISLR, and categorization of these techniques in form of taxonomy. Section 3 presents the related work on these techniques. Finally, conclusion and future scope is given in Sect. 4.

2 Feature Extraction in ISL and Its Techniques

Feature extraction is the most fundamental step in ISL after image pre-processing. This stage is very crucial for ISLR system performance. Feature extraction addresses to the process of finding the most promising and informative features set to improve accuracy and efficiency of the data to be tested [27]. Features are also referred to as descriptors. Selection of appropriate features and feature extraction methods is the most critical decision for the success of system. They are two types of descriptors, i.e., local and global features. Global features define the image as a whole to generalize the entire object. Global descriptors [14, 19, 28] are normally used for image retrieval, object recognition and classification such as shape matrices, invariant moments (Hu, Zernike), HOG and Co-HOG. On the counterpart local features defines the image patches and texture in an image patch. The local descriptors [20, 21, 29] are used for object recognition and identification [30] like SIFT, SURF. Various techniques such as Zernike moments, Hu moments, HOG, SIFT, ED, FD, DWT, ANN, CNN, Fuzzy,

GA have been employed for feature extraction in vision-based gesture recognition system [24, 25, 31–35].

Description of the commonly used techniques are as follows:

1. **Zernike moments**: These are a set of moments produced on an orthogonal basis defined on the unit circular disk. They characterize a basic determined arrangement of distinct properties that can be utilized for shape extraction and object detection. Mathematically, Zernike moments are in order '*n*' with repetition '*m*' for continuous image function $f(x, y)$, where x, y are the polar coordinates on unit disk for Eq. (1) of order set. Feature vector A_{nm} size will be increased with an increase in order.

$$A_{nm} = \left(\frac{n+1}{\pi}\right) \int \int_{x^2+y^2\leq 1}^{\pi} V_{nm}(x, y) f(x, y) dx dy. \tag{1}$$

2. **Contour moments**: These moments extract features from image based on the spatial moments. Contour moments are based on central features. These moments are invariant to translation. To make them scale independent, they are scaled on the basis of object area. The normalized contour moments can be calculated as Eq. (2).

$$nu_{ji} = \frac{mu_{ji}}{m_{00}^{\frac{(i+j)}{2+1}}} \tag{2}$$

Where nu_{ji} are the normalized $(i + j)$ th order moment feature, mu_{ji} are the $(i + j)$ th order spatial moments where $(i, j) \in 0,1,2,3$.

3. **Hu moments**: They are based on set of 7 ordeation whereas first 6 moments are independent to scaling, translation, orientation and reflection. These moments also provide invariance to the infinite image resolution. But these techniques are dependent on orientation of image and background illumination. Equation (3) shows Hu moments at order 1, similarly it is derived up-to 7th orders.

$$Hu[0] = \eta WX + \eta XW \tag{3}$$

where ηWX and ηXW denote $(i + j)$th order moments at coordinates (i, j), respectively.

4. **Euclidean distance (ED)**: The ED is the straight-line distance between two pixels with linear transformation. They give feature vector for stretching, compression and flipping of image without affecting the original image. Large number of eigenvalue depicts the accurate gesture. Euclidean distance can be evaluated in the form of Eq. (4).

$$E.D = \sqrt{\sum_{m}^{n} E_{v1(n)} - E_{v2(n)}} \tag{4}$$

Where $E_{v1(n)}$ and $E_{v2(n)}$ denotes the eigenvector set of t>he test image and database image respectively.

5. **Fourier Descriptor (FD)**: These techniques use Fourier transform for encoding the shape of 2D object where every (*u, v*) point in the boundary map to a complex number $(u + iv)$. It provides smooth and simplified boundaries by using inverse transformation, which also recovered the original shape.
6. **Discrete Wavelet Transform (DWT)**: These wavelets are discretely sampled. The main fact behind wavelet is that they can be integrated to zero so they can wave up and down around the axis. In comparison to Fourier transform it can capture both frequency and location information. Decomposition of multi-resolution images is done on the basis of wavelet coefficient and scaling which makes it invariant to orientation.
7. **Scale Invariant Feature Transform (SIFT)**: This works on local features at different scales without affecting scaling, rotation or translation of the image. SIFT is also partially invariant to illumination changes with a level of tolerance for view point. Due to its advantage of low probability mismatch it allows to extract accurate object detection with its location and pose. It can also give better recognition rate for visual recognition system for small databases.
8. **Speeded Up Robust Feature (SURF)**: It is an improved version of SIFT with a capacity of computing distinctive features quickly commonly used for object detection, image registration and classification. Compared to SIFT it is faster and more robust against image transformation. SURF algorithm is consisting of three components, first component is interest point detection, second one is local neighborhood description and third is matching.
9. **Histogram Oriented Gradient Descriptor (HOG)**: HOG is based on occurrences of gradient orientation in local portion of the image. It is independent of illumination and pre-processing of image. It is computed on dense grid using overlapping of cells for normalization that improves its accuracy. Also, geometric and photometric transformation does not affect the original image as it operates on cells.
10. **Genetic Algorithm (GA)**: GA is a search-based optimization technique which mimics the process of natural selection. In GA images are taken as a pool or population on which operation of mutation, crossover and selection of fittest are applied. Because of its powerful optimization technique, it is used for image enhancement and segmentation. GAs are basically the evolutionary based algorithms.
11. **Fuzzy Logic**: Fuzzy sets and fuzzy logic has a capability to handle roughness and uncertainty in data. It represents the vagueness and imprecise information of images. Fuzzy logic substitutes the need of image segmentation as it can handle both image smoothing, filtering and noise removal itself.
12. **Neural Network**: Neural network are a set of algorithms designed to recognize patterns, the way human brain operates. In neural network, input is taken as x_i which has some weight w_{ij} to form net input, where *i, j* is the input layer. The net input signal for threshold value is calculated by shown in Eq. (5):

$$net_j = \sum_{i}^{n} x_i w_{ij} \tag{5}$$

Further neural network is broadly divided into two categories artificial neural network (ANN) and convolution neural network (CNN) which are commonly used for feature extraction and recognition. ANNs focuses on the boundaries and identify different feature which are invariant to translation, rotation, shear, scale, orientation, illumination and stretch while CNNs are specially designed for natural feature extraction process because of shift variance functionality.

13. **Hybrid**: Hybrid techniques are those techniques that integrates two or more above discussed methods.

After the exhaustive examination of above used methods in literature we had made following observations.

1. Techniques i-ii-iii make use of statistical parameters, so we had collectively put them in statistical category.
2. Techniques iv-v-vi are based on shape extraction phenomenon which invariant to translation; hence they are categorized into shape transform based technique.
3. Statistical and shape transform techniques are categorized as Content based image retrieval (CBIR) as they can extract features based on content of image like shape, texture and color.
4. Techniques from vii–xii are invariant to illumination, outplace the use of image pre-processing and can recognize large number of gestures hence they are classified under soft computing techniques.
5. Hybrid techniques are the fusion of CBIR and soft computing technique to improve the recognition rate and makes system more efficient.

A taxonomy of these techniques on the basis of above observations has been made (see Fig. 2).

3 Related Work

A summative review work of the techniques discussed in previous section used for SLR system in feature extraction process has been presented in this section.

3.1 CBIR

Statistical: Zernike moments requires lower computation time compared to regular moments [36, 37]. In [33] these moments are used for extraction of mutually independent shape information while [38] has used this approach on Tamil scripts to overcome the loss in information redundancy of geometric mean. Oujaoura et al.

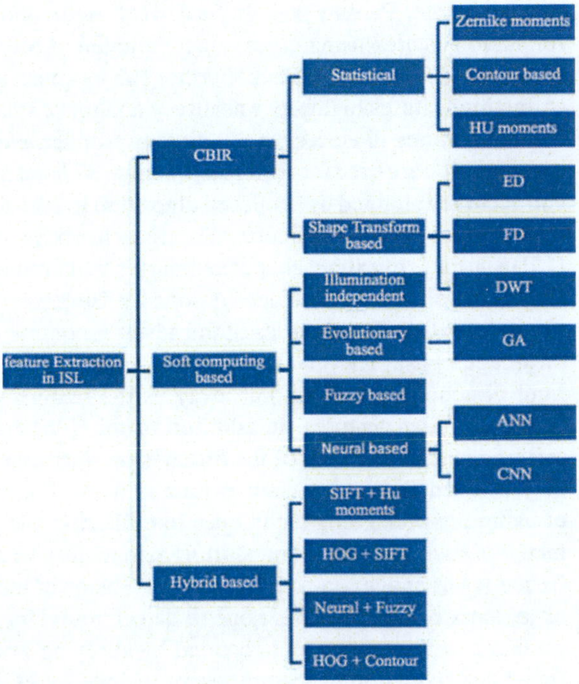

Fig. 2 Taxonomy of feature extraction techniques

[36, 37] has used this method to find luminance and chrominance characteristic of image for locating the forged area up-to order 5. In addition to this [32] used this to extract hand shape. But recognition rate lacks in similar structures such as (M, N) and (C, L). Contour moments are further used to extracts features from image based on the boundaries of an object. In [34, 39] these techniques have been used to represent moments based on statistical distribution like variance, area, average and finger-tips. Convexity defect detection and convex hull formation is used for extraction. The proposed system will work under ideal illuminated condition and the accuracy obtained is 94%. Most apparently used statistical technique is Hu moments as it can calculate central moments also. Rokade and Jadav [40] used these to describe, characterize and quantify the shape of an object in an image. Fourier descriptors are used in [40] to generate projection vectors. Further thirteen features were extracted from each sign language. Hu invariants moments are applied to extract geometrical moments of hand region [41]. Although computation of these feature vector is easy here, but recognition rate is less efficient. These features are invariant to shape and angles but variant to background and illumination.

Shape based: These techniques are based on phenomenon that without any change in shape of image we can extract the accurate features. Khan and Ibraheem [42] determines active finger count by evaluating the ED distance between palm and wrist. As a result, feature vector of Finger projected distance (FPD) and finger base angle (FBA) are computed. But the features selection depends on orientation and rotation angle. Pansare and Ingle [14], Singha and Das [43] proposed a system for static gestures using eigenvalue weighted. Accuracy achieved by the proposed system is 97% on 24 gestures. Further ED has been also used with convex- hull for extracting features to improve accuracy and make system more reliable [44]. Several distinct features like eccentricity, fingertip finder, elongatedness, rotation and pixel segmentation are used for feature extraction. 37 hand gestures are used for recognition and accuracy attained by proposed algorithm in real-time environment is 94.32%. To improve recognition rate Fourier descriptor has been used. Kishore and Rajesh Kumar [15] uses these to extract shape boundary's with minimum loss of shape information. Classification of trained images is done by using train fuzzy inference system. While to extract external boundaries of the gestures contour extraction technique is applied on images [45]. The main 220 coefficients of Fast Fourier Transform (FFT) for limit directions were then put away as the feature vector resulting in recognition of even similar gestures. In addition to this [46] frames were extracted from the reference video and each of the frame is pre-processed individually. All the features of processed frames are then extracted using Fourier descriptor method. Instead of using pre-processing techniques like filtering and segmentation of hand gesture, methods such as scaling and shifting parameters were extracted based on high low frequency of images up-to 7th level [47]. Fusion of morphological process and canny edge operator with DWT is done to detect boundary pixels of the hand image [48]. Further features vector set is created by applying Fourier descriptors on each pixel frame and reduction of feature vector is done by PCA. Proposed system concludes that more the number of training samples more accuracy is attained, i.e., 96.66%. Further [6] has also used DWT to overcome the limitations of device-based and vision-based approach. These feature extraction techniques lack for large database in terms of accuracy and efficiency [45]. They also cannot perform well in cluttered background [48] and are variant to illumination changes.

3.2 Soft Computing Based Feature Extractions

Soft computing is an emerging approach in field of computing that gives a remarkable ability of learning in the atmosphere uncertainty. It is a collection of methodologies such as neural network, fuzzy logic, probabilistic reasoning methods and evolutionary algorithms to achieve traceability, robustness, and low computation cost. It has been observed that these techniques perform well on a large database and with vague in images [49, 50]. Soft computing has also been successfully applied in applied other fields like optimization [51], VRP [52–54], and pattern recognition [55]. It is aimed for extracting the relevant features automatically.

Illumination independent based feature extraction such as SIFT, SURF and HOG are commonly used for ISL recognition. Dardas et al. [56] has used SIFT algorithm with bag-of-feature model to extract vectors from each image. Sharing feature concept is used to speed up the testing process and accuracy achieved by the system is up-to 90%. Gurjal and Kunnur [57] works on low-resolution images in real-time environment. Pandita and Narote [58] developed improved SIFT method to compute edges of an image [59] extracts distinct features and feature matching using SIFT which results in robustness to noise. Further an improved version of SIFT, i.e., SURF has been [60] used with affine invariant algorithm, which is partially invariant to viewpoint changes in image, results in a computation efficient system. Yao and Li [61] uses SURF with various sized filters for fast convolution to make system less sensitive to computational cost. Also, the system shows high performance in images having noisy background. Multi-dimensional SURF is used to reduce the number of local patches yielding a much faster convergence speed of SURF cascade [62]. In addition to describe appearance and shape of local object within an image HOG is used in [63, 64]. Tripathi and Nandi [65] works on continuous gesture recognition by storing 15 frames per gesture in database.

Hamda and Mahmoudi [66] uses HOG for vision-based gesture recognition. Reddy et al. [35] extracts global descriptors of image by local histogram feature descriptor (LHFD). Evolutionary algorithm is now marking a trend in field of HCI, hence in ISL recognition they are very convenient for use. These are very useful when feature vector is large [67, 68] uses GA with a feedback linkage from classifier. In addition to an improved GA working directly on pixels has been applied in [69] resulting in a better recognition rate. To reduce computation time fuzzy rule system has been further used for ISL recognition. Fang et al. [70] used fuzzy decision tree for large, noisy system to reduce the computational cost due to large recognized classes. Kishore et al. [71] has used Sugeno fuzzy inference system for recognition of gesture by generating optimum fuzzy rules to control the quality of input variables. fuzzy c-means clustering has been used to recognize static hand gestures in [49]. Verma and Dev [50] uses fuzzy logic with finite state machine (FSM) for hand gesture recognition by grouping data into clusters. Nölker and Ritter [72] describe GREFIT algorithm based on ANN to detect continuous hand gesture from gray-level video images. This approach works well for blur images; however, it requires high frame rate with vision acquisition. In addition to this [73] uses ANN algorithm for selfie-based video continuous ISL recognition system for embedding it into smartphones. [74] develops three novel methods (NN-GA, NN- EA and NN-PSO) for effective recognition of gestures in ISL. The NN has been optimized using GA, EA and PSO. Experimental results conclude that NN-PSO approach outperforms the two other methods. Huang et al. [75] uses CNNs for automating construction of pool with similar local region of hand. Yang and Zhu [76] applied CNNs to directly extracts images from video. Ur Rehman et al. [77], Li et al. [78], Beena et al. [79] automatic clustering of all frames for dynamic hand gesture is done by CNNs. Three max-pooling layers, two fully connected layers and one SoftMax layer constitutes the model.

Disadvantage of Soft computing technique: Although these techniques provides accuracy, but feature vector size is large. So, it requires feature extraction approaches resulting in high time complexity.

3.3 Hybrid Technique for Feature Extraction

Fusion of soft computing based and CBIR bases techniques are also employed in literature to have advantages of both techniques. Sharath Kumar and Vinutha [80] integrates SURF and Hu moments to achieve high recognition rate with less time complexity. Agrawal [32] embeds SIFT and HOG for robust feature extraction of images in cluttered background and under difficult illumination. Singha and Das [13] uses Haar like features for skin and non-skin pixel differentiation and HOG is used for feature vector extraction. Dour and Sharma [81] a neural-fuzzy fusion is done to reduce complexity of similar gesture recognition. To improve efficiency a multiscale oriented histogram within addition to contour directions is used for feature extraction [9]. This integration of approaches makes system memory efficient with high recognition rate of 97.1%. Hybrid approaches develops efficient and effective system, but implementation is complex.

Based on the analysis, Table 1 summarizes some of the selected articles for feature extraction in ISL. The first column enlists the paper and the second column represents the technique used. The gestures recognized, advantages and disadvantages of the technique are discussed in column three, four and five respectively. The last column discusses the accuracy achieved by the adopted technique. Most of the recent work (64%) is devoted to use of CBIR technique in ISL recognition. Among them statistical and shape transform technique are commonly used approaches. Some of the soft computing techniques such as HOG, ANN, Fuzzy, etc., have also been used for dynamic gesture recognition.

4 Conclusion and Future Direction

Vision-based ISL is a boon for deaf-mute people to express their thoughts and feelings. The accurate recognition of gesture in ISL depends upon feature extraction phase. Owing to different orientation of hands, background, light conditions, etc., there exists various feature extraction techniques in ISL. A lot of research is yet being going in this area. However, to our best knowledge no efforts have been done to provide a systematic review of the work after 2010. We have attempted to bridge the gap by reviewing some of the significant feature extraction techniques in this area. A taxonomy of various techniques, categorizing them into three board groups namely: CBIR, soft computing and hybrid is also developed. A comparative table of recent work is also presented.

From the previous work, hybrid and soft computing appears as promising for real-time gesture recognition; however, CBIR methods are cost effective for static gesture.

Table 1 Comparison of ISL feature extraction techniques

Paper	Feature extraction technique	Gesture's	Advantage	Disadvantage	Accuracy (%)
[43]	Euclidean distance	24	Less time complexity, recognize double-handed gesture, differentiate skin color	Only static images have been used.	97
[14]	Euclidean distance	24	On video sequence, recognize single and double-handed gestures accurately	Works only in ideal lighting conditions	96.25
[45]	Fourier Descriptors	15	Differentiated similar gestures	Large dataset	96.15
[47]	Fourier Descriptor	46	Dynamic gestures	Dataset of 130,000 is used	92.16
[48]	DWT	52	Considers dynamic gesture	Simple background, large dataset	81.48
[6]	DWT	24	Increase adaptability in background complexity and illumination	Less efficient for similar gestures	90
[70]	Fuzzy logic	90	Invariant to scaling, translation and rotation	Can't work in real-time system	96
[74]	ANN	22	No noise issue, data normalization is easily done	–	99.63
[81]	Fuzzy + neural	26	High recognition rate for single and double-handed gestures	Accuracy lacks for similar gestures	97.1

After extensive review of recently used techniques, some of the significant gaps to be filled for future work in this area are as follows: Firstly, it has been observed that the main focus is on developing complex and accurate techniques, but an effective and efficient technique is the required. Secondly, it has been observed that proposed techniques lack accuracy for similar gestures. As a result, there is still a potential of improvement in the techniques used. Third, although the recent techniques work well for different background, light conditions, orientation of hands, etc., for lesser gestures but loses efficiency for large databases. So, there is scope of further work to make them efficient for large databases. Finally, it has been also observed that proposed techniques achieve high accuracy for static gestures but should also be able to recognize dynamic gestures, sentences and phrases efficiently.

References

1. Rahaman MA, Jasim M, Ali MH, Hasanuzzaman M (2003) Real-time computer vision-based Bengali sign language recognition. In: 2014 17th international conference computer information technology ICCIT 2014, pp 192–197
2. Zhang L-G, Chen Y, Fang G, Chen X, Gao W (2004) vision-based sign language recognition system using tied-mixture density HMM. In: Proceedings of the 6th international conference on Multimodal interfaces (ICMI '04), pp 198–204
3. Garg P, Aggarwal N, Sofat S (2009) Vision based hand gesture recognition. World Acad Sci Eng Technol 49(1):972–977
4. Ren Y, Gu C (2010) Real-time hand gesture recognition based on vision. In: International conference on technologies for e-learning and digital entertainment, pp 468–475
5. Ibraheem NA, Khan RZ (2012) Vision based gesture recognition using neural networks approaches: a review. Int J Human Comput Interact (IJHCI) 3(1):1–14
6. Ahmed W, Chanda K, Mitra S (2017) Vision based hand gesture recognition using dynamic time warping for indian sign language. In: Proceeding of 2016 international conference information science, pp120–125
7. Juneja, S, Chhaya Chandra PD, Mahapatra, SS, Bahadure NB, Verma S (2018) Kinect Sensor based Indian sign language detection with voice extraction. Int J Comput Sci Inf Secur (IJCSIS) 16(4)
8. Ren Y, Xie X, Li G, Wang Z, Member S (2018) Hand gesture recognition with multiscale weighted histogram of contour direction normalization for wearable applications. IEEE Trans Circuits Syst Video Technol 28:364–377
9. Joy J, Balakrishnan K, Sreeraj M (2019) SignQuiz: a quiz based tool for learning fingerspelled signs in indian sign language using ASLR. IEEE Access 7:28363–28371
10. Mittal A, Kumar P, Roy PP, Balasubramanian R, Chaudhuri BB (2019) A modified- LSTM model for continuous sign language recognition using leap motion. IEEE Sens J 19
11. Cheok MJ, Omar Z, Jaward MH (2019) A review of hand gesture and sign language recognition techniques. Int J Mach Learn Cybern 10:131–153
12. Rautaray SS, Agrawal A (2012) Vision based hand gesture recognition for human computer interaction: a survey. Artif Intell Rev 43:1–54
13. Singha J, Das K (2013) Recognition of Indian sign language in live video. Int J Comput Appl 70:17–22
14. Pansare JR, Ingle M (2016) Vision-based approach for american sign language recognition using edge orientation histogram. 2016 Int Conf Image Vis Comput ICIVC 86–90
15. Kishore PVV, Rajesh Kumar P (2012) A video based Indian sign language recognition system (INSLR) using wavelet transform and fuzzy logic. Int J Eng Technol 4(5):537
16. Hore S, Chatterjee S, Santhi V, Dey N, Ashour AS, Balas VE, Shi F (2017) Indian sign language recognition using optimized neural networks. In Inf Technol Intell Transp Syst pp 553–563
17. Suharjito, WF, Kusuma GP, Zahra A (2019) Feature Extraction methods in sign language recognition system: a literature review. In: 1st 2018 Indonesian association for pattern recognition international conference (INAPR), pp 11–15
18. Narang S, Divya Gupta M (2015) Speech feature extraction techniques: a review. Int J Comput Sci Mob Comput 43:107–114

19. Pavlovic VI, Sharma R, Huang TS (1997) Visual interpretation of hand gestures for human-computer interaction: a review. IEEE Trans Pattern Anal Mach Intell 19:677–695
20. Marcel S (2002) Gestures for multi-modal interfaces: a review, technical report IDIAP-RR 02–34
21. Ping Tian D (2013) A review on image feature extraction and representation techniques. Int J MultimediaUbiquitous Eng 8(4):385–396
22. Wiryana F, Kusuma GP, Zahra A (2018) Feature extraction methods in sign language recognition system: a literature review. In: 2018 Indonesian Association for Pattern Recognition International Conference (INAPR), pp 11–15
23. Yasen M, Jusoh S (2019) A systematic review on hand gesture recognition techniques, challenges and applications. PeerJ Comput Sci 5:e218
24. Pisharady PK, Saerbeck M (2015) Recent methods and databases in vision-based hand gesture recognition: a review. Comput Vis Image Underst 141:152–165
25. Bhavsar H, Trivedi J (2017) Review on feature extraction methods of image based sign language recognition system. Indian J Comput Sci Eng 8:249–259
26. Kusuma GP, Ariesta MC, Wiryana F (2018) A survey of hand gesture recognition methods in sign language recognition. Pertanika J Sci Technol 26:1659–1675
27. Fei L, Lu G, Jia W, Teng S, Zhang D (2019) Feature extraction methods for palmprint recognition: a survey and evaluation. IEEE Trans Syst Man Cybern Syst 49:346–363
28. Tuytelaars T Mikolajczyk K (2008) Local invariant feature detectors: a survey. Found Trends® in Comput Graph Vis 3(3):177–280
29. Chaudhary A, Raheja JL, Das K, Raheja S (2011) A survey on hand gesture recognition in context. Adv Comput 133:46–55
30. Juan, L, Gwon L (2007) A comparison of sift, pca-sift and surf. Int J Sign Proc Image Proc Pattern Recogn 8(3):169–176
31. Athira PK, Sruthi CJ, Lijiya A (2019) A signer independent sign language recognition with co-articulation elimination from live videos: an indian scenario. J King Saud Univ Comput Inf Sci 0–10
32. Agrawal SC, Jalal AS, Bhatnagar C (2012) Recognition of Indian sign language using feature fusion. In 2012 4th international conference on intelligent human computer interaction (IHCI), pp 1–5
33. Li S, Lee MC, Pun CM (2009) Complex Zernike moments features for shape-based image retrieval. IEEE Trans Syst Man, Cybern Part ASyst Humans 39:227–237
34. Kakkoth SS (2018) Real time hand gesture recognition and its applications in assistive technologies for disabled. In: 2018 fourth international conference computer communication control automatically, pp 1–6
35. Reddy DA, Sahoo JP, Ari S (2018) Hand gesture recognition using local histogram feature descriptor. In: Proceeding 2nd international conference trends electronic informatics, ICOEI 2018, pp 199–203
36. Oujaoura M, El Ayachi R, Fakir M, Bouikhalene B, Minaoui B (2012) Zernike moments and neural networks for recognition of isolated Arabic characters. Int J Comput Eng Sci 2:17–25
37. Zhao Y, Wang S, Zhang X, Yao H (2013) Robust hashing for image authentication using zernike moments and local features. IEEE Trans Inf Forensics Secur 8:55–63
38. Sridevi N, Subashini P (2012) Moment based feature extraction for classification of handwritten ancient Tamil Scripts. Int J Emerg Trends 7:106–115
39. Haria A, Subramanian A, Asokkumar N, Poddar S (2017) Hand gesture recognition for human computer interaction. Procedia Comput Sci 115:367–374
40. Rokade YI, Jadav PM (2017) Indian sign language recognition system. Int J Eng Technol 9:189–196
41. Dardas NH, Georganas ND (2011) Real-time hand gesture detection and recognition using bag-of-features and support vector machine techniques. IEEE Trans Instrum Meas 60:3592–3607
42. Khan R, Ibraheem NA (2014) Geometric feature extraction for hand gesture recognition. Int J Comput Eng Technol (IJCET) 5(7):132–141

43. Singha J, Das K (2013) Indian sign language recognition using eigen value weighted euclidean distance based classification technique. Int J Adv Comput Sci Appl 4:188–195
44. Islam M, Siddiqua S, Afnan J (2017) Real time hand gesture recognition using different algorithms based on american sign language. In: 2017 IEEE International Conference Imaging, Vision and Pattern Recognition, pp 1–6
45. Shukla, P, Garg A, Sharma K, Mittal A (2015) A DTW and fourier descriptor based approach for indian sign language recognition. In: 2015 third international conference on image information processing (ICIIP). IEEE, pp 113–118
46. Badhe PC, Kulkarni V (2016) Indian sign language translator using gesture recognition algorithm. In: 2015 IEEE international conference computer graph visualization information security. CGVIS 2015, pp 195–200
47. Kumar N (2017) Sign language recognition for hearing impaired people based on hands symbols classification. In: 2017 international conference on computing, communication and automation (ICCCA). IEEE, pp 244–249
48. Prasad MVD, Kishore PVV, Kiran Kumar E, Anil Kumar D (2016) Indian sign language recognition system using new fusion based edge operator. J Theor Appl Inf Technol 88:574–558
49. Korde SK, Jondhale KC (2008) Hand gesture recognition system using standard fuzzy C-means algorithm for recognizing hand gesture with angle variations for unsupervised users. In: Proceeding 1st international conference on emerging trends in engineering, technology. (ICETET) 2008, pp 681–685
50. Verma R, Dev A (2009) Vision based hand gesture recognition using finite state machines and fuzzy logic. In: 2009 international conference on ultra modern telecommunications work, pp 1–6
51. Jang JSR, Sun CT, Mizutani E (1997) Neuro-fuzzy and soft computing-a computational approach to learning and machine intelligence [Book Review]. IEEE Trans Autom Control 42(10):1482–1484
52. Bansal S, Goel R, Mohan C (2014) Use of ant colony system in solving vehicle routing problem with time window constraints. In: Proceedings of the second international conference on soft computing for problem solving, pp 39–50
53. Bansal S, Katiyar V (2014) Integrating fuzzy and ant colony system for fuzzy vehicle routing problem with time windows. Int J Comput Sci Appl (IJCSA) 4(5):73–85
54. Goel R, Maini R (2017) Vehicle routing problem and its solution methodologies: a survey. Int J Logistics Syst Manage 28(4):419–435
55. Singh V, Misra AK (2017) Detection of plant leaf diseases using image segmentation and soft computing techniques. Inf Process Agric 4(1):41–49
56. Dardas N, Chen Q, Georganas ND, Petriu EM (2010) Hand gesture recognition using bag-of-features and multi-class support vector machine. In: 2010 IEEE international symposium on haptic audio visual environment, pp 1–5
57. Gurjal P, Kunnur K (2012) Real time hand gesture recognition using SIFT. Int J Electron Electr Eng 2(3):19–33
58. Pandita S, Narote SP (2013) Hand gesture recognition using SIFT ER. Int J Eng Res Technol (IJERT) 2(1)
59. Mahmud H, Hasan MK, Tariq AA, Mottalib MA (2016) Hand gesture recognition using SIFT features on depth image. In: Proceedings of the ninth international conference on advances in computer-human interactions (ACHI), pp 359–365
60. Pang Y, Li W, Yuan Y, Pan J (2012) Fully affine invariant SURF for image matching. Neurocomputing 85:6–10
61. Yao, Y, Li, C-T (2012) Hand posture recognition using surf with adaptive boosting. In: British Machine Vision Conference Workshop, pp 1–10
62. Li J, Zhang Y (2013) Learning SURF cascade for fast and accurate object detection. In: Proceedings of the IEEE computer society conference on computer vision and pattern recognition, pp 3468–3475
63. Tavari NV, Deorankar AV (2014) Indian sign language recognition based on histograms of oriented gradient. Int J Comput Sci Inf Technol 5(3):3657–3660

64. Chaudhary A, Raheja JL (2018) Optik Light invariant real-time robust hand gesture recognition. Opt Int J Light Electron Opt 159:283–294
65. Tripathi K, Nandi NBGC (2015) Continuous indian sign language gesture recognition and sentence formation. Procedia Comput Sci 54:523–531
66. Hamda M, Mahmoudi A (2017) Hand gesture recognition using kinect's geometric and hog features. In: Proceedings of the 2nd international conference on big data, cloud and applications, ACM, p 48
67. Cerrada M, Vinicio Sánchez R, Cabrera D, Zurita G, Li C (2015) Multi-stage feature selection by using genetic algorithms for fault diagnosis in gearboxes based on vibration signal. Sens (Basel, Switzerland) 15(9):23903–23926
68. Ibraheem NA, Khan RZ (2014) Novel algorithm for hand gesture modeling using genetic algorithm with variable length chromosome. Int J Recent and Innov Trends Comput Commun 2(8):2175–2183
69. Kaluri R, Reddy CP (2016) A framework for sign gesture recognition using improved genetic algorithm and adaptive filter. Cogent Eng 64:1–9
70. Fang G, Gao W, Zhao D (2004) Large vocabulary sign language recognition based on fuzzy decision trees. IEEE Trans Syst Man, Cybernet-Part A: Syst Humans 34(3):305–314
71. Kishore PVV, Rajesh Kumar P (2014) A video based indian sign language recognition system (INSLR) using wavelet transform and fuzzy logic. Int J Eng Technol 4:537–542
72. Nölker C, Ritter H (2002) Visual recognition of continuous hand postures. IEEE Trans Neural Networks 13:983–994
73. Rao GA, Kishore PVV (2018) Selfie video based continuous Indian sign language recognition system. Ain Shams Eng J 9(4):1929–1939
74. Hore S, Chatterjee S, Santhi V, Dey N, Ashour AS, Balas VE, Shi F (2017) Indian sign language recognition using optimized neural networks. Adv Intell Syst Comput 455:553–563
75. Huang J, Zhou W, Li H, Li W (2015) Sign language recognition using 3D convolutional neural networks. In: 2015 IEEE international conference on multimedia expo, pp 1–6
76. Yang S, Zhu QX (2018) Video-based chinese sign language recognition using convolutional neural network. In: 2017 9th IEEE international conference on communication software networks, ICCSN 2017. 2017-Janua, pp 929–934
77. Ur Rehman MZ, Waris A, Gilani SO, Jochumsen M, Niazi IK, Jamil M, Farina D, Kamavuako EN (2018) Multiday EMG-based classification of hand motions with deep learning techniques. Sensors (Switzerland) 18:1–16
78. Li J, Huai H, Gao J, Kong D, Wang L (2019) Spatial-temporal dynamic hand gesture recognition via hybrid deep learning model. J Multimodal User Interfaces 13:1–9
79. Beena MV, Namboodiri MA, Dean PG (2017) Automatic sign language finger spelling using convolution neural network: analysis. Int J Pure Appl Math 117(20):9–15
80. Sharath Kumar YH, Vinutha V (2016) Hand gesture recognition for sign language: a skeleton approach. Adv Intell Syst Comput 404:611–623
81. Dour G, Sharma S (2016) Recognition of alphabets of indian sign language by Sugeno type fuzzy neural network. Pattern Recognit Lett 30:737–742

Feature-Based Supervised Classifier to Detect Rumor in Social Media

Anamika Joshi and D. S. Bhilare

Abstract Social media is the most important and powerful platform for sharing information, ideas, and news almost immediately. With this, it also attracted antisocial elements for spreading and distributing rumors that is unverified information. Malicious and intended misinformation spread on social media has a severe effect on societies, people and individuals, especially in case of real-life emergencies such as terror strikes, riots, earthquakes, floods, war, etc. Thus, to minimize the harmful impact of rumor on society, it will be better to detect it as early as possible. The objective of this research and analysis is to develop a modified rumor detection model targeted for the proliferation of any malicious rumors related to any significant events. It is achieved through a binomial supervised classifier. The classifier uses a combination of explicit and implicit features to detect rumors. Our enhanced model significantly achieved it with 85.68% accuracy.

Keywords Rumor detection · Social media data analysis · Classification · Feature-based supervised model

1 Introduction

Social media has opened a new door for useful and versatile group communication. People have uncontrolled reach and span much more than ever before. It is a very useful platform for sharing information, ideas, and news. People have the power to spread information and news almost instantly. It affects almost all aspects of life. Social media like Twitter is mostly used and is an important source of news especially at the time of emergency [1]. Twitter could be extremely helpful during an emergency, but it could be as harmful when misinformation is rapidly spread during an emergency or crisis [2] and [3].

One immense pro and con about social media is that it spreads widely news that is not verified and confirmed. Misinformation spread on social media fast and

A. Joshi (✉) · D. S. Bhilare
School of Computer Science, Devi Ahilya University, Indore, MP, India
e-mail: anajoshi4@gmail.com

V. Singh et al. (eds.), *Computational Methods and Data Engineering*, Advances in Intelligent Systems and Computing 1227,
https://doi.org/10.1007/978-981-15-6876-3_5

they have caused harm from financial losses, virus Ebola scares, riots, disorder, etc. Misinformation particularly at the time of emergency may cause disturbance or unrest situation. Like Mass exodus of people of northeast India in 2012 [4] and [5], the riots in Muzaffarnagar in 2013 [6] and in Jammu and Kashmir in 2017 [7]. That's why nowadays most of the time during emergency, government, or law enforcement agencies stop internet and social media services to maintain law and order. For example, in Haryana at the time of Ram Rahim case [8] and recently in Jammu and Kashmir after the removal of article 370 [9].

Information whether true or not, verified or not, pass through social media rapidly. The recourse to beat back a rumor or misinformation is either to spread correct and authentic information instead of or classify rumors as true or false. This classification of rumors will drastically reduce the amount of data a vigilance service has to examine and act on.

The rest of the research work is arranged as follows. In Sect. 2, the overview is presented of the related work and in Sect. 3 we have described our proposed rumor detection model, especially explicit and implicit features that significantly contributing to rumor detection. In Sect. 4, we explain our evaluation and experiment results and the conclusion of the research and analysis is at the end with future work.

2 Related Work

Social media network like Twitter is being progressively used by professionals, organizations, and individuals as a primary source of information to find out the current affairs [10–12]. In spite of the rising potential of Twitter as a primary source of information, its tendency is to spread misinformation that is rumors and its impact attracted several researchers [13, 14]. Researchers studied, analyzed, and developed the ways to detect and classify rumors so that the end-users can get accurate and verified information. With this, we can also lessen the impact of the misinformation on individuals, organizations, and society.

Rumor detection problem is a classification problem and most of the rumor detection models are based on supervised learning. The main key factor of the classification model is feature extraction. Most of the classifiers based on explicit features. The existing extracted features for detection of rumor can be grouped into the following groups:

- The user-based properties.
- The content-based properties.
- The propagation-based properties.
- The linguistic (implicit) properties.

The recognized research work in the field of rumor detection is as shown in Table 1.

Most of the research works are based on explicit features. Some of the researchers also include some implicit features like linguistic features, internal and external consistency, etc. But by analyzing and including some more implicit features like sentiment or viewpoint of messages and replies we can enhance the accuracy and efficiency of a classifier to detect rumor.

Table 1 Important research works in rumor detection

Recognized research work	Data source	Contribution	Classifiers
Yang et al. [15]	Chinese micro-blogging platform Sina Weibo	It is a double approach feature having a client-based and location-based properties. The client-based properties give evidence that which software was used to send the message. The location-based properties give details about geographical location of the message where the relevant event happened or not	SVM
Castillo et al. [16]	Twitter	Message-based, user-based, topic-based (section of tweets having URL, hashtags, etc.), and propagation-based properties Initiate work in this direction	Bayesian networks, SVM classifiers, and decision trees based on J48
Kwon et al. [17]	Twitter	Proposed temporal, structural and linguistic features	Random forest, logistic regression, decision tree, and SVM
Liu et al. [18]	Twitter	Extend [16] and [15] work with verification features. These features include "source credibility, source identification, source diversity, source and witness location, and event propagation and belief identification".	Random forest, Decision Trees and SVM

(continued)

Table 1 (continued)

Recognized research work	Data source	Contribution	Classifiers
Zhang et al. [19]	Twitter	Identified and mentioned implicit properties contain properties like "popularity orientation, internal and external/consistency, sentiment polarity and opinion of comments, social influence, opinion retweet influence, and match overall degree of messages", etc.	SVM
Zhiwei Jin et al. [20]	Twitter	Focused on specific topic that is "2016 US President Election related rumors" Detect rumors in Twitter by comparing them with verified rumor articles	TF-IDF and BM25, Word2Vec and Doc2Vec, Lexicon matching
Kwon et al. [21]	Studied properties stability over a specified timeline and report the possible structural and temporal properties that detect rumors at a later stage as they are not available in early stage of rumor propagation. In variance, the user and dialectal features are better substitutes when we want to identify rumor as quickly as possible		

3 Methodology

Our rumor detection system is based on both explicit and implicit features. It is designed to detect rumors that are related to any noteworthy event that may have a sizeable impact on the society, also we want to detect it at the earliest, especially important during an emergency. To facilitate early detection, we have used explicit and implicit features that are based on users and content data and metadata. We not only include some new combinations of implicit and explicit features that significantly contribute to rumor detection, but also analyze replies of the messages resulting in more accurate results. These additional features will enhance the authenticity and efficiency of our diagnosis.

3.1 Problem Statement

Rumor detection can be handled as a classification problem. For it we have used a supervised classifier. We required a large labeled dataset for supervised classification. The classification job is to classify the label or class for a given unlabeled point. Formally, a classifier is a function or model that classifies the class label point for a given Input. To generate the supervised classifier, we need a training data set that is correctly class labeled points. After designing the model, we can test this model on testing data set then it is ready to predict the class or label for any new point.

Considering the above requirement, we can define our rumor detection system as follows. We take a set of news events $E = \{e_1;\ldots; e_n\}$, where each event e_i is related with a set of Tweet messages $T_i = \{t_{i,1}; \ldots; t_{i,m}\}$. So, a rumor detection model M is a function $R^{mxf} \rightarrow R^f \rightarrow \{1; 0\}$ that combines a feature matrix for all news event Tweet messages to a f-dimensional feature vector of the related event and, then maps it to a binary class: rumor (1) or non-rumor (0).

3.2 Proposed Model

To detect the proliferation of a rumor, we have used a binomial supervised classifier. Supervised classification methods or model are methods that find out the association between independent variables and a dependent variable. Dependent variable is also known as target variable. Generally, the classification models describe features, patterns, or classification rules that are concealed in the dataset. These classification rules help to predict the value of the dependent variable by knowing the values of the independent variables. The models that predict categorical (discrete, unordered) class labels are known as classifiers. We have built a classification model to classify a rumor as true or false. Figure 1, shows the methodology we have used to design our rumor detection model.

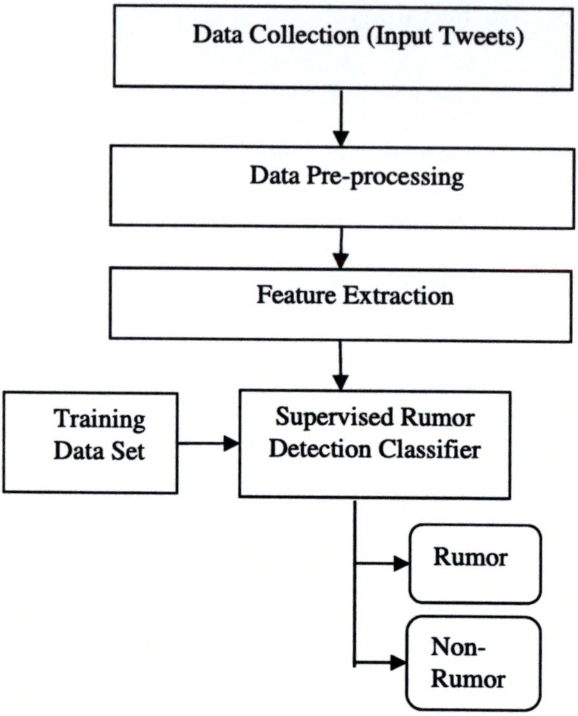

Fig. 1 The framework of modified rumor detection model

3.3 Feature Extraction

Our rumor detection model is designed for rumors that are associated with newsworthy events. In this case, we have to deal with unseen rumors emerge during the newsworthy events, one does not know in advance, and the particular keywords related to a rumor are yet to be identified. To deal with this type of rumors, a classifier based on generalized patterns is used to identify rumors during emerging events. In this work we studied and extracted features related to event-based rumors.

The two essential components of a message are content and users. By examining these two aspects of a message we identified salient characteristics of the rumors. The properties may be extracted from elementary attributes of user or content or may be generated by mining and analyzing the linguistic, belief, opinion, or sentiment of user and its message. In this way, we can further divide all the features into two groups they are explicit features and implicit features. By examining related work in these fields [15, 22 and 23] and by examining publicly available Tweeter data and metadata, we have identified overall 32 features. We examined the significance of each of these features in rumor detection and found that not all were significantly contributing. We have used Wald z-statistic and the associated p-values to measure the contribution of each feature in rumor detection. Small p-values indicate statistical significance, which means there is a significant relationship between a feature and the

Table 2 Explicit features of rumor detection model

Category	Name	Description
Explicit user-based features	Reliability	To identify as to whether Twitter has verified the user's account or not
	Has description	To identify whether a personal description or self-summary has been given by the user or not
	Has Profile URL	To identify if the user has revealed a profile URL or not
	Has image	Whether the user has profile image or not
	Influence	Number of followers
Explicit content-based features	Time span	To assess the time interval between the posting of the message and registration of the user
	Has URLs	To access if the message has URL that is actually pointing to an external source or not

outcome. After removing all insignificant features, we were left with 7 explicit and 7 implicit features, which is a total of 14 features. The contributions of individually of the 14 features from the two categories are explained in detail below.

Explicit Features

Explicit features are features those extracted from fundamental characteristics of the user or its content. In this model, we found seven explicit features that are significantly contributing to the outcome of our model are described in Table 2.

Implicit Features

Implicit features or properties are extracted by mining the message content and user information. They are extracted by examining the linguistic, opinion, sentiment, belief, or viewpoint of tweets and user information. In this model, we found seven implicit features that are significantly contributing to the outcome of our model they are described in Table 3.

Rumor detection problem is modeled as a binomial classification problem. Most of the research work is modeled on explicit properties of text messages, user, propagation, and other metadata [15, 22–26]. But such explicit properties sometimes could not differentiate among rumor and normal messages. It has been observed that implicit features like replies of the mass are very useful to detect rumors. People frequently give mixed opinions such as support or deny in response to message. We could enhance the accuracy of the existing rumor detection model by including some implicit features like replies to the Tweet etc.

Thus, we hypothesize: H1—The implicit features are effective, and explicit and implicit features jointly give a more significant contribution to detect rumor on online social media than only explicit features.

Table 3 Implicit features of rumor detection system

Category	Name	Description
Implicit content-based Features	Exaggeration of message	Refers to the sentimental polarity of message. Usually, the contents of rumors are exaggerated and generally use extreme words
	Acceptance of message	States to the level of acceptance of the message. To measure the acceptance, we analyzed the responses and replies to the tweets. Usually, the content of rumors receives large number of doubtful, inquiring and uncertain replies
	Formality of message	Measures the formality (or the informality) of a message. Each Tweet is checked for abbreviations and emoticons and then grouped into formal and informal
	Linguistic inquiry and word count (LIWC)	Find the presence of opinion, insight, inferring, and tentative words. Based on presence and absence tweets could be classified into found or not found
Implicit user-based features	Originality	Measures the originality of the user's message. It is a ratio of the total number of original tweets, to the total number of retweets
	Role in social media	It is a ratio of followers and followees of a Twitter account
	Activeness	Measures the activeness of a user on Twitter from joining

4 Design of the Experiment

4.1 Experimentation Platform

To implement rumor detection model, we have used the following two platforms:

R programming language with IDE RStudio: *R* programming language is an open-source, highly extensible software package to perform statistical data analysis. *R* provides a wide range of machine learning, statistical, classification, clustering, and graphical techniques. We have used *R* to extract Twitter data and metadata, for data preprocessing, for feature extraction and to test their significance and finally to test the fitness of the model.

Table 4 The details of annotated dataset

Event	Rumors	Nonrumors	Total
Sydney siege	522 (42.8%)	699 (57.2%)	1221
German wings crash	238 (50.7%)	231 (49.3%)	469
Total	760	930	1690

Weka: It is a platform-independent, open-source, and easily useable software package written in java. It is an assembly of machine learning algorithms that are used for data mining tasks.

To assess the performance of five classifiers *logistic regression (LR), Naive Bayes (NB), Randon Tree (RT), linear support vector machine (SVM), and J48,* we have used Weka.

4.2 Data Collection and Dataset

Data source: As the rumor detection problem requires public opinions and reactions, we are using Twitter as a data source.

Dataset: As rumor detection is a classification problem and we are using supervised binomial classification; we need a reliable annotated dataset. So, we are working on a subset of publicly accessible datasets that is the PHEME dataset of rumors and nonrumors [27]. These are extracted from Twitter. The tweets are in English and associated with different events that had caught the attention of people and contained rumors. To create a generalized model, we have used dataset from separate events one for training and other for testing:

1. **Sydney siege**: On December 15, 2014, in an incident a gunman held hostage 8 employees and 10 customers of Lindt Chocolate Cafe which was located at Martin Place in Sydney, Australia.
2. **Germanwings plane crash**: On March 24, 2015, all passengers and crew were dead when a plane from Barcelona to Dusseldorf had crashed on the French side of Alps. It was judged after an investigation that the plane was deliberately crashed by a co-pilot.

The details of annotated dataset used in rumor detection model are shown in Table 4.

4.3 Result Analysis and Evaluation

There four main objectives to evaluate and examine the rumor detection model are as follows:

1. To measure the accuracy of our model, at which it forecasts the rumor.
2. To measure the significance of each property or feature in rumor detection.
3. To measure the contribution of the explicit features, implicit features, and explicit-implicit features together in rumor detection.

The results of logistic regression of explicit-implicit features for rumor detection are shown in Table 5. Column A presents the results for "Explicit Features", Column B presents the results for "Implicit Features" and Column C provides the results for combined "Explicit-Implicit" Features.

The table shows the estimate that is a coefficient. The binary logistic regression coefficients calculate the variation in the log odds of the result for a one-unit increase in the independent variable. The Wald z-statistic and significance stars give statistical significance for individual independent variables. We could easily find in the table that all the features or properties are considerably contributing to rumor detection. The findings of the regression models are that if micro-blogging is associated with absence of profile picture, description, profile url, url; non-reliable users, lower followings, and lower time span between message posting and registration, there are higher chances that the message is a rumor. Similarly, higher sentimental polarity and opinions in the message, lower acceptability of message, higher formality, lower

Table 5 Results of logistic regression of explicit-implicit features model

Variable names	Explicit features model A		Implicit features model B		Explicit-implicit features model C	
	Estimates	Z-Value	Estimates	Z-Value	Estimates	Z-Value
(Intercept)	10.986	12.002***	−1.140	−3.710***	18.578	6.153***
Has_Profile_Img	−5.1005	−10.176***			−9.163	−6.226***
Has_Description	−2.970	−7.962***			−4.736	−5.272***
Reliable	−1.210	−2.937**			−2.310	−2.266*
Influence	−2.268	−6.204***			−3.874	−4.256***
Has_Profile_URL	−2.673	−7.061***			−3.760	−4.341***
Has_URL	−1.190	−3.471***			−2.154	−3.068**
Time. Span	−3.207	−7.635***			−4.048	−4.790***
Exaggeration			1.331	7.471***	2.522	3.302***
Acceptance			−0.870	−4.899***	−3.511	−4.109***
Formality			1.620	8.927***	1.763	2.573*
LIWC			1.567	8.760***	2.409	3.524***
Activeness			2.208	11.963***	2.897	4.054***
RoleSM			−1.465	−6.317***	−3.919	−4.320***
Originality			−1.454	−7.986***	−2.411	−3.536***
McFadden R2	0.845		0.498		0.966	

***, **, * significant at 1%, 5%, 10%, respectively

originality, high activeness, and higher disjointed connections are associated with high chances of rumor.

Absence of profile image was found to be the most significant explicit feature for rumor detection with the highest coefficient value. Activeness of user and non-acceptance of messages were found to be most significant implicit features for rumor output. By including the "Implicit" features in the model, the McFadden $R2$ value has increased from 0.845 to 0.966. It means that we get better-fitted model by including the "implicit" features for rumor detection.

In the above result, we can see that all the features or properties are considerably contributing to rumor detection. The proposed model is well fitted and there is considerable improvement in the model after including implicit features as we have in hypothesis. The improvement is shown in Fig. 2. But for the complete study we have designed and trained five different classifiers on these sets of significant features using the following methods: *logistic regression (LR), Naive Bayes (NB), Randon Tree (RT), linear support vector machine (SVM), and J48*. Out of these, we have selected the classifier that has given the best result. To measure the performance of our classifier we have used four standard prediction qualities measures, they are:

1. Overall accuracy and error rate: these measures the overall performance of the classifier.
2. Precision, Recall, $F1$: Those measure the class level performance of the classifier.
3. AUC value and ROC curve: That measures the performance of the model by evaluating the tradeoffs between true-positive rate and false-positive rate.
4. Kappa-statistics: IT is a non-parametric test-based metric. It is a measure of the agreement between the predicated and the actual classifications.

The comparison of all five classifiers and results are given in Table 6 and Fig. 3.

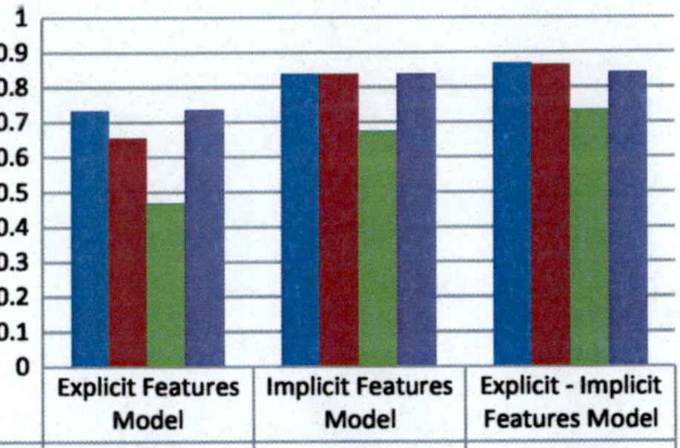

	Explicit Features Model	Implicit Features Model	Explicit - Implicit Features Model
Accuracy	0.733	0.838	0.868
F score	0.656	0.838	0.866
Kappa Statistics	0.469	0.675	0.735
AUC	0.736	0.838	0.842

Fig. 2 Evaluation metrics for explicit-implicit features models

Table 6 Performance result of five classifiers

	Logistic	SVM	Naive Bayes	Random tree	J48
Accuracy (%)	85.6838	85.4701	84.6154	79.0598	82.0513
Precision	0.875	0.872	0.863	0.813	0.845
Recall	0.857	0.855	0.846	0.791	0.821
F measure	0.855	0.853	0.845	0.787	0.818
ROC area	0.960	0.856	0.945	0.793	0.837
Kappa statistic	0.7144	0.7101	0.6931	0.5825	0.6422

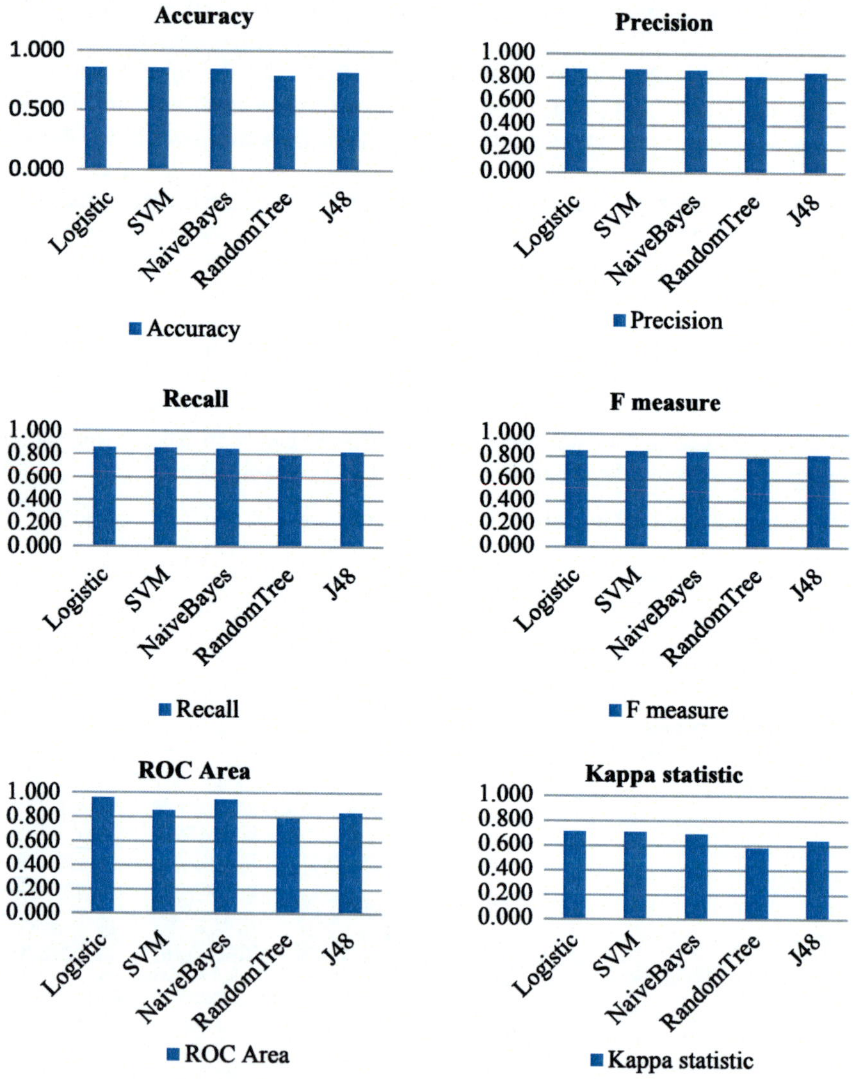

Fig. 3 Performance results of five classifiers

From the above observation and analysis, we finally conclude that the new modified rumor detection model has been effective in getting a significant improvement in accuracy, precision, recall, and F-score with a very low value of false-positive rate.

5 Conclusion and Future Scope

Rumor that is unverified information can cause severe impact on the individual or society especially at the time of emergency. This study conducts the investigation of social media networks like Twitter messages for rumors. The rumor detection in social media network has fascinated lots of attention in current years due to their impact on the prevailing socio-political situation in the world. Most of the previous works focused on the explicit features of users and messages. To effectively distinguish rumors from normal messages, we need more deep analysis of data and metadata. In this study, we proposed a rumor detection method combining both explicit features and user-content-based implicit features. The study found that all explicit and implicit features are significant for rumor detection in online social networks.

We have proposed this rumor detection model for Twitter. Further one can extend this model to other social media platforms. There will always be scope to add more features to any such investigation so that ill-intentioned rumors could be detected more effectively and as early as possible. This will help to make society more livable.

References

1. AlKhalifa HS, AlEidan RM (2011) An experimental system for measuring the credibility of news content in Twitter. Int J Web Inf Syst 7(2):130–151
2. Sivasangari V, Pandian VA, Santhya R (2018) A modern approach to identify the fake news using machine learning. Int J Pure Appl Math 118(20):3787–3795
3. Mitra T, Wright GP, Gilbert E (2017) A parsimonious language model of social media credibility across disparate events. In: Proceedings of the 2017 ACM conference on computer supported cooperative work and social computing, pp 126–145
4. Northeasterners' exodus in India underlines power of social media, Aug 18, 2012, Available: http://articles.latimes.com/2012/aug/18/world/la-fgindia-social-media-20120819
5. Social media and the India exodus, BBC World News, Available: http://www.bbc.com/news/world-asia-india-19292572
6. Social media being used to instigate communal riots, says HM Rajnath Singh, Nov 5, 2014, Available: http://www.dnaindia.com/india/report-socialmedia-being-used-to-instigate-communal-riots-rajnath-singh-2032368
7. J&K Bans Facebook, WhatsApp And Most Social Media from Kashmir Valley Indefinitely, Apr 26, 2017, Available: http://www.huffingtonpost.in/2017/04/26/jandk-bansfacebook-whatsapp-and-most-social-media-from-kashmirv_a_22056525/
8. Mobile internet services suspended, trains cancelled, govt offices closed ahead of Dera chief case verdict, The Times of India, Aug 24, 2017, Available: https://timesofindia.indiatimes.com/india/mobile-internet-services-suspended-trains-cancelled-govt-offices-closed-ahead-of-dera-chief-case-verdict/articleshow/60210295.cms

9. Article 370 and 35(A) revoked: how it would change the face of Kashmir, The Economic Times, Aug 5 2019, Available: https://economictimes.indiatimes.com/news/politics-and-nation/article-370-and-35a-revoked-how-it-would-change-the-face-of-kashmir/articleshow/70531959.cms
10. Kwak H, Lee C, Park H, Moon S (2010) What is twitter, a social network or a news media? In: Proceedings of the 19th international conference on world wide web, ACM, pp 591–600
11. Stassen W (2010) Your news in 140 characters: exploring the role of social media, in journalism. Global Media J-Afr Ed 4(1):116–131
12. Naaman M, Boase J, and Lai CH (2010) Is it really about me? message content in social awareness streams. In: Proceedings of the 2010 ACM conference on computer supported cooperative work, pp 189–192
13. Sivasangari V, Mohan AK, Suthendran K Sethumadhavan M (2018) Isolating rumors using sentiment analysis. J Cyber Secur Mob 7(1 & 2)
14. Yavary A, Sajedi H (2018) Rumor detection on twitter using extracted patterns from conversational tree. In: 4th international conference on web research (ICWR), IEEE
15. Yang F, Liu Y, Yu X, Yang M (2012) Automatic detection of rumor on sina weibo. In: Proceeding of the ACM SIGKDD workshop on mining data semantics, p 13
16. Castillo C, Mendoza M, Poblete B (2013) Predicting information credibility in timesensitive social media. Internet Res 23(5):560–588
17. Kwon S, Cha M, Jung K, Chen W, Wang Y (2013) Prominent features of rumor propagation in online social media. In: 2013 IEEE 13th international conference on data mining (ICDM), pp 1103–1108
18. Liu X, Nourbakhsh A, Li Q, Fang R, Shah S (2015) Real-time rumor debunking on twitter. In: Proceedings of the 24th ACM international conference on information and knowledge management, pp 1867–1870
19. Zhang Q, Zhang S, Dong J, Xiong J, Cheng X Automatic detection of rumor on social network. Springer International Publishing Switzerland, pp 113–122
20. Jin Z, Cao J, Guo H, Zhang Y, Wang Y, Luo J (2017) Detection and Analysis of 2016 US presidential election related rumors on Twitter. Springer International Publishing AG 2017, Springer, pp 230–239
21. Kwon S, Cha M, Jung K (2017) Rumor detection over varying time windows. PLOS ONE 12(1)
22. Wu K, Yang S, Zhu KQ (2015) false rumors detection on sina weibo by propagation structures. In: IEEE international conference of data engineering
23. Tolosi L, Tagarev A, Georgiev G (2016) An analysis of event-agnostic features for rumour classification in twitter, the workshops of the tenth. International AAAI conference on web and social media, Social Media in the Newroom: Technical Report WS-16–19
24. Ratkiewicz J, Conover M, Meiss M, Goncalves B, Patil S, Flammini A, Menczer FM (2011) Detecting and tracking political abuse in social media. In: Proceedings of ICWSM, WWW, pp 249–252
25. Sun S, Liu H, He J, Du X (2013) Detecting event rumors on sina weibo automatically. In: Web technologies and applications. Springer, pp 120–131
26. Seo E, Mohapatra P, Abdelzaher T (2012) Identifying rumors and their sources in social networks. SPIE defense security and sensing, international society for optics and photonics
27. Zubiaga A, Liakata M, Procter R (2016) Learning reporting dynamics during breaking news for rumour detection in social media. Pheme: computing veracity—the fourth challenge of big data

K-harmonic Mean-Based Approach for Testing the Aspect-Oriented Systems

Richa Vats and Arvind Kumar

Abstract Testing is an important activity of software development and lot of effort can be put on the testing of softwares. In turn, the cost of development of software can be increased. The development cost is increased due to execute large number of test cases for testing the software. So, optimizing the test cases is also a challenging problem in field of software testing. The optimized test cases can either reduce the development cost or ensure the timely delivery of software. In present time, the paradigm shifted from OOP system to AOP sysetm. In AOP, less number of work is reported on testing process. Hence, in this work, KMH approach is applied to optimize the test cases. The performance of KMH is evaluated using two cases studies. It reveals that KHM is efficient approach for testing the AOP system.

Keywords Aspect-oriented system · Testing · K-harmonic approach · Object-oriented system · Data flow diagram

1 Introduction

Software testing is the process to test the working of the software product. Prior to delivery of the software product, it should bet test whether the software is as per user needs or not. The different test cases are developed to accomplish the software testing process. A software comprises of different modules and each module consists of different supporting functions. The logic of the software is described through different function. To differentiate the functions from program logic, aspect-oriented programming is applied [1, 2]. The main objective of the AOP is to make the program more modular. AOP is a new programming language having advantage over object-oriented programming in terms of code scattering, tangling, etc. It can be written using

R. Vats (✉) · A. Kumar
SRM University Delhi-NCR, Sonepat, Haryana, India
e-mail: ritzi1606@gmail.com

A. Kumar
e-mail: k.arvind33@gmail.com

V. Singh et al. (eds.), *Computational Methods and Data Engineering*,
Advances in Intelligent Systems and Computing 1227,
https://doi.org/10.1007/978-981-15-6876-3_6

AspectJ language an extension of Java language. AOP can also be written using other language viz AspectCan extension of c, Aspect C++ an extension of C++, CaserJ and HyperJ [3]. Moreover, it is stated that the object-oriented programming can crosscut a single repository in multiple components and it is one of major drawback of OOP [4]. But, the AOP can address the crosscutting problems of OOP and separate the functions in terms of aspect. A lot of work is reported on testing process using OOP. But, lack of work is presented using AOP for testing process. Testing in aspect-oriented system is an early phase and it is an important activity of SDLC. This process contains the large number of test cases. Executing the each test case can increase the testing time. Further, the delivery of software can be affected. So, the main task of testing process is to arrange the test cases in optimum manner. A small set of test cases can be executed to examine the behavior of software instead of entire test cases. Hence, the objective of this paper is to address the testing process of AOP using meta-heuristics algorithm. In this work, k-harmonic mean (KMH) approach is used determine the optimal subset test cases [5]. Using KMH approach, all test cases are divided into different clusters and from each cluster selects few test cases to check the behavior of software. The performance of KMH approach is evaluated using two cases studies. These case studies are ATM system and library management system. The detailed description of these case studies along with other necessary components are discussed in Sect. 4. It is noticed that KMH efficiently works with aspect-oriented system.

Rest of paper is organized as Sect. 2 presents the related works on the aspect-oriented system. Section 3 describes the proposed KMH approach and its steps. Section 4 demonstrates the experimental results of our study using two case studies. The wok is concluded in Sect. 5.

2 Related Works

This section describes the work reported aspect-oriented programing. Raheman et al. presented a review on aspect-oriented programs [6]. In this review, different perspective and challenges are discussed in context of aspect-oriented programming. Furthermore, dependence graph and complexity are also discuss of aspect programing. To reduce the test cases for aspect-oriented programming, Joyti and Hooda applied a fuzzy c-means algorithm [7]. In this work, authors consider online banking system to evaluate the performance of fuzzy c-means algorithm. It is stated that FCM algorithm obtains state-of-the-art results.

A review on aspect-oriented system is presented in [8]. Chandra and Singhal discussed the impact of data flow testing and unit testing in context of object-oriented and aspect-oriented programing [9]. In this work, point-cut-based coverage and interprocedural data flow analysis are presented.

To address the cost and effort issues of testing, Dalal and Hooda presented a hybrid approach for testing the aspect-oriented programs [10]. The proposed approach is the combination of genetic algorithm and fuzzy c-means. The proposed approach is validated using well-known heating kettle problem. It is observed that proposed approach obtains better results. To address remodularization problem, Jitemder and Chabbra adopted the harmony search-based algorithm for object-oriented system [11]. In this study, structural aspect of the software system is considered to evaluate the performance of harmony search-based algorithm. It is noticed that proposed algorithm is one of competitive and efficient algorithm for structural aspect of software system.

Assuncao et al. [12] explored different strategies for integration testing using aspect-oriented programs. In this study, three approaches are considered for integration testing. These approaches are traditional approach, GA-based approach, and multi-objective approach. Simulation results stated that multi-objective approach provides more convenient results for integration testing instead of traditional and GA-based approaches. Boudaa et al. [13] developed an aspect-oriented model for context-aware-based service applications. The proposed model is the combination of MDD and AOM. AOM contains different awareness context logic, called ContexttAspect. It is observed the combination of MDD and AOM successfully overcomes the pitfalls of earlier approaches.

Ghareb and Allen presented the different metrics to measure the development of aspect-oriented systems [14]. Dalal and Hooda explored the prioritized genetic algorithm to test the aspect-oriented system [15]. The traditional banking system example is considered to evaluate the performance of proposed algorithm. It is stated that prioritized GA provides more efficient results than random order and unprioritized GA algorithm.

Sangaiah et al. [16] explored the cohesion metrics in context of reusability for aspect-oriented system. Further, in this work, a relationship is established between the cohesion metrics and reusability. Authors developed PCohA metrics to measure the package-level cohesion in aspect-oriented systems. The proposed metrics is validated using theoretical as well as experimentally.

Kaur and Kaushal developed fuzzy logic-based approach to assess external attributes using package-level internal attributes [17]. The proposed approach is validated using external attributes. Results stated that proposed approach provides quality results. Singhal et al. applied harmony search algorithm to prioritize the test cases for aspect-oriented systems [18]. The benchmark problems are considered to evaluate the performance of harmony search algorithm.

These problems are implemented using AspectJ. It is observed that proposed approach provides better results over random and non-prioritization approaches.

3 Proposed Methodology

K-harmonic means is a popular algorithm that can be applied to obtain the optimum cluster centers [5]. Many researchers have been applied KHM algorithm to solve diverse optimization problems such as clustering, feature selection, dimension reduction, outlier detention, and many more [19, 20]. KHM algorithm is superior than k-means algorithm because it is not sensitive the initial cluster centers. Whereas the performance of k-means algorithm depends on initial cluster centers. The steps of KHM algorithm is highlighted using Algorithm 1.

Algorithm 1: Steps of KHM algorithm	
Step 1:	Compute the initial cluster centers in random order.
In this work, applicability of KHM algorithm is explored to determine the reduced set of test cases for aspect oriented programing	
Step 2:	Compute the value of objective function using Eq. 1. $\mathrm{KHM}(X, C) = \sum_{i=1}^{M} \frac{k}{\sum_{j=1}^{k} \frac{1}{x_i - c_j^p}}$ (1)
Step 3:	Compute the membership function $m(c_j/x_i)$ for each cluster centers using Eq. 2. $(c_j/x_i) = \frac{x_i - c_j^{-p-2}}{\sum_{j=1}^{k} x_i - c_j^{-p-2}}$ (2)
Step 4:	Compute the weight of each data instances using Eq. 3. $w(x_i) = \frac{\sum_{j=1}^{k} x_i - c_j^{-p-2}}{\left(\sum_{j=1}^{k} x_i - c_j^{-p}\right)^2}$ (3)
Step 5:	Compute the cluster centers again using membership function and weight. $c_j = \frac{\sum_{i=1}^{n} m(c_j/x_i) w(x_i) x_i}{\sum_{i=1}^{n} m(c_j/x_i) w(x_i)}$ (4)
Step 6:	Repeat steps 2–5, untill optimized clusters are not obtained
Step 7:	Obtain optimized clusters

3.1 Steps of the Proposed Algorithm

This section discusses the KHM algorithm using aspect-oriented programing for test cases optimization. The aim of KHM algorithm is to obtain reduced set of test cases. The steps of the proposed methodology are listed as

Algorithm 1: Steps of proposed algorithm for reduction of test cases Input: Set of test cases Output: Reduced set of test cases	
Step 1:	Design the activity diagram for a given project using UML.
Step 2:	Construct the control flow graph (CFG) using activity diagram of the project.
Step 3:	Compute the sequential, aspect and decision nodes from the control flow graph
Step 4:	Compute the cyclomatic complexity using the CFG.
Step 5:	Determine the independent paths in the given CFG.
Step 6:	Compute the cost of independent paths.

(continued)

(continued)

Step 7:	Apply the K-harmonic algorithm to determine the closeness of test cases using clustering method
Step 8:	Determine the optimal test cases from clusters based on the minimum closeness criteria with respect to cluster centers
Step 9:	Evaluate the performance of the proposed algorithm using efficiency parameter

4 Results and Discussion

This section discusses the simulation results of the proposed KHM algorithm using two cases studies. To validate the proposed algorithm, ATM system and library system are considered. Furthermore, activity diagram and control flow graphs are designed for both of case studies. The performance of proposed algorithm is evaluated using efficiency rate.

4.1 *Case Study 1: ATM System*

This subsection considers the ATM system case study to validate the proposed KHM algorithm. In initial step of the algorithm, activity diagram is developed for ATM system. Further, the control flow graph is designed with the help of activity diagram. The working of the proposed algorithm is started by determining sequential nodes, decision nodes, and aspect nodes. For ATM case study, authentication, withdrawal, dispense cash, etc., are described as aspects. The step by step working of proposed algorithm is given below.

Step 1: The activity diagram of the ATM system is illustrated in Fig. 1. This diagram consists of sequential nodes, aspect nodes, and decision nodes. The number of sequential nodes, decision nodes, and aspect nodes for ATM system is given as

$$\text{Sequential nodes} : 1, 2, 5, 14, 15, 19, 20, 21, 22, 23, 24, 26$$
$$\text{Aspect Nodes} : 3, 6, 8, 9, 10, 12, 13, 17, 18$$
$$\text{Decision Nodes} : 4, 7, 11, 16,$$

Step 2: In second step, cyclomatic complexity is computed. The cyclomatic complexity for ATM system is defined as

$$\text{Cyclomatic Complexity} = 30 - 26 + 2 * 6 = 16$$

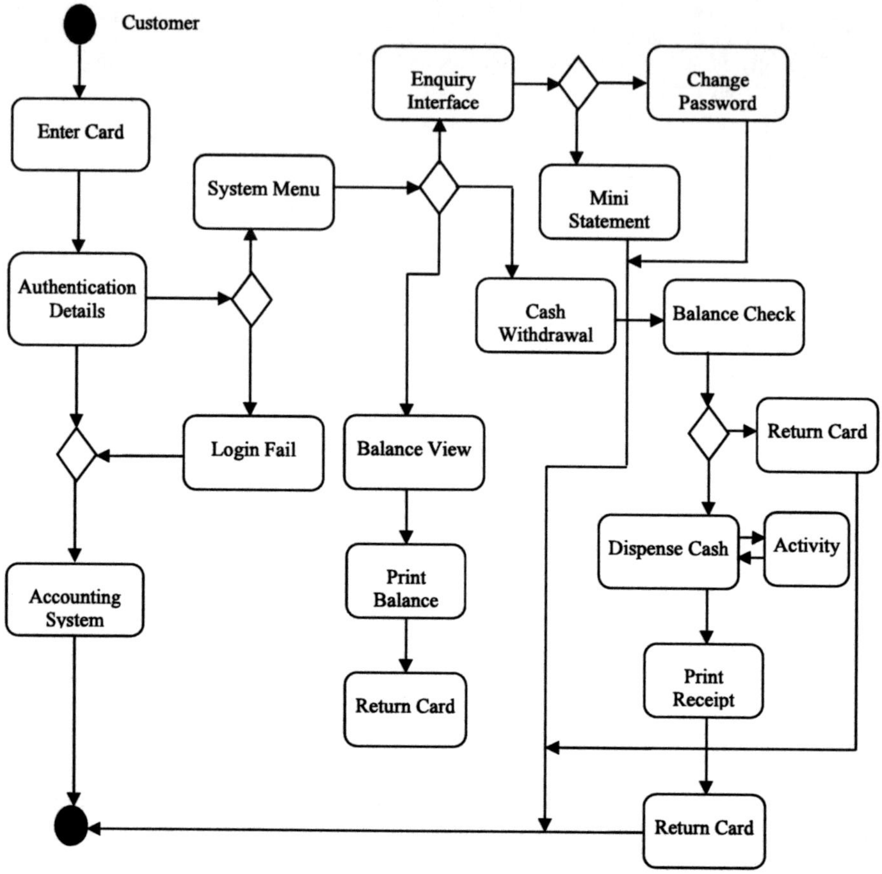

Fig. 1 Activity diagram of ATM system

Step 3: In this step, independent paths are computed. The independent paths in ATM system are listed as

$TC1 = 1 \rightarrow 2 \rightarrow 3 \rightarrow 4 \rightarrow 5 \rightarrow 25 \rightarrow 26$

$TC2 = 1 \rightarrow 2 \rightarrow 3 \rightarrow 4 \rightarrow 5 \rightarrow 25 \rightarrow 3 \rightarrow 4 \rightarrow 5 \rightarrow 25 \rightarrow 26$

$TC3 = 1 \rightarrow 2 \rightarrow 3 \rightarrow 4 \rightarrow 6 \rightarrow 7 \rightarrow 8 \rightarrow 22 \rightarrow 23 \rightarrow 24 \rightarrow 26$

$TC4 = 1 \rightarrow 2 \rightarrow 3 \rightarrow 4 \rightarrow 6 \rightarrow 7 \rightarrow 9 \rightarrow 14 \rightarrow 15 \rightarrow 16 \rightarrow 18 \rightarrow 26$

$TC5 = 1 \rightarrow 2 \rightarrow 3 \rightarrow 4 \rightarrow 6 \rightarrow 7 \rightarrow 9 \rightarrow 14 \rightarrow 15 \rightarrow 16 \rightarrow 19 \rightarrow 20 \rightarrow 21 \rightarrow 26$

$TC6 = 1 \rightarrow 2 \rightarrow 3 \rightarrow 4 \rightarrow 6 \rightarrow 7 \rightarrow 10 \rightarrow 11 \rightarrow 13 \rightarrow 26$

$TC7 = 1 \rightarrow 2 \rightarrow 3 \rightarrow 4 \rightarrow 6 \rightarrow 7 \rightarrow 10 \rightarrow 11 \rightarrow 12 \rightarrow 26$

Step 4: In step 4, the cost of each path is computed. The cost of path is described in terms of number of nodes presented in path. The cost of path for ATM system is given as

TC1 = 1 → 2 → 3 → 4 → 5 → 25 → 26; Cost = 7
TC2 = 1 → 2 → 3 → 4 → 5 → 25 → 3 → 4 → 5 → 25 → 26; Cost = 11
TC3 = 1 → 2 → 3 → 4 → 6 → 7 → 8 → 22 → 23 → 24 → 26; Cost = 11
TC4 = 1 → 2 → 3 → 4 → 6 → 7 → 9 → 14 → 15 → 16 → 18 → 26;
Cost = 12
TC5 = 1 → 2 → 3 → 4 → 6 → 7 → 9 → 14 → 15 → 16 → 19 → 20 → 21 → 26;
Cost = 15
TC6 = 1 → 2 → 3 → 4 → 6 → 7 → 10 → 11 → 13 → 26;
Cost = 10
TC7 = 1 → 2 → 3 → 4 → 6 → 7 → 10 → 11 → 12 → 26; Cost = 10

Step 5: In this step, k-harmonic algorithm is applied to obtain optimal test cases. The cost of test cases is given as input to KHM algorithm and number of clusters is set to 2. The output of KHM algorithm is mentioned below. (Fig. 2)

Cluster centre 1 = 9
Cluster centre 2 = 12.25

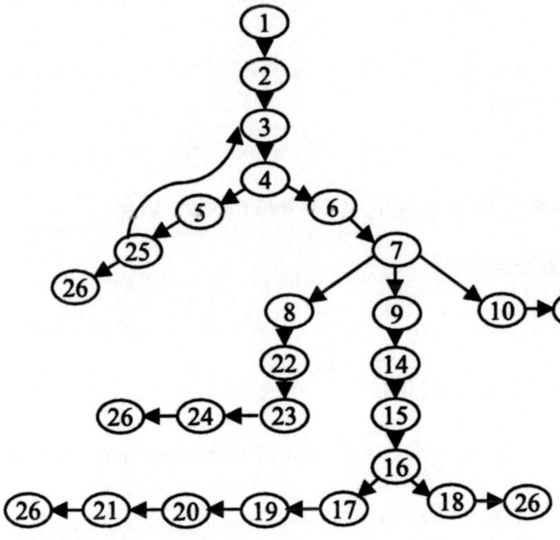

Fig. 2 Control flow graph of ATM system

KMH algorithm divides the test cases in two clusters, and the optimal center of these clusters are 9 and 12.5, respectively. Furthermore, test cases are assigned to clusters bases on minimum Euclidean distance. In turn, TC1, TC6, and TC7 are assigned to cluster 1, whereas, TC2, TC3, TC4, TC5 are allocated to cluster 2.

Step 6: In step 6, optimum test cases are selected using the outcome of the step 5. To determine the optimum test case, Manhattan distance is computed between cluster centers and corresponding data. The optimum test cases are selected based on the minimum distance between cluster centers and data. In cluster 1, three test cases are allotted. These test cases are TC1, TC6, and TC7. The corresponding cost of these test cases are 7, 10, and 10.

For cluster 1:

$$\mathrm{TC1} = 9 - 7 = 2$$
$$\mathrm{TC6} = 9 - 10 = 1$$
$$\mathrm{TC7} = 9 - 10 = 1$$

For cluster 2:

$$\mathrm{TC2} = 12.25 - 11 = 1.25$$
$$\mathrm{TC3} = 12.25 - 11 = 1.25$$
$$\mathrm{TC4} = 12.25 - 12 = 0.75$$
$$\mathrm{TC5} = 12.25 - 15 = 2.75$$

So, the minimum value for cluster 1 is 1, while the minimum value for cluster 2 is 0.75.

Step 7: This step computes the efficiency of the proposed KMH algorithm to compare old test cases over new test cases.

$$\text{First} = (1 - \text{no.of test clusters}/\text{total no.of test cases}) * 100$$
$$\text{First} = \left(1 - \frac{2}{7}\right) * 100 = 72\%$$
$$\text{Cluster 1} = (-\text{minimum difference}/\text{total sum of test cases within cluster}) * 100$$
$$\text{First} = \left(1 - \frac{1}{27}\right) * 100 = 96.29\%$$
$$\text{Cluster 2} = (1 - \text{minimum difference}/\text{total sum of test cases within cluster}) * 100$$
$$\text{First} = \left(1 - \frac{0.75}{49}\right) * 100 = 98.46\%$$

4.2 Case Study 2: Library System

This subsection describes the library system case study to evaluate the efficiency of proposed KHM algorithm. In initial step of the algorithm, activity diagram is developed for library system. Further, the control flow graph is designed with the help of activity diagram. The working of the proposed algorithm is strated by determining sequential nodes, decision nodes, and aspect nodes. For ATM case study, log in details, user validation, return book, etc., are defined as aspects. The step by step working of proposed algorithm is given below.

Case Study 2: Library System Step 1: The activity diagram of library system is illustrated in Fig. 3. This diagram consists of sequential nodes, aspect nodes, and decision nodes. The number of sequential nodes, decision nodes, and aspect nodes for ATM system is given as

$$\text{Sequential nodes}: 1, 4, 12, 13, 14, 15, 16, 18, 23, 24$$
$$\text{Aspect Nodes}: 2, 5, 7, 8, 10, 11, 19, 21, 22$$

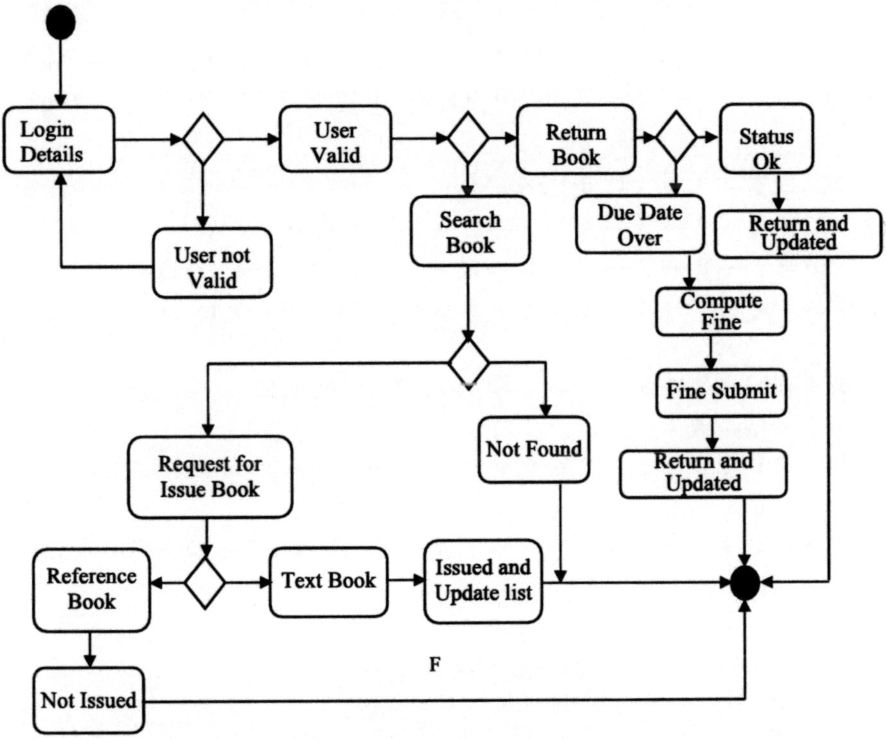

Fig. 3 Illustrates the activity diagram of library system

$$\text{Decision Nodes} : 3, 6, 9, 17, 20$$

Step 2: In second step, cyclomatic complexity is computed. The cyclomatic complexity for library system is defined as

$$\text{Cyclomatic Complexity} = 28 - 24 + 2 * 5 = 14$$

Step 3: In this step, independent paths are computed. The independent paths in library system are listed as

$$\begin{aligned}
&\text{TC1} = 1 \rightarrow 2 \rightarrow 3 \rightarrow 4 \\
&\text{TC2} = 1 \rightarrow 2 \rightarrow 3 \rightarrow 4 \rightarrow 2 \rightarrow 2 \rightarrow 3 \rightarrow 4 \\
&\text{TC3} = 1 \rightarrow 2 \rightarrow 3 \rightarrow 5 \rightarrow 6 \rightarrow 8 \rightarrow 9 \rightarrow 11 \rightarrow 12 \rightarrow 13 \\
&\text{TC4} = 1 \rightarrow 2 \rightarrow 3 \rightarrow 5 \rightarrow 6 \rightarrow 8 \rightarrow 9 \rightarrow 10 \rightarrow 14 \rightarrow 15 \rightarrow 16 \rightarrow 13 \\
&\text{TC5} = 1 \rightarrow 2 \rightarrow 3 \rightarrow 5 \rightarrow 6 \rightarrow 7 \rightarrow 17 \rightarrow 19 \rightarrow 13 \\
&\text{TC6} = 1 \rightarrow 2 \rightarrow 3 \rightarrow 5 \rightarrow 6 \rightarrow 7 \rightarrow 17 \rightarrow 18 \rightarrow 20 \rightarrow 22 \rightarrow 23 \rightarrow 13 \\
&\text{TC7} = 1 \rightarrow 2 \rightarrow 3 \rightarrow 5 \rightarrow 6 \rightarrow 7 \rightarrow 17 \rightarrow 18 \rightarrow 20 \rightarrow 21 \rightarrow 24 \rightarrow 13
\end{aligned}$$

Step 4: In step 4, the cost of each path is computed. The cost of path is described in terms of number of nodes presented in path. The cost of path for library system is given as

$$\begin{aligned}
&\text{TC1} = 1 \rightarrow 2 \rightarrow 3 \rightarrow 4; \text{Cost} = 4 \\
&\text{TC2} = 1 \rightarrow 2 \rightarrow 3 \rightarrow 4 \rightarrow 2 \rightarrow 2 \rightarrow 3 \rightarrow 4; \text{Cost} = 7 \\
&\text{TC3} = 1 \rightarrow 2 \rightarrow 3 \rightarrow 5 \rightarrow 6 \rightarrow 8 \rightarrow 9 \rightarrow 11 \rightarrow 12 \rightarrow 13; \text{Cost} = 10 \\
&\text{TC4} = 1 \rightarrow 2 \rightarrow 3 \rightarrow 5 \rightarrow 6 \rightarrow 8 \rightarrow 9 \rightarrow 10 \rightarrow 14 \rightarrow 15 \rightarrow 16 \rightarrow 13; \\
&\qquad \text{Cost} = 12 \\
&\text{TC5} = 1 \rightarrow 2 \rightarrow 3 \rightarrow 5 \rightarrow 6 \rightarrow 7 \rightarrow 17 \rightarrow 19 \rightarrow 13; \\
&\qquad \text{Cost} = 9 \\
&\text{TC6} = 1 \rightarrow 2 \rightarrow 3 \rightarrow 5 \rightarrow 6 \rightarrow 7 \rightarrow 17 \rightarrow 18 \rightarrow 20 \rightarrow 22 \rightarrow 23 \rightarrow 13; \\
&\qquad \text{Cost} = 12 \\
&\text{TC7} = 1 \rightarrow 2 \rightarrow 3 \rightarrow 5 \rightarrow 6 \rightarrow 7 \rightarrow 17 \rightarrow 18 \rightarrow 20 \rightarrow 21 \rightarrow 24 \rightarrow 13; \\
&\qquad \text{Cost} = 12
\end{aligned}$$

Step 5: In this step, k-harmonic algorithm is applied to obtain optimal test cases. The cost of test cases is given as input to KHM algorithm and number of clusters is set to 2. The output of KHM algorithm is mentioned below. (Fig. 4)

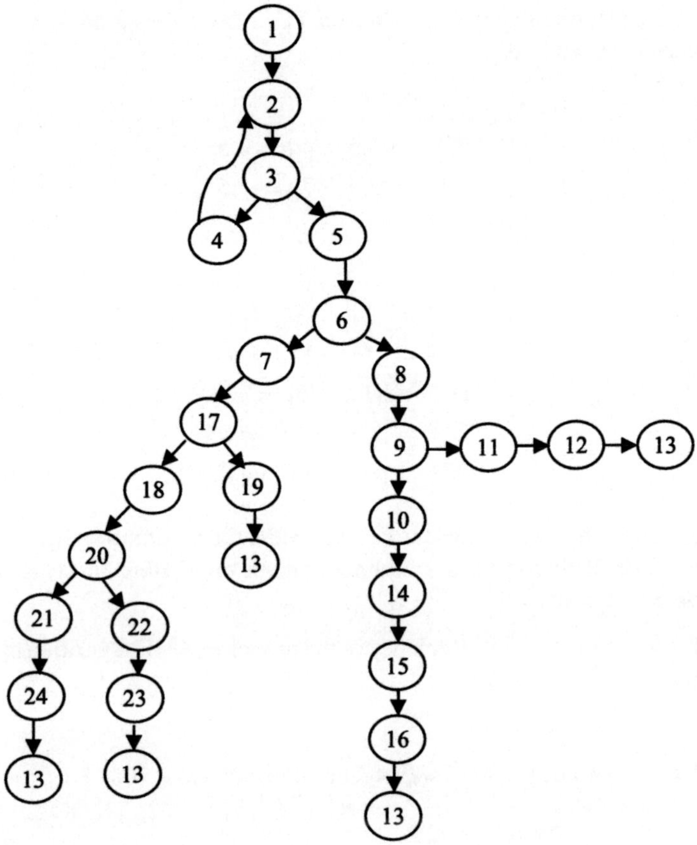

Fig. 4 Control flow graph of library system

$$\text{Cluster centre } 1 = 5.5$$
$$\text{Cluster centre } 2 = 12$$

KMH algorithm divides the test cases in two clusters, and the optimal center of these clusters are 5.5 and 12, respectively. Furthermore, test cases are assigned to clusters bases on minimum Euclidean distance. In turn, TC1 and TC2 are assigned to cluster 1, whereas, TC3, TC4, TC5, and TC6 are allocated to cluster 2.

Step 6: In step 6, optimum test cases are selected using the outcome of step 5. To determine the optimum test case, Manhattan distance is computed between cluster centers and corresponding data. The optimum test cases are selected based on the minimum distance between cluster centers and data. In cluster 1, three test cases are allotted. These test cases are TC1, TC6, and TC7. The corresponding cost of these test cases are 7, 10, and 10.

For cluster 1:

$$\text{TC1} = ||5.5 - 4|| = 0.5$$
$$\text{TC2} = ||5.5 - 7|| = 1.5$$

For cluster 2:

$$\text{TC3} = ||11 - 10|| = 1$$
$$\text{TC4} = ||11 - 12|| = 1$$
$$\text{TC5} = ||11 - 9|| = 2$$
$$\text{TC6} = ||11 - 12|| = 1$$
$$\text{TC7} = ||11 - 12|| = 1$$

So, the minimum value for cluster 1 is 0.5, while the minimum value for cluster 2 is 1. But in cluster 2, four test cases obtain minimum value. Here, test case is selected on first come first basis.

Step 7: This step computes the efficiency of the proposed KMH algorithm to compare old test cases over new test cases.

$$\text{First} = (1 - \text{no.of test clusters/total no.of test cases}) * 100$$
$$\text{First} = \left(1 - \frac{2}{7}\right) * 100 = 72\%$$
$$\text{Cluster 1} = (1 - \text{minimum difference/total sum of test cases within cluster}) * 100$$
$$\text{First} = \left(1 - \frac{0.5}{11}\right) * 100 = 95.45\%$$
$$\text{Cluster 2} = (1 - \text{minimum difference/total sum of test cases within cluster}) * 100$$
$$\text{First} = \left(1 - \frac{1}{55}\right) * 100 = 98.18\%$$

5 Conclusion

In this work, KHM-based algorithm is proposed to reduce the number of test cases in aspect-oriented programming. The performance of proposed algorithm is tested over two case studies, i.e., ATM system and library system. Both of cases studies are explored through activity diagrams and further, the control flow graphs are designed to determine independent paths. In this study, seven test cases are designed for each case study. The KHM algorithm is applied on the cost of independent paths. It is observed that KHM algorithm obtains significant results for both of cases studies. It is concluded that only two test cases can be executed to test the entire systems. The proposed algorithm provides more than ninety-five percent efficiency rate for both of case studies.

References

1. Laddad R (2010) AspectJ in action. Manning publication, vol II
2. Sommerville (2009) Software engineering, 8th ed. Pearson
3. Chauhan N (2012) Software testing: principles and practices, 5th ed. Oxford University Press
4. Harman M (2014) The current state and future of search based software engineering. In: IEEE international conference on software engineering
5. Zhang B, Hsu M, Dayal U (1999) K-harmonic means-a data clustering algorithm. Hewlett-packard labs technical report HPL-1999–124 55
6. Raheman SR, Maringanti HB, Rath AK (2018) Aspect oriented programs: issues and perspective. J Electr Syst Inf Technol 5(3):562–575
7. Jyoti SH (2017) Optimizing software testing using fuzzy logic in aspect oriented programming. Int Res J Eng Technol 04(04):3172–3175
8. Jyoti SH (2017) A systematic review and comparative study of existing testing techniques for aspect-oriented software systems. Int Res J Eng Technol 04(05):879–888
9. Chandra A, Singhal A (2016) Study of unit and data flow testing in object-oriented and aspect-oriented programming. In: 2016 international conference on innovation and challenges in cyber security (ICICCS-INBUSH). IEEE
10. Dalal S, Hooda S (2017) A novel technique for testing an aspect oriented software system using genetic and fuzzy clustering algorithm. In: 2017 International conference on computer and applications (ICCA). IEEE
11. Chhabra JK (2017) Harmony search based remodularization for object-oriented software systems. Comput Lang Syst Struct 47:153–169
12. Assunção W, Klewerton G et al (2014) Evaluating different strategies for integration testing of aspect-oriented programs. J Braz Comput Soc 20(1):9
13. Boudaa B et al (2017) An aspect-oriented model-driven approach for building adaptable context-aware service-based applications. Sci Comput Program 136:17–42
14. Ghareb MI, Allen G (2018) State of the art metrics for aspect oriented programming. In: AIP conference proceedings, vol. 1952, no. 1. AIP Publishing (2018)
15. Dalal S, Susheela H (2017) A novel approach for testing an aspect oriented software system using prioritized-genetic algorithm (P-GA). Int J Appl Eng Res 12(21):11252–11260
16. Kaur PJ et al (2018) A framework for assessing reusability using package cohesion measure in aspect oriented systems. Int J Parallel Program 46(3):543–564
17. Kaur PJ, Kaushal S (2018) A fuzzy approach for estimating quality of aspect oriented systems. Int J Parallel Program 1–20
18. Singhal A, Bansal A, Kumar A (2019) An approach for test case prioritization using harmony search for aspect-oriented software systems. In: Ambient Communications and Computer Systems. Springer, Singapore, pp 257–264

19. Kumar Y, Sahoo G (2015) A hybrid data clustering approach based on improved cat swarm optimization and K-harmonic mean algorithm. AI Communications 28(4):751–764 (2015)
20. Kumar Y, Sahoo G (2014) A hybrid data clustering approach based on cat swarm optimization and K-harmonic mean algorithm. J Inf Comput Sci 9(3):196–209

An Overview of Use of Artificial Neural Network in Sustainable Transport System

Mohit Nandal, Navdeep Mor, and Hemant Sood

Abstract The road infrastructure is developed to provide high mobility to road users, but, at present, the rapidly growing population and number of registered vehicles led to traffic congestion all around the world. Traffic congestion causes air pollution, increases fuel consumption, and costs many hours to the road users. The establishment of new highways and expanding the existing ones is an expensive solution and sometimes may not be possible everywhere. The better way is to detect the vehicle location and accordingly guiding the road users to opt for a fast route. Nowadays, Artificial Neural Network (ANN) is used for detecting vehicle location and estimation of the vehicle speed on the road. Route forecasting and destination planning based on the previous routes are missing elements in Intelligent Transport System (ITS). The GPS application in new generation mobiles provides good information for prediction algorithms. The objective of this study is to discuss the ANN technique and its use in transportation engineering. The paper also gives an overview of the advantages and disadvantages of ANN. Regular maintenance within the urban road infrastructure is a complex problem from both techno-economic and management perspectives. ANN is useful in planning maintenance activities regarding road deterioration.

Keywords Intelligent transport system · Artificial neural network · Traffic congestion

M. Nandal · H. Sood
Civil Engineering Department, NITTTR, Chandigarh, India

N. Mor (✉)
Civil Engineering Department, Guru Jambheshwar University of Science and Technology, Hisar, India
e-mail: navdeepmor12345@gmail.com

V. Singh et al. (eds.), *Computational Methods and Data Engineering*,
Advances in Intelligent Systems and Computing 1227,
https://doi.org/10.1007/978-981-15-6876-3_7

1 Introduction

Every year 1.35 million people die as a result of road accidents throughout the world and between 20 and 50 million suffer from injuries. Road traffic accidents lead to considerable economic losses to individuals, their families, and the nation too. About 3% of Gross Domestic Product (GDP) of most of the countries is wasted in road crashes. Although, the target of the 2030 agenda for Sustainable Development is to cut down the number of accidents and injuries by 50% (MORTH). One of the primary reasons behind a huge number of road accidents is traffic congestion. Real-time and accurate prediction of the arrival of a public transit vehicle is very important as passengers can accordingly plan their trips resulting in time and resource management [1]. The quality of data being processed and real-time situations will improve the output of prediction models. Most industry-leading engines are proprietary and algorithms are highly enhanced and refined by depending heavily on historic and crowdsourced data. The whole road network is considered as a graph where nodes denote points or intersections, and edges denote road segments. Various physical parameters like a number of stops along the route, speed limits, the distance between adjacent stops, historical average speed data, real-time crowdsource traffic data including traffic signals, and actual travel times are considered while modeling of data.

The weight is assigned to parameters on the basis of historical data. The algorithms based on this data will provide an acceptable bus Estimated Time of Travel (ETA) without a complex prediction model. Generally, the prediction models follow a certain pattern and if certain data (traffic signal malfunction, road crash, and speed limit) is not present then less accurate ETA prediction will be done.

Open Source Routing Machine used a prediction model known as Contraction Hierarchies [2]. The model is very effective but also time-consuming while updating real-time traffic data. Uber made use of OSRM model for ETA at pick up locations which were later modified and known as "Dynamic Contraction Hierarchies". This model updates applicable segments when a real-time traffic update comes. The model improved the pre-processing time and provided an almost accurate ETA. Artificial Neural Network is used for ETA prediction in many public transit applications by obtaining multiple sources of data from system administrating vehicle scheduling, tracking, and operations. The centralized server contains the data and conducts its management and processing of business functions. The algorithms are used for this processing which is required to be fast and responsive to provide quick updates to passengers in case of delay or change in schedule.

In recent years, Artificial Intelligence (AI) has engaged consideration of many researchers from different branches of pattern recognition, signal processing, or time series forecasting [3]. Artificial Intelligence is generally encouraged by biological processes that involve learning from past experiences. The primary work of AI methods is based on learning from experimental data and transmitting human knowledge into analytical models. This paper evaluates the meaning of ANN, the structure of ANN, its applications in transportation engineering, summarizes characteristics of ANN, and examines the interface of its techniques.

2 Definition of Artificial Neural Network

ANN is a computational model which stands for Artificial Neural Network. The working of ANN is based on the structure and function of a biological neural network. The human body is having 10 billion to 500 billion neurons [4]. A cell body, dendrites, an axon forms a biological neuron. The arrangement of neurons in between layers is known as the architecture of net. The architecture of ANN consists of an input layer, hidden layers, and an output layer. The structure of an Artificial Neural Network is affected by the transfer of information. The elements which process the information are called neurons. Difference between ANN and Biological Neural Network:

i. The processing speed of ANN is very fast as compared to Biological Neural Network. Cycle time is the time consumed in processing a single piece of information from input to output.
ii. ANN is having only a few kinds of processing units while the biological neuron network consists of more than a hundred processing units.
iii. Knowledge in biological neural networks is versatile while knowledge in Artificial Neural Networks is replaceable.
iv. The human brain has better error correction.

The neuron with n input is used to determine the output as given in Eq. 1:

$$a = f\left(\sum_{i=1}^{n} w_i p_i + b\right) \tag{1}$$

where

p_i value of ith input
w_i value of ith weight
b bias
f activation function of neuron.

Generally, the activation function "f" will be of the following types:

i. Linear function:

$$f(x) = x$$

ii. Threshold function or heavy side step function:

$$f(x) = \begin{Bmatrix} 1 \text{ if } x > 0 \\ 0 \quad \text{else} \end{Bmatrix}$$

iii. Sigmoid function:

$$f(x) = \tan h(x) \text{ or } f(x) = \frac{1}{1 + e^{-x}}$$

The heavy side step function is used in the output layer to generate the final decision and linear function, and sigmoid function is used in the first two layers, i.e., input layer and hidden layer. The number of Processing Elements (PE) in the input layer is the same as the number of input variables that are used to determine the required output [5]. The PE in the output layer determines the variables to be forecasted. The relation between input and output layer depends on the complication of one or several intermediate layers of processing elements known as hidden layer. The most important property of ANN is its ability of mapping non-linear relations between variables explaining the model's behavior.

3 Structure of ANN

ANNs consist of various nodes that act like genetic neurons of the human brain [6]. Interaction between these neurons is assured by the links which are connected with these neurons. The input data is hence received by nodes and executes the operation on the data which is passed on to other neurons. The final output at an individual node is termed as "node value". Every link is capable of learning as they are associated with the same weight. If output generated by using ANN is good, there is no need of adjusting the weights, but if the overall output of the model generated is poor, then confidently weight should be altered to improve the results. The diagram of ANN is given in Fig. 1.

Usually, the two different types of ANN are as follows:

(a) Feed Forward ANN: The flow of information is unidirectional in this network. It does not involve any feedback loops [7]. The accuracy of output will be increased by using a greater number of hidden layers.
(b) Feedback ANN: The flow of information is allowed in both directions means feedback loops are available in this ANN. The implementation of this network is complex. This network is also known as Recurrent or Recursive Network (Table 1).

4 Machine Learning in Artificial Neural Network

Various types of machine learning techniques that are being used in ANN [2] are as follows:

(1) Supervised Learning: This learning technique contains the training data which means both input and output are available to us. The value of output is checked

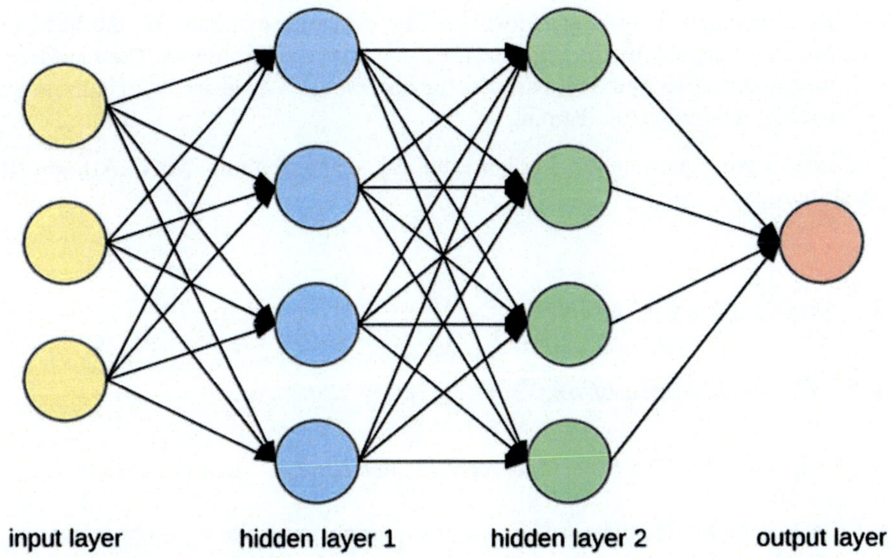

Fig. 1 Artificial neural network

Table 1 Advantages and disadvantages of ANN

Sr. No.	Advantages	Disadvantages
1.	ANN simulates the naive mechanisms of brain and permits external input and output to allocate proper functioning	ANN requires long training and has problems with multiple solutions
2.	ANNs have various ways of learning which depends on adjusting the strength of connections between the processing units	ANN does not cover any basic internal relations and there is no up-gradation of knowledge about the process
3.	ANN can use different computing techniques. The accuracy of output will be changed by altering the hidden units	
4.	Programming is installed in ANN at once, then the only requirement is to feed data and train it	There is no proper guidance for using the type of ANN for a particular problem
5.	ANN can estimate any multivariate non-linear function	

by putting different values of training data. Naive Bayes algorithm is used in supervised learning. Example: Exit Poll.

(2) Unsupervised Learning: This learning technique is used in the absence of data set. It contains only input value based on which it performs clustering. The ANN modifies the weight to achieve its own built-in criterion. *K* mean algorithm is used in unsupervised learning. Most of the machines come under this category only.

(3) Reinforcement Learning: It involves the formation of policy on the basis of reward or penalty by the action of the agent on the environment. The reinforcement learning technique is based on the observations. Q-learning algorithms are used in reinforcement learning.

These algorithms can be implemented by using Python, MATLAB, or R-programming.

5 Applications of ANNs

5.1 General Applications

a. Aerospace: ANN can be used for fault detection in aircraft, or in autopilot aircrafts.
b. Electronics: ANN can be used for chip failure analysis, code sequence prediction, and IC chip layout.
c. Military: ANN is used in tracking the target, weapon orientation, and steering.
d. Speech: ANN is used in speech recognition and speech classification.
e. Medical: ANN can be used in EEG analysis, cancer cell analysis, and ECG analysis.
f. ***Transportation**: ANN can be used in vehicle scheduling, brake system diagnosis, and driverless cars.
g. Software: ANN is used for recognition patterns like optical character recognition, face recognition, etc.
h. Telecommunications: ANN can be used in different applications in different ways like image and data compression.
i. Time Series Prediction: ANN can be used to predict time as in the case of natural calamities and stocks.

5.2 Applications of ANNs in Transportation Engineering

1. **Traffic Forecasting**: The forecasting of traffic parameters is done in order to manage local traffic control system [8]. Different approaches can be used while forecasting based on statistics, i.e., the method may vary in generating an output of the network or manner in which forecasting task is identified. Identification involves the parameters which are passed into the input of the network. The parameters that can be used are the speed of travel, length of trip, and traffic flow. Finally, after training of data, the network produces the next values of input variables to obtain the output.
2. **Traffic Control**: Various computation methods on the basis of ANN can be used in the design of road traffic control devices, and traffic management systems.

ANN involves the blending of historical data with the latest parameters of road condition for an explanation of control decisions. The algorithms can be developed in order to enhance the efficiency of traffic control by around 10%. ANN is also effective in establishing an optimal time schedule for a group of traffic signals.

3. **Evaluation of Traffic Parameters**: Traffic situation can be mapped with higher accuracy in the case of Intelligent Transportation System (ITS) by making use of Origin & Destination (O-D) matrices [9]. Work report in many areas has been developed to determine the O-D matrix in the absence of complete data by using different measuring devices. The major problem for Intelligent Transportation System functioning is to detect the location of a traffic accident which may disturb the equilibrium of the managed traffic system.
4. **Maintenance of Road Infrastructure**: The primary concern in the maintenance of road infrastructure is the restoration of pavement. Neural networks are enforced for predicting pavement performance and condition, maintenance, and management strategies.
5. **Transport Policy and Economics:** ANN can be used in the appraisal of the significance of transport infrastructure expansion. The composition of the neural network is proposed for developing the order of carrying out objectives of expansion by considering the resources of investment.
6. **Driver Behavior and Autonomous Vehicles**: Decision making of driver's awareness of road condition and judgment is governed by many factors where conventional modeling methods are not applicable. ANN can be used in developing a vehicle control system for driverless movement and ensuring the safety of the driver. This development involves the position of the driver, his ability of driving, and the position of conquering the traffic situations which is dangerous while driving.
7. **Pattern Recognition**: ANN is useful in automatic detection of any road accident, identification of cracks in bridge or pavement structure, and processing image for traffic data collection.
8. **Decision Making**: ANN is useful in making a decision whether a new road is to be constructed or not, how much money should be assigned to rehabilitation and maintenance activities, and which bridge or road segment is required to be maintained and whether to divert traffic to some other route in case of accident situation.
9. **Weather Forecasting**: ANN consists of certain tools that can be used to inform weather conditions to the driver for planning a suitable route.

5.3 Important Previous Research in Transportation Engineering Using ANNs

ANN has been used worldwide in transportation engineering. Some of the important studies where ANN has been used are discussed below:

Amita et al. [1] developed a time prediction model for bus travel using ANN and regression model. The model provided real-time bus arrival data to the passengers. For the analysis, the authors took input data like time delays, dwell-time, and the total distance traveled by bus at each stop. The authors concluded that ANN model is better than the regression model in terms of accuracy. Faghri and Hua [3] evaluated applications of Artificial Neural Network in transportation engineering. The paper summarizes the characteristics of Artificial Neural Network in different fields and its comparison with a biological neural network. The authors performed a case study for forecasting trip routes by making use of two Artificial Neural Network models and one traditional method in order to represent the potential of ANNs in transportation engineering. The authors compared the methods and concluded that ANN is highly capable of forecasting trip routes for transportation engineering operations than other methods of artificial intelligence. Pamuła [5] summarized the application of ANN in Transportation Research. The author discussed the various examples of road traffic control, prediction of traffic parameters, and transport policy and economics. The author concluded that the Feed Forward multilayer neural network is the most commonly used network in transportation research. Štencl and Lendel [6] discussed the applications of Artificial Intelligence (AI) techniques in Intelligent Transportation Systems. The authors concluded that traditional is expensive and time-consuming in the field of AI and use of ANN method is appropriate because it can solve multivariate non-linear functions easily. Behbahani [10] compared four ANN techniques, i.e., Probabilistic Neural Network (PNN), Extreme Learning Machine (ELM), Multilayer Perceptron (MLP), and Radial Basis Function (RBF) in forecasting accident frequency on urban road network. The authors concluded ELM as the most efficient method in prediction models based on different measures, i.e., Root Mean Square Error (RMSE), and Mean Absolute Error (MAE). Moreover, the authors found ELM as the fastest algorithm and most accurate method for prediction of the road accident location. Gurmu et al. [8] developed an ANN model for accurate prediction of bus travel time and to provide real-time information to passengers using GPS data. The authors took a unique set input–output values for offline training of ANN model. The authors analyzed the performance of ANN on the basis of robustness and prediction accuracy and concluded that ANN had better results in both aspects.

6 Conclusions and Recommendations

Data analysis in prediction of maintenance of the road, location of traffic congestion, black spots are very complex phenomena, but ANN is considered to be a useful tool in analyzing the data by performing data clustering. The Artificial Neural Network depicts the overall connection of the system along with numeric weights, which can be adjusted on the basis of the input unit, hidden layers, output unit, and experience. One of the important advantages of Artificial Neural Network is varying the topology of hidden layers in order to improve the final result. ANN has a wide range of application which includes traffic forecasting, traffic control, etc. This paper summarizes the concept of Artificial Neural Network specifically for Transportation Infrastructure Systems (TIS). The paper has demonstrated various advantages of ANN and the core advantage of this technique is its ability to solve complicated problems in the field of transportation engineering.

The ANN is capable of providing a good solution for increased congestion so it should be used in urban areas for developing traffic signals and finding an appropriate schedule plan for public transport. The ANN can also be used by individual drivers in optimizing their routes. The automobile companies should make use of ANN for guiding its customers regarding vehicle safety during the service life of the product. The use of ANN can be made by highway authorities in finalizing the decision regarding road infrastructure rehabilitation.

References

1. Amita J, Singh JS, Kumar GP (2015) Prediction of bus travel time using artificial neural network. Int J Traffic Transp Eng 5(4):410–424
2. Data Flair Homepage. https://data-flair.training/blogs/artificial-neural-network/
3. Faghri A, Hua J (1992) Evaluation of artificial neural network applications in transportation engineering. Transp Res Rec 71–79
4. Experion technologies Homepage. https://www.experionglobal.com/predicting-vehicle-arrivals-in-public-transportation-use-of-artificial-neural-networks/
5. Pamuła T (2016) Neural networks in transportation research–recent applications. Transp Prob 111–119
6. Štencl M, Lendel V (2012) Application of selected artificial intelligence methods in terms of transport and intelligent transport systems. Period Polytech Transp Eng 40(1):11–16
7. Dougherty M (1995) A review of neural networks applied to transport. Transp Res Part C: Emerg Technol 3(4):247–260
8. Gurmu ZK, Fan WD (2014) Artificial neural network travel time prediction model for buses using only GPS data. J Public Transp 17(2):3–14
9. Abduljabbar R, Dia H, Liyanage S, Bagloee S (2019) Applications of artificial intelligence in transport. An overview. Sustainability 11(1):189–197
10. Behbahani H, Amiri AM, Imaninasab R, Alizamir M (2018) Forecasting accident frequency of an urban road network. A comparison of four artificial neural network techniques. J Forecast 37(7):767–780

Different Techniques of Image Inpainting

Megha Gupta and R. Rama Kishore

Abstract Image inpainting was generally done physically by artists to remove deformity from works of art and photos. To cover the area of target or missing data from a signal, utilizing encompassing details and restructure signal is the fundamental job of image inpainting algorithms. We have considered and audited numerous distinct algorithms available for image inpainting and clarified their methodology. This paper includes various works in the branch of image inpainting and will help beginners who want to work and develop the image inpainting techniques.

Keywords PDE · Image inpainting · Exemplar-based inpainting · Structural inpainting · Texture synthesis · Neural network-based inpainting

1 Introduction

Inpainting is the craft of reestablishing lost pieces of a picture and reproducing them dependent on the foundation data. This must be done in an imperceptible manner. The word inpainting is taken from the old specialty of reconstructing the image by expert image restorers in exhibition halls and so forth. Digit Image Inpainting attempts to impersonate this procedure and do the inpainting through algorithms. Figure 1 demonstrates a case of this tool where a building is supplanted by appropriate data from the image in a perceptibly conceivable manner. Through an automatic process, algorithm does this such that the image looks "sensible" to humans. Information that is covered up entirely by the object to be expelled can't be restored by any algorithm. Subsequently, the target for the image inpainting isn't to recreate the first image, yet to restore image so that it has impressive similarity with the primary image.

M. Gupta (✉)
USICT, Guru Gobind Singh Indraprastha University, Dwarka, Delhi, India
e-mail: meghagupta0820@gmail.com

R. Rama Kishore
Guru Gobind Singh Indraprastha University, Dwarka, Delhi, India
e-mail: ram_kish@yahoo.com

V. Singh et al. (eds.), *Computational Methods and Data Engineering*,
Advances in Intelligent Systems and Computing 1227,
https://doi.org/10.1007/978-981-15-6876-3_8

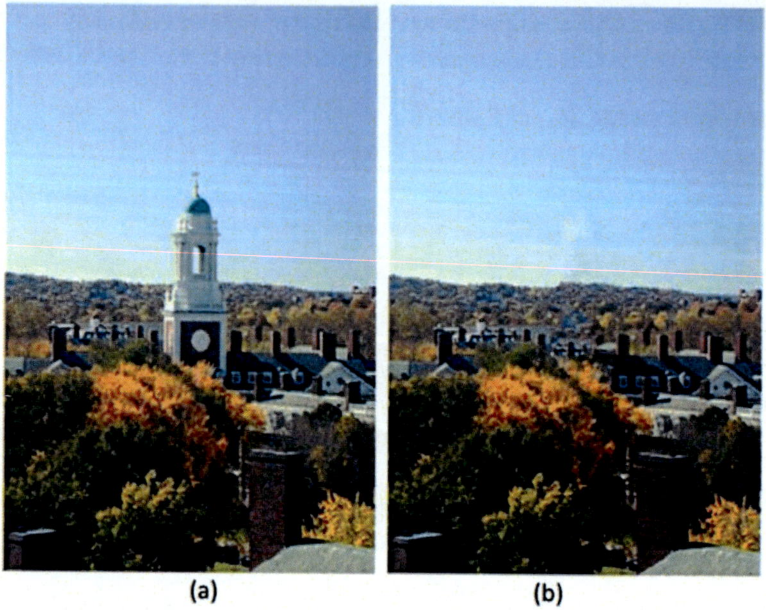

Fig. 1 Demonstration of image inpainting to remove the large unwanted object. **a** Original image. **b** Unwanted building removed by using image inpainting

Image inpainting has many benefits, like restoring photos. In truth, the term inpainting has been obtained from the specialty of restoring falling apart photos and artistic creations by skilled restorers in exhibition halls and so on. A very long time back, individuals were taking care of their pictorial works cautiously. With age, their work gets harmed and scratched. Clients would then be able to utilize the tool to expel the deformities from the photograph. Another utilization of image inpainting is in making embellishments by expelling undesirable elements from the image. Undesirable things may extend from mouthpieces, ropes, individual and logos, stepped dates and content, and so forth in the image. During the transmission of images to a system, there might be a few pieces of an image that may go absent. These parts would then be able to be reproduced utilizing image inpainting. There has been a good amount of research on the most proficient method to utilize image inpainting in various fields.

2 Literature Review

These days, there are various procedures of image inpainting are accessible. A few procedures have been utilized by the specialists for advanced image inpainting are being listed below in broad classes:

- Partial Differential Equation(PDE) based inpainting.
- Texture Synthesis-based inpainting.
- Hybrid inpainting.
- Exemplar-based inpainting.
- Deep generative model-based inpainting.

2.1 Partial Differential Equation (PDE) Based Inpainting

The first PDE-based methodology presented by Bertalmio [1]. Pixels on edges are likewise not secured because it utilizes the idea of the isophotes and propagation operation as shown in Fig. 2. The essential issue with this strategy is that it obscures the impact of diffusion operation and so replication of huge texture isn't executing great. Total Vibrational model is put forwarded by Chang and Shen and utilizes Aniso-tropic diffusion along with Euler–Lagrange equation [2–4]. Out of the TV model, a new algorithm is displayed dependent on the Curvature Driven Diffusion model that incorporates curvature appertain to isophotes.

Andrea et al. [5] utilized changed Cahn–Hilliard equation to accomplish quick binary image inpainting. This altered equation is used to do inpainting of binary shapes, content reparation, street interpolation, and high-resolutions. The equation does the best processing when the end client determines the inpainting space, their methodology can likewise be utilized for interpolating uncomplicated streets and different scenes where a client characterized inpainting part isn't doable. Using a unique two-step procedure, the strategy can inpaint over huge separation in a repeatable manner. Albeit different arrangements, inclusive of broken connections, might be conceivable numerically, the method can provide continual computation by first performing diffused yet ceaseless connection, and afterwards utilizing this as new information for later inpainting with sharp transitions among white and dark districts. With regards to binary image inpainting, the changed Cahn–Hilliard equation has

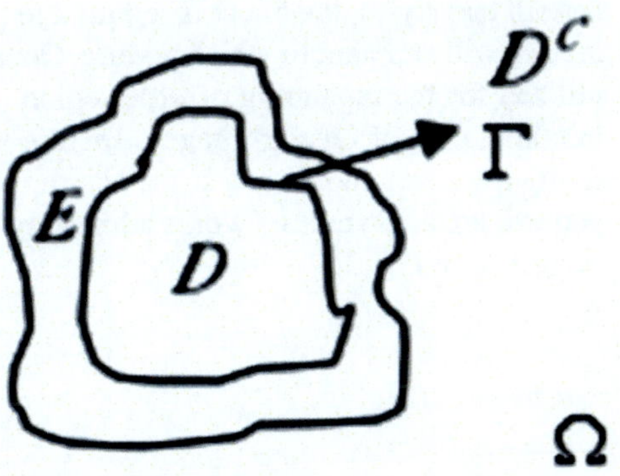

Fig. 2 Shows the diffusion method used in PDE-based image inpainting algorithms

displayed a significant reduction in the calculation when contrasted along with other PDE-based inpainting techniques. Quick numerical methods present for the Cahn–Hilliard equation is additionally a less productive calculation with generally enormous datasets.

2.2 *Texture Synthesis-Based Image Inpainting*

Approaches based on PDE are appropriate for fixing little imperfections, content overlays, and so on. In any case, PDE procedure typically comes up short whenever applied to zones having regular patterns and when applied to a textured field. This disappointment is a direct result of the accompanying explanation:

1. For the most part gradients of high intensity in grains might be deciphered inaccurately as edges moreover erroneously conveyed into the area to be painted.
2. In inpainting based on PDE, the data utilized is just the boundary condition that is available inside a small circle near the object to be inpainted. Therefore, it is incapable to calculate structured and textured objects or ordinary areas from such minute information.

In this technique, holes are fixed by testing and replicating nearby pixels [6, 7]. The central issue between textured based algorithms is the means by which they keep up the progression among hole's pixel and novel picture pixels. This technique is just working for chosen images, not with all images. Yamauchi et al. introduced algorithms that texture in various brilliance conditions and working environment for multi resolutions [8].

The quick synthesizing algorithm shown in [9], utilizes image quilting (sewing little clumps of pre-existing images). Every texture-based technique is diverse as far as their ability to create texture for various color variations, statistical, and gradients features. Texture synthesis-based inpainting technique does not deal truly well for regular images. These techniques don't deal with edges and limits productively. In certain text styles, the client is required to provide information about texture, which texture will supplant to which texture. Consequently, these particular techniques are utilized for the inpainting of little region [10]. These algorithms experience issues in taking care of regular images. Texture synthesis procedures might be applied in fixing digitized photos and if the improper region wants to be filled in any regular pattern, texture synthesis works admirably. Procedure of texture synthesis method is described in Fig. 3.

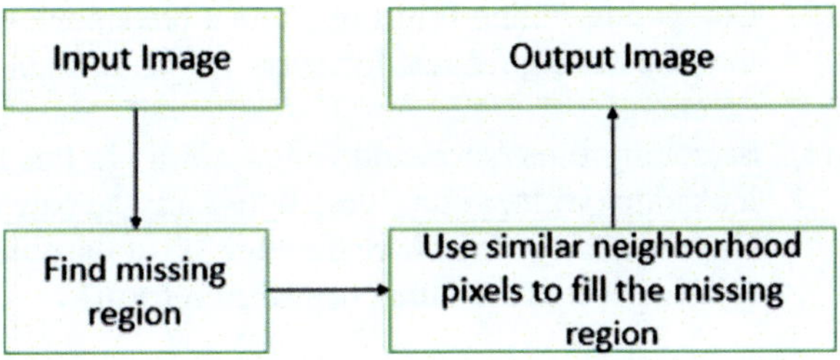

Fig. 3 Texture synthesis method

2.3 *Hybrid Inpainting*

Criminisi et al. [11] attempted to fuse structure with textural inpainting utilizing an exceptionally canny rule, by which texture get inpainted in the direction of isophote as per its quality. Unfortunately, it was constrained for limited structures, frequently bringing about discontinuities at the place where texture meet. Structural inpainting attempts to uphold a smoothness earlier still protects isophotes. They characterize the textural image from the disintegration "u = us + ut", "us" is the structural component and "ut" is the textural component [12–14]. It is redefining of the high texture region in the past subsection, by comprising genuine colors of pixels. Bertalmio et al. [15] applied inpainting to textures and structures as per the information portrayed before. The advantage of isolating texture against structure is in the capacity to inpaint each sub-image independently. The texture of sub-image inpainted utilizing an exemplar-based technique that duplicates pixels instead of patches [16]. This texture inpainting may not produce a remarkable arrangement, along with may not recreate the sample attributes clearly.

2.4 *Exemplar-Based Image Inpainting*

This technique for image Inpainting is an effective way to deal with reproducing huge objective areas. Exemplar-based Inpainting method repeatedly synthesis the goal section by most analogous area in the source section. The exemplar-based method makes samples from the finest identical areas out of the known area, whose likeness is calculated by certain metrics, and pastes into the goal areas in the misplaced area. Essentially it involves two elementary phases: In the primary phase priority assignment work is completed and the latter phase comprises the selection of the finest matching area [17–19].

Normally, an exemplar-based inpainting algorithm incorporates the accompanying fundamental advances:

(1) Setting the Target Region, where the underlying lost parts are obtained and illustrated with suitable information.

(2) Computing Filling Priorities, here a predefined operation is utilized to process the dispatching request for every single unfilled pixel in the start every filling iteration.
(3) Searching Illustration and Compositing, in this utmost comparative model, is looked out of the source area Φ to form the patch.
(4) Updating Image Info, here the edge $\delta\Omega$ of the objective region Ω and the needed data to process the filling needs are refreshed.

For the exemplar-based image Inpainting, many algorithms are created. For example, Criminisi built up a proficient and basic way to fill the required information by the boundary of the area which is required and where the quality of the isophotenear the missing area was substantial, after that we utilized the sum of squared difference (SSD) [20] to choose the most similar patch among applicant source patches. In this algorithm of criminisi order of filling the area is dictated by the priority-based system. Wu [16, 21] presented a model, cross isophotes exemplar-based model utilizing the local texture data and the cross-isophote diffusion information which chose dynamic size belonging to exemplars (Sun et al.) [22]. Fills the obscure data utilizing the process of texture propagation but before that, it made fundamental curves on its own which inpaint the missing structure. Hung (Baudes et al. 2005) [23] utilized Bezier curves and the structure formation to build the missing information on the edges. By curve filling operation we utilize reconnecting contours and structure information to inpaint damaged areas. A patch-based image synthesis process and Resolution preparing process. Duval et al. (2010) [24] gave ideas of sparsity on the patch level in order to do modeling of the patch priority and representation.

In comparison with methodologies that are diffusion-based, exemplar-based methodologies accomplish impressive outcomes in reiterative structures regardless of even if they are targeted on the enormous areas or not. A large portion of exemplar-based algorithms takes up the avid methodology, so these algorithms experience the ill effects of the regular issues of the avid algorithm, being the filling order (to be specific need) is demanding. Exemplar-based inpainting will create great outcomes just if the missing area comprises of smooth texture and structure [25, 26].

Jin et al. in 2015 [27] presented a methodology for situation-aware patch-based image inpainting, here textural descriptors are utilized to manage and quicken the quest for similar (candidate) patches. Top-down parting technique isolates the image into many distinctive size squares as per their specific situation, obliging in this way for the quest of candidate patches to no local image areas of similar details. This methodology can be utilized to boost the processing and execution of essentially any patched-based technique of inpainting. We use this way with the Markov Random Field (MRF) [28] to deal with the purported global image inpainting. Earlier, where MRF e codes from the earlier information about the consistency of neighboring image patches.

2.5 Deep Generative Model-Based Inpainting

Ongoing deep learning-based methodologies have demonstrated encouraging outcomes for difficult work of inpainting for substantially imperfect information in an image. These techniques can produce valuable textures and image structures, however, frequently make twisted structures or hazy textures in-reliable with encompassing territories. This is for the most part because of the ineffectualness of convolutional neural networks in unequivocally obtaining or duplicating data from far off spatial areas. Then again, conventional patch and texture synthesis methodologies are especially reasonable during the time it needs to get textures out of the encompassing locale.

Yang et al. in 2017 [29] brought together a unified feed-forward generative system. It has a unique contextual attention layer to inpaint the image. Their presented system has two phases. In the primary stage, quick inpainting takes place because the simple dilated convolution network is trained to recreate loss quickly without including all details. In the second stage, contextual attention is applied. The information of known patches is to be utilized while conventional filters will process the created patches, and it is the main logic behind the contextual attention layer [30, 31]. It is planned and executed with convolution for calculating and putting the generated matching patches with already known relevant patches, and check applicable patches and deconvolution to remake the produced patches with contextual patches. Spatial engendering layer energizes spatial coherency of attention. Which results in network hallucinating novel contents, in parallel they have a convolutional pathway and relevant attention pathway. To get the end result two pathways accumulated and sustained into a single decoder. The entire system is skilled from start to finish with reproduction losses and two Wasserstein GAN losses [1, 32–34], one examiner checks the global image while another checks the local patch of the missing area.

Another work of jiahui et al. presented a novel deep learning-based image inpainting framework to finish pictures with free-form masks and inputs. Their presented framework depends on gated convolutions gained from a huge number of images without extra labeling efforts. The proposed gated convolution fix the issue of vanilla convolution that treats all input information pixels as legitimate ones, sums up partial convolution by giving a learnable powerful feature selection system for each channel at each spatial area over all layers. Besides, as freestyle masks may show up at any place in images with any shapes, global and local GANs intended for a solitary rectangular mask are not appropriate. To this end, they likewise present a novel GAN loss, named SN-PatchGAN [35], by applying phantom standardized discriminators on thick image patches. It is basic in the formulation, quick and stable in training. Demir et al. in 2018 [36] presented a generative CNN model and a training method for the subjective and voluminous hole filling issue. The generator system takes the tainted image and attempts to remake the fixed image. They used the ResNet [32, 37, 38] design as their generator model with a couple of changes. During the training, they utilize the dual loss to acquire reasonable looking results. The key purpose of their work is to make a novel discriminator network that joins G-GAN structure with

PatchGAN approach which they call PGGAN. Karras et al. represented a training procedure for generative antagonistic networks. The key thought is to advance both the generator and discriminator continuously: beginning from a low resolution, they include new layers model progressively increase fine details as training advances. This speeds the training and balances it with great efficiency, enabling them to create pictures of extraordinary quality [20, 30]. They likewise proposed a basic method to expand the variety in produced pictures, Additionally, they portray a few usage subtleties that are significant for decreasing undesirable competition between the generator and discriminator.

Kamyar et al. in 2019 [34] explain image inpainting in below-mentioned procedures, edge generation, and image accomplishment. Edge formation is exclusively centered around hallucination operation on the edges of the target areas. The image completion network utilizes the hallucination operation on edges along with that guesses RGB pixel's strength of the target areas. The two phases pursue an inimical framework [10, 39] to guarantee that the hallucination operation on edges along with the RGB pixel's intensity potential are visibly unchanging. The two networks assimilate losses dependent on deep features to implement perceptually reasonable outcomes [40].

Wei et al. in 2019 proposed an approach that fuses a deep generative model with the operation that searches for analogous patches. This technique firstly skills a "U-Net" [8, 41, 42] generator utilizing the Pix2Pix [9, 43] method its engineering is like VAE-GAN along with the generator produces a rough image and its patch of the missing area is hazy semantic info. Their technique looks for a comparable patch in a huge facial picture data set by using this coarse image. At long last, Poisson blending [37, 44, 45] is utilized to associate with analogous patch and the coarse image. The mix of the two techniques resolves their separate weaknesses, which are the hazy outcomes and the absence of earlier semantic data when utilizing the method to look analogous patches.

3 Comparative Study

On the basis of a detailed analysis of all the papers, we have presented the comparative study of all the approaches used in Image Inpainting. We divide them into two categories. First, traditional approaches generally fail to give results on the large missing areas. The second category of deep neural network-based approaches these algorithms are faster and provide better results but they take more time in training.

Below are Tables 1 and 2 in which the merits and demerits of all the approaches are shown.

Table 1 Merits and demerits of traditional inpainting algorithms

Methods	Authors	Merits	Demerits
Diffusion based inpainting	Sapiro et al. [38]	It can inpaint uncomplicated and limited region	Image information is compromised
PDE-based inpainting	Bertalmio et al. [1]	Better performance and structural arrangement is maintained	Subject of inpainting occupying large region gets obscure
Texture synthesis method	Grossauer et al. [46]	Blurring issue for sizeable subject of inpainting is resolved	This method is not useful for subjects which are rounded and have broad damaged area
Hybrid inpainting	Li et al. [47]	Even with smoother results, it maintains linear subject's structure and image's texture	With disproportionate patch size and broad scratched regions, results are boxlike
Exempler based inpainting	Criminisi et al. [11]	Commendable outcomes as it maintains great information of the image intact	Failed image inpainting results spill over different areas in the image

Table 2 Merits and demerits of deep neural network algorithms

Methods	Authors	Merits	Demerits
Deep generative model with convolution neural network	Zhu et al. [48]	Their approach exploit semantics learned from large-scale dataset to fill contents in non-stationary images	Does not work well enough with curved structures
Patch-based inpainting with generative adversial network	Demir et al. [49]	Produces visually and quantitatively better results	Training time is high and slight blurriness is present
Progressive growing of Gans	Karras et al. [50]	The training is stable in large resolutions	Need improvement in curved structure
Edge connect: generative image inpainting with adversarial edge learning	Nazeri et al. [34]	Images with numerous uneven missing areas can be inpainted	Problem in processing edges around high texture region in the image

4 Conclusion

Image inpainting has an exceptionally wide application and research area. It has a significant job in reconstructing lost or decayed regions in an image and expelling an unfortunate or undesirable part from the image. There are numerous methods developed for the same reason with many benefits and shortcomings. In this paper, diverse image inpainting techniques are considered. Various advances received for image inpainting in different techniques are explained. Their advantages and drawbacks are discussed in a nutshell. For various strategies, specialists performed experiments on pictures of various scenarios. At present, algorithms do not function admirably enough with curved structures, which can be enhanced. Training time of generative based calculations is high which has the degree to lessen it in not so distant future.

5 Future Work

Computerized inpainting techniques expect to robotize the procedure of inpainting, and to achieve this it needs to put a limit on the end-user cooperation. Yet, the cooperation from a user which is difficult to dispense is with the determination of the inpainting area since that relies upon the decision of the user, but intelligent suggestions can be provided. Presently, it requires improvement to inpaint the curved structures. The inpainting technique can be further used for the expulsion of moving bodies from a video by tracing and inpainting the moving bodies in real-time. The inpainting algorithm can likewise be stretched out for mechanized recognition and elimination of content in recordings. The video recordings many times include dates, titles, and other extra things that are not required. The above process can be done automatically without user cooperation.

References

1. Bertalmio M, Vese L, Sapiro G (2000) Simultaneous structure and texture image inpainting. IEEE Trans Image Process 12(8)
2. Bertalmio M, Vese L, Sapiro G, Osher S (2003) Simultaneous structure and texture image inpainting In IEEE Trans. Image Process 12(8):882–889
3. Yang C, Lu X, Lin Z, Shechtman E, Wang O, Li H (2017) High-resolution image inpainting using multi-scale neural patch synthesis. In: Proceedings of the IEEE computer vision pattern recognition, pp 6721–6729
4. Zhu X, Qian Y, Zhao X, Sun B, Sun Y (2018) A deep learning approach to patch-based image inpainting forensics. Signal Process Image Commun 67:90–99
5. Andrea L, Bertozzi L, Selim E, Alan G (2007) Inpainting of binary images using the Cahn–Hilliard equation. In IEEE Trans Image Process 16(1)
6. Liu Y, Caselles V (2013) Exemplar-based image inpainting using multiscale graph cuts In IEEE Trans. Image Process 22(5):1699–1711

7. Guillemot C, Meur O (2014) In image inpainting: overview and recent advances. IEEE Signal Process Mag 31(1):127–144
8. Meur O, Gautier J, Guillemot C (2011) Examplar-based inpainting based on local geometry In: Proceedings of the 18th IEEE international conference image process, pp 3401–3404
9. Li Z, He H, Tai H, Yin Z, Chen F (2015) Color-direction patchsparsity-based image inpainting using multidirection features In IEEE Trans Image Process 24(3):1138–1152
10. Ruzic T, Pizurica A (2019) Context-aware patch-based image inpainting using Markov random field modeling In IEEE Trans Image Process 24
11. Criminisi A, Perez P, Toyama K (2003) Object removal by exemplar based inpainting. In: Proceedings of the conference computer vision and pattern recognition, Madison
12. Kumar V, Mukherjee J, Mandal Das S (2016) Image inpainting through metric labelling via guided patch mixing. IEEE Trans Image Process 25(11):5212–5226
13. Arias P, Facciolo G, Caselles V, Sapiro G (2011) A variational framework for exemplar-based image inpainting In Int. J Comput Vis 93(3):319–347
14. Meur O, Ebdelli M, Guillemot C (2013) Hierarchical super resolution-based inpainting. IEEE Trans Image Process 22(10):3779–3790
15. Cai K, Kim T (2015) Context-driven hybrid image inpainting In IET Image Process. 9(10):866–873
16. Barnes C, Shechtman E, Finkelstein A, Goldman D (2009) PatchMatch: a randomized correspondence algorithm for structural image editing. ACM Trans Graph 28
17. Bertalmio M, Bertozzi A, Sapiro G (2001) Navier-Stokes, fluid dynamics, and image and video inpainting In Proceedings of the IEEE international conference computer vision pattern recognition, pp 1355–1362
18. Ulyanov D, Vedaldi A, Lempitsky V (2018) Deep image prior. In: Proceedings of the IEEE computer vision pattern recognition, pp 9446–9454
19. Pawar A, Phatale A (2016) A comparative study of effective way to modify moving object in video: using different inpainting methods. In 10th international conference on intelligent systems and control
20. He K, Sun J (2014) Image completion approaches using the statistics of similar patches". IEEE Trans Pattern Anal Mach Intell 36(12):2423–2435
21. Darabi S, Shechtman E, Barnes C, Goldman D, Sen P (2012) Image melding: combining inconsistent images using patch-based synthesis. ACM Trans Graph 31(4):82:1–82:10
22. Ram S, Rodríguez J (2016) Image super-resolution using graph regularized block sparse representation. In: Proceedings of the IEEE Southwest symposium analysis interpretation, pp 69–72
23. Xie J, Xu L, Chen E (2012) Image denoising and inpainting with deep neural networks. In: Proceedings of the 25th international conference neural information processing systems (NIPS), pp 341–349
24. Duval V, Aujol J, Gousseau Y (2010) On the parameter choice for the non-local meansSIAM. J Imag Sci 3:1–37
25. Li F, Pi J, Zeng T (2012) Explicit coherence enhancing filter with partial adaptive elliptical kernel. IEEE Signal Process Lett 19(9):555–558
26. Grossauer H, Scherzer O (2003) Using the complex Ginzburg-Landau equation for digital inpainting in 2D and 3D In: Proceedings of the 4th international conference scale space methods computer vision, pp 225–236
27. Siegel S, Castellan N (1998) Nonparametric statistics for the behavioral sciences, 2nd edn. McGraw-Hill, New York, NY, USA
28. Wang J, Lu K, Pan D, He N, Bao B (2014) Robust object removal with an exemplar-based image inpainting approach. Neurocomputing 123:150–155
29. Ram S, Rodríguez J (2014) Single image super-resolution using dictionary-based local regression. In: Proceedings of the IEEE Southwest symposium on image analysis interpretation, pp 121–124
30. Huang J, Kang S, Ahuja N, Kopf J (2014) Image completion using planar structure guidance. ACM Trans Graph 33(4):129:1–129:10

31. Criminisi A, Pérez P, Toyama K (2004) Region filling and object removal by exemplar-based image inpainting. IEEE Trans Image Process 13(9):1200–1212
32. Yu J, Lin Z, Yang J, Shen X, Lu X, Huang T (2018) Generative image inpaining with contextual attention. In: Proceedings of the IEEE computer vision pattern recognition, pp 5505–5514
33. Cham T, Shen J (2001) Local inpainting models and TV inpainting. SIAM J Appl Math 62:1019–1043
34. Nazeri, Kamyar & Ng, Eric & Joseph, Tony & Qureshi, Faisal & Ebrahimi, Mehran. (2019). EdgeConnect: Generative Image Inpainting with Adversarial Edge Learning.
35. Lee J, Choi I, Kim M (2016) Laplacian patch-based image synthesis. In: Proceedings of the IEEE computer vision pattern recognition, pp 2727–2735
36. Ballester C, Bertalmio M, Caselles V, Sapiro G, Verdera J (2011) Filling-in by joint interpolation of vector fields and gray levels. IEEE Trans Image Process 10(8):1200–1211. August 2001
37. Li P, Li S, Yao Z, Zhang J (2013) Two anisotropic fourth-order partial differential equations for image inpainting. IET Image Process 7(3):260–269
38. Bertalmio M, Saprio G, Caselles V, Ballester C (2000) Image inpainting In: Proceedings of the 27th annual conference on computer graphics and interactive technique, pp 417–424
39. Deng L, Huang T, Zhao X (2015) Exemplar-based image inpainting using a modified priority definition. PLoS ONE 10(10):1–18
40. Ram S (2017) Sparse representations and nonlinear image processing for inverse imaging solutions. Ph.D. dissertation, Dept. Elect. Comput. Eng., Univ. Arizona, Tucson, AZ, USA
41. Jin K, Ye J (2015) Annihilating filter-based low-rank Hankel matrix approach for I age inpainting. IEEE Trans Image Process 24(11):3498–3511
42. Buyssens P, Daisy M, Tschumperle D, Lezoray O (2015) Exemplar-based inpainting: technical review and new heuristics for better geometric reconstructions. IEEE Trans Image Precess 24(6):1809–1824
43. Ding D, Ram S, Rodriguez J (2018) Perceptually aware image inpainting. Pattern Recogn 83:174–184
44. Ogawa T, Haseyama M (2013) Image inpainting based on sparse representations with a perceptual metric. EURASIP J Adv Signal Process 2013(179):1–26
45. Abbadeni N (2011) Computational perceptual features for texture representation and retrieval. IEEE Trans Image Process 20(1):236–246
46. Grossauer H, Pajdla T, Matas J (2004) A combined PDE and texture synthesis approach to inpainting. In: Proceedings of the European conference on computer vision, vol 3022. Berlin, Germany: Springer, pp 214–224
47. Mansoor A and Anwar A (2010) Subjective evaluation of image quality measures for white noise distorted images. In: Proceedings of the 12th international conference advanced concepts for intelligent vision systems, vol 6474, pp 10–17
48. Li X (2011) Image recovery via hybrid sparse representations: a deterministicannealing approach In IEEE. J Sel Topics Signal Process 5(5):953–962
49. Demir U, Gozde U (2018) Patch-based image inpainting with generative adversarial networks. Comput Vis Pattern Recognit
50. Karras T, Aila T, Laine S, Lehtihen J (2018) Progressive growing of GANs for improved quality, stability, and variation. In: International conference on learning representations
51. Meur O, Gautier J, Guillemot C (2012) Super-resolution-based inpainting. In: Proceedings of the 12th European conference on computer vision, pp 554–567

Web-Based Classification for Safer Browsing

Manika Bhardwaj, Shivani Goel, and Pankaj Sharma

Abstract In the cyber world, attack (phishing) is a big problem and it is clear from the reports produced by the Anti-Phishing Working Group (APWG) every year. Phishing is a crime that can be committed over the internet by the attacker to fetches the personal credentials of a user by asking their details like login/credit or debit card credentials, etc. for financial gains. Since 1996 phishing is a well-known threat and internet crime. Research community is still working on phishing detection and prevention. There is no such model/solution existed that can prevent this threat completely. One useful practice is to make the users aware of possible phishing attacking sites. The other is to detect the phishing site. The objective of this paper is to analyze and study existing solutions for phishing detection. The proposed technique uses logistic regression to correctly classify whether the given site is malicious or not.

Keywords Phishing · Cybercrime · Logistic regression · TF-IDF

1 Introduction

Phishing is an illegitimate act started since 1996 in which the offender transmits the mocked messages via those occur to arrive from prominent authenticated and authorized organization or trademark, prompting to enter secret account information,

M. Bhardwaj (✉) · P. Sharma
ABES Engineering College, Ghaziabad, India
e-mail: manika.bhardwaj@abes.ac.in

P. Sharma
e-mail: pankaj.sharma@abes.ac.in

S. Goel
Bennett University, Greater Noida, India
e-mail: shivani.goel@bennett.edu.in

V. Singh et al. (eds.), *Computational Methods and Data Engineering*,
Advances in Intelligent Systems and Computing 1227,
https://doi.org/10.1007/978-981-15-6876-3_9

e.g. password of bank account, phone number, username, address, and more. For the novice and experienced computer users, the open, unidentified and uncontrolled internet infrastructure enables a tremendous platform for cyber-attacks, which grants severe security exposures. Across the various attacks upon cybersecurity, phishing has been paid special attention because of its contrary impact upon the economy.

1.1 Phishing Scenario

Nowadays, phishing can be possible with little or no technical skills and with an insignificant cost. This cyberattack can be launched from anywhere in the world [1]. There are so many methods or techniques due to which a phishing site exactly looks like a legitimate site. To create a phishing site is not a difficult task for the attacker. However, at the end attacker rely on the URL to redirect their victims to the trap. A general phishing scenario is shown in Fig. 1.

The steps followed by user to attempt a phishing attack are as follows:

1. Spoofed mails are sent by attackers to targets that run a fake version of targeted sites. This type of email generally contains important information in which the user immediately acts. For example, he/she provides some information to the bank otherwise his/her account will be locked.
2. Users are directed to a similar-looking login web page through the fraudulent link sent by the attacker via email.
3. The victim enters all his/her personal details on the fraudulent website, which is then tracked by the attacker.
4. Finally, with the help of that personal information which was entered by victim user, the attacker attempts the fraud and earn money from that.

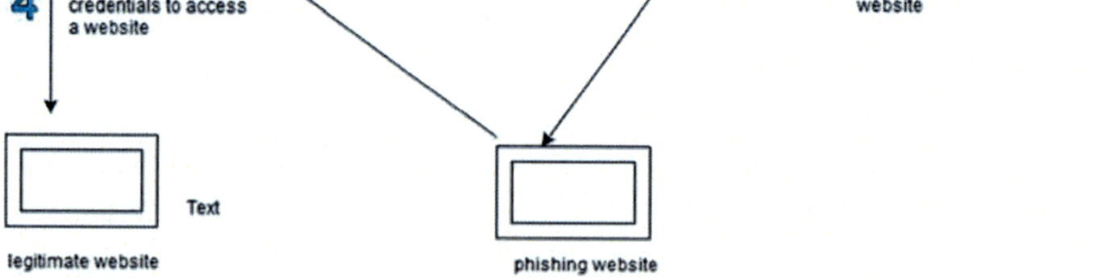

Fig. 1 A phishing scenario

1.2 Current Phishing Statistics

From the year phishing was started, anti-phishing societies like PhishTank and APWG recorded a high amount of phishing attacks. Anti-Phishing Working Group in 2017 released 'Global Phishing Survey for 2016' data which showed that there were minimum 255,065 distinct phishing attacks across the globe. This represented an increase of over ten percent on 230,280 site count received in the year 2015 [2]. According to a report produced by APWG in 2018, 138,328 number of phishing sites were detected in fourth quarter (Q4). In Q3, the site count was 151,014, Q2 had 233,040 and Q1 had 263,538 sites. These are shown in Fig. 2.

The confirmed phishing sites count has declined as 2018 proceeded. Detection of harmful phishing sites has become harder because the phishers are using obfuscated phishing URLs which includes multiple redirections [2]. There is an urgent need to find an appropriate solution for phishing detection by a researcher.

Solutions to phishing can be phishing detection, phishing prevention or trained users on phishing related activities. This paper analyzed phishing detection because it was observed by several researchers that phishing detection is cheaper than phishing prevention. Logistic regression was used by the authors to detect phishing and it has reported an accuracy of 100%.

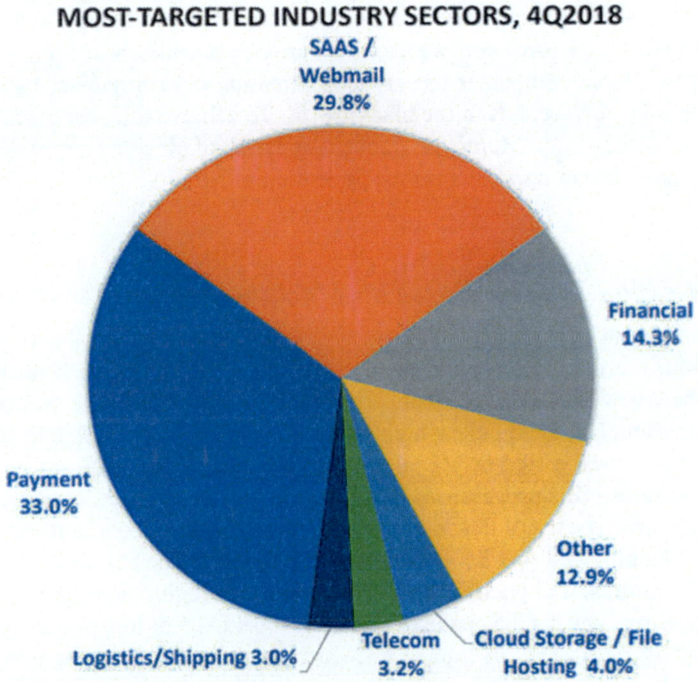

Fig. 2 Last quarter of 2018 (APWG 2018)

The rest of the paper is organized as follows: The related work is discussed in Sect. 2. The proposed work and methodology used are given in Sect. 3. Results are presented in Sect. 4. Conclusion and future scope are given in Sect. 5.

2 Related Works

Detection systems for phishing can be broadly classified into two categories: User awareness-based or software-based. Software-based detection can be done either based on lists or machine learning (ML) algorithms.

2.1 List Based Detection Systems

This type of system usually consists of two types of lists, blacklists, and whitelists to detect phishing and legitimate web pages. A blacklist is a list that consists of fraudulent IP addresses, URL's and domains. Whitelist contains a list of legitimate sites. The method was developed to advice the users on web by automatically updating the whitelist of legitimate websites [3]. Blacklist are the lists that are updated frequently but protection against zero-hour is attacked not provided [4]. Google safe browsing API and PhishNet are some companies which provide blacklist based phishing detection systems. These companies use an approximation matching learning algorithm to validate if the URL exists in the blacklist [5]. To achieve this, the blacklist needs to be updated frequently. Moreover, frequent updation of blacklist requires excessive resources. So, a better approach is the application of ML.

2.2 Detection Systems Based on Machine Learning

ML methods are the most popular methods for phishing detection. Detecting malicious websites is basically a classification problem. To classify, a learning-based detection system has to be built which required trained data with lots of features in it. Researchers used different ML algorithms to classify the URL on the basis of different features. This process was performed on client-side for detecting phishing of web pages [6]. The list of IP addresses, malicious URLs, and domains were created named as blacklist and this list needed regular updates. The authors selected eight basic URL features, six hyperlink specific features, one login form feature, three web identity features, and 1 CSS feature (F16). Datasets used in this paper were PhishTank phishing pages and Openphish in which they used 2141 webpages. Permissible datasets were extracted from various sources like Alexa, payment gateway, top banking websites. 1981 webpages were considered for training and testing. With the use of ML algorithms, they achieved a true positive rate of 99.39%.

PhishDef, a classification system performed a proactive classification of phishing URLs by applying the AROW algorithm with the lexical features [7]. For phishing URL's following dataset were used PhishTank, Malware Patrol, Yahoo Directory and Open Directory. For finding legitimate URL, random genuine URL was taken by only utilizing lexical features. PhishDef reduced the latency while loading page and avoided dependence upon remote servers. By implementing the AROW algorithm, PhishDef scored high accuracy in classification, even with noisy data and consumed lesser system resources thereby reducing hardware requirements.

To achieve good generalization ability and high accuracy a risk minimization principle was designed. The classification was done based on neural network which used a stable and simple Monte Carlo algorithm [8]. The main advantage of using neural network is generalization, nonlinearity, fault tolerance and adaptiveness. Overfitting is also one of the problems faced in neural networks. The main advantage of using this approach was that it was not dependent on third parties, detection is performed in real-time and the accuracy was improved 97.71%.

Varshney et al. [1] and Khonji et al. [4] analyzed the classification of several schemes of web phishing detection. Phishing detection was broadly classified into the categories- search engine based, DNS based, whitelist and blacklist based, visual similarity-based and heuristic (proactive phishing URL) based and ML-based. This paper also gave a comparative study of each and every scheme on the basis of their capabilities, novelty and accuracy. The existing machine learning based approaches extract features from sources like a search engine, URL and third-party services. Examples of third-party examples are whois record, Domain Name service, website traffic, etc. The extraction of third-party features is a complicated and time-consuming process [9].

Sahingoz et al. [10] proposed an anti-phishing system that is real-time. It compared the result of seven classification algorithms with URL, NLP, word count, Hybrid features. Out of all seven algorithms used with the only NLP features, random forest gave the best accuracy of 97.98%. Language independence was the major advantage of the proposed approach. Big size of dataset was used for both legitimate and phishing data. Real-time execution was used, independent from third-party services.

Artificial Neural Network (ANN) with 1 input layer and 2 hidden layers were used for phishing detection [11]. Number of neurons used in input layer was 30 and 18 neurons were used in hidden layer. Different URL features were used for training of the data like URL length, prefix or suffix, @ symbol, IP address, subdomain length and with the help of all these features, the approach reached the accuracy of 98.23%.

Recurrent neural networks were used for phishing detection by Bahnsen et al. [12]. The lexical and statistical features of URL were used as an input to the random forest and its performance was compared with the recurrent neural network [i.e. long/short term memory (LSTM)]. Here the dataset used was PhishTank with approximately 1 million phishing URL and the result accuracy achieved was 93.5% and 98.7% for random forest and LSTM, respectively.

Jeeva et al. (2016) utilized association rule mining for the detection of phishing URLs [13]. For training, the algorithm used features such as count of the host URL characters, slashes (/) in the URL, dots (.) in the host name of the URL, special characters, IP addresses, unicode in URL, transport layer security, subdomain, specific keywords in the URL, top-level domain name, count of dots in the path of the URL and presence & count of hyphen in the hostname of the URL. The a priori and predictive apriori algorithms were used for extraction of rules for phishing URL. Both of these algorithms generated different rules. The analysis indicated that the apriori was quite faster than the predictive apriori.

Blum Aaron et al. [14] produce a dynamic and extensible system to detect phishing by exploring the algorithm named confidence weighted classification. In this, the confidence weighted parameter is used to improve the overall accuracy of the model which leads to 97% classifying accuracy on emerging phishing.

Table 1 demonstrates the comparative study of various ML-based systems for phishing detection. URL feature is the most important feature which was used with every detection method.

3 Proposed Work

In this paper, first of all we collect the database of various URL including phishing and legitimate URL. Then we extract the feature of URLs as shown in Fig. 3 in which the structure of the URL is described which consists of protocol, second-level domain, top-level domain, subdomain, etc. For phishing, the attacker majorly used the combination of top-level domain and second-level domain for the creation of Phishing URL. Once, we identified the features we trained our model, and finally our model predicts that the given URL is Phishing URL or Legitimate URL. If the given URL is Phishing then our model returns True otherwise it returns False for the Legitimate URL.

Figure 4 shows the workflow diagram of the proposed approach.

Logistic Regression (LR) is used for phishing detection. It is ML classification algorithm in which observations are assigned to a distinct set of classes. In linear regression, the output values with continuous numbers. In logistic regression, a probability value is returned using a logistic sigmoid function. It is then mapped to two or more distinct classes. The probability is found as the success or failure of an event. LR is used when the dependent variable is binary in nature, i.e. it takes the values 0 or 1, True or False, Yes or No.

$$\text{logit}(p) = b_0 + b_1 X_1 + b_2 X_2 + b_3 X_3 + \cdots + b_1 X_k \quad (1)$$

where p is the probability indicating whether the characteristic of interest is present or not. The logit transformation is defined as given as Eqs. 2 and 3:

Table 1 Comparative study of ML-based phishing detection systems in reverse chronological order

Project	Feature used	Dataset used		Algorithm(s) used with accuracy %
		Phishing	Legitimate	
Sahingoz et al. [10]	URL, NLP, word count, hybrid	Ebbu2017	Google	Random forest (97.98), Naive Bayes, kNN ($n = 3$), K-star, Adaboost, decision tree
Ferreira et al. [11]	URL	University of California's ML and Intelligent Systems Learning Center	Google	ANN (98.23)
Babagoli et al. [15]	Wrapper	UCI dataset	–	Nonlinear Regression model based on harmony search (92.8) and SVM (91.83)
Feng et al. [8]	URL	UCI repository	Millersmile's, Google's searching operators	Novel Neural Network (Monte Carlo Algorithm) (97.7)
Bahnsen et al. [12]	URL Based Feature	PhishTank	–	Machines (SVM), *k*-means and density-based spatial clustering (98.7)
Jain and Gupta [6]	URL based, login form based, hyperlink specific features CSS based	PhishTank, Openphish	Alexa, Payment gateway, top banking websites	Random forest, SVM, Naïve-based, logistic regression, neural networks (99)
Varshney et al. [1]	URL	PhishTank, Castle Cops	Millersmile's, Yahoo, Alexa, Google, NetCraft	
Jeeva and Rajsingh [13]	URL	PhishTank	Millersmile's, Yahoo, Alexa, Google, NetCraft	Association rule mining (93)
Akinyelu and Adewumi [16]	URL	Nazario	Ham corpora	Random forest (99.7)
Khonji et al. [4]	URL, model based, hybrid features	PhishTank	Google	–

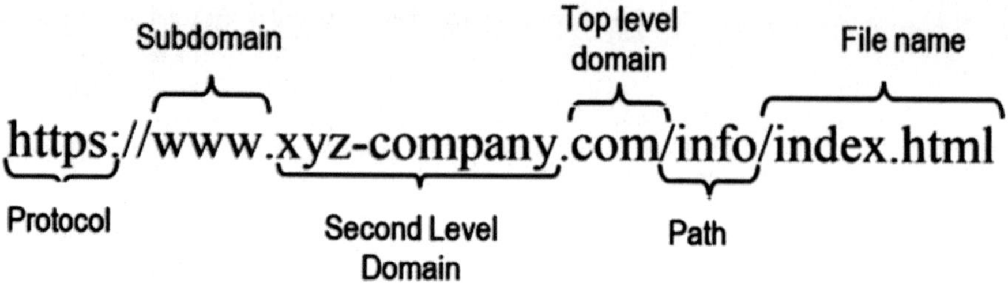

Fig. 3 Structure of URL [11]

Fig. 4 Work flow diagram

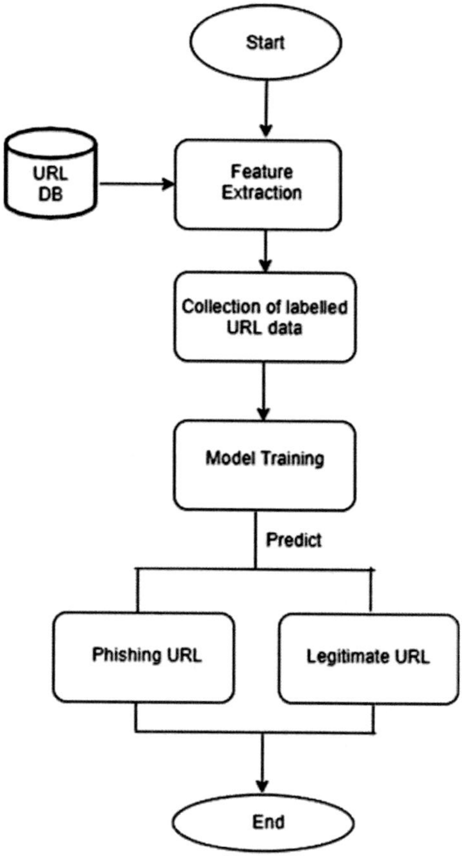

$$\text{odds} = \frac{p}{1-p} = \frac{\text{probability of presence of characteristic}}{\text{probability of absence of characteristic}} \quad (2)$$

And

$$\text{logit}(p) = \ln\left(\frac{p}{1-p}\right) \tag{3}$$

In ordinary regression, parameters are chosen which minimize the sum of squared errors In LR the parameters that maximize the likelihood of observing the sample values are chosen in estimation.

3.1 Methodology

Text data requires initial preparation before it can be used by phishing detection algorithm. Tokenization is used to remove stop words while parsing the text. The words are then encoded as integers or floating-point values so that these can be input to an ML algorithm. The process is called vectorization or feature extraction. The scikit-learn library is used for tokenization and feature extraction.

TFidVectorizer was used for converting a collection of raw documents to a matrix of TF-IDF features. Word counts are a good starting point but are very basic, but later it was shifted to word frequencies. It was due to the fact that counts of commonly occurring words like 'the' may not be very meaningful in the encoded vectors. The frequency is called the TF-IDF weight or score is given in Eq. (1). TF-IDF is an acronym that stands for 'Term Frequency—Inverse Document Frequency'.

- **Term Frequency**: It is used to count how many times a given word appears within a document, $tf_{t,d}$. Since this value may be very large for stop words like is are, the, a, and, log base 10 is taken to reduce the effect of very large frequency of common words.
- **Inverse Document Frequency**: This downscales words that appear a lot across documents. It is calculated using log base 10 of the term (N/df_t) where N is the total number of documents in the corpus or dataset and df_t is the number of documents in which the term t appears. IDF increases the weight for rare terms and reduces the weight of common words. This is important to incorporate correct score as per the informedness of the term instead of frequency only.

$$w_{t,d} = \left(1 + \log tf_{t,d}\right) \cdot \log \frac{N}{df_t} \tag{4}$$

TF-IDF weights $w_{t,d}$ are word frequency scores that are highlighting the most useful words, e.g. frequently occurring in one document but not in many documents.

Documents are tokenized using the TFidfVectorizer. Then TF-IDF weights are calculated for each token and new documents are encoded.

After that, the URL data is loaded in csv format. Then a tokenizer is created to split the URL, remove repetitions of words and 'Com'. After tokenization a model is built using logistic regression and the model is trained. The model is tested for accuracy after training. The dataset used for phishing URLs is PhishTank [17].

4 Results

URL	PHISH
http://www.cheatsguru.com/pc/thesims3ambitions/requests/	False
http://www.sherdog.com/pictures/gallery/fighter/f1349/137143/10/	False
http://www.mauipropertysearch.com/maui-meadows.php	False
https://www.sanfordhealth.org/HealthInformation/ChildrensHealth/Article/	False
http://strathprints.strath.ac.uk/18806/	False
http://th.urbandictionary.com/define.php?term=politics&defid=1634182	False
http://moviesjingle.com/auto/163.com/index.php	True
http://rarosbun.rel7.com/	True
http://www.argo.nov.edu54.ru/plugins/system/applse3/54e9ce13d8baee9569663325 7b33b2b5/	True
http://tech2solutions.com/home/wp-admin/includes/trulia/index.html	True
http://www.zeroaccidente.ro/cache/modlogin/home/37baa5e40016ab2b877fee2f0c9 21570/realinnovation.com/css/menu.js	True

The accuracy of classification of phishing is 100% which is calculated as:

$$\text{Accuracy} = \frac{\text{TP} + \text{TN}}{\text{TP} + \text{TN} + \text{FP} + \text{FN}} \tag{5}$$

where

TP True Positive,
FP False Positive,
TN True Negative,
FN False Negative.

5 Conclusion and Future Work

In this paper, the comparative analysis of phishing detection techniques has been done. The conclusion is that phishing detection is a much better way than user training solutions and phishing prevention. Moreover, in terms of hardware requirement and for managing a password it was found that detection technique is comparatively inexpensive. On the basis of features, methodology and accuracy, this paper contributes towards relative study of several phishing detection schemes. A solution using logistic regression is proposed for detecting phishing URL which has reported ac accuracy of 100%.

References

1. Varshney G et al. (2016) A survey and classification of web phishing detection schemes. Secur Commun Netw 9(18) https://doi.org/10.1002/sec.1674
2. https://www.antiphishing.org/
3. Jain AK, Gupta BB (2018) Towards detection of phishing websites on client–side using machine learning based approach. Telecommun Syst 68(4):687–700
4. Khonji M, Iraqi Y, Jones A (2013) Phishing detection: a literature survey. IEEE Commun Surv Tutor 15(4):2091–2121
5. https://developers.google.com/safe-browsing/v4/
6. Jain AK, Gupta BB (2016) A novel approach to protect against phishing attacks at client side using auto updated white-list. EURASIP J Inf Secur. Article 9
7. Le A, Markopoulou A, Faloutsos M (2011) Phishdef: URL names say it all. Proc IEEE INFOCOM 2011:191–195
8. Feng F et al (2018) The application of a novel neural network in the detection of phishing websites. J Ambient Intell Humaniz Comput. https://doi.org/10.1007/s12652-018-0786-3
9. Whittaker C et al. (2010) Large scale automatic classification of phishing pages. In: Report NDSS Symposium
10. Sahingoz OK et al (2019) Machine learning based phishing detection from URLs. Expert Syst Appl 117:345–357
11. Ferreira RP et al (2018) Artificial neural network for websites classification with phishing characteristics. Soc Netw 7:97–109
12. Bahnsen AC et al. (2017) Classifying phishing URLs using recurrent neural networks. In: Proceedings: 2017 APWG symposium on electronic crime research (eCrime) https://doi.org/10.1109/ecrime.2017.7945048
13. Jeeva SC, Rajsingh EB (2016) Intelligent phishing URL detection using association rule mining. Hum Centric Comput Inf Sci. Article 10
14. Blum A, Wardman B, Solorio T (2010) Lexical feature based phishing URL detection using online learning. In: Proceedings of the 3rd ACM workshop on security and artificial intelligence, AISec 2010, Chicago, Illinois, USA, 8 Oct 2010. https://doi.org/10.1145/1866423.1866434
15. Babagoli M, Aghababa MP, Solouk V (2018) Heuristic nonlinear regression strategy for detecting phishing websites. Soft Comput 12:1–13
16. Akinyelu AA, Adewumi AO (2014) Classification of phishing email using random forest machine learning technique. J Appl Math
17. PhishTank. Verified phishing URL Accessed 24 July 2018. https://www.phishtank.com/

A Review on Cyber Security in Metering Infrastructure of Smart Grids

Anita Philips, J. Jayakumar, and M. Lydia

Abstract In the era of digitizing electrical power networks to smarter systems, there arises an increased demand of security solutions in various components of the Smart Grid networks. The traditional and general security solutions applicable to hardware devices, network elements and software applications are no longer able to provide comprehensive readymade alternatives to secure the systems. As the scalability of the system increases, component-wise security solutions are essential for end-to-end security. Considering this current scenario, in this paper, the key management techniques, particularly the lightweight Key Management Systems (KMS) methodologies that have been proposed in the past are reviewed in the context of Advanced Metering Infrastructure (AMI) of the Smart Grid Systems.

Keywords Smart grid · Cyber security · Advanced metering infrastructure · Key management systems · Lightweight KMS solutions

1 Introduction

The European Technology defines "a Smart Grid (SG) as an electricity network that can intelligently integrate the actions of all users connected to it—generators, consumers and those that do both, in order to efficiently deliver sustainable, economic and secure electricity supply". A Smart grid in short is an electric system that is more efficient, reliable, resilient and responsive. It aims for better electricity delivery by use of advanced technologies to increase the reliability and efficiency of the electric grid, from transmission to distribution.

A. Philips (✉)
Department of Electrical and Electronics Engineering, Karunya University, Coimbatore, India
e-mail: anucornee@yahoo.com

J. Jayakumar · M. Lydia
Department of Electrical and Electronics Engineering, SRM University, Delhi NCR, Sonepat, Haryana, India

V. Singh et al. (eds.), *Computational Methods and Data Engineering*, Advances in Intelligent Systems and Computing 1227,
https://doi.org/10.1007/978-981-15-6876-3_10

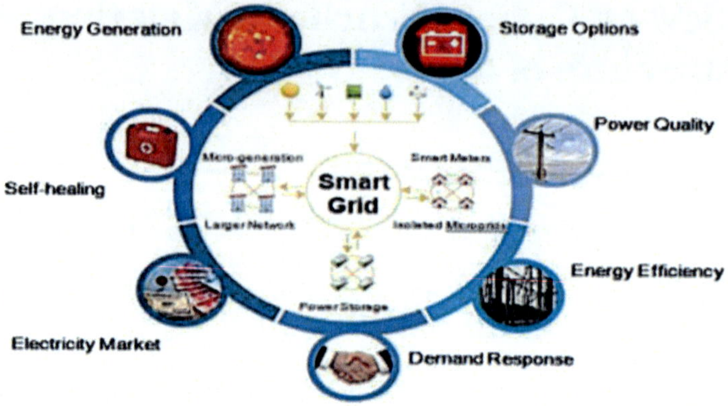

Fig. 1 Smart grid framework [1]

SG includes automation and controllable power devices in the whole energy value chain from production to consumption. In particular, the computing and two-way communication capabilities of the SG aids to exchange real-time information between utilities and consumers, thus achieving the desirable balance of energy supply and demand.

Hence, the SG incorporates many technologies such as advanced metering, network communication, distributed generation and storage, integration with renewable energy sources, Internet of Things (IoT) enabled devices etc. The framework of Smart Grid systems involve energy generation, energy storage, electricity market, power quality and demand-response balance (Fig. 1).

2 Cyber Security in Smart Grids

Upgrading the power grid to Smarter Grid will present many new security challenges which need to be dealt with before the deployment and implementation. The increasingly sophisticated nature and speed of the attacks, especially of the cyber domain is alarming.

Due to the gravity of these threats, the Federal Energy Regulatory Commission (FERC) policy statement on the SG states that cybersecurity is essential to the operation of the SG and that the development of cybersecurity standards is a key priority. In the SG, the physical power system is integrated and tightly coupled with the cyber system.

Therefore, in the case of any attack in either domain may have an impact on the other domain and lead to potential cascading failures. [Black-outs, financial losses etc.]

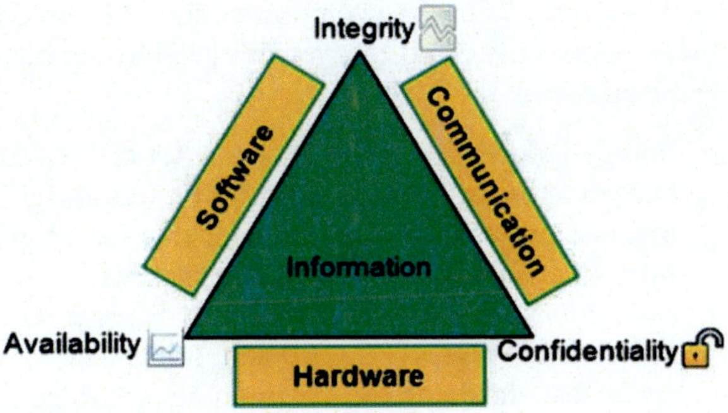

Fig. 2 CIA triad

2.1 Cyber Security Requirements

The simple definition of Cyber Security could be stated as "Cybersecurity is the practice of protecting systems, networks, and programs from digital attacks". Major areas of cyber security could be listed as application security, information security, disaster recovery and network security.

The term 'Information security' is defined by NIST as: "A condition that results from the establishment and maintenance of protective measures that enable an enterprise to perform its mission or critical functions despite risks posed by threats to its use of information systems."

A combination of deterrence, predict and prevent, early detection, recovery, and remedial measures that should form part of the business's risk management methods are some of the protective measures for information security. Information security comprises of three core principles:

- Confidentiality—Only authorized parties can access computer-related assets.
- Integrity—Modifications can be permitted only for authorized parties or through authorized ways.
- Availability—Assets are accessible to authorized parties at appropriate times.

Together these principles, the "CIA triad," as explained in Fig. 2, provide reliable access to appropriate information for the authorized people, applications, and machines. CIA (Confidentiality, Integrity, and Availability) ensures the security of information, and it is obvious that breaking "CIA" leads to a sequence of cyber threats.

2.2 Cyber Attack Models, Threats & Challenges

Some of the common cyber-attacks could be classified as DoS/DDoS attacks, MitM attacks, false data injection, malware attacks, brute force, replay attacks, supply chain attacks. With the evolution of the SG, the process of developing the standards for security protocols were initiated by various authorities. The efforts can be summarized as:

- Energy Independence and Security Act of 2007 (EISA)—The National Institute of Standards and Technology [NIST], was assigned to develop a framework that includes protocols and model standards for information management to achieve interoperability of SG devices and systems.
- June 2008—The US Department of Energy (DOE) published its "Metrics for Measuring Progress Toward the Implementation of the Smart Grid" which states that standards for the smart electrical grid must incorporate seven major characteristics namely:
  - Facilitate active participation by end users
  - Availability of generation and storage options
  - Enable new products, services, and markets
  - Provide power quality for the range of applications
  - Optimize asset utilization and operating efficiency
  - Anticipate and act to system failures in a self-healing manner
  - Resilience against physical and cyber-attack and natural disasters
- January 2010—NIST released the framework and roadmap for SG Interoperability Standards, Release 1.0
- September 2010—NIST released the guidelines for SG cyber security.

The communication networks in SG bring increased connectivity along with increased security vulnerabilities and challenges. As millions of digital devices are inter-connected via communication networks throughout critical power entities, in a hugely scalable infrastructure, cyber-security emerges to be critical issue.

2.3 Security Solutions

The cyber security solutions as proposed by Aloul et al. [2], for the SG are implicit deny policy to grant explicit access permissions, malware protection on embedded systems, Network Intrusion Detection System [IDS] technologies, periodic vulnerability assessments, Virtual Private Network [VPN] architecture and authentication protocols. In [3], the authors discuss methods for accurate and secure information sharing across the SG domain, and insists on cyber-physical system security. In general, the security solutions are to be used in combination, for addressing the existing and future cyber-attacks.

As found in [4], SEGRID (Security for Smart Electricity GRIDs) is a collaboration project, funded by the EU under the FP7 program. The main objective here is to protect SGs from cyber attacks by applying the risk-management approach to a no. of SG use cases for end-to-end security solution.

Fig. 3 Iterative process of SEGRID

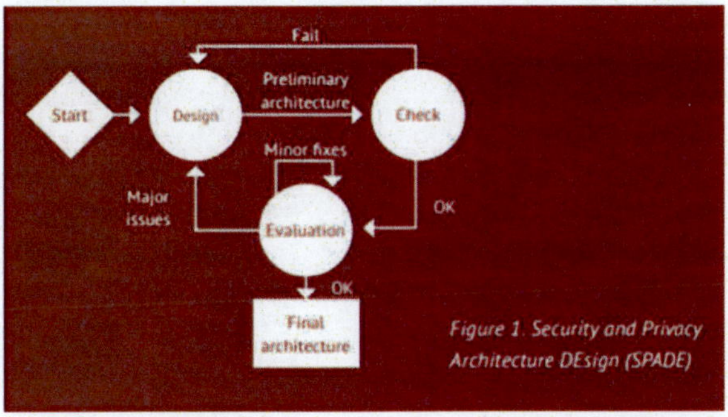

The iterative phases of the Security and Privacy Architecture Design SPADE, namely design, check and evaluation are performed repeatedly to achieve the desired security requirements as found in Fig. 3. As the SG is a system involving multiple stakeholders, this risk assessment method is more suitable for establishing secured architecture.

3 Advanced Metering Infrastructure

Advanced Metering Infrastructure (AMI) is the most crucial part of SG and aids for the efficiency, sustainability and reliability of the system. Therefore the cyber threats that are possible in the AMI has a huge impact on the reliable and efficient operation of SG.

In [5], the components of AMI are discussed, as the AMI is comprised of smart meters, data collector, and communications network. AMI transmits the user's electricity consumption information to the meter data management system (MDMS) or other management systems [6]. The main drawback of implementing security scheme in AMI is stated in [7] as the limited memory and low computational ability of the smart meters, and the scalability of AMI being a huge network, consisting of millions of meters.

In general, the communication overhead and computational abilities utilized for encryption schemes and key management increases with the degree of encryption as explained in [8]. The need for lightweight authentication protocols arises due to reasons of long key sizes, ciphers and certificates, maintenance of Public Key Infrastructure (PKI), keeping track of Certificate Revocation Lists and timers.

Therefore, in AMI which consists of limited capability components like smart meters, lightweight key management techniques are more appropriate to be used.

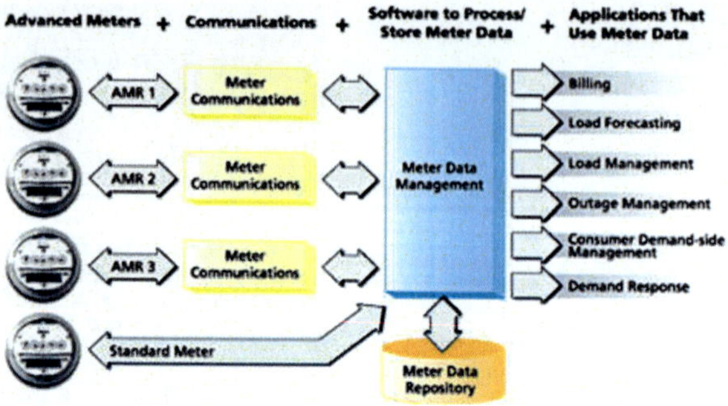

Fig. 4 Components of AMI [9]

3.1 *Ami Components and Benefits*

The primary goals of the Advanced Metering Infrastructure can be summarized as:

- The real time date about energy usage is provided to the utility companies.
- Based on the Time of Use [ToU] prices, the consumers will be able to make informed choices of power consumption.
- Peak shaving option can be provided where the demand for electricity can be reduced during the periods of expensive electricity production.

The smart meter network aids in establishing two-way communication link between the power utility and consumers, thereby increasing the risk of exposing the AMI communication architecture to cyber-attacks. Therefore, the AMI network of the SG is more vulnerable to many cyber-attacks. This may lead to poor system performance and incorrect energy estimations, which affects the stable state of the grid.

The AMI comprises of the components as below and are shown in Fig. 4: Smart Meters, Communication Network, Meter Data Acquisition System (Data concentrators) and Meter Data Management System (MDMS).

The two-way information flow in the Advanced Metering Infrastructure, between the utility data centre and consumers aids for the energy demand response efficiency. The benefits of the AMI can be depicted in Fig. 5.

3.2 *Attack Models in Ami*

The scalability of the AMI Communication network varies from hundreds to thousands of smart meter collector devices, each in turn serving thousands of smart meters. This gives rise to a magnitude of vulnerabilities that has an impact on system operations resulting in physical, cyber and economical losses. According to Wei et al. [11], some of the physical and cyber attacked targeted towards AMI are depicted as in Table 1.

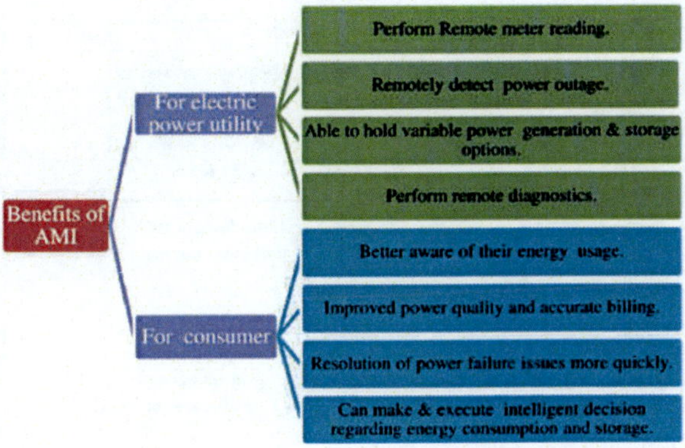

Fig. 5 Benefits of AMI [10]

Table 1 Attacks targeted towards AMI [12]

Attack type		Attack target	
		Smart meter	AMI communication network
Physical		1. Meter manipulation 2. Meter spoofing and energy fraud attack	1. Physical attack
Cyber	Availability	1. Denial of service (DoS)	1. Distributed denial of service (DDoS)
	Integrity	1. False data injection attack (FDIA)	2. False data injection attack (FDIA)
	Confidentiality	1. De-pseudonymizalion attack 2. Man-in-the-middle Attack 3. Authentication attack 4. Disaggregation attack	1. WiFi/ZigBec attack 2. Internet attack 3. Data confidentiality attack

In [12], a new Denial of Service (DoS) attack, a puppet attack is discovered in the AMI network. Here, the normal network nodes are selected as puppets and specific attack information is sent through them which results in huge volume of route packets. This is in turn result in network congestion and DoS attack happens.

3.3 Security Solutions in Ami

Specifically, cyber security plays a crucial role in the AMI of SG because it has a direct impact on real time energy usage monitoring, as AMI offers bidirectional flow of crucial power related information across the components.

AMI is an integrated system of Smart meters, Communications networks and Data management systems that enables bidirectional flow of information between

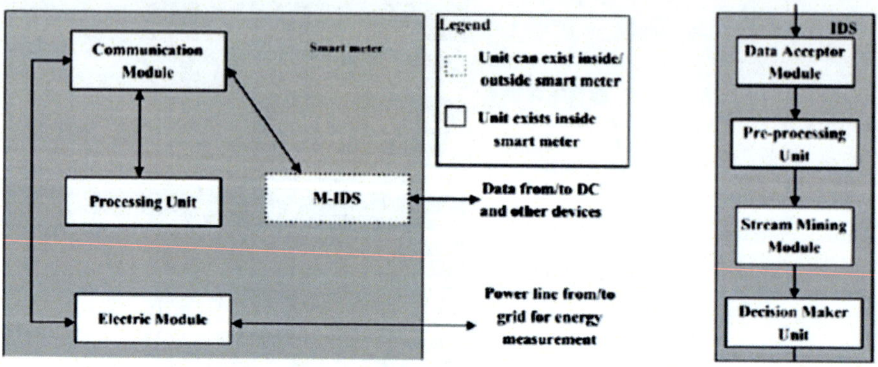

Fig. 6 Smart Meter with IDS [13]

power utilities and consumers. So, the AMI is the critical component of SG which enables the two-way communication path from the appliances of the consumer to the electric utility control centre. Hence, the operational efficiency and reliability of SG hugely relies on the security and stability of AMI system. In addition to the security solutions like authorization, cryptography and network firewall, the mechanisms like IDS [Intrusion Detection system] are to be used in combination. In [13], it is recommended to use anomaly-based IDS using data mining which is analyzed for each component of the AMI using MOA (Massive Online Analysis). The design structure of this security mechanism including IDS is shown in Fig. 6.

The DoS attacks explained in [12], is detected and prevented using a distributed method, and the attacker is isolated using link cut-off mechanism. The Wireless Sensor Networks (WSN) features like multi-hop, wireless communications are utilized to disconnect the attacker nodes from the neighbor nodes.

4 Key Management in AMI

For providing security in data communication, the fundamental technique used is the cryptographic key management. The data flow diagram for secured communications using cryptographic keys are depicted in Fig. 7. In general, the cryptography algo-

Fig. 7 Communication with cryptographic keys

Table 2 Comparison of KMS features [14]

Feature/algorithm	Hash	Symmetric	Asymmetric 1
No. of Keys	0	1	2
NI5T recommended key length	256 bits	128 bits	2048 bits
Commonly used	SHA	AES	RSA
Key management/sharing	N/A	Big issue	Easy & secure
Effect of key compromise	N/A	Loss of both sender & receiver	Only loss for owner of asymmetric key
Speed	Fast	Fast	Relatively slow
Complexity	Medium	Medium	High
Examples	SHA-224, SHA-256, SHA-384 or SHA-512	AES, Blowfish, Serpent, Two fish, 3DES, and RC4	RSA, DSA, ECC, Diffie-Hellman

rithms are classified based on the number of cryptographic keys used namely Hash Functions, Symmetric-Key, and Asymmetric-Key.

The comparative features of the key management mechanisms are illustrated in Table 2.

Key management systems (KMS) are an important part of AMI that facilitates secure key generation, distribution and rekeying. Lack of proper key management in AMI may result in possible key acquisition by attackers, and hence may affect secure communications.

The general goals of secure cryptographic key management in AMI of SGs includes:

- Enabling the energy control systems to withstand cyber-attacks.
- Ensuring secure communications for the smart meters within the advanced metering infrastructure.

AMI must support application level end-to-end security by establishing secure communication links between the communicating components, which requires the implementation of encryption techniques. This implementation of encryption techniques in an AMI requires effective and scalable methods for managing encryption keys. Hence, security solutions for SG can be delivered by using proper key management systems in the AMI using encryption based techniques.

5 Analysis on Lightweight KMS Approaches

Some of the lightweight KMS approaches available in literature are discussed below:

Abdallah and Shen in [15] proposes a lightweight security and privacy preserving scheme. The scheme is based on forecasting the electricity demand for the houses in a neighborhood area network (NAN) and it utilizes the lattice-based N-th degree Truncated Polynomial Ring (NTRU) cryptosystem to reduce the computation complexity. Only if the overall demand of a group of homes in the same neighborhood needs to be changed, the messages are exchanged, thereby reducing communication complexity and the computation burden becomes light. The two phases Initialization and Message exchange phases establishes the connection between different parties and organizes the electricity demand.

The Initialization phase works with Key Generation where the encryption public and private keys and signing keys are generated by the Trusted Authority (TA) for the main Control Centre (CC) and the Building Area Network (BAN) gateway, Demand Forecast where a forecasting function is applied for each Home Area Network (HAN) and aggregated in the BAN along with a backup value, Electricity Agreement where exchange of Agreement request and response messages happens between the CC and BAN gateway thereby guaranteeing the required electricity share to the HANs.

The Exchange Message phase initially supplies each HAN with specific electricity share based on the previously calculated amounts and if Demand change occurs, encrypted demand message using BAN's public key is sent from the HAN to the BAN gateway where the new amount is computed, and if Price change occurs, the revised price message is broadcast to all connected BANs signed by CC's public key which is accepted after signature verification and forwarded by the BAN gateway to connected HANs with its own signing keys which in turn is accepted after signature verification and validity check.

The security requirements are satisfied in this scheme, as the connection is established in two different steps, i.e. CC to BAN gateways and BAN to HAN, the customer's privacy is preserved when any adversary intercepts the exchanged messages at any point, confidentiality and authentication are guaranteed because of the use of public keys for CC and BANs, as the messages are signed and hashed the message integrity is assured, DoS attacks can be identified and the malicious HANs could be blocked if the BAN gateway receives abnormal number of messages from a malicious HAN.

As the demand messages from HANs are sent only when the electricity share changes, and only one billing message is sent to the CC for the whole BAN, when compared to traditional methods there is significant reduction in number of messages and thus the communication overhead.

Also, in this scheme, the computation operations as calculated based on NTRU cryptosystem proves that there is remarkable decrease in computation time. The computation overhead of this protocol is shown in Table 3.

In [8], Ghosh et al. proposes a lightweight authentication protocol for the AMI in SG specifically between the BAN gateway and HAN smart meters. Regardless of the limited memory and processor capabilities of the smart meter devices, the sensitive information related to the individual meter readings need to be protected. It works in two phases namely pre-authentication protocol and authentication protocol.

The pre-authentication phase exchanges two messages, the first message is the identifier of the HAN smart meter and it is delivered to the BAN gateway, on receiving this the BAN gateway generates its public key, applies its master secret key to create the HAN smart meter's private key and it is conveyed as the second message of acknowledgement is sent to the HAN SMs from the BAN gateway.

In the Authentication phase, three messages are exchanged, the first message is the encrypted variable using the BAN gateway's public key from HAN SM along with its identifier, the second message is the variable calculated using the pairing based cryptography and applying hash function from the BAN gateway, the HAN SM authenticates the BAN gateway if the bilinear pairing properties are equal on both sides, the third message is sent from the HSN SM comprising of the period of validity, sequence number and the signed session key which is compared with its session key at the BAN gateway to authenticate the HAN SM.

In this scheme, the application of one-directional hash function prevents the Replay attacks, zero knowledge password protocol prevents the impersonation attacks, the combination of security policies prevent the Man-in-the-middle attacks, different combination of variables prevent the Known Session key attacks, individual generation of keys at both ends prevent key control attacks.

Lesser number of hash functions in the BAN side and mutual authentication using just one encryption-decryption step and one sign-verify step assures less computational costs. The reduced computational costs of the protocol is summarized in Table 4.

Simulation results show a comparatively lesser communication overhead and average delay over Elliptic Curve Digital Signature Algorithm (ECDSA).

As in [16], a lightweight authentication protocol is proposed for the two-way device authentication of the Supervisory node [SN] and control node [CN] in the SG by Qianqian Wu, Meihong Li. This scheme is based on the shared security key which is embedded in the device chip and random number to authenticate the identity of SN and CN. The use of certificates and third-party services are avoided in this

Table 3 Computational overhead of proposed protocol [15]

	Computation Overhead
Traditional	$810 * T_E + 810 * T_D + 810 * T_S + 810 * T_V$
Proposed	$90 * T_E + 90 * T_D + 90 * T_S + 90 * T_V$

Table 4 Computational costs of proposed protocol [8]

	Proposed protocol
HAN side	$5T_h + 2T_{\text{exp}} + 1T_{\text{bm}} + 1T_{\text{mul}} + 1T_{\text{sub}}$
BAN side	$4T_h + 1T_{\text{exp}} + 1T_{\text{bm}} + 3T_{\text{mul}} + 1T_{\text{sub}} + 1T_{\text{add}}$

method, and a symmetric cryptographic algorithm and hash operation are adopted. This scheme works in three phases namely system initialization, device certification and device key updation.

During the system initialization, the shared key is stored in a dedicated chip added to both SN and CN, and the identification of the devices can be found directly though IP addresses. Device Certification consists of random number generation and device identity authentication. Here, SN generates a random number and sends request to CN which authenticates based on the corresponding shared key from the local chip. Then CN generates random number and sends the response to SN which in turn similarly authenticates based on the corresponding shared key in the local chip. Further, CN decrypts the data packets received from SN and Integrity checks, validation and verification steps are carried out, if these steps fail authentication fails.

For the subsequent device certification processes, Device key updation is done based on key update cycle and the latest key may be used for communication.

Embedded key in the device prevents Man-in-the-middle attacks and key leakage possibility is eliminated, random number generation prevents Replay attacks, Message Digest is calculated to ensure data integrity, one-way hash protocol improves computing speed.

George et al. in [17] proposes a hybrid encryption scheme for unicast, multicast and broadcast communication in AMI, guaranteeing forward and backward security. During the initial network establishment phase, the identities of smart meters [SM], center station [CS] and public, private key pairs are delivered by the certification authority CA.

In the Unicast communication, the steps involved are Handshake process in which identities of CS and SM are verified with PKC and certificate issues by CA, Session Key Generation in which session key is generated by CS and sent to SM using PKC, Message encryption in which the message is encrypted using the session key and PKC, and key refreshing is done after every session.

In the Multicast communication, during the group key generation, the session keys generated for each SM are combined to create group session key and are decrypted by the SM of respective group and the acknowledgement is sent to CS. Message encryption is done using the group key and SMs will decrypt using the public-private key pair generated by CA and key refreshing is done when new SMs are added or removed in the network.

In the Broadcast communication, in broadcast key generation, a common symmetric key is generated based on session keys of SMs belonging to the broadcast, and message encryption is done using the broadcast key and key refreshing is done periodically.

Implementation results shows reduced execution time for multicast and broadcast communications as the computation is carried out in the highly equipped utility servers on the CS side. The execution time for different modes are shown in Table 5. The flexible key updation process of this proposed scheme ensures the confidentiality, authenticity and integrity requirements of AMI are satisfied.

In [18], Yan et al. propose a lightweight authentication and key agreement scheme to provide mutual authentication and key agreement without a trust third party. The

Table 5 Execution time for different modes [17]

Mode of communication	Execution time (ms)	
	CS (Linux PC)	SM (Raspberry Pi)
Unicast	1.38	65.134
Multicast	25.91	30.575
Broadcast	29.13	30.202

scheme works in four phases namely Registration, Authentication and key agreement, key refreshment and multicast key generation.

During the Registration, the embedded password and id of the smart meter SM is submitted to the BAN gateway which personizes the SM using the one-way hash function. In Authentication and key agreement phase, the SM and BAN gateway authenticate each other and generate the session key. The session key is refreshed in a short-term or long-term process.

If multicast communication is required, the BAN gateway sends a message to the SM using the symmetric secret key to join the group, and after verifying the identity of the SM can start communicating.

As mutual authentication is established with a secret shared key between the SM and BAN gateway, replay attacks, Man-in-the-middle attacks and impersonation attacks are prevented. The computation complexity is reduced as only hash functions and exclusive OR operations are performed. Performance evaluation shows lesser communication overhead.

Rizzetti et al. in [19] propose a secure multicast approach using WMN and symmetric cryptography for lightweight key management in SGs.

Gateway Multihop Application Multicast scenario is assumed in which application layer is used but packet filtering is at the link layer. The multicast messages from the gateway to the WMN nodes need to be acknowledged. The gateway acts a Key distribution centre KDC for the shared keys to each WMN nodes. So all messages sent are signed by the sender node's private key.

All smart meters are treated as WMN nodes. First, the initiator node [SM] and the responder [GW] are authenticated to each other. Initiator generates a nonce value, sends the symmetric key encrypted data along with the hash of the certificate. As the symmetric key is lighter, the computation requirements are at the minimum.

Security analysis shows the prevention of Replay attacks, MiM attacks and perfect forward secrecy is achieved.

New key management schemes are proposed by Benmalek et al. in [20] based on individual and batch rekeying using multi-group key graph structure to support unicast, multicast and broadcast communications.

In the initialization between MDMS and smart meters, individual keys are established for unicast communications, and for mulicast communications, these keys are used to generate the multi-group key graph, and for broadcast communications, the group key is generated by MDMS and transmitted to SMs.

In verSAMI scheme, Group key management is achieved through Multi-group key graph structure instead of LKH Logical Key Hierarchy protocol, to enable managing only one set of keys and thereby reduce cost. Instead of using separate LKH operation for each DR [demand response] group, the key graph technique allows multiple groups to share a new set of keys.

In verSAMI + scheme, One Way Function [OFT] trees are adopted to allow reducing number of rekeying messages.

In batch verSAMI and verAMI + schemes, membership changes are handled in groups during batch rekeying intervals instead of individual rekeying operations.

Security analysis assures strong forward and backward secrecy and the batch keying schemes prevent out-of-sync problem. Detailed performance analysis and comparative studies show less storage and communication overheads.

In [21], Mahmood et al. has proposed a ECC based lightweight authentication scheme for SG communications.

The scheme works in three phases namely the Initialization, Registration and Authentication phases.

In the initialization phase, the trusted third part TA, using the one-way hash functions, generate the secret public and private key pair.

During registration, the user sends the id to TA where it deduces the corresponding key and sends back to the user for registering.

Each node needs to be authenticated in order to communicate with each other. The sender sets a time-stamp while transmitting, the receiver checks if the time-stamp to be within a specific threshold, then determines the shared session key and sends a challenge message. The challenge message is determined on checking the freshness of the time-stamp. So successful exchange of shared session key aids for secure communication. Security analysis shows perfect forward secrecy, and as mutual authentication is achieved, Replay attacks, Privileged Insider attacks, Impersonation, MiM attacks are prevented. Performance analysis shows substantially less computation costs and reduced memory overhead and communication overhead.

6 Research Challenges

With the recent advances in the SG and an equal growth in cyber-attack capabilities, robust threat/attack detection mechanisms are to be in place. With a focus on early detection of possible attacks, the need of the hour is to establish end-to-end security solutions achieved through component-wise mechanisms. However, with an emphasised role of AMI in the SG networks combined with its limited memory and computation capabilities, more research works need to carried out to seek accurate information and network security of AMI communications. Lightweight KMS approaches prove to be promising in this case, but more comprehensive security architectures are essential.

7 Conclusion

This review analyzed the cyber security issues found in the SG networks along with the attack models, threats and security solutions. The emphasis is on the Metering Infrastructure AMI of SGs, as it forms the crucial component for the successful operation of the system. Then, the KMS techniques are examined in detail in the context of SGs and in particular, the lightweight key management schemes available for SG communications are analyzed. It is observed that the lightweight approach is appropriate to be used in SG components which require lesser computational operations than the traditional schemes. Symmetric keys are generally used owing to its reduced key length.

However, the particular scheme is decided based on factors like type of communication security goals, the devices where the scheme to be deployed, techniques used for generating secret keys and whether trusted authorities are required for initialization.

References

1. Jain A, Mishra R (2015) Changes & challenges in smart grid towards smarter grid. In: 2016 international conference on electrical power and energy systems (ICEPES), INSPEC Accession Number: 16854529
2. Aloul F, Al-Ali AR, Al-Dalky R, Al-Mardini M, El-Hajj W (2012) Smart grid security: threats, vulnerabilities and solutions. Int J Smart Grid Clean Energy 1(1)
3. Kotut L, Wahsheh LA (2016) Survey of cyber security challenges and solutions in smart grids. In: 2016 cybersecurity symposium (CYBERSEC)
4. Fransen F, Wolthuis R, Security for smart electricity GRIDs How to address the security challenges in smart grids. A publication of the SEGRID Project www.segrid.eu, info@segrid.eu
5. Xu J, Yao Z (2015) Advanced metering infrastructure security issues and its solution: a review. Int J Innov Res Comput Commun Eng 3(11)
6. Mohamed N, John Z, Sam K, Elisa B, Kulatunga A (2012) Cryptographic key management for smart power grids. Cyber Center Technical Reports
7. Parvez I, Sarwat AI, Thai MT, Srivastava AK (2017) A novel key management and data encryption method for metering infrastructure of smart grid
8. Ghosh D, Li C, Yang C (2018) A lightweight authentication protocol in smart grid. Int J Netw Secur 20(3):414–422
9. https://electricenergyonline.com/energy/magazine/297/article/Conquering-Advanced-Metering-Cost-and-Risk.htm
10. Rohokale VM, Prasad R (2016) Cyber security for smart grid—the backbone of social economy. J Cyber Secur 5:55–76
11. Wei L, Rondon LP, Moghadasi A, Sarwat AI (2018) Review of cyber-physical attacks and counter defense mechanisms for advanced metering infrastructure in smart grid. In: IEEE/PES transmission and distribution conference and exposition (T&D), April 2018
12. Yi P, Zhu T, Zhang Q, Wua Y, Pan L (2015) Puppet attack: a denial of service attack in advanced metering infrastructure network. J Netw Comput Appl
13. Faisal MA, Aung Z, Williams JR, Sanchez A (2012) Securing advanced metering infrastructure using intrusion detection system with data stream mining. In: PAISI 2012, LNCS 7299. Springer-Verlag, Berlin Heidelberg, pp 96–111, pp 96–111

14. https://www.cryptomathic.com/news-events/blog/differences-between-hash-functions-symmetric-asymmetric-algorithms
15. Abdallah A, Shen X (2017) Lightweight security and privacy preserving scheme for smart grid customer-side networks. IEEE Trans Smart Grid 8(3)
16. Wu Q, Li M (2019)A lightweight authentication protocol for smart grid. IOP Conf. Ser Earth Environ Sci 234:012106
17. George N, Nithin S, Kottayil SK (2016) Hybrid key management scheme for secure AMI communications. Procedia Comput Sci 93:862–869
18. Yan L, Chang Y, Zhang S (2017) A lightweight authentication and key agreement scheme for smart grid. Int J Distrib Sens Netw Vl 13(2)
19. Rizzetti TA, da Silva BM, Rodrigues AS, Milbradt RG, Canha LN (2018) A secure and lightweight multicast communication system for Smart Grids. EAI Endorsed Trans Secur Saf
20. Benmalek M, Challal Y, Derhab A, Bouabdallah A (2018) VerSAMI: versatile and scalable key management for smart grid AMI systems. Comput Netw
21. Mahmood K, Chaudhry SA, Naqvi H, Kumari S, Li X, Sangaiah AK (2017) An Elliptic Curve Cryptography based on lightweight authentication scheme for smart grid communication. Future Gener Comput Syst

On Roman Domination of Graphs Using a Genetic Algorithm

Aditi Khandelwal, Kamal Srivastava, and Gur Saran

Abstract A Roman dominating function (RDF) on a graph G is a labelling $f: V \rightarrow \{0, 1, 2\}$ such that every vertex labelled 0 has at least one neighbour with label 2. The weight of G is the sum of the labels assigned. Roman domination number (RDN) of G, denoted by $\gamma_R(G)$, is the minimum of the weights of G over all possible RDFs. Finding RDN for a graph is an NP-hard problem. Approximation algorithms and bounds have been identified for general graphs and exact results exist in the literature for some standard classes of graphs such as paths, cycles, star graphs and $2 \times n$ grids, but no algorithm has been proposed for the problem for exact results on general graphs in the literature reviewed by us. In this paper, a genetic algorithm has been proposed for the Roman domination problem in which two construction heuristics have been designed to generate the initial population, a problem specific crossover operator has been developed, and a feasibility function has been employed to maintain the feasibility of solutions obtained from the crossover operator. Experiments have been conducted on different types of graphs with known optimal results and on 120 instances of Harwell–Boeing graphs for which bounds are known. The algorithm achieves the exact RDN for paths, cycles, star graphs and $2 \times n$ grids. For Harwell–Boeing graphs, the results obtained lie well within bounds.

Keywords Roman domination · Genetic algorithm · Roman domination number

A. Khandelwal (✉) · K. Srivastava · G. Saran
Dayalbagh Educational Institute, Dayalbagh, Agra 282005, India
e-mail: aditikhandelwal8193@gmail.com

K. Srivastava
e-mail: kamal.sri@dei.ac.in

G. Saran
e-mail: gursaran.db@gmail.com

V. Singh et al. (eds.), *Computational Methods and Data Engineering*, Advances in Intelligent Systems and Computing 1227,
https://doi.org/10.1007/978-981-15-6876-3_11

1 Introduction

The Roman domination problem (RDP) is inspired by an article 'Defend the Roman Empire!' by I. Stewart in the year 1999 [1]. The problem is of interest from the point of view of both history and mathematics. The Roman domination problem is formally defined as: Let $G(V, E)$ be a graph with $|V|$ vertices and $|E|$ edges. Let $N(u) = \{u \in V | uv \in E\}$ be the neighbourhood of u. A Roman dominating function (RDF) $f : V \rightarrow \{0, 1, 2\}$ is a map such that $\forall\ u \in V$, if $f(u) = 0$, then $\exists\ v \in N(u)$ with $f(v) = 2$. In other words, map f assigns labels $\{0, 1, 2\}$ to each vertex of V such that every vertex with 0 label has at least one neighbour with label 2. Let Φ denotes all RDFs defined on G.

Then the Roman domination number (RDN) of G is $\gamma_R(G) = \min\limits_{\forall f \in \Phi} w(f)$, where $w(f) = \Sigma f(v), \forall\ v \in V$, denotes the weight of G for the RDF f. Thus, the objective of the RDP for a graph G is to find an RDF with minimum weight. It is an NP-hard problem [2]. It has various applications in the field of server placement and assignment [3]. Throughout, we will denote the Roman domination number G by $\gamma_R(G)$.

Various theoretical results and bounds have been proved in [3–10]; however, it remains unexplored from metaheuristic point of view. The problem was introduced by I. Steward (1999). A lower bound for the Roman domination number for any graph G with $n = |V|$ is $\frac{2n}{\Delta+1}$ is proved in [4], where Δ is the maximum degree of G. Cockayne et al. [5] proved a probabilistic upper bound $\frac{2+\ln\left(\frac{1+\delta(G)}{2}\right)}{1+\delta(G)}$ for the RDN on general graphs and optimal RDN values for cycles $\left(\gamma_R(C_n) = \lceil \frac{2n}{3} \rceil\right)$, paths $\left(\gamma_R(P_n) = \lceil \frac{2n}{3} \rceil\right)$, complete n- partite graphs, graphs which contain a vertex of degree n-1 $(\gamma_R(S_n) = 2)$, $2 \times n$ grid graphs $\left(\gamma_R\left(G_{2,n}\right) = n + 1\right)$ and isolate-free graphs.

Note that the domination number, $\gamma_R(G)$, of G is defined as the minimum cardinality of the dominating set $S \subseteq V$ such that every vertex in $V - S$ is adjacent to at least one vertex in S. Cockayne et al. [5] have also given results on the relation between $\gamma_R(G)$ and $\gamma(G)$. They have proved that for any graph G, $\gamma(G) \leq \gamma_R(G) \leq 2\gamma(G)$.

Mobaraky and Sheikholeslami [6] have given lower and upper bounds on RDN with respect to girth and diameter of G. Favaron et al. [7] have proved that for connected graphs with $n \geq 3$, $\gamma_R(G) + \frac{\gamma(G)}{2} \leq n$. Chambers et al. [3] have improved the bounds and have proved that $\gamma_R(G) \leq \frac{4n}{5}$ and for graphs with $\delta(G) \geq 2$ and $n \geq 9$, $\gamma_R(G) \leq \frac{8n}{11}$, where $\delta(G)$ is the minimum degree of graph G.

Shang and Hu [2] have given some approximation algorithms for the RDP. Liu and Chang first established an upper bound on graphs with minimum degree at least 3 and on big claw-free and big net-free graphs [8] and later proved that RDP is NP-hard for bipartite graphs and NP-complete for chordal graphs [9]. Later Liedloff et al. [10] have shown that RDN for interval graphs and cographs can be computed in linear time.

In the literature reviewed by us, no heuristic has been designed to solve this problem. Therefore, we propose a genetic algorithm (GA) for the Roman domination problem which involves designing two new construction heuristics for the initial population and a problem specific crossover operator for the iterative phase. Experiments conducted on instances with known optimal RD values show that our GA is capable of achieving these values. Further, for other instances, results obtained by GA lie well within the known bounds.

1.1 Organization of the Paper

The rest of the paper is organized as follows. Section 2 describes GA for RDP. Implementation details of GA for RDP are present in Sect. 3. Section 4 is dedicated to experiments and their results. This is followed by the conclusion in Sect. 5.

2 Genetic Algorithm for RDP

Genetic algorithm (GA) mimics the process of natural evolution to generate solutions of optimization problems. Inspired by Darwin's principle of natural evolution and introduced by John Holland, as the name suggests, works on the principle of natural genetics and natural selection [11]. It uses artificial construction of search algorithms that need minimal information but provides us with vigorous results. The process starts with generating an initial population of feasible solutions. Then the fitness of the solutions obtained is evaluated depending on the underlying objective of the problem. Further, the selection procedure generates an intermediate population that helps to retain good solutions. Then the genetic crossover operator generates new solutions from those solutions that were selected on the application of selection operator. To maintain diversity among population individuals, the mutation operator alters the solutions obtained after crossover. The solutions of the initial population are replaced by better solutions of this new population. The GA continues on this new population until certain termination criterion is met. The adaptation of GA for the RDP is outlined in Fig. 1. The implementation details are presented in Sect. 3.

The GA proposed for RDP starts with generating initial population *pop* (Step 2) consisting of *ps* number of solutions (here a solution refers to a RDF) using construction heuristics which are detailed in Sect. 3.2. The objective function then computes the fitness of each population individual and stores the minimum RDN obtained as *bestcost* in Step 3. Then in Step 5, an intermediate population *interPop* is generated by applying tournament selection operator on *pop*. This helps to retain the solutions that perform better in terms of their RDN. Among remaining individuals in *interPop* a problem specific crossover operator is applied, with probability 0.25, to obtain the child population *childPop* (Step 6). The solutions in *childPop* undergo a feasibility checking procedure and are repaired accordingly. The *childPop* acts as

Pseudocode of GA for Roman Domination Problem (RDP)	
Step 1:	Initialize *ps*
Step 2:	Generate *pop*
Step 3:	*bestcost* = least *cost* obtained so far
Step 4:	while *termination criteria*
Step 5:	*interPop*← Tournament(*pop*)
Step 6:	*childPop*←Crossover (*interPop*) after being checked for feasibility
Step 7:	*pop* = *childPop*
Step 8:	Update *bestcost*
Step 9:	end while

Fig. 1 Pseudocode of RDP

the initial population for the next generation. The *bestcost* is updated if the least cost solution in *pop* is smaller than current *bestcost*. Steps 5-8 are repeated until *max_iter* generations are completed. The algorithm may terminate if there is no improvement in the *bestcost* for 100 consecutive generations.

3 Implementation Details of RDP

This section gives the implementation details of the problem based on the algorithm in Fig. 1.

3.1 *Solution Representation*

Each solution in the population is represented in the form of an array of length $n = |V|$ as shown in Fig. 2. For the graph shown in the same figure, the numbers in parenthesis are the labels assigned to the vertices, whereas those inside the circles are vertex identifiers. If solution array is represented by s, then $s[i]$ is the label assigned to vertex i as per the condition of RDF. Clearly, weight of G corresponding to this assignment is $\sum_{i=1}^{|V|} s[i]$, which will be referred to as the *cost* of the solution s, throughout the paper.

3.2 *Initial Population*

The initial population is generated using two construction heuristics specially designed for RDP described below. In the context of RDP, randomly generated solutions do not serve the purpose as they are infeasible in general and putting penalties on them and then improving them consumes a lot of computational time. First heuristic

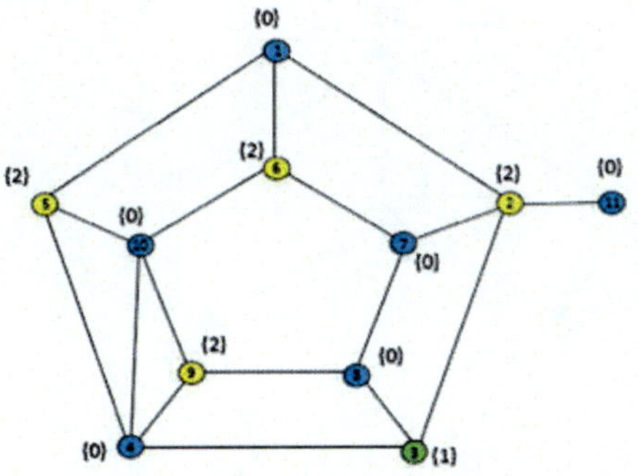

$s \rightarrow$

0	2	1	0	2	2	0	0	2	0	0

Fig. 2 Representation of a solution

$H1$ is a greedy heuristic but also has some random features, whereas second heuristic $H2$ generates solutions based on the degree of vertices.

Heuristic 1 ($H1$)

This heuristic begins with picking one random vertex u from a set *unvisited* which initially is V and assigns it label 2 and then assigns 0 to all its unvisited neighbours, thus satisfying the condition for RDF. From the remaining vertices, another vertex is chosen at random and the same process is repeated until all the vertices are labelled (Fig. 3).

Figure 4 illustrates the generation of a solution using $H1$. The heuristic begins with constructing a set of unassigned vertices $unvisited = \{1, 2, 3, 4, 5, 6, 7, 8, 9, 10, 11\}$. Vertex 6 is chosen randomly and is labelled $f(6) = 2$. Then $N(6) = \{1, 7, 10\}$. As indicated in Step 6 of the algorithm, vertices 1, 7 and 10 are assigned 0 and

Procedure *H1*:

Step 1: $unvisited \leftarrow V$, (the set of vertices)
Step 2: **while** ($unvisited \neq \phi$)
Step 3: select vertex u from *unvisited* randomly
Step 4: $f(u) = 2$
Step 5: $unvisited = unvisited \setminus \{u\}$
Step 6: $f(v) = 0 \ \forall \ v \in N(u)$
Step 7: $unvisited = unvisited \setminus N(u)$
Step 8: if ($|unvisited| = 1$)
Step 9: $f(u) = 1$; $u \in unvisited$
Step 10: **end**
Step 11: **return** $f(V)$

Fig. 3 Heuristic $H1$

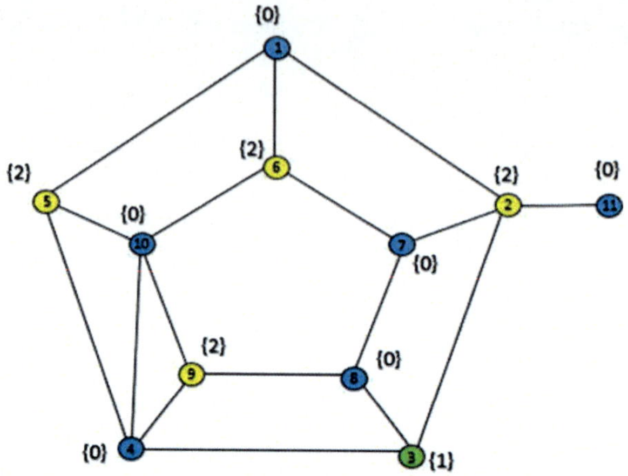

Fig. 4 Graphical representation of solution generated by $H1$ with RDN = 9

unvisited is updated as {2,3,4,5,8,9,11}. Now let vertex 2 is chosen at random and the process continues until *unvisited* is left with just one vertex or no vertex which is {3} in this example. This vertex is labelled 1 as it satisfies the condition given in Step 8 of the algorithm.

Heuristic 2 (*H*2)

The heuristic begins with placing the vertices in descending order of their degrees in a set *unvisited*. Ties are broken randomly. The first vertex of the set, i.e. having the highest degree, is labelled 2 and all its neighbours are labelled 0. All the labelled vertices are removed from V. From the remaining *unvisited* vertices, the one with the highest degree is again picked to be labelled and the same process continues until the vertex set V is exhausted (Fig. 5).

Procedure *H2*:

Step 1: *unvisited* ← V, (the set of vertices)
Step 2: sort *unvisited* in descending order of degree
Step 3: while (*unvisited* $\neq \phi$)
Step 4: u = first vertex from *unvisited*
Step 5: $f(u) = 2$
Step 6: *unvisited* = *unvisited* \ $\{u\}$
Step 7: $f(v) = 0 \ \forall \ v \in N(u)$
Step 8: *unvisited* = *unvisited* \ $N(u)$
Step 9: if (|*unvisited*| = 1)
Step 10: $f(u) = 1$; $u \in$ *unvisited*
Step 11: end
Step 12: return $f(V)$

Fig. 5 Heuristic $H2$

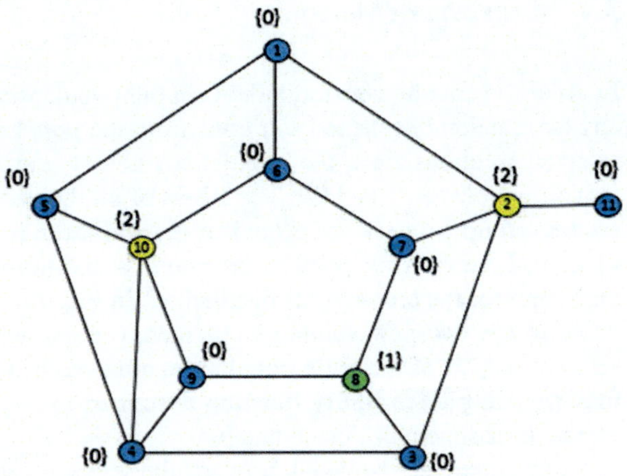

Fig. 6 Solution generate by $H2$ with RDN = 5

Figure 6 illustrates the generation of a solution using $H2$. The heuristic begins with constructing a set of unassigned vertices *unvisited* = {1, 2, 3, 4, 5, 6, 7, 8, 9, 10, 11}. This is then sorted based on descending order of degree of each vertex as *unvisited* = {2, 10, 4, 1, 3,9 ,7 ,5, 6, 8, 11}. The vertices 2, 4 and 10 have the same degree, and they are randomly permuted and placed in the *unvisited* set. A similar random permutation is done for vertices 1, 3, 5, 6, 7, 8 and 9 as they all have degree 3.

Vertices are now picked from *unvisited* and labelled. The first vertex in *unvisited* is picked and labelled as $f(2) = 2$. Now $N(2) = \{1, 3, 7, 11\}$. As given in the above algorithm, they are all labelled 0 and *unvisited* is updated as *unvisited* = {10, 4, 9, 5, 6, 8}. The next vertex is picked and labelled, and the process continues until *unvisited* = {8}. This vertex is labelled 1 as it satisfies the condition given in Step 9 of the algorithm.

Heuristic $H1$ is used to generate *ps* -1 solutions and heuristic $H2$ contributes only one solution in the initial population since it generates a unique solution for a graph though breaking the ties may provide more solutions.

3.3 Selection

The binary tournament operator [12] is applied for selection to ensure that good solutions participate in the crossover to obtain child population. In this, each solution participates in two tournaments and the one with a better objective value wins. These selected solutions form the new population *interPop* which undergoes crossover in the next step. This type of binary tournament selection ensures that a solution can have at most two copies in the population in contrast to Roulette wheel selection which creates multiple copies of good solutions. This may cause premature convergence after getting stuck in local minima.

3.4 Crossover Operator

In order to generate new individuals in the population, a crossover is performed on any two randomly selected solutions from the *pop*. Let $s1$ and $s2$ be two randomly selected solutions from the population. Let $r1$ and $r2$ are two numbers selected randomly between 1 and $|V|$. The labels of all the vertices lying between $r1$ and $r2$ are picked up from $s1$ and stored in $v1$, and those in $s2$ are stored in $v2$. The sets $v1$ and $v2$ are then swapped to get two new solutions $c1$ and $c2$, respectively. The algorithm for the crossover is detailed out in Fig. 7.

From the two new solutions obtained, the one with a better objective value is selected as part of the child population, after ensuring that the solution obtained is feasible using a feasibility function described in the next section. The process of crossover is shown in Figs. 8 and 9.

After crossover, the labels between the selected vertices are swapped (Fig. 9.).

Procedure: *Crossover*(s1, s2)

Step 1: $\{r1, r2\} \leftarrow$ two random vertices from V
Step 2: $v1 \leftarrow$ set of labels between $r1$ and $r2$ in $s1$
Step 3: $v2 \leftarrow$ set of labels between $r1$ and $r2$ in $s2$
Step 4: swap v1 and v2 to generate child solutions c1 and c2
Step 5: *feasibility*($c1$,$c2$)
Step 6: *child*=min{weight($c1$), weight ($c2$)}

Fig. 7 Crossover operator

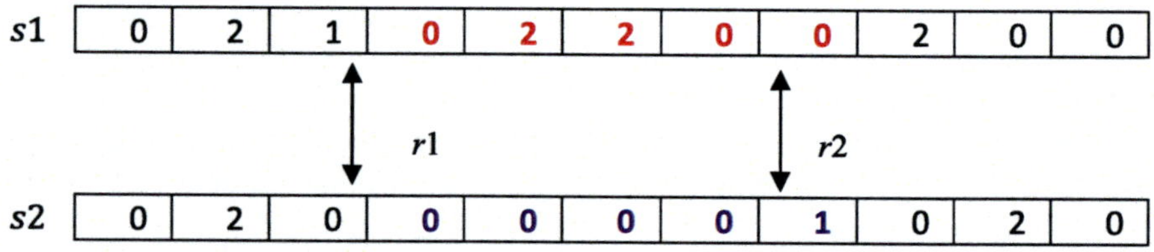

Fig. 8 Solution after crossover

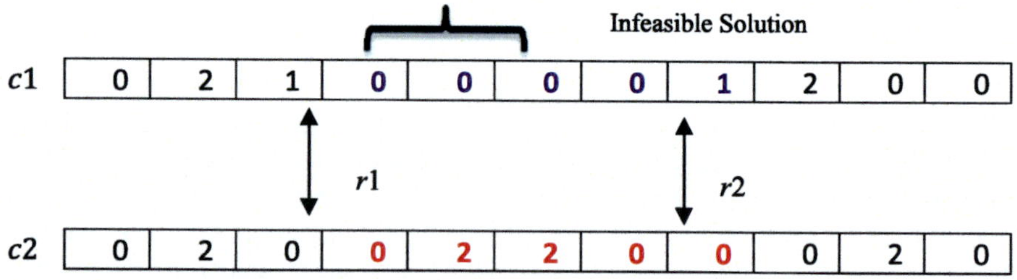

Fig. 9 Solution after crossover

Procedure: *Feasibility* (s)

Step 1: Z set of all $v \in V$ such that $s(v) = 0$
Step 2: for $\forall\, u \in Z$ do
Step 3: if $\nexists\, w \in N(u)$ such that $s(w) = 2$
Step 4: $s(u) \leftarrow 1$
Step 5: end for

Fig. 10 Feasibility function

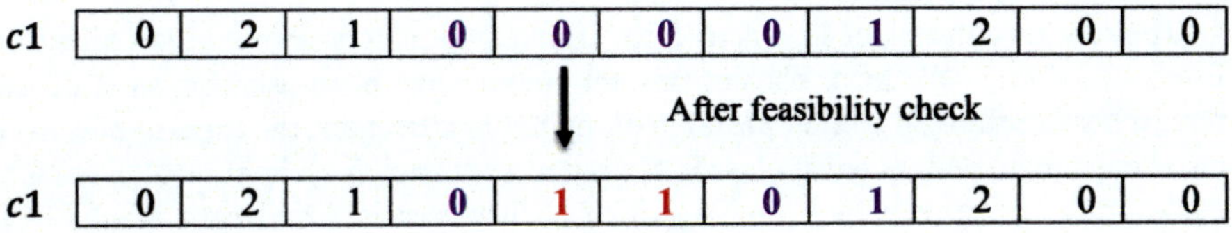

Fig. 11 Solution after attaining feasibility with RDN = 8

Feasibility Check

When new solutions are produced by crossover operator, it is quite possible that they do not belong to the feasible region. In other words, the labelling does not conform to the definition of Roman domination function. Thus, every solution generated by crossover undergoes a feasibility check which not only identifies infeasible solutions but also transforms them into a feasible solution with minimal changes in the labelling. This procedure is quite simple since a solution is infeasible if there is a vertex with label 0 but none of its adjacent vertices are labelled 2. The conversion to feasible solution just requires that such vertices have their labels replaced from 0 to 1. The algorithm for the feasibility check and subsequent conversion to a feasible solution is outlined in Fig. 10.

For an illustration, solution shown in Fig. 11 is an infeasible solution obtained after crossover in Fig. 9. Each vertex with a label 0 is checked for its neighbour having label 2. Since vertex 5 and 6 are labelled 0 but have no neighbour with label 2, they make the solution infeasible. The feasibility function relabels these vertices from 0 to 1 to make the solution feasible. It is worth to mention here that this conversion increases the *cost* of the solution.

3.5 Termination Criteria

We have set the termination criteria of GA as follows: the iterative steps of GA are executed for 1000 generations or if there is no improvement in the *bestcost* obtained in the process for 100 consecutive generations whichever is earlier.

4 Experiments and Results

This section is devoted to describe the experiments which were conducted for the proposed GA for RDP and the results obtained. The metaheuristic was implemented in C++, and experiments were conducted on Ubuntu 16.04 LTS machine with Intel (R) Core(TM) i5-2400 CPU @3.10_4 GHz and 7.7 GiB of RAM. The experiments were carried out in two phases. Initial experiments were conducted on a representative set to tune GA parameters, and the second phase helps to analyse the performance of the proposed algorithm.

The test set consists of instances from Harwell–Boeing sparse matrix collection listed in Table 3. We have chosen this set since it has been used by most of the researchers working in similar problem domains [13]. Besides, the experiments have also been conducted on some classes of graphs with known optimal results, namely cycle graphs (C_n), in which the degree of each vertex is 2 and path graphs (P_n) which is defined as a sequence of vertices of $V(G)$ such that for each $i = 1$ to $k - 1, (u_i, u_{i+1}) \in E(G)$. Star graphs ($S_n$), which is a complete bipartite graph $K_{1,k}$, and grid graphs ($G_{2,n}$) or lattice graph, a graph Cartesian product of path graphs $P_m \otimes P_n$ on m and n vertices, are also taken. These graphs are listed in Table 1, in which the entries in the second column show the range of the graph sizes considered for the experiments.

4.1 Representative Set

The representative set consists of 11 HB graphs which are ash85, bcspwr03, bcsstk01, bcsstk05, bus494, can24, can73, can715, dwt162, dwt245, ibm32, two random graphs denoted as R_{n_b}, where b is the edge density, one star graph (S_n), two $2 \times n$ grid graphs ($G_{2,n}$), one cycle (C_n) and one path (P_n) graph. The number of vertices ranges from 24 to 715 in the representative set.

Table 1 Class of graphs with known optimal Roman domination number

Graphs	$n = \|V\|$	Optimal results [5]
$2 \times n$ grid graphs	32–1000	$n + 1$
Star graphs	16–500	2
Path graphs	16–500	$\lceil \frac{2n}{3} \rceil$
Cycle graphs	16–500	$\lceil \frac{2n}{3} \rceil$

4.2 Tuning the Parameter Population Size (ps)

To determine the population size, experiments were conducted by taking $ps = n/2$, $n/4$ and $n/6$, on the representative set. Each instance of the test set was used to provide objective values and time taken by the GA for 30 trials. Two-way ANOVA with repetition with 5% level of significance was used to analyse the data, and no significant difference was found among the mean objective values of populations. However, it was observed that with population size $n/2$, GA performed quite well with respect to time. Thus, based on time analysis, $n/2$ was taken to be the population size for further experimentation.

4.3 Final Experiments

After setting the population size ps, experiments were conducted on the test set to validate the proposed algorithm. The results obtained show that optimal values were attained for the instances of the classes of graphs listed in Table 1. For the instances with unknown optimal RDN, the values attained lie within the bounds. These experiments show that our GA is capable of achieving satisfactory results.

The final experiments to validate the designed GA were first conducted on instances with known optimal results in the literature. Cycle graph C_n and path graph P_n were tested for $16 \leq n \leq 500$. Optimal RDN for cycle $\gamma_R(C_n) = \lceil \frac{2n}{3} \rceil$ and $\gamma_R(P_n) = \lceil \frac{2n}{3} \rceil$ which are achieved by our GA.

For star graphs S_n, the optimal value of 2 is readily achieved by our GA in each test instance run for graphs tested for $16 \leq n \leq 500$.

For $2 \times n$ grid graphs $G_{2,n}$, the known optimal RDN is $n + 1$ [5]. The algorithm is tested for n up to 500. The optimal RDN for these is sometimes present in the initial population itself, and the results, given bold in Table 2, are obtained quickly by the algorithm.

We also tested the proposed algorithm on Harwell–Boeing Graphs. Though the optimal results remain unknown, upper and lower bounds for any graph of order n are given [5]. We have used these bounds to compute upper bound (UB) and lower bound (LB) given in Table 3. The RDN for all the instances of HB graphs, tested on the GA for RDP, lies well within these bounds. The results obtained are listed in Table 3.

Table 2 Results for grid graphs $G_{2,n}$

Graphs	$n = |V|$	Optimal values	RDN
$G_{2,16}$	32	17	**17**
$G_{2,16}$	56	29	**29**
$G_{2,40}$	80	41	**41**
$G_{2,102}$	204	103	**103**
$G_{2,234}$	468	235	**235**
$G_{2,288}$	576	289	**289**
$G_{2,310}$	620	311	**311**
$G_{2,344}$	688	345	**345**
$G_{2,378}$	756	379	**379**
$G_{2,426}$	852	427	**427**
$G_{2,466}$	932	467	**467**
$G_{2,500}$	1000	501	**501**

Table 3 Results for HB graphs

Graphs	n	LB	UB	RDN	Graphs	n	LB	UB	RDN
can24	24	5	17	9	ash292	292	41	280	43
pores_1	30	6	22	13	can_292	292	16	259	85
ibm32	32	5	22	15	dwt_310	310	56	301	122
bcspwr01	39	13	35	28	gre_343	343	76	336	185
bcsstk01	48	8	38	24	dwt_361	361	80	354	135
bcspwr02	49	14	44	40	str_200	363	14	315	177
curtis54	54	6	40	28	dwt_419	419	64	408	161
will57	57	10	48	19	bcsstk06	420	30	394	123
dwt_59	59	19	55	39	bcsstm07	420	32	396	156
impcol_b	59	6	43	22	impcol_d	425	53	411	229
can_61	61	4	38	18	bcspwr05	443	88	435	306
bfw62a	62	5	43	31	can_445	445	68	434	177
bfw62b	62	9	51	38	nos5	468	40	447	177
can_62	62	17	57	44	west0479	479	24	442	234
bcsstk02	66	14	44	28	bcsstk020	485	88	476	241
dwt_66	66	22	62	33	mbeause	492	2	9	8
dwt_72	72	28	69	49	bus494	494	98	486	340
can_73	73	16	66	39	mbeacxc	496	2	12	9

(continued)

Table 3 (continued)

Graphs	n	LB	UB	RDN	Graphs	n	LB	UB	RDN
steam3	80	13	70	18	mbeaflw	496	2	12	10
ash85	85	17	77	48	dwt_503	503	40	480	177
dwt_87	87	13	76	39	lns_511	511	78	500	115
can_96	96	21	89	33	gre_512	512	113	505	300
nos4	100	28	95	52	pores_3	532	106	524	284
gent113	113	8	88	62	fs_541_1	541	90	531	122
gre_115	115	3	107	68	dwt_592	592	78	579	213
bcspwr03	118	23	110	88	steam2	600	42	574	94
arc130	130	2	7	7	west0655	655	33	618	222
hor_131	131	22	398	158	bus662	662	132	654	417
lns_131	131	20	120	94	shl_200	663	3	225	48
bcsstk04	132	5	88	25	nnc666	666	78	651	189
west0132	132	11	111	60	fs_680_1	680	97	668	256
impcol_c	137	18	124	71	bus685	685	105	674	205
bcsstk22	138	34	132	80	can_715	715	13	612	211
can_144	144	19	131	48	nos7	729	208	724	419
bcsstk05	153	12	131	47	fs_760_1	760	63	738	233
can_161	161	35	154	58	mcfe	765	13	657	92
dwt_162	162	36	155	69	bcsstk19	817	148	808	388
west0167	167	15	148	69	bp_0	822	6	558	399
mcca	180	5	118	67	bp_1000	822	5	574	560
fs_183_1	183	3	80	70	bp_1200	822	5	513	369
gre_185	185	41	178	83	bp_1400	822	5	513	363
can_187	187	37	179	70	bp_1600	822	5	520	344
dwt_193	193	12	165	40	bp_200	822	5	541	411
will199	199	28	187	80	bp_400	822	5	529	381
impcol_a	207	31	196	115	bp_600	822	5	522	363
dwt_209	209	24	194	79	bp_800	822	5	520	359
gre_216a	216	48	209	113	can_838	838	52	808	294
dwt_221	221	36	211	80	dwt_878	878	175	870	301
impcol_e	225	12	191	75	orsirr_2	886	126	874	411
can_229	229	41	220	99	gr_30_30	900	200	893	335
dwt_234	234	46	226	140	dwt_918	918	141	907	362

(continued)

Table 3 (continued)

Graphs	n	LB	UB	RDN	Graphs	n	LB	UB	RDN
nos1	237	94	234	150	jagmesh1	936	267	931	457
saylr1	238	95	235	146	nos2	957	382	954	509
steam1	240	22	221	51	nos3	960	106	944	222
dwt_245	245	37	234	126	west0989	989	56	956	202
can_256	256	6	175	85	jpwh_991	991	123	977	361
nnc261	261	30	246	135	dwt_992	992	976	110	199
lshp265	265	75	260	124	saylr3	1000	285	995	555
can_268	268	14	233	83	sherman1	1000	285	995	495
bcspwr04	274	34	260	179	sherman4	1104	315	1099	489

5 Conclusion

In this paper, we have described a genetic algorithm for the Roman domination problem for general graphs. For this problem, we have designed two construction heuristics to provide feasible initial solutions. In order to generate new solutions, a crossover operator has been designed. An important feature of the algorithm is the feasibility function that keeps a check on the feasibility of solutions obtained after crossover. The algorithm achieves the exact Roman domination number for those classes of graphs for which optimal results were known. For Harwell–Boeing instances with known bounds, the RDN obtained is within bounds.

As future work, single solution-based metaheuristic and other population-based metaheuristics can be designed for the improvement of results. New crossover operators can also be designed to improve the performance of the GA. Techniques can also be designed for other variants of the problem.

References

1. Stewart I (1999) Defend the Roman Empire! Sci Am 281(6):136–139
2. Shang W, Hu X (2007) The Roman domination problem in unit disk graphs. In: International conference on computational science (3), LNCS, vol. 4489, Springer, pp 305–312
3. Chambers EW, Kinnersley W, Prince N, West DB (2009) Extremal problems for Roman domination. SIAM J on Disc Math 23(3):1575–1586
4. Cockayne EJ, Grobler PJP, Grundlingh WR, Munganga J, van Vuure JH (2005) Protection of a graph. Util Math 67:19–32
5. Cockayne EJ, Dreyer PA Jr, Hedetniemi SM, Hedetniemi ST (2004) Roman domination in graphs. Disc Math 278:11–22
6. Mobaraky BP, Sheikholeslami SM (2008) Bounds on Roman domination numbers of graphs. Matematiki Vesnik 60:247–253
7. Favaron O, Karami H, Khoeilar R, Sheikholeslami SM (2009) Note on the Roman domination number of a graph. Disc Math 309:3447–3451

8. Liu CH, Chang GJ (2012) Upper bounds on Roman domination numbers of graphs. Disc Math 312:1386–1391
9. Liu CH, Chang GJ (2013) Roman domination on strongly chordal graphs. J Comb Optim 26:608–619
10. Liedloff M, Kloks T, Liu J, Peng SL (2005) Roman domination over some graph classes. LNCS 3787:103–114
11. Deb K (2008) Multiple-objective optimization using evolutionary algorithms. Wiley, New York
12. Jain P, Saran G, Srivastava K (2016) On minimizing vertex bisection using a memetic algorithm. Inf Sci 369:765–787
13. Torres-Jimenez J, Izquierdo-Marquez I, Garcia-Robledo A, Gonzalez-Gomez A, Bernal J, Kacker RN (2015) A dual representation simulated annealing algorithm for the bandwidth minimization problem on graphs. Inf Sci 303:33–49

General Variable Neighborhood Search for the Minimum Stretch Spanning Tree Problem

Yogita Singh Kardam and Kamal Srivastava

Abstract For a given graph G, minimum stretch spanning tree problem (MSSTP) seeks for a spanning tree of G such that the distance between the farthest pair of adjacent vertices of G in tree is minimized. It is an NP-hard problem with applications in communication networks. In this paper, a general variable neighborhood search (GVNS) algorithm is developed for MSSTP in which initial solution is generated using four well-known heuristics and a problem-specific construction heuristic. Six neighborhood strategies are designed to explore the search space. The experiments are conducted on various classes of graphs for which optimal results are known. Computational results show that the proposed algorithm is better than the artificial bee colony (ABC) algorithm which is adapted by us for MSSTP.

Keywords General variable neighborhood search · Artificial bee colony · Minimum stretch spanning tree problem

1 Introduction

Finding the shortest paths between pairs of vertices in a graph has always been a problem of interest due to its applications. The minimum stretch spanning tree problem (MSSTP) consists of finding spanning tree of a graph such that the vertices in the tree remain as close as possible. For a given undirected connected graph $G = (V, E)$, where $V(G) = \{v_1, v_2, \ldots, v_n\}$ is the set of vertices and $E(G) = \{(u, v) : u, v \in V(G)\}$ is the set of edges, the MSSTP is defined formally as follows:

Let

$$\phi(G) = \{ST : ST \text{ is a spanning tree of } G\}$$

Y. S. Kardam (✉) · K. Srivastava
Dayalbagh Educational Institute, Dayalbagh, Agra 282005, India
e-mail: yogita.singh27@gmail.com

K. Srivastava
e-mail: kamal.sri@dei.ac.in

V. Singh et al. (eds.), *Computational Methods and Data Engineering*, Advances in Intelligent Systems and Computing 1227,
https://doi.org/10.1007/978-981-15-6876-3_12

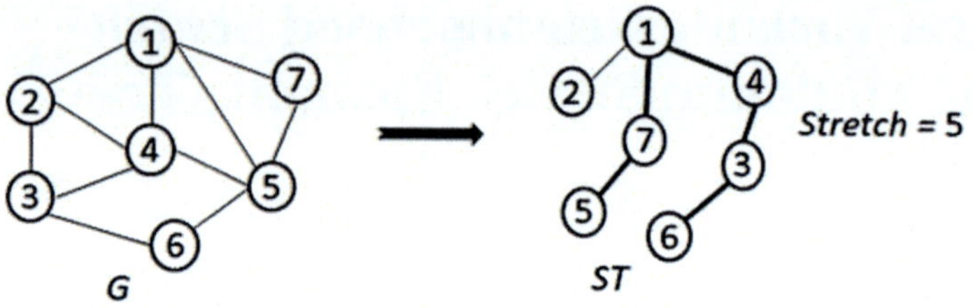

Fig. 1 *Stretch* in spanning tree *ST* of a graph *G*

Then, MSSTP is to find a spanning tree $ST^* \in \phi(G)$ such that

$$Stretch\left(G, ST^*\right) = \min_{\forall ST \in \phi(G)} \{Stretch(G, ST)\}$$

where

$$Stretch(G, ST) = \max_{\forall (u,v) \in E(G)} d_{ST}(u, v)$$

Here, $d_{ST}(u, v)$ is the distance (path length) between u and v in ST.

Let $(u, v) \in E(G)$, then a path between the vertices u and v in ST is termed as critical path if $d_{ST}(u, v)$ is maximum over all pairs of adjacent vertices of G, i.e., $d_{ST}(u, v) = Stretch(G, ST)$. Note that a spanning tree can have more than one critical path. Throughout this paper, a solution S to the problem is a spanning tree of the input graph G and $Stretch(S)$ or $Stretch$ refers to the objective value corresponding to that solution.

Figure 1 shows the $Stretch$ in the spanning tree ST of a given graph G. Here, $d_{ST}(1, 5) = 2$, $d_{ST}(2, 3) = 3$, $d_{ST}(2, 4) = 2$, $d_{ST}(4, 5) = 3$, $d_{ST}(5, 6) = 5$ and for the remaining edges of G it is 1. Therefore, $Stretch$ is 5 as it is the maximum distance between any two adjacent vertices of G in ST and the critical path is (5, 7, 1, 4, 3, 6) (shown with the bold edges).

MSSTP is NP-hard problem for general graphs. For this problem, exact methods and approximation algorithms are adopted by the researchers in the literature; however, it remains unstudied from metaheuristic point of view, which is perhaps one of the most useful approaches to deal with such a problem. Thus, our main focus in this paper is to design and implement a widely used metaheuristic, namely general variable neighborhood search (GVNS) algorithm, a single solution base metaheuristic, which guides the search procedure by changing the pre-defined neighborhoods in systematic manner. It is a variant of variable neighborhood search (VNS) algorithm which was first proposed in 1997 by Mladenovic and Hansen [1]. GVNS combines variable neighborhood descent (VND) and reduced VNS (RVNS) where the first one is entirely deterministic and the second one is stochastic [2]. GVNS balances between diversification (RVNS) and intensification (VND) in search space by changing the neighborhoods both in deterministic and in random way [3]. This motivated us to design GVNS for MSSTP.

In this paper, a GVNS algorithm is designed for MSSTP in which five construction heuristics are considered for generating initial solution. These construction heuristics are well-known procedures for obtaining a spanning tree of a given graph and are adapted as per the requirements of MSSTP. Neighborhoods play a vital role in the functioning of VNS as the solution in VNS is searched through the solution space by moving from one neighborhood to another in some specific manner. Therefore, six different neighborhood strategies balancing between diversification and intensification are developed for MSSTP which are based on subtree replacement and cycle exchanges, respectively. The performance of GVNS is tested by conducting experiments on a class of graphs with known optimal results. Though this problem has not been dealt with by the metaheuristic community so far, an artificial bee colony (ABC) algorithm, proposed recently for a similar relevant problem on weighted graphs, has been adapted for MSSTP for comparison purposes. Computational experiments show the effectiveness of GVNS as it outperforms the ABC.

Rest of the paper is organized as follows. Section 2 contains the work related to MSSTP. The algorithm proposed for the problem and its methods is explained in Sect. 3. Section 4 discusses the experiments conducted on particular classes of graphs using the proposed algorithm for MSSTP and compares the results of GVNS with the results obtained from ABC after implementing it for MSSTP. Section 5 concludes the paper.

2 Related Work

MSSTP is a special case of generalized tree t-spanner problem which was first introduced in [4] in order to develop a technique for constructing network synchronizers by using a relationship between synchronizers and the structure of t-spanner over a network. As tree spanners have been of great use in distributed systems, network design and communication networks, this has led to a number of problems related to tree spanners [5]. In the literature, tree t-spanner problem has been extensively studied while MSSTP has been tackled with very few approaches. In [6], graph theoretic, algorithmic and complexity issues pertaining to tree spanners have been studied and various results have been proved on weighted and unweighted graphs for different values of t. More recent work on tree t-spanner is the ABC algorithm for weighted graphs [7] which has been claimed to outperform the only existing metaheuristic (genetic algorithm) proposed for the problem so far. In a technical report [8], MSSTP is dealt by restricting the input graphs to grids, grid subgraphs and unit disk graphs. The optimal results have also been proved for some standard graphs such as Petersen graph, complete k-partite graphs, and split graphs [9]. Since these results help in validating the metaheuristic designed by us, therefore we list these results in the following section. However, to the best of our knowledge no metaheuristic for general graphs is available in the literature for MSSTP.

2.1 Known Optimal Results for Some Classes of Graphs [9]

1. **Petersen Graph** $(\boldsymbol{Pt})$: The optimal result is 4 for this graph.
2. **Cycle Graph** $(\boldsymbol{C_n})$: The optimal is $n-1$ for these graphs, where $n \geq 3$.
3. **Wheel Graph** $(\boldsymbol{W_n})$: The optimal value for this class is 2 for $n \geq 4$.
4. **Complete Graph** $(\boldsymbol{K_n})$: The optimal $Stretch$ is 2 for $n \geq 3$.
5. **Split Graph** $(\boldsymbol{S_n})$: It is a connected graph with set of vertices $V = X \cup Y$, where X is a clique (no two vertices in it are adjacent) and Y is an independent set. For these graphs, optimal is 2, if $\exists$ a vertex $x \in X$ such that the degree of every vertex $y \in Y \backslash Nbr_G(x)$ is 1 and it is 3, otherwise.
6. **Complete** $\boldsymbol{k}$**-Partite Graph** $(K_{n_1,n_2,\ldots,n_k})$ For $k \geq 3$

$$Stretch = \begin{cases} 2, & \text{if}\, n_1 = 1. \\ 3, & \text{otherwise.} \end{cases} \tag{1}$$

and For $k = 2$

$$Stretch = 3, n_1, n_2 \geq 2$$

7. **Diamond Graph** $\left(\boldsymbol{K_{1,n-2,1}}\right)$: It is a complete tripartite graph with partite sets $\{P_1, P_2, P_3\}$ with $|P_1|=|P_3| = 1$ and $|P_2| = n-2$, where $n \geq 4$ and is an even number. For this class of graphs, optimal result is 2 for $n \geq 4$.
8. **Triangular Grid** $(\boldsymbol{T_n})$: The optimal of these graphs is $\lceil \frac{2n}{3} \rceil + 1$, where $n \geq 1$.
9. **Rectangular Grid** $(\boldsymbol{P_m \times P_n})$: The optimal result is $2\lfloor \frac{m}{2} \rfloor + 1, 2 \leq m \leq n$ for this class of graphs.
10. **Triangulated Rectangular Grid** $\left(\boldsymbol{TR_{m,n}}\right)$: The optimal is m, $2 \leq m \leq n$ for the graphs of this class.

3 Proposed Algorithm: General Variable Neighborhood Search for Minimum Stretch Spanning Tree Problem (GVNS-MSSTP)

The GVNS proposed for the MSSTP is sketched in Algorithm 1. It starts by generating an initial solution S (Step 2) using a construction heuristic described in Sect. 3.1. The S_{best} maintains the best solution found at any step of the algorithm. Step 7 performs the Shake procedure. It is done by generating a neighbor S' of S randomly in the neighborhood NBD_i (explained in Sect. 3.2) of S. In Step 8, a local minimum solution S'' is obtained from S' using VND method. S_{best} is updated if the $Stretch$ of S'' is better than that of S_{best} (Steps 9-11). Now, $Stretch$ of the two solutions S and S'' is compared (Step 12) and S is replaced by S'' if it is an improvement over S (Step 13). In this case, i is set to 1 (Step 14), i.e., the new solution will be explored starting again with the first neighborhood and if S'' fails to improve S in the current neighborhood, then search is moved to the next neighborhood (Step 16). Steps 7–17 are repeated until all the neighborhoods (1 to max_*nbd*) are explored. The search continues till the stopping criterion is met, i.e., *iter* reaches to the maximum number of iterations max_*iter*.

Algorithm 1 General Variable Neighborhood Search Algorithm for MSSTP (GVNS-MSSTP)

1: Initialize number of neighborhoods (max_nbd) and number of iterations (max_iter)
2: $S \leftarrow$ generate initial solution
3: $S_{best} \leftarrow S$
4: **while** $iter \leq max\_iter$ **do**
5:   $i \leftarrow 1$
6:   **while** $i \leq max\_nbd$ **do**
7:     $S' \leftarrow \mathrm{NBD}_i(S)$ ⊳ Shaking
8:     $S'' \leftarrow \mathrm{VND}(S', max\_nbd)$
9:     **if** $Stretch(S'') < Stretch(S_{best})$ **then**
10:       $S_{best} \leftarrow S''$
11:     **end if**
12:     **if** $Stretch(S'') < Stretch(S)$ **then** ⊳ Neighborhood Change
13:       $S \leftarrow S''$
14:       $i \leftarrow 1$
15:     **else**
16:       $i \leftarrow i + 1$
17:     **end if**
18:   **end while**
19: $iter \leftarrow iter + 1$
20: **end while**
21: **return** S_{best}

Algorithm 2 explains the procedure VND used in GVNS-MSSTP to find a local minimum after exploring all the neighborhoods of a given solution. It starts by finding a best neighbor S_1' of solution S_1 in its jth neighborhood using function FindBestNbr (Step 3). Then, neighborhood is changed accordingly by comparing the solutions S_1 and S_1' (Steps 4-9). S_1 keeps improving in a similar way until all the neighborhoods of S_1 are explored.

Algorithm 2 VND(S_1, *max_nbd*)

```
j ← 1
while j ≤ max_nbd do
    S_1' ← FindBestNbr(S_1, j)
    if Stretch(S_1') < Stretch(S_1) then        ▷ Neighborhood Change
        S_1 ← S_1'
        j ← 1
    else
        j ← j + 1
    end if
end while
return S_1
```

Algorithm 3 presents the function FindBestNbr used in VND. In Steps 5–13, neighbor S_2'' of S_2 keeps generating in neighborhood NBD_k of S_2 till the improved neighbor is found. The complete process (from Steps 4 to 19) is repeated if thus improved solution S_2 keeps improving the original solution S_2'.

Algorithm 3 FindBestNbr(S_2, k)

```
1: flag_1 ← 0
2: flag_2 ← 0
3: while flag_1 ≠ 1 do
4:     S_2' ← S_2
5:     while flag_2 ≠ 1 do
6:         S_2'' ← NBD_k(S_2)
7:         if Stretch(S_2'') < Stretch(S_2) then
8:             S_2 ← S_2''
9:             flag_2 ← 0
10:        else
11:            flag_2 ← 1
12:        end if
13:    end while
14:    if Stretch(S_2) < Stretch(S_2') then
15:        flag_1 ← 0
16:        flag_2 ← 0
17:    else
18:        flag_1 ← 1
19:    end if
20: end while
21: return S_2'
```

The different construction heuristics and the neighborhood strategies used in GVNS are discussed as follows.

3.1 Initial Solution Generation

The initial solution is generated using a construction heuristic selected randomly from the five construction heuristics. These heuristics are based on the well-known algorithms for finding the spanning trees of a graph and are explained below.

1. ***Random_Prim***: This heuristic constructs a spanning tree of a given graph using a random version of Prim's algorithm in which at every step the edges are picked randomly.
2. ***Random_Kruskal***: The idea of this heuristic is based on another well-known spanning tree algorithm, namely Kruskal's algorithm that forms a spanning tree of a graph based on the weights associated with edges. As the underlying graph is unweighted, the selection of edges is completely random.
3. ***Random_Dijkstra's***: This heuristic implements Dijkstra's algorithm by considering unit weight on each edge. The vertices from set V are added to the set U (initially empty) on the basis of their distance from a fixed vertex u chosen randomly. At every step of the algorithm, a vertex not included in U and having minimum distance from u is obtained and added to U. The entire process is repeated until all vertices of V are included in U.
4. ***Max_degree_BFS***: This heuristic uses the well-known BFS algorithm to construct a spanning tree of a given graph. It explores all the vertices of the graph starting from a maximum degree vertex as root. The neighbor vertices are also traversed in decreasing order of their degrees. The process continues until all vertices are traversed. Preferring to visit higher degree vertices helps in keeping the neighbors close and hence may lead to a spanning tree with lower $Stretch$.
5. ***Random_BFS***: Spanning tree is produced as a result of BFS with neighbors being visited randomly.

3.2 Neighborhood Strategies

We have designed six neighborhood strategies for generating a neighbor of a given solution. These strategies are detailed out below.

1. **Method1 ($\mathbf{NBD}_1$)**: In this method, neighbors of solutions are generated based on the cycle exchange. An edge $(u, v) \in E(G)\backslash E(ST)$ is selected randomly and added to ST creating a cycle C. Now, $(u', v') \in C\backslash(u, v)$ is picked up randomly and removed from C resulting in a neighbor ST'. This method helps in diversification as the edges to be added and deleted are chosen randomly (see Algorithm 4).

Figure 2 illustrates this process. An edge (3, 4) belonging to G is added to its spanning tree ST which forms a cycle in ST. Now the edge (2, 5) appearing in this cycle is removed from ST producing a neighbor ST'.

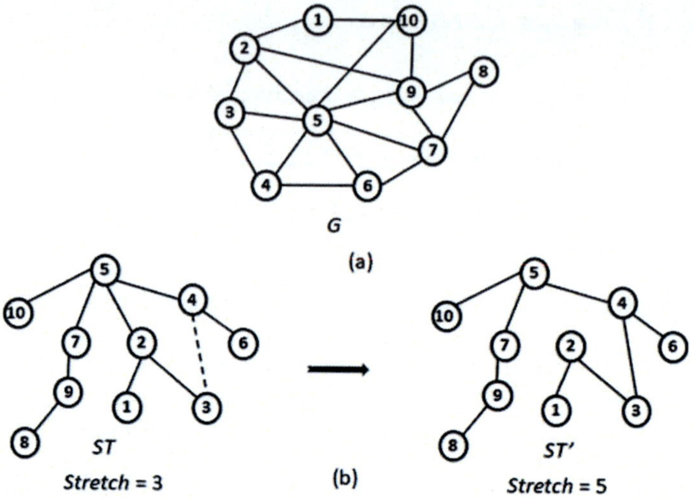

Fig. 2 **a** Graph G and its **b** Spanning tree ST with its neighbor ST' obtained from NBD_1

Algorithm 4 $\text{NBD}_1(ST)$

1: $(u, v) \leftarrow$ select randomly such that $(u, v) \in E(G) \backslash E(ST)$
2: $ST \leftarrow ST \cup (u, v)$
3: $C \leftarrow$ cycle obtained by adding edge (u, v) to ST
4: $(u', v') \leftarrow$ select randomly such that $(u', v') \in E(C)$
5: $ST' \leftarrow ST \backslash (u', v')$
6: **return** ST'

2. **Method2 ($\textbf{NBD}_2$)**: This method generates a neighbor of a spanning tree by replacing its subtree with another subtree of the graph (see Algorithm 5). Initially, a critical path CP in ST is selected randomly and a subgraph G' of G induced by the vertices of CP is formed. A spanning tree PT of G' is then generated using the heuristic *Random_Prim* described in Sect. 3.1. Now with the help of partial tree PT and the given ST, a neighbor ST' is obtained by adding those edges of ST to PT which are not in CP. This method favors the intensification as one of the critical paths is chosen for the replacement and hence may provide an improved solution.

This procedure is explained in Fig. 3. ST in Fig. 3a shows a spanning tree of G in Fig. 2a which has two critical paths $\{(1, 2, 3, 4, 5, 10), (2, 3, 4, 5, 7, 9)\}$ corresponding to $Stretch$ 5. Now, a path (1, 2, 3, 4, 5, 10) (shown in colored) is selected randomly and a subgraph G' of G is produced from its vertices. A spanning tree PT of G' is created using *Random_Prim* (Fig. 3b). This PT is transformed into a complete spanning tree ST' by adding those edges (shown with the dotted lines) to it from ST which are not in (1, 2, 3, 4, 5, 10) (Fig. 3c).

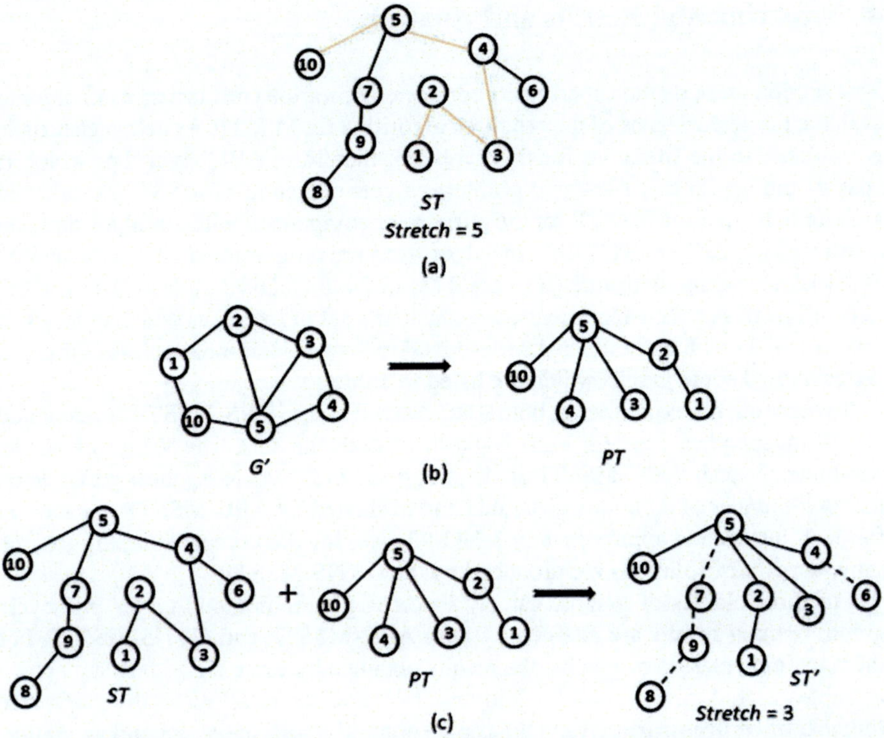

Fig. 3 **a** Spanning tree ST of G in Fig. 2a and its **b** subgraph G' induced by the vertex set of critical path CP and its spanning tree PT obtained using *Random_Prim*, **c** neighbor ST' of ST obtained from ST and PT using NBD_2

Algorithm 5 $NBD_2(ST)$

$CP \leftarrow$ critical path in ST selected randomly

$V' \leftarrow$ vertices in CP

$E(CP) \leftarrow$ edges in CP

$G' \leftarrow$ subgraph of G with vertex set V'

$PT \leftarrow$ spanning tree of G' generated using *Random_Prim*

$ST' \leftarrow PT \cup (u, v), \forall (u, v)$ such that $(u, v) \in E(ST) \setminus E(CP)$

return ST'

The remaining methods NBD_3 to NBD_6 are similar to NBD_2, where the partial tree PT is generated using the heuristics *Random_Kruskal*, *Random_Dijkstra's*, *Max_degree_BFS*, and *Random_BFS*, respectively.

4 Experimental Results and Analysis

This section presents the experiments conducted on various test instances in order to evaluate the performance of the proposed algorithm for MSSTP. As no metaheuristic is available in the literature for this problem; therefore, ABC algorithm which is state of the art for tree t-spanner problem (a generalization of MSSTP) is adapted and implemented for MSSTP for the purpose of comparison with our algorithm and is referred as ABC-MSSTP. Both the algorithms are programmed in C++ on ubuntu 16.04 LTS machine with Intel(R) Core(TM) i5-2400 CPU@3.10 × 4 GHz and 7.7 GiB of RAM. For the experiments, we consider a set of instances which consists of graphs with known optimal results (described in Sect. 2.1). For each class of this set, we generated some graphs which are listed in Table 1.

To carry out the experiments, both ABC-MSSTP and GVNS-MSSTP are executed for 10 independent runs for each instance of both the sets. The values of all the parameters used in ABC-MSSTP are kept as given in [7]. As using these parameters, the total number of solutions generated and evaluated by ABC-MSSTP in each run for each instance is approximately 3 lakh; hence, for the comparison purpose, the same number of solutions are produced by the GVNS-MSSTP.

For three classes of graphs, namely Petersen graph, diamond graph, and cycle graph, optimal results are attained by both ABC-MSSTP and GVNS-MSSTP. For the remaining classes of graphs, the results obtained by these algorithms are shown in Tables 2, 3, 4, 5, 6, 7, and 8. Columns '$|V|$' and 'optimal' mean the graph size (number of vertices in graph) and the known optimal results of corresponding graphs, respectively, in all the tables. Columns 'ABC-MSSTP' and 'GVNS-MSSTP' show the minimum $Stretch$ while columns 'Avg-ABC' and 'Avg-GVNS' show the average $Stretch$ obtained by ABC-MSSTP and GVNS-MSSTP, respectively, over 10 runs.

Table 1 Graphs with known optimal results

Graphs	Size	# instances
Petersen Graph (Pt)	$\lvert V\rvert = 10$	1
Diamond Graph $(K_{1,n-2,1})$	$4 \le \lvert V\rvert \le 120$	10
Cycle Graph (C_n)	$5 \le \lvert V\rvert \le 150$	10
Wheel Graph (W_n)	$5 \le \lvert V\rvert \le 150$	10
Complete Graph (K_n)	$5 \le \lvert V\rvert \le 100$	10
Split Graph (S_n)	$10 \le \lvert V\rvert \le 50$	10
Complete k-Partite Graph ($K_{n_1,n_2,\ldots,n_k}$)	$8 \le \lvert V\rvert \le 50$	10
Triangular Grid (T_n)	$10 \le \lvert V\rvert \le 136$	10
Rectangular Grid ($P_m \times P_n$)	$6 \le \lvert V\rvert \le 1080$	18
Triangulated Rectangular Grid $(TR_{m,n})$	$12 \le \lvert V\rvert \le 1500$	18

Table 2 Comparison of results obtained by GVNS-MSSTP and ABC-MSSTP for wheel graphs

Graphs	$\|V\|$	Optimal	ABC-MSSTP	GVNS-MSSTP	Avg-ABC	Avg-GVNS
W_5	5	2	2	***2***	2.0	2.0
W_7	7	2	2	***2***	2.0	2.0
W_{10}	10	2	3	***2****	3.0	2.1
W_{15}	15	2	3	***2****	3.0	2.0
W_{20}	20	2	3	***2****	3.1	2.0
W_{30}	30	2	4	***2****	4.0	2.3
W_{50}	50	2	5	***2****	5.0	2.6
W_{70}	70	2	6	***2****	6.0	3.4
W_{100}	100	2	7	***2****	7.3	3.8
W_{150}	150	2	9	***2****	9.3	4.0

Table 3 Comparison of results obtained by GVNS-MSSTP and ABC-MSSTP for complete graphs

Graphs	$\|V\|$	Optimal	ABC-MSSTP	GVNS-MSSTP	Avg-ABC	Avg-GVNS
K_5	5	2	2	***2***	2.0	2.0
K_7	7	2	2	***2***	2.0	2.0
K_9	9	2	2	***2***	2.2	2.0
K_{10}	10	2	2	***2***	2.5	2.0
K_{15}	15	2	4	***2****	4.0	2.0
K_{20}	20	2	4	***2****	4.0	2.1
K_{25}	25	2	4	***2****	4.5	2.0
K_{30}	30	2	5	***2****	5.8	2.1
K_{50}	50	2	6	***2****	6.5	2.0
K_{100}	100	2	9	***2****	9.2	2.8

Table 4 Comparison of results obtained by GVNS-MSSTP and ABC-MSSTP for split graphs

Graphs	$\|V\|$	Optimal	ABC-MSSTP	GVNS-MSSTP	Avg-ABC	Avg-GVNS
S_{10}	10	2	2	***2***	2.0	2.0
S_{12}	12	2	2	***2***	2.0	2.0
S_{12}	12	2	2	***2***	2.0	2.0
S_{15}	15	2	2	***2***	2.0	2.0
S_{15}	15	2	2	***2***	2.5	2.0
S_{20}	20	2	2	***2***	2.2	2.0
S_{35}	35	2	3	***2****	3.8	2.1
S_{35}	35	2	4	***2****	4.0	2.3
S_{50}	50	2	4	***2****	4.0	2.1
S_{50}	50	2	4	***2****	5.0	2.0

Table 5 Comparison of results obtained by GVNS-MSSTP and ABC-MSSTP for complete *k*-partite graphs

Graphs	$\|V\|$	Optimal	ABC-MSSTP	GVNS-MSSTP	Avg-ABC	Avg-GVNS
$K_{3,2,3}$	8	3	3	***3***	3.0	3.0
$K_{5,3,4,6}$	18	3	4	***3****	4.0	3.0
$K_{2,2,2,2,2,2,2,2,2,2}$	20	3	4	***3****	4.0	3.0
$K_{7,5,9,2}$	23	3	4	***3****	4.2	3.0
$K_{2,3,7,4,9}$	25	3	4	***3****	4.6	3.0
$K_{5,10,15}$	30	3	4	***3****	4.9	3.3
$K_{3,3,3,3,3,3,3,3,3,3}$	30	3	5	***3****	5.7	3.2
$K_{5,5,5,5,5,5,5}$	35	3	5	***3****	5.9	3.4
$K_{7,7,7,7,7,7,7}$	49	3	6	***3****	6.6	3.2
$K_{10,10,10,10,10}$	50	3	6	***3****	6.6	3.0

Table 6 Comparison of results obtained by GVNS-MSSTP and ABC-MSSTP for triangular grids

Graphs	$\|V\|$	Optimal	ABC-MSSTP	GVNS-MSSTP	Avg-ABC	Avg-GVNS
T_3	10	3	3	***3***	3.0	3.0
T_4	15	4	4	***4***	4.0	4.0
T_5	21	5	5	***5***	5.0	5.0
T_6	28	5	5	***5***	5.0	5.0
T_7	36	6	6	***6***	6.0	6.0
T_8	45	7	7	***7***	7.0	7.0
T_9	55	7	7	***7***	7.8	7.4
T_{10}	66	8	8	***8***	8.9	8.9
T_{11}	78	9	9	***9***	9.6	9.2
T_{15}	136	11	13	***11****	13.3	12.9

For the graphs, shown in Tables 2, 3, 4, 5, and 6, GVNS-MSSTP attains optimal values for all the instances (shown in italics); however, ABC-MSSTP attains optimal for few instances. For rectangular grids (see Table 7) and triangulated rectangular grids (see Table 8), GVNS-MSSTP attains optimal for all the instances of size ≤ 50. In particular, GVNS-MSSTP is able to achieve optimal in 85 cases out of 107, whereas ABC-MSSTP attains optimal values in 50 cases. The proposed algorithm is better than ABC-MSSTP in 47 cases (shown in bold with asterisk sign) and obtains same results in 58 cases (shown in bold). For the remaining 2 instances, ABC-MSSTP is better than GVNS-MSSTP. The 'mean_Dev' from optimal for all the instances is 13.22% in case of GVNS-MSSTP, whereas this value is 50.79% in case of ABC-MSSTP. However, the mean $Stretch$ values obtained by the two algorithms are almost similar for these instances.

Table 7 Comparison of results obtained by GVNS-MSSTP and ABC-MSSTP for rectangular grids

Graphs	$\|V\|$	Optimal	ABC-MSSTP	GVNS-MSSTP	Avg-ABC	Avg-GVNS
$P_2 \times P_3$	6	3	3	***3***	3.0	3.0
$P_2 \times P_5$	10	3	3	***3***	3.0	3.0
$P_2 \times P_{10}$	20	3	3	***3***	3.0	3.0
$P_5 \times P_{10}$	50	5	7	***5****	8.0	6.7
$P_9 \times P_{11}$	99	9	11	**11**	12.4	11.6
$P_2 \times P_{50}$	100	3	5	***3****	6.8	5.8
$P_4 \times P_{25}$	100	5	9	***5****	9.8	8.8
$P_5 \times P_{20}$	100	5	11	**9***	11.0	10.6
$P_{10} \times P_{10}$	100	11	13	**13**	13.0	13.0
$P_8 \times P_{13}$	104	9	11	**11**	12.8	12.3
$P_7 \times P_{15}$	105	7	11	**11**	12.2	11.8
$P_8 \times P_{120}$	960	9	33	**33**	36.0	34.6
$P_{10} \times P_{100}$	1000	11	37	**33***	38.6	33.6
$P_{20} \times P_{50}$	1000	21	45	**29***	46.2	36.2
$P_{25} \times P_{40}$	1000	25	43	45	46.0	47.8
$P_{30} \times P_{34}$	1020	31	45	47	46.4	48.2
$P_{15} \times P_{70}$	1050	15	45	**27***	45.6	32.2
$P_{12} \times P_{90}$	1080	13	41	**23***	43.2	35.4

A statistical comparison of results of GVNS-MSSTP and ABC-MSSTP on all the instances is also done using paired two-sample t-test with 5% level of significance which shows that there is a significant difference between the mean values of $Stretch$ of both the algorithms. From the results, it can be seen that in most of the cases GVNS-MSSTP performs better than ABC-MSSTP in terms of minimum as well as average value of $Stretch$. This comparison is also shown in Fig. 4 for some classes of graphs which clearly indicate the superiority of GVNS-MSSTP over ABC-MSSTP.

5 Conclusion

In this paper, a general variable neighborhood search (GVNS) is proposed for MSSTP which uses the well-known spanning tree algorithms for generating initial solution. Six problem-specific neighborhood techniques are designed which help in an exhaustive search of the solution space. Extensive experiments are conducted on various types of graphs in order to assess the performance of the proposed algorithm. Further, the results are compared with the adapted version of ABC initially

Table 8 Comparison of results obtained by GVNS-MSSTP and ABC-MSSTP for triangulated rectangular grids

Graphs	$\|V\|$	Optimal	ABC-MSSTP	GVNS-MSSTP	Avg-ABC	Avg-GVNS
$TR_{3,4}$	12	3	3	***3***	3.0	3.0
$TR_{4,4}$	16	4	4	***4***	4.0	4.0
$TR_{4,5}$	20	4	4	***4***	4.1	4.0
$TR_{4,6}$	24	4	5	***4****	5.0	4.9
$TR_{5,5}$	25	5	5	***5***	5.0	5.0
$TR_{5,7}$	35	5	6	***5****	6.0	5.8
$TR_{3,15}$	45	3	5	***3****	5.3	4.4
$TR_{5,10}$	50	5	7	***5****	7.0	7.0
$TR_{5,15}$	75	5	8	**7***	8.4	8.1
$TR_{10,15}$	150	10	13	**13**	13.8	13.6
$TR_{11,15}$	165	11	14	**13***	14.8	13.6
$TR_{20,25}$	500	20	26	**26**	27.8	27.2
$TR_{15,40}$	600	15	27	**26***	29.3	28.2
$TR_{20,30}$	600	20	29	**29**	30.4	29.8
$TR_{8,120}$	960	8	27	**15***	28.5	25.4
$TR_{33,40}$	1320	33	46	**41***	47.3	46.8
$TR_{35,40}$	1400	35	44	**43***	48.4	46.8
$TR_{30,50}$	1500	30	49	**45***	50.6	49.8

proposed for the tree t-spanner problem in the literature. Effectiveness of GVNS-MSSTP is clearly indicated through the results obtained by the two approaches in a majority of instances.

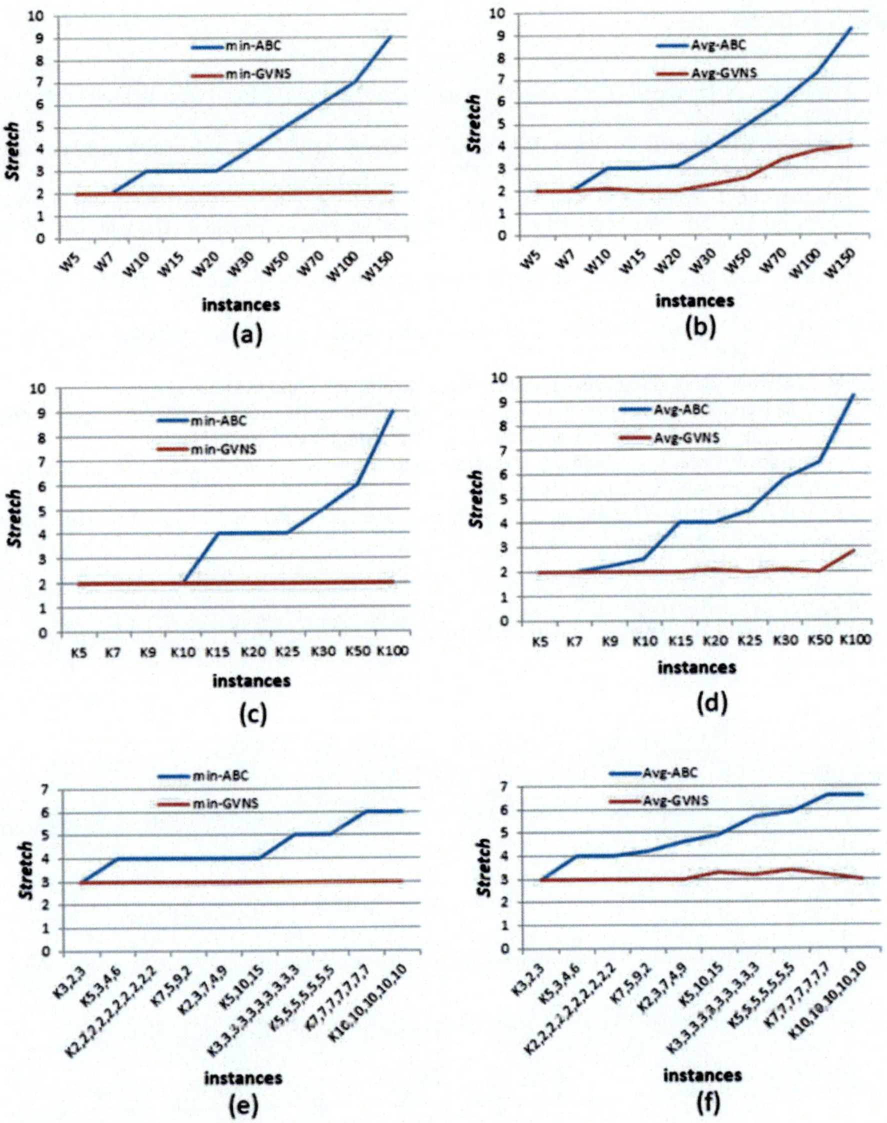

Fig. 4 Comparison of minimum *Stretch* values in (**a**), (**c**), (**e**) and average *Stretch* values in (**b**), (**d**), (**f**) obtained by ABC-MSSTP and GVNS-MSSTP over 10 runs for the instances of wheel graphs, complete graphs and complete *k*-partite graphs, respectively

References

1. Mladenovic N, Hansen P (1997) Variable neighborhood search. Comput Oper Res 24(11):1097–1100
2. Hansen P, Mladenovic N, Perez JAM (2010) Variable neighbourhood search: methods and applications. Ann Oper Res 175(1):367–407
3. Sanchez-Oro J, Pantrigo JJ, Duarte A (2014) Combining intensification and diversification strategies in VNS. An application to the Vertex Separation problem. Comput Oper Res 52:209–219
4. Peleg D, Ullman JD (1989) An optimal synchronizer for the hypercube. SIAM J Comput 18(4):740–747
5. Liebchen C, Wunsch G (2008) The zoo of tree spanner problems. Discrete Appl Math 156(5):569–587
6. Cai L, Corneil DG (1995) Tree spanners. SIAM J Discrete Math 8(3):359–387
7. Singh K, Sundar S (2018) Artificial bee colony algorithm using problem-specific neighborhood strategies for the tree t-spanner problem. Appl Soft Comput 62:110–118
8. Boksberger P, Kuhn F, Wattenhofer R (2003) On the approximation of the minimum maximum stretch tree problem. Technical report 409
9. Lin L, Lin Y (2017) The minimum stretch spanning tree problem for typical graphs. arXiv preprint arXiv:1712.03497

Tabu-Embedded Simulated Annealing Algorithm for Profile Minimization Problem

Yogita Singh Kardam and Kamal Srivastava

Abstract Given an undirected connected graph G, the profile minimization problem (PMP) is to place the vertices of G in a linear layout (labeling) in such a way that the sum of profiles of the vertices in G is minimized, where the profile of a vertex is the difference of its labeling with the labeling of its left most neighbor in the layout. It is an NP-complete problem and has applications in various areas such as numerical analysis, fingerprinting, and information retrieval. In this paper, we design a tabu-embedded simulated annealing algorithm for profile reduction (TSAPR) for PMP which uses a well-known spectral sequencing method to generate an initial solution. An efficient technique is employed to compute the profile of a neighbor of a solution. The experiments are conducted on different classes of graphs such as T^4-trees, tensor product of graphs, complete bipartite graphs, triangulated triangle graphs, and a subset of Harwell–Boeing graphs. The computational results demonstrate an improvement in the existing results by TSAPR in most of the cases.

Keywords Tabu search · Simulated annealing · Profile minimization problem

1 Introduction

Various methods in numerical analysis involve solving systems of linear equations which require to perform operations on the associated sparse symmetric matrices [1]. In order to reduce the computational effort as well as the storage space of such matrices, it is needed to rearrange the rows and columns of these matrices in such a way that the nonzero entries of the matrix should be as close to the diagonal as possible. With this objective, the profile minimization problem (PMP) was proposed

Y. S. Kardam (✉) · K. Srivastava
Dayalbagh Educational Institute, Dayalbagh, Agra 282005, India
e-mail: yogita.singh27@gmail.com

K. Srivastava
e-mail: kamal.sri@dei.ac.in

V. Singh et al. (eds.), *Computational Methods and Data Engineering*,
Advances in Intelligent Systems and Computing 1227,
https://doi.org/10.1007/978-981-15-6876-3_13

Fig. 1 Computing profile of a matrix M

$$\begin{array}{c} \\ 1 \\ 2 \\ 3 \\ 4 \\ 5 \\ 6 \end{array} \begin{array}{c} \begin{array}{cccccc} 1 & 2 & 3 & 4 & 5 & 6 \end{array} \\ \begin{bmatrix} 0 & 1 & 1 & 0 & 0 & 0 \\ 1 & 0 & 0 & 0 & 0 & 1 \\ 1 & 0 & 0 & 1 & 1 & 0 \\ 0 & 0 & 1 & 0 & 0 & 1 \\ 0 & 0 & 1 & 0 & 0 & 1 \\ 0 & 1 & 0 & 1 & 1 & 0 \end{bmatrix} \end{array} \begin{array}{c} j - r_j \\ (1-1=0) \\ (2-1=1) \\ (3-1=2) \\ (4-3=1) \\ (5-3=2) \\ (6-2=4) \end{array}$$

$$Profile(M)=0+1+2+1+2+4=10$$

[2]. Besides this original application, this linear ordering problem also has relevance in fingerprinting, archeology, and information retrieval [3].

Let M be a symmetric matrix of order n, then the profile of M ($Profile(M)$) is defined as $\sum_{j=1 \text{ to } n} (j - r_j)$. Here r_j denotes the index of first encountered nonzero entry in row j; i.e., $Profile(M)$ is the sum of profiles of each row of a matrix M and if $r_j > j$ for any jth row, then the profile of that row is 0.

Figure 1 shows the computation of profile of a matrix M of order 6. PMP for a matrix M seeks a permutation matrix Q such that $M' = Q \cdot M \cdot Q^{\mathrm{T}}$ has minimum profile. Note that here Q is the identity matrix of order n whose columns need to be permuted.

The PMP for a matrix M can be transformed into a graph-theoretic optimization problem by considering the nonzero entries of M as the edges of the graph G and the permutation of columns and rows with the swapping of labels of vertices in G. This relation can be seen in [2]. Based on this relation, PMP for graphs can formally be defined as follows:

Let $G = (V, E)$ be an undirected graph, where V and E are the set of vertices and edges, respectively. The layout of G is a bijective function $\Psi : V \rightarrow \{1, 2, \ldots, n\}$. Let $\Omega(G)$ denotes the set of all layouts of G, then the profile of G for a layout Ψ is $\mathrm{pf}(\Psi, G) = \sum_{u \in V} \left(\Psi(u) - \min_{v \in \overline{N_G(u)}} \Psi(v)\right)$, where $\overline{N_G}(u) = \{u\} \cup \{v \in V : (u, v) \in E\}$. PMP is to find a layout $\Psi^* \in \Omega(G)$ such that $pf(\Psi^*, G) = \min_{\forall \Psi \in \Omega(G)} pf(\Psi, G)$.

In Fig. 2, a graph layout Ψ is shown which corresponds to the matrix M given in Fig. 1 where each row of M maps to a vertex in this layout. The vertices A to F in the layout are labeled from 1 to 6. The profile of each vertex is the difference between its labeling and the labeling of its neighbor which is at the left to it in the layout and has minimum label value. As vertex A has no neighbor to its left, therefore $pf(A) = 0$. Vertex B has only one neighbor to its left with label 1, so $pf(B) = 1$. In the same way, profile of the remaining vertices is computed. The sum of profiles of each vertex yields the profile of G, i.e., 10.

In the literature, PMP is tackled using several approaches. In this paper, a tabu-embedded simulated annealing algorithm for profile reduction (TSAPR) for minimizing the profile of graphs is proposed that embeds the features of tabu search (TS) in the simulated annealing (SA) framework with a good initial solution generated

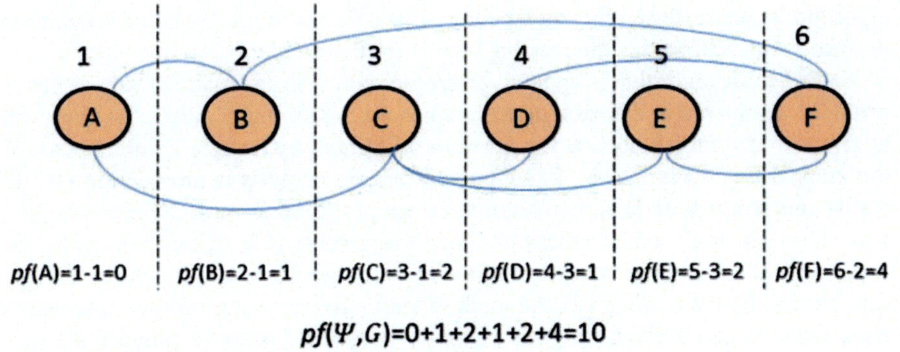

Fig. 2 Computing profile of a layout Ψ of a graph G

from spectral sequencing (SS) method [4]. SA algorithm [5, 6] is a local search method which accepts the worse solutions with some probability in order to avoid local optima. On the other hand, TS [7] prevents the process of getting stuck into local optima by a systematic use of memory. Both the techniques provide high-quality solutions and are broadly applied to a number of NP-hard combinatorial optimization problems in the literature. Experiments are conducted on different classes of graphs, and it is observed that the proposed algorithm gives better results than that of the existing approaches. The basic idea of TSAPR is to build a globally good solution using SS and to improve this solution locally using SA. A TS is also incorporated to avoid cycles of the same solutions and to explore more promising regions of the search space. The proposed TSAPR algorithm provides solutions which are not only comparable with the state-of-the-art metaheuristic but also improves the solutions of some of the instances tested by us.

The rest of the paper is organized as follows. In Sect. 2, existing approaches for the problem are given. Section 3 is dedicated to the proposed algorithm. The experimental results obtained from TSAPR are discussed in Sect. 4. Finally, the paper is concluded in Sect. 5.

2 Existing Approaches

The PMP is an NP-complete problem [8] which was introduced in fifties due to its application in numerical analysis. Since then, this problem has gained much attention of the researchers. In [9], a direct method to obtain a numbering scheme to have a narrow bandwidth is presented which uses level structures of the graph. An improved version of this method is given in [10] which reverses the order of the numbering obtained in [9]. Some more level structure-based approaches are developed in [11] and [12] which provide comparable results in significantly lower time. In [13], an algorithm to generate an efficient labeling for profile and frontal solution schemes is designed which improves the existing results for the problem.

In [14], a SA algorithm is applied for profile reduction and claimed to have better profile as compared to the existing techniques for the problem. A spectral approach is given in [15] which obtains the labeling of a graph by using a Fiedler vector of the graph's Laplacian matrix. In support of this, an analysis is provided in [10] to justify this strategy. In [16], two algorithms are proposed for profile reduction. The first algorithm is an enhancement of Sloan's algorithm [13] in solution quality as well as in run time. The second one is a hybrid algorithm which reduces the run time further by combining spectral method with a refinement procedure that uses a modified version of Sloan's algorithm. In [17], different ways to enhance the efficiency and performance of Sloan's algorithm are considered by using supervariables, weights, and priority queue. A hybrid approach that combines spectral method with Sloan's algorithm is also examined. Profiles of triangulated triangle graphs have been studied by [18]. Another algorithm for reducing the profile of arbitrary sparse matrices is developed by [19]. Exact profiles of products of path graphs, complete graphs, and complete bipartite graphs are obtained by [20]. A systematic and detailed survey of the existing approaches for PMP is given in [21]. A scatter search metaheuristic has been proposed by [3] which uses the network structure of the problem for profile reduction. In this, path relinking is used to produce new solutions and a comparison is done with the best heuristics of literature, i.e., RCM [10] and SA [14]. An adaptation of tabu search for bandwidth minimization problem (TS-BMP) and the two general-purpose optimizers are also used for comparison. Recently, a hybrid SA algorithm HSAPR is designed for this problem [22]. Besides the heuristic approaches, researchers from combinatorics community have attempted to find/prove exact optimal profiles of some classes of graphs [20, 23, 24].

3 Tabu-Embedded Simulated Annealing Algorithm (TSAPR) for Minimizing Profile of Graphs

This section first describes an efficient method to compute the profile of a neighbor of a given solution (layout). Then, the proposed algorithm TSAPR designed for PMP is explained in detail.

3.1 Efficient Profile Computation of a Neighbor of a Solution

Each time when a neighbor of a solution is produced, it is needed to compute its profile. Since in the proposed algorithm a neighbor Ψ' of a solution Ψ is generated by swapping the label of a vertex u with the label of a vertex v in Ψ and as the computing profile of each vertex again in the Ψ' is expensive, so instead of computing it for each vertex, it is evaluated only for the swapped vertices and for the vertices which are adjacent to them. It helps in reducing the computational efforts and is done using the following gain function $g_{\Psi}(u, v)$:

$$g_{\Psi}(u, v) = CB_{\Psi}(u, v) - CB_{\Psi'}(u, v)$$

where $CB_{\Psi}(u, v)$ is the contribution of vertices u and v in profile evaluation of a solution Ψ and is defined as:

$$CB_{\Psi}(u, v) = \sum_{r \in \overline{N_G}(u)} pos_{\Psi}(r) + \sum_{s \in \overline{N_G}(v/u)} pos_{\Psi}(s)$$

where

$$pos_{\Psi}(r) = \Psi(\mathrm{r}) - min_{p \in \overline{N_G}(r)} \Psi(p)$$

and

$$\overline{N_G}(v/u) = \overline{N_G}(v) - \overline{N_G}(u) = \left\{y \in \overline{N_G}(v) \wedge y \notin \overline{N_G}(u)\right\}$$

3.2 TSAPR Algorithm

The TSAPR algorithm designed for PMP is outlined in Algorithm 1. With initial values of maximum number of iterations max_iter, number of neighbors nbd_size, and the initial temperature t (Step 1), TSAPR starts with generating an initial solution Ψ using a well-known spectral sequencing method [4] (Step 3). In the iterative procedure (Steps 5–23) of the algorithm, initially the nbd_size number of neighbors is generated of Ψ with the help of a randomly selected vertex v using a function $Neighbor\_gain$ (given in Algorithm 2) which returns a triplet containing the vertices u, v and the gain value corresponding to these vertices (Step 9). Note that here the vertex u is the vertex which gives maximum gain in profile when swapped with vertex v in the solution Ψ. The key idea of tabu search is used by keeping the record of visited solutions in $visited$ in each iteration. Thus, repeated moves are forbidden during an iteration that helps in avoiding cycles (Steps 2-6 of Algorithm 2). From these nbd_size number of triplets, the one giving the maximum gain $(x^*, y^*, gain_{\Psi'}(x^*, y^*))$ is selected (Steps 11 and 12) which is used to decide the Ψ' for the next iteration. If for a randomly generated number ρ (generated with a uniform distribution between 0 and 1), either $\exp(gain_{\Psi'}(x^*, y^*)/t) > \rho$ or the gain is found positive [25], then the neighbor Ψ' corresponding to the vertices x^* and y^* is considered for the next iteration (Steps 13 and 14). Then, the temperature is reduced using a geometric cooling schedule for the next iteration (Step 21). In $global\_best$, the record of the best solution obtained so far is maintained throughout this procedure. The process continues until $count$ exceeds the max_iter.

4 Results and Discussion

With an aim to test the efficiency of the proposed TSAPR algorithm for PMP, the experiments are conducted on various graphs. Also, the results are analyzed and compared with scatter search algorithm, HSAPR algorithm and with the existing best-known results for the problem. For the experiments, a machine with Windows 8 operating system with 4 GB of RAM and with an intel(R) Core (TM) i3-3110 M CPU 2.40 GHz is used and the algorithm is coded in MATLAB R2010a. For the comparison purpose, TSAPR is run on the same machine for all the instances of test set. The experiments are performed on different kinds of data sets which are described in the following subsection.

Algorithm 1 Tabu Embedded Simulated Annealing Algorithm for Profile Reduction (TSAPR)

```
Set the values of maximum number of iterations max_iter, number of neighbors
  nbd_size and the initial temperature t
count ← 1
Ψ ← solution generated using Spectral Sequencing
global_best ← Ψ
while (count ≤ max_iter) do
    Set visited(u, v) = 0 for each u, v ∈ Ψ
    for i ← 1 to nbd_size do
    v ← a vertex in Ψ selected randomly
    [v, u, gain_value_i] ← Neighbor_gain(Ψ, v)
    end for
    Neighbor_gain_best ← max_{i=1,...,nbd_size}{gain_value_i}
    (x*, y*) ← best favorable pair corresponding to Neighbor_gain_best
    if Neighbor_gain_best > 0 or exp(Neighbor_gain_best)/t > ρ then
        Ψ' ← swap(Ψ, x*, y*)*
    if Ψ' is better than global_best then
    global_best ← Ψ'
    count ← 0
    end if
    Ψ ← Ψ'
    end if
    t ← t × α
    count ← count + 1
end while
return global_best
```

Algorithm 2 $Neighbor\_gain(\Psi, p)$

for $i \leftarrow 1$ to $n = |V|$ **do**
    if $visited(p, i) = 0$ **then**
        $\Psi' \leftarrow swap(\Psi, p, i)*$
        $g_{\Psi}(p, i) = CB_{\Psi}(p, i) - CB_{\Psi'}(p, i)$
        $visited(p, i) \leftarrow 1$
    end if
end for
return $\left(p, i^{*}, max_{i=1,\ldots,n} g_{\Psi}(p, i)\right)$ ▷ i^{*} corresponds to the vertex which gives max gain
$*$ $swap(\Psi, r, s)$ returns a neighbor of Ψ obtained after swapping the labels of any two vertices r and s in Ψ

4.1 Test Set

Two types of graphs are considered for the experiments which are given below:

1. **Graphs with known optimal results**: It consists of trees with Diameter 4 (T^4 graphs) [23], complete bipartite graphs [24] and tensor product of graphs [20].
   (a) T^4 Graphs: This set contains 91 instances with $10 \leq |V| \leq 100, 9 \leq |E| \leq 99$ [3].
   (b) Complete bipartite graphs: This set has 98 instances with $4 \leq |V| \leq 142, 3 \leq |E| \leq 5016$.
   (c) Tensor product of graphs: This set consists of 10 instances with $45 \leq |V| \leq 180, 144 \leq |E| \leq 9450$.
2. **Graphs with unknown optimal results**: This class contains triangulated triangle (TT_l) graphs and Harwell–Boeing (HB) graphs.
   (a) Triangulated triangle graphs: This set has 4 instances with $21 \leq |V| \leq 1326$ and $44 \leq |E| \leq 3825$ [18].
   (b) Harwell–Boeing graphs: This set consists of 35 instances with $|V| \leq 443$, a subset of HB graphs which are used in [3].

4.2 Tuning of Parameters

The different parameters, namely initial temperature t, number of neighbors nbd_size, cooling rate α of the proposed algorithm are tuned by conducting the experiments on a subset of 10 instances of HB graphs.

1. **Initial temperature** (t): This is an important parameter of SA algorithm. In order to decide initial temperature t, we have conducted experiments with $t = 50$, 60,

70, 80, 90, 100. The choice of initial temperature is based on the concept that amount of exploration directly depends on the temperature as higher temperature means a higher probability of accepting bad solutions in the initial iterations. A statistical comparison of the profiles of 10 instances, done by ANOVA and Tukey's HSD test shows that there is no significant difference among all the 6 values of t. Thus, the value of t is set 100 with the aim of maximum exploration of solution space.

2. **Number of neighbors (*nbd_size*)**: After setting the temperature t, *nbd_size* number of neighbors of a solution are generated. This parameter is set by taking $nbd\_size = |V|/2, |V|/4$ and $|V|/6$. The *nbd_size* depends on number of vertices, since more neighbors need to be explored for a larger graph for a better exploitation of a solution. The experimental results obtained by setting these values of *nbd_size* are compared statistically using ANOVA but no significant difference in quality is observed among them. The Tukey's HSD test shows that these are not significantly different pairwise also. Thus, $nbd\_size = |V|/6$ is set as the average computation time is the least for this over the 10 instances.
3. **Cooling rate (α)**: For updating the temperature, geometric cooling schedule is used which starts with an initial temperature t and after each iteration the current temperature is decreased by a factor of α using $t \leftarrow t \times \alpha$. To tune the value of α, experiments are conducted by taking $\alpha = 0.90.\ 0.93, 0.96, 0.99$. In order to investigate the performance difference among these values, ANOVA is used which shows that there is a significant difference in the profiles obtained for these values of α. For pairwise comparison Tukey's HSD test is used, and it is found that $\alpha = 0.90, 0.96$ and $\alpha = 0.90, 0.99$ are significantly different from each other. Table 1 shows the average profiles, average deviation of profiles from the best-known values so far and the average computation time over 30 runs for each value of α on the instances of representative set. Since the average of profile values is less for $\alpha = 0.96$, so for the final experiments this value of α is used.
4. **Maximum number of iterations (*max_iter*)**: The algorithm terminates after performing maximum number of iterations *max_iter*. Initially, the value of this parameter is set very high so that a large number of good solutions can be explored. From the experiments, it is observed that after $\max\_iter = 200$ the profile value becomes constant (Fig. 3).

The final experiments are conducted using these values of control parameters in the proposed algorithm TSAPR.

Table 1 Experimental results for different values of α

Cooling rate α	0.90	0.93	0.96	0.99
Average profile value	1269.1	1257.3	1245.5	1250.2
Average deviation from best (%)	4.99	4.43	3.86	4.17
Average time (in s)	1621.4	2344.4	3722.2	4576.2

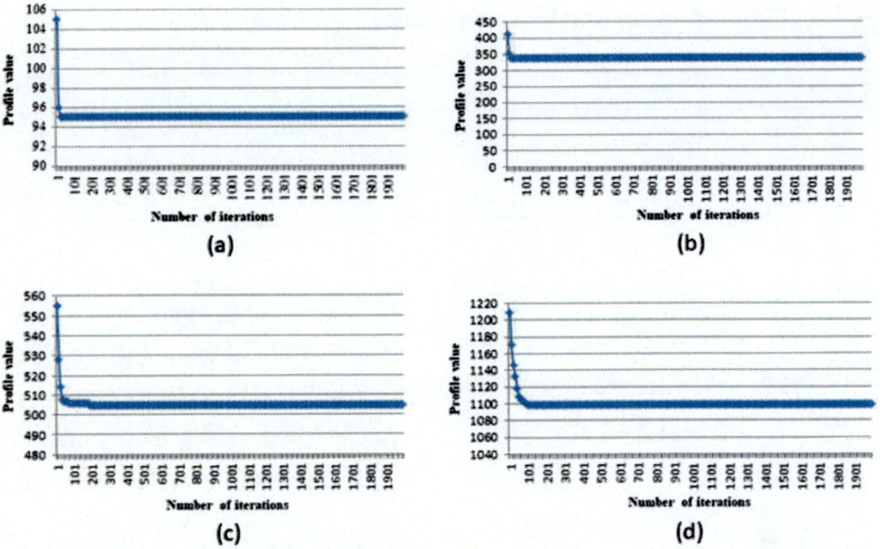

Fig. 3 Number of iterations versus profile graphs of **a** can24, **b** can61, **c** ash85, and **d** can96 by TSAPR

4.3 *Final Experiments*

For T^4 trees and complete bipartite graphs, the optimal results are achieved by TSAPR algorithm as well as by the HSAPR algorithm for all the instances of both the graphs. For tensor product of graphs also both the algorithms attain optimal results for all the 10 instances which are shown in Table 2.

For triangulated triangle graphs, the performance of TSAPR algorithm is same as that of HSAPR algorithm. Table 3 shows the results obtained by TSAPR on the instances of this class of graphs. From the table, it can be seen that TSAPR not only attains the profile given by Guan and Williams (GW) [18] but for TT_{10} and TT_{20} (marked with asterisk) is able to lower the profile further. Figure 4a, b shows the

Table 2 Known optimal results for the instances of tensor product of graphs

Graphs	#vertices	#edges	Optimal
$K_5 \otimes K_9$	45	720	946
$P_5 \otimes K_9$	45	144	428
$K_7 \otimes K_7$	49	882	1143
$K_6 \otimes K_9$	54	1080	1385
$P_6 \otimes K_9$	54	360	544
$P_8 \otimes K_8$	64	392	609
$K_8 \otimes K_{10}$	80	2520	3101
$P_{10} \otimes K_{10}$	100	810	1251
$K_{10} \otimes K_{15}$	150	9450	11052
$P_{12} \otimes K_{15}$	180	2310	3514

Table 3 TSAPR results on triangulated triangle graphs

Graphs	#vertices	#edges	GW	TSAPR
TT_5	21	44	72	72
TT_{10}	66	165	401	**400***
TT_{20}	231	630	2585	**2583***
TT_{50}	1326	3825	34,940	34,940

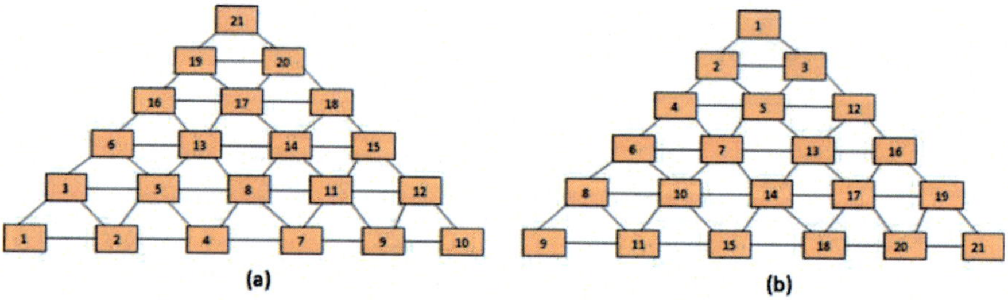

Fig. 4 **a** TSAPR ordering and **b** GW ordering for TT_5, Profile = 72

labeling patterns obtained from TSAPR and GW ordering scheme, respectively, for the graph TT_5.

To examine how TSAPR works on a wide range of graphs, it is also applied to a subset of graphs from the Harwell–Boeing matrix collection. Table 4 shows the results obtained from the scatter search algorithm [3] and those obtained by HSAPR algorithm [22] on HB graphs. The results of the proposed TSAPR algorithm are shown in last column. The best-known results [3] so far for these instances are shown in column "best known." In last column, the values in bold show that TSAPR either achieves the same results as that of best or improves the results of scatter search or of the HSAPR, whereas the values in bold with asterisk sign show the improvement of TSAPR over all the existing methods in the literature.

Figures 5, 6, 7, 8, and 9 show the change in profile values of some HB graphs before and after applying the proposed algorithm. Here, each dot represents a nonzero entry in the adjacency matrix of a given graph. The correctness of the algorithm can also be seen from these spy graphs as the nonzero entries are coming closer to the diagonal by applying TSAPR.

5 Conclusion

In this paper, a hybrid algorithm for reducing the profile of graphs is presented that combines TS and SA algorithms. On one hand, the initial solution is produced by spectral sequencing method for SA that helps in accelerating the search process, and

Table 4 Comparison of TSAPR with best-known results

Graphs	#vertices	best known	Scatter Search	HSAPR	TSAPR
can24	24	95	95	95	**95**
bcspwr01	39	82	82	83	83
bcsstk01	48	460	466	462	**461**
bcspwr02	49	113	113	113	**113**
dwt59	59	214	223	214	**214**
can61	61	338	338	338	**338**
can62	62	172	172	178	**174**
dwt66	66	127	127	127	**127**
bcsstk02	66	2145	2145	2145	**2145**
dwt72	72	147	151	150	**150**
can73	73	520	520	523	523
ash85	85	490	490	491	491
dwt87	87	428	434	428	**428**
can96	96	1078	1080	1083	**1082**
nos4	100	651	651	653	**651**
bcsstk03	112	272	272	272	**272**
bcspwr03	118	434	434	434	**433***
bcsstk04	132	3154	3159	3195	**3162**
can144	144	969	969	969	**969**
bcsstk05	153	2191	2192	2196	**2195**
can161	161	2482	2482	2534	**2509**
dwt162	162	1108	1286	1117	**1117**
can187	187	2184	2195	2163	**2163**
dwt193	193	4355	4388	4308	**4270***
dwt198	198	1092	1092	1097	**1096**
dwt209	209	2494	2621	2604	**2567**
dwt221	221	1646	1646	1668	**1644***
can229	229	3928	4141	3961	**3953**
dwt234	234	782	803	820	820
nos1	237	467	467	467	**467**
dwt245	245	2053	2053	2119	**2115**
can256	256	5049	5049	5041	**4969***
can268	268	5215	5215	5005	**4936***
plat362	362	9150	10,620	8574	**8489***
bcspwr05	443	3076	3354	3121	**3121**

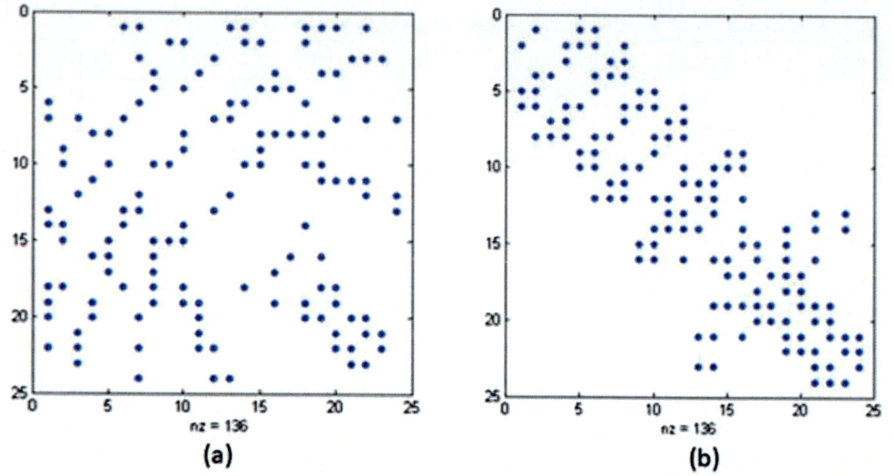

Fig. 5 Sparsity graph of can24 **a** before applying TSAPR with profile = 238 and **b** profile reduced to 95 after applying TSAPR

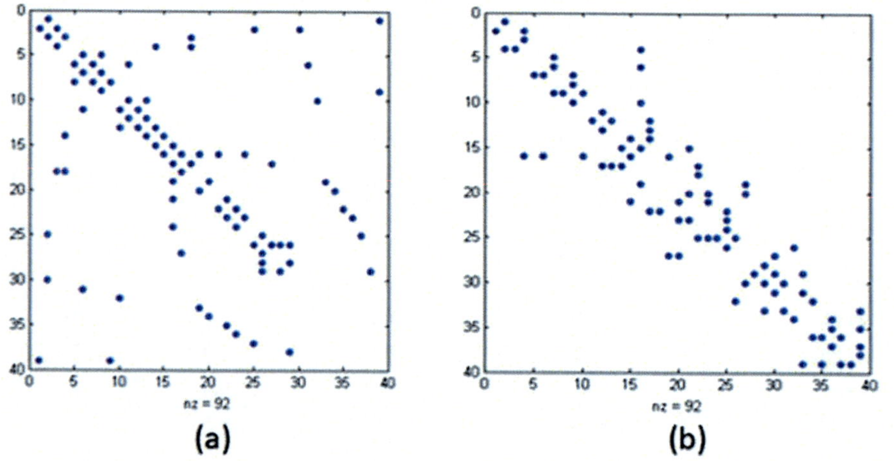

Fig. 6 Sparsity graph of bcspwr01 **a** before applying TSAPR with profile = 292 and **b** profile reduced to 83 after applying TSAPR

on the other hand, TS uses memory structure to tabu the visited solutions for a short-term period in order to explore the search space. Hence, a balance between exploration and exploitation is maintained by using this hybrid approach. The performance of the proposed algorithm is assessed on some standard benchmark graphs. The results show that the algorithm is able to attain not only the best profiles known so far in majority of cases but also improves the profiles further in some of the cases. For future work, some intelligent local improvement operators can be designed which cannot

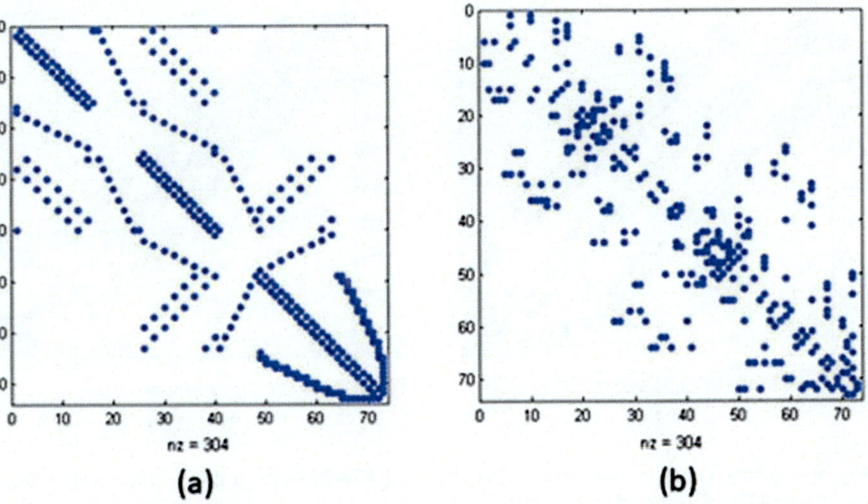

Fig. 7 Sparsity graph of can73 **a** before applying TSAPR with profile = 797 and **b** profile reduced to 523 after applying TSAPR

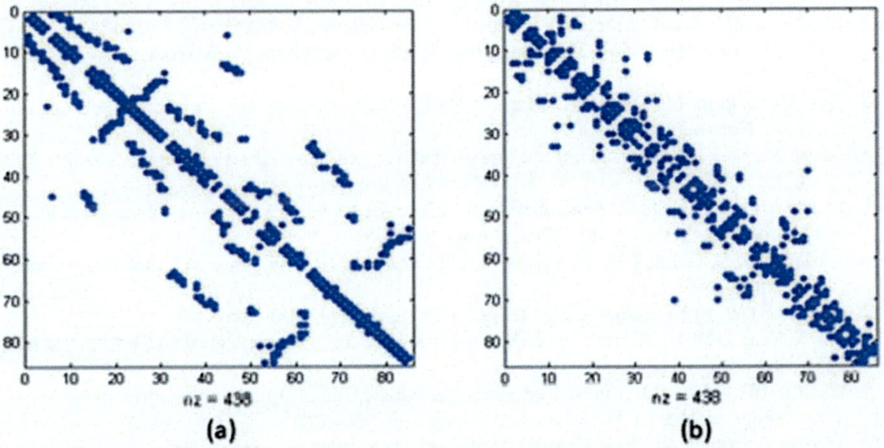

Fig. 8 Sparsity graph of ash85 **a** before applying TSAPR with profile = 1153 and **b** profile reduced to 491 after applying TSAPR

only accelerate the entire procedure but can also enhance the quality of solutions. The test suite can also be enriched further by considering more benchmark graphs.

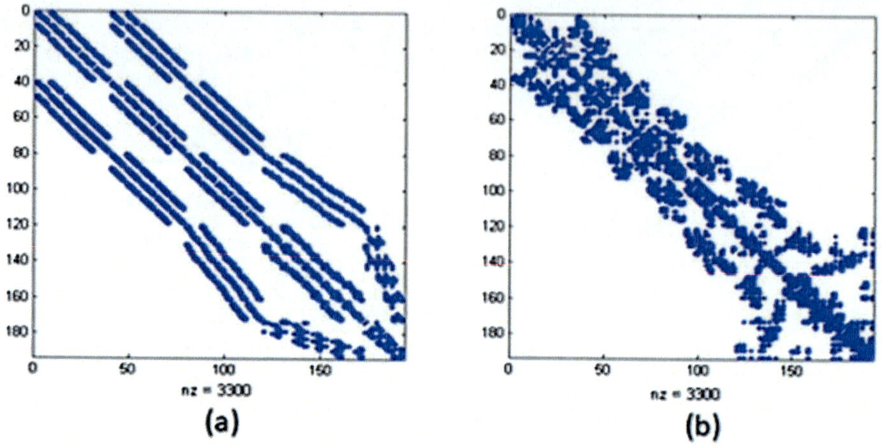

Fig. 9 Sparsity graph of dwt193 **a** before applying TSAPR with profile = 7760 and **b** profile reduced to 4270 after applying TSAPR

References

1. Saad Y (2003) Iterative methods for sparse linear systems. SIAM 82
2. Diaz J, Petit J, Serna M (2002) A survey of graph layout problems. ACM Comput Surv 34:313–356
3. Oro JS, Laguna M, Duarte A, Marti R (2015) Scatter search for the profile minimization problem. Networks 65(1):10–21
4. Juvan M, Mohar B (1992) Optimal linear labelings and eigenvalues of graphs. Discrete Appl Math 36:153–168
5. Cerny V (1985) A thermodynamical approach to the traveling salesman problem: an efficient simulation algorithm. J Optim Theory Appl 45:41–51
6. Kirkpatrick S, Gelatt CD, Vecchi MP (1983) Optimization by simulated annealing. Science 220(4598):671–680
7. Glover F (1989) Tabu search: Part I. ORSA J Comput 1(3):190–206
8. Lin Y, Yuan J (1994) Minimum profile of grid networks in structure analysis. J Syst Sci Complex 7:56–66
9. Cuthill EH, Mckee J (1969) Reducing the bandwidth of sparse symmetric matrices. In: Proceedings of 24th ACM National conference, pp 157–172
10. George A, Pothen A (1997) An analysis of spectral envelope reduction via quadratic assignment problems. SIAM J Matrix Anal Appl 18(3):706–732
11. Gibbs NE, Poole Jr WG, Stockmeyer PK (1976) An algorithm for reducing the bandwidth and profile of a sparse matrix. SIAM J Numer Anal 13(2):236–250
12. Lewis JG (1982) Implementation of the Gibbs-Poole-Stockmeyer and Gibbs-King algorithms. ACM Trans Math Softw 8:180–189
13. Sloan SW (1986) An algorithm for profile and wavefront reduction of sparse matrices. Int J Numer Meth Eng 23(2):239–251
14. Lewis RR (1994) Simulated annealing for profile and fill reduction of sparse matrices. Int J Numer Meth Eng 37(6):905–925
15. Barnard ST, Pothen A, Simon H (1995) A spectral algorithm for envelope reduction of sparse matrices. Numer Linear Algebra Appl 2(4):317–334
16. Kumfert G, Pothen A (1997) Two improved algorithms for envelope and wavefront reduction. BIT Numer Math 37(3):559–590

17. Reid JK, Scott JA (1999) Ordering symmetric sparse matrices for small profile and wavefront. Int J Numer Meth Eng 45(12):1737–1755
18. Guan Y, Williams KL (2003) Profile minimization on triangulated triangles. Discrete Math 260(1–3):69–76
19. Ossipov P (2005) Simple heuristic algorithm for profile reduction of arbitrary sparse matrix. Appl Math Comput 168(2):848–857
20. Tsao YP, Chang GJ (2006) Profile minimization on products of graphs. Discrete Math 306:792–800
21. Bernardes JAB, Oliveira SLGD (2015) A systematic review of heuristics for profile reduction of symmetric matrices. In: Procedia Comput Sci ICCS, 221–230
22. Kardam YS, Srivastava K, Sharma R (2017) Minimizing profile of graphs using a hybrid simulating annealing algorithm. Electron Notes Discrete Math 63:381–388
23. Lin Y, Yuan J (1994) Profile minimization problem for matrices and graphs. Acta Mathematicae Applicatae Sinica 10(1):107–112
24. Lai YL, Williams K (1999) A survey of solved problems and applications on bandwidth, edgesum, and profile of graphs. J Graph Theory 31(2):75–94
25. Metropolis N, Rosenbluth AW, Rosenbluth MN, Teller AH, Teller E (1953) Equation of state calculations by fast computing machines. J Chem Phys AIP 21(6):1087–1092

Deep Learning-Based Asset Prognostics

Soham Mehta, Anurag Singh Rajput, and Yugalkishore Mohata

Abstract In this highly competitive era, unpredictable and unscheduled critical equipment failures result in a drastic fall in productivity and profits leading to loss of market share to more productive firms. Due to Industry 4.0 and large-scale automation, many plants have been equipped with complex, automatic and computer-controlled machines. The conventional run-to-failure maintenance system, i.e., repairing machines after complete failure lead to unexpected machine failures where the cost of maintenance and associated downtime is substantially high, especially for unmanned, automatic equipment. Ineffective maintenance systems have a detrimental effect on the ability to produce quality products that are competitive in the market. In this context, an effective asset prognostics system which accurately estimates the Remaining Useful Life (RUL) of machines for pre-failure maintenance actions assumes special significance. This paper represents a deep learning approach based on long short-term memory (LSTM) neural networks for efficient asset prognostics to save machines from critical failures. The effectiveness of the proposed methodology is demonstrated using the NASA-CMAPSS dataset, a benchmark aero-propulsion Engine maintenance problem.

Keywords Asset prognostics · Remaining Useful Life (RUL) · Long short-term memory (LSTM) neural network · NASA-CMAPSS

1 Introduction

In this age of Industrial 4.0 and Internet of Things, many plants are equipped with complex, computer-controlled and automatic equipment. However, to reduce further capital expenditures, companies use simple maintenance regimes which lead to costly failures of complex, automatic equipment, high downtime and large inventories of equipment spare parts, leading to a drastic increase in operating costs. The decision

S. Mehta (✉) · A. S. Rajput · Y. Mohata
Department of Industrial Engineering, Pandit Deendayal Petroleum University, Gandhinagar, Gujarat, India
e-mail: mehtasoham9@gmail.com

V. Singh et al. (eds.), *Computational Methods and Data Engineering*, Advances in Intelligent Systems and Computing 1227,
https://doi.org/10.1007/978-981-15-6876-3_14

of not investing in an effective maintenance system to save capital in the short term leads to increased operating costs in terms of costly failures, increased unexpected downtime and reduced productivity, thereby consuming higher capital in the longer time horizons. In this context, a maintenance regime consisting of an effective asset prognostic system is of utmost significance to minimize operating costs and maximize productivity. Maintenance systems can broadly be classified into three main categories—(1) corrective system, (2) preventive system and (3) predictive system [1, 2]. In corrective maintenance system, maintenance is done as to identify, isolate and rectify a fault such that equipment or the system can be restored to operational condition. It involves repairing the equipment after the failure of the equipment. This causes an increased system downtime and increased inventory of spare parts. This type of system cannot be used in a product layout-based manufacturing organization as it can lead to downtime of complete manufacturing unit.

Preventive maintenance system is maintenance activities carried out on systems on scheduled basis in order to keep the system up and running. It has some benefits like less unplanned downtime and reduced risk of injury but on the other side, it has some major drawbacks like unnecessary corrective actions leading to large economic losses in terms of manpower, equipment and other resources.

Predictive maintenance system involves asset prognosis to determine Remaining Useful Life (RUL) of machines and implementing appropriate maintenance actions before the failure of the equipment. This leads to reduction in equipment repair and downtime costs as the equipment is saved from critical failures.

Asset prognostics involves accurate estimation of the RUL of equipment. RUL of equipment or a system is a technical term used to describe the length of equipment useful life from current time to the equipment or system failure [3, 4]. A known RUL can directly help in operation scheduling and reduction of resources utilized. An accurate RUL prediction can aid an organization in saving equipment from critical failure and help it in achieving the goal of zero system downtime or equipment failure [4]. It can further lead to decreased inventory level of spare parts of equipment, increased service life of parts, improved operator safety, negligible or no emergency repair, provides better product quality and at last, an increased production capacity. Thus, an effective asset prognosis system can be of great help for the organization to become lean.

The proposed research work focuses on the development of a deep learning-based asset prognostics system which utilizes long short-term memory (LSTM) neural networks for determining the RUL of machines.

2 Literature Review

Models for RUL prediction can broadly be categorized into two types—physics driven model and data-driven model [3, 5, 6]. Physics-driven models use domain knowledge, component properties and physical behavior of system to generate mathematical model for failure mechanisms. Whereas data-driven models provide a better approximation about equipment or the system failure based on historical and live sensor data. This approach can easily be applied when historical performance data is available or the data is easy to collect. It involves usage of different machine learning algorithms in order to learn degradation patterns [7]. This approach is much more feasible and capable of learning the degradation process if sufficient data is available. Meanwhile, it is more difficult to establish precise physics-driven model for complex systems comprising of sophisticated machines.

In the literature, many researchers have utilized techniques such as Hidden Markov Models, Kalman Filter, Support Vector Machine (SVM), Dynamic Bayesian Networks, and Particle Filter [7–9]. However, getting accurate and timely estimates of RUL is still a challenging task due to the changing dynamic conditions. Deep learning is one of the emerging fields of study that may give the best solution to this problem. One of the major advantages of deep learning is that domain knowledge is not required to obtain the feature set whereas feature engineering is the most important step of machine learning-based methods. Hence, efforts need to be undertaken in the selection and design of deep learning-based techniques rather than feature engineering. The researcher generally tries to use an optimum feature set that reduces run time and improves prediction performance while preserving the RUL causation factors. A larger feature set does not guarantee an improved performance as features that do not correlate with the target variable acts as noise, thereby reducing the performance. Hence an optimum feature set is chosen.

A deep neural network (DNN) consists of numerous layers of neurons such that every neuron in the current layer is connected to every neuron in the preceding and succeeding layer. Hence, another advantage offered by DNN is that for the same number of independent variables, a large number of parameters are computed at each layer of the network as compared to traditional machine learning-based methods. This helps deep neural networks learn highly complex functions thereby giving out highly accurate predictions.

System data acquired by the sensors is in chronological order with a certain time rate. Hence, it is imperative to harness this sequential data as there is information in the sequence of data itself. In order to accurately predict the RUL, it becomes imperative to determine the degradation that has occurred till date. Hence, the problem involves sequential modeling. Machine learning-based techniques do not take recital all information build due to the progressive sequence over time. Other techniques like Hidden Markov Models (HMM), Kalman Filtering are widely used by the researchers. However, both techniques assume that the probability to move from one state to another is constant, i.e., the failure rate of the system/component is constant [5]. This acutely limits the ability of the model to generate accurate RUL predictions.

This paper proposes a deep learning-based asset prognostic system which utilizes long short-term memory (LSTM) neural networks to obtain highly accurate RUL predictions of machines using machine sensor data. The main benefit of using LSTM over other methods is that they are able to learn the long-term dependencies in the time series. The proposition is demonstrated, discussed and tested by creating a LSTM model on the NASA C-MAPSS dataset, a health monitoring dataset of aero-propulsion engines provided by NASA.

3 LSTM Deep Learning

Deep learning is a branch of machine learning that uses deep neural networks (DNN) to make predictions and/or classifications by determining the relationship between dependent and independent variables. A neural network with more than two hidden layers is referred to as deep neural network (DNN).

In a DNN, the input layer of neurons receives the input data, hidden layer performs computations and the output layer of neurons gives the output. The neurons are connected to each other via links to transmit signals. Each link has weight and each neuron has bias. By using methods like Gradient Descent [1], Error Backpropagation [10], the weights and biases can be modified in order to obtain the desired output for the given input. Training of the neural networks is carried out by adjusting the weights and biases connected to each neuron. Training of neural networks is the key process of estimating the relationship between the independent and dependent variable. This estimated relationship is used for predicting the variable of interest.

A machine's RUL is affected by multiple factors like Pressure, Temperature, vibrations. Since sensor data consists of operating conditions of machine at a specific time, and the RUL is in time context (remaining cycles/hours), the data is of multivariate time series format. ANN assigns a weight matrix to its input and then produces an output, completely forgetting the previous input. Hence, as information flows only once through ANN and previous input data is not retained, ANN do not perform suitably when sequential data is to be processed. The time-context problem can be resolved by implementing recurrent neural networks (RNN).

RNN is a special form of ANN with a recurring connection to itself. This allows the output at the current iteration to be again used as an input for the next iteration for a more contextual prediction. Hence, recurrent neural network (RNN) behaves as if it has "memory" to generate output according to data processed previously [11], and giving it a sense of time context. Figure 1 shows the schematic representation of a recurrent neural network. The X_i represents the ith input and h_i represents the ith

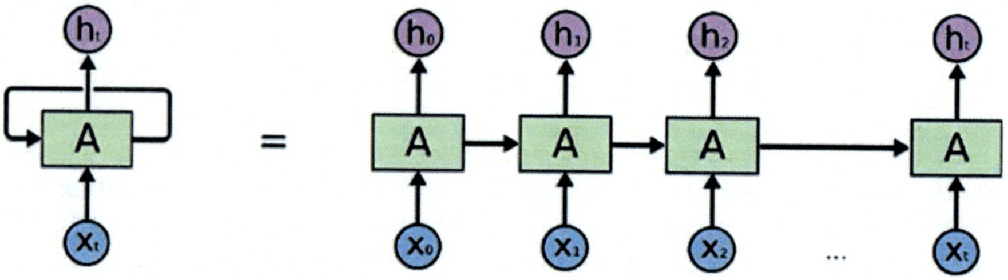

Fig. 1 Recurrent neural network

output. For the prediction of h_i, all the previous inputs and outputs starting from X_c and h_0 are used.

However, many times RNN suffer from the problem of Vanishing Gradients and Exploding Gradients which make the model less effective for problems involving long-term dependencies, i.e., learning the relationships of inputs that are far apart in the time series [11].

The long short-term memory (LSTM) is a special type of recurrent neural network that can solve the problem of long-term dependencies by using a memory cell controlled by three gates—input gate, forget gate and output gate. The forget gate determines whether the historical data should be removed or not from the memory cell, i.e., the data point is removed once it becomes obsolete, the input gate determines which input data should be added to the memory cell and the output gate determines which part of the memory cell should be given as output [12]. Hence, the relevant past information is stored in the memory cell. In addition to the current datapoint, LSTM make use of the relevant past information stored in the memory cell to make highly accurate and contextual predictions.

Degradation of machines is a gradual process that occurs over a period of time. In order to determine the Remaining Useful Life of a machine at the current cycle, it is imperative to know the degradation that has happened till date to obtain accurate predictions. This information is stored in the memory cell. LSTM make use of this stored information in addition to the current datapoints to understand the context and deliver highly accurate predictions.

This paper proposes the utilization of LSTM neural networks for asset prognostics to obtain highly accurate RUL estimates. The proposed method is demonstrated and tested on the NASA—CMAPSS dataset for estimation of RUL of 100 aero-propulsion engines.

4 Data Description

To test the adequacy of the proposed method, an experiment was carried on the NASA-C MAPSS (Commercial Modular Aero-Propulsion System Simulation) dataset. The intent of this experiment was to predict the RUL of 100 aero-propulsion engines. The NASA C-MAPSS dataset consists of four sub-datasets with different operating and fault conditions. Each dataset is divided into two parts—one consisting of data that is used for training the model (Training set) while the other is used to determine the accuracy of the model (Test set). All the datasets are of multivariate time series format.

Each aero-propulsion engine has a different time series with different life-cycles, sensor readings and operating conditions—the data is from a fleet of engines which are of the same type. All engines start with varying degrees of initial wear which is unknown to the user. Each time series has a different length, i.e., each engine has different number of time cycles until the engine is considered damaged.

The experiment is performed on the FD001 dataset. The dataset is a 20,631 × 26 matrix where 26 represents the number of features in the dataset. The first feature is the engine ID, the second, third and fourth features are the operational settings of the engines which are input by the engine operator and the 6th to 26th features are the readings of the 21 sensors installed on the engines. Each row in the dataset represents the state the engine during that operating time cycle. Hence 20,631 are the data records of the engines for each of the 26 fields such that each data record is collected within a single time cycle. Figure 2 represents variation in the sensor readings over the life cycle for sensors 3 & 4.

In the training set, at the last cycle, the engine cannot be operated further and is considered damaged (Fig. 3).

Figure 3 depicts the life cycle of engines in the training data. However, in the given testing set, the time series terminates before the complete degradation of the engine, i.e., the engine is still in normal operating condition at the last time cycle. Figure 4 shows the life cycle of the engines in the given test dataset.

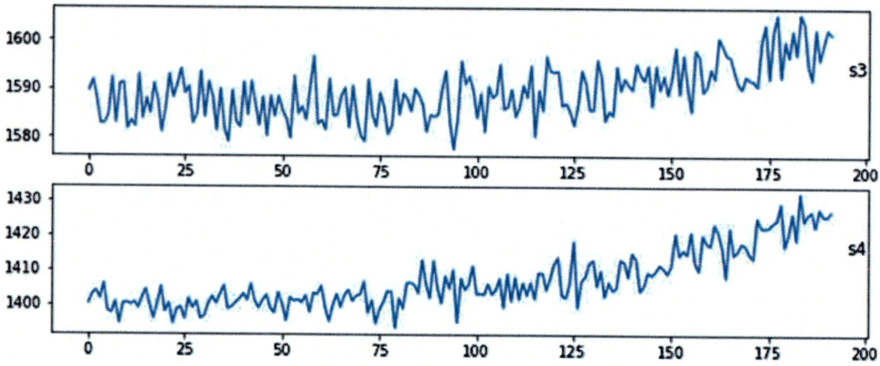

Fig. 2 Variation of Sensor 3 and 4 readings for machine id 1

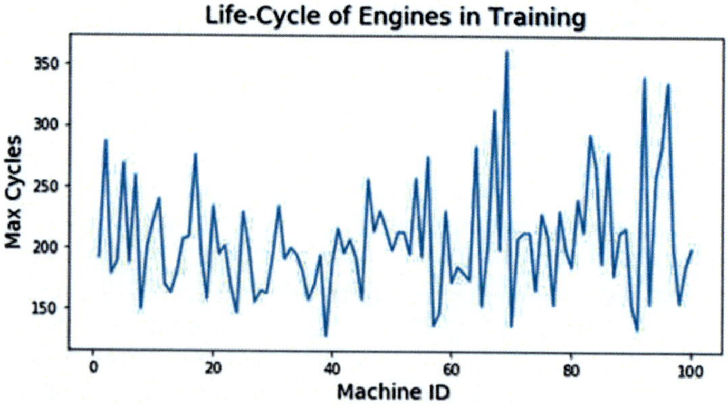

Fig. 3 Life cycle of engines in training set

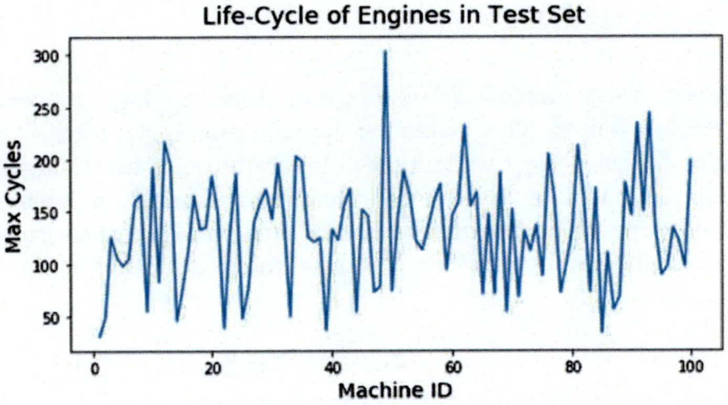

Fig. 4 Life cycle of engines in test set

The objective is to determine the Remaining Useful Life in the test dataset, i.e., the number of functional cycles that the engine will continue to work properly until failure. The predicted RUL should not be overestimated because it can lead to system failure or fatal accidents. An underestimation may be tolerated to some extent depending on the available resources and criticality of the conditions being estimated.

For evaluating the accuracy of Remaining Useful Life estimations, the Root Mean Square Error (RMSE) metric is used.

5 Data Processing

5.1 Data Labeling

Generally, for a brand-new machine, the degradation is negligible and does not affect the RUL much as the machine is in a healthy state. The degradation occurs after the machine has been operated for some cycles or period of time and initial wear has developed.

Hence, for initial cycles, the RUL of the engine is estimated at a constant maximum value and then it decreases linearly to 0. Hence, the target vector is piecewise linear.

For the estimation of the initial RUL value, the approach used in [13] was used. Since the minimum engine life cycle is 127 cycles, and the average life cycle is 209, the knee in the curve should occur at around 105th cycle. Experimentation was done by training the model on different values of initial RUL. Initial RUL value of 125 worked the best.

5.2 Data Normalization

Since different sensors measure different entities, there were huge differences in the range of sensor readings. This makes the normalization important as it brings all the values in the same range without much loss of information. There are various normalization techniques available such as minmax normalization, Logarithmic, Z-score normalization and linear normalization. Minmax and Z-score normalization are the most commonly used one. Minmax Normalization is done using the following Eq. (1)

$$\boldsymbol{X_i} = \frac{X_i - \min(X_i)}{\max(X_i) - \min(X_i)} \tag{1}$$

where, $\boldsymbol{X_i}$ is the normalized data, X_i is the original data, max (X_i) is the maximum value of the given X_i column and min (X_i) is the minimum value of the given X_i column.

Here, in the proposed approach Z-Score normalization of the input data is carried out according to Eq. (2).

$$Z = \frac{X_{i-u}}{\sigma_i} \tag{2}$$

In (2), X_i represents the original ith sensor data. μ_i is the mean of the ith sensor data from all engines, and σ_i is the corresponding standard deviation. The training dataset is used for calculating the mean and standard deviation. Using the calculated mean and standard deviation, both the training and testing data are normalized. This is done to ensure the same transformation is applied on both the training and testing set.

5.3 Data Conversion

Since LSTM have a temporal dimension, the two-dimensional data was reshaped into three dimensions—(samples, time steps, features). Samples represent the number of sequences/time series. Time step represents how many previous cycles should be taken into account to make future predictions. Features denote the fields (operating settings and sensor data) indicating the operating state of the engine. A low value of time steps will not capture long-term dependencies while a high value of time steps will reduce the weight assigned to more recent data. Hence, hyperparameter tuning is done to determine the optimum value of time steps.

Sensors having zero standard deviation, i.e., almost constant value for all the cycles were discarded as they do not have much impact on the output.

6 Model Architecture

The model proposed in the method consists of an input ANN layer, two LSTM layers where each layer has 64 neurons and an output ANN layer. The cost function used was Mean Squared Error. RMSprop optimizer was used which can adaptively adjust the learning rate. A Normal kernel initializer was used, i.e., the initialized random weights were normally distributed. This ensures that the weights given for starting the model training are uniformly and independently distributed over the given range here (0, 1).

7 Experiment Implementation

This paper utilizes the Keras library for the implementation of the long short-term memory neural network model. Keras is a neural network Application Programming Interface (API) based on Tensorflow and Theano.

Keras is coded in Python programming language and can be used for the implementation of different types of neural networks, cost functions and optimizers. After creating the LSTM model using Keras, the data was split into training data (9/10) and validation data (1/10) to initiate the training process. After each epoch (one iteration through the dataset), the cost function values of the training and validation data decreased. However, if the experiment is run for too many epochs, the model may over fit. Once the validation error stabilizes, the experiment is stopped. The predicted and actual RUL is given in Fig. 4. The test data is given as an input to the trained model and the RUL estimates are obtained. In order to reduce the effect of local optima, the experiment is carried out thrice and the average RMSE is determined.

8 Experimental Results and Comparison

Using the RUL predictions and the actual RUL values, the RMSE of the model is computed. An RMSE of 14.359 was obtained.

The experimental results of the proposed method are compared with other reported results (Table 1). Based on the comparison made in the table, it can be clearly seen that the proposed model outperforms all the other algorithms used by different researchers so far. Figure 5 represents the graph for the predicted and the actual values of the RUL for the machine Id 15 during training. In most of the cases, overestimation has been eliminated by the model which ensure that in any case the machine will not lead to any system failure due to the prediction error. Similar results are also shown in Fig. 6.

Table 1 Comparison of the proposed method with LSTM using other methods in terms of root mean square error

Methods	FD001
MLP [14]	37.5629
SVR [14]	20.9640
RVR [14]	23.7985
CNN [14]	18.4480
Proposed method (LSTM)	14.359
Improvement	22.165

Improvement = (1 − CNN/LSTM)

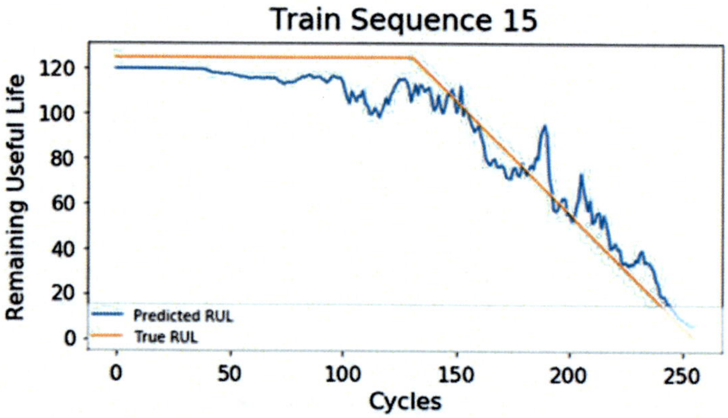

Fig. 5 Predicted and Actual RUL comparison for Training set machine Id 15

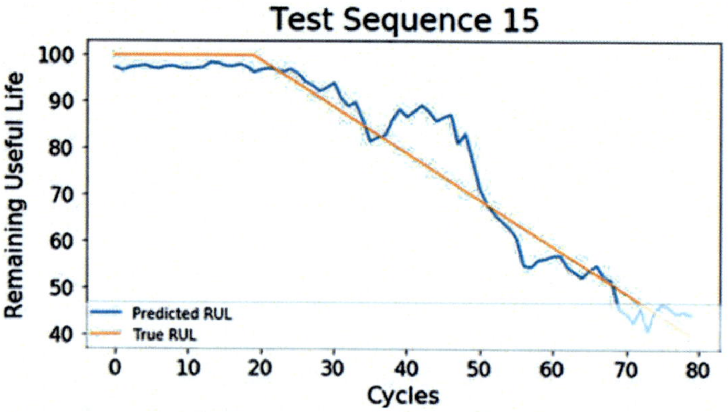

Fig. 6 Predicted and Actual RUL comparison for Test set machine Id 15

9 Conclusion

This paper proposes a deep learning-based asset prognostics system which utilizes long short-term memory neural networks. The LSTM can solve the problem of long-term dependencies. The efficiency of the suggested method is demonstrated using the NASA C-MAPSS dataset for estimating RUL of 100 aero-propulsion engines. The results are compared to the state of the art methods like Support Vector Regression, Relevance Vector Regression, Multi-Layer Perceptron and Convolutional Neural Network. The comparisons show that the proposed method performs better than the aforementioned methods in terms of RMSE values.

In the future scope of work, the factors causing machine degradation can be identified by performing statistical tests on the data. Based on the results, the manufacturing process can be adjusted to keep process parameters of degradation causing features to optimum values, thereby decreasing the machine degradation and hence increasing the lifespan of the asset.

References

1. Ruder S (2016) An overview of gradient descent optimization algorithms. ArXiv:1609.04747
2. Yam R, Tse P, Li L, Tu P (2001) Intelligent predictive decision support system for condition-based maintenance. Int J Adv Manuf Technol 17:383–391
3. Zheng S, Ristovski K, Farahat A, Gupta C (2017) Long short-term memory network for remaining useful life estimation. In: 2017 IEEE international conference on prognostics and health management (ICPHM). IEEE, pp 88–95
4. Listou Ellefsen A, Bjørlykhaug E, Æsøy V, Ushakov S, Zhang H (2019) Remaining useful life predictions for turbofan engine degradation using semi-supervised deep architecture. Reliab Eng Syst Saf 183:240–251
5. Elsheikh A, Yacout S, Ouali M (2019) Bidirectional handshaking LSTM for remaining useful life prediction. Neurocomputing 323:148–156
6. Deutsch J, He D (2018) Using deep learning-based approach to predict remaining useful life of rotating components. IEEE Trans Syst Man Cybern Syst 48:11–20
7. Pektas A, Pektas E (2018) A novel scheme for accurate remaining useful life prediction for industrial IoTs by using deep neural network. Int J Artif Intell Appl 9:17–25
8. Sikorska J, Hodkiewicz M, Ma L (2011) Prognostic modelling options for remaining useful life estimation by industry. Mech Syst Signal Process 25:1803–1836
9. Li X, Zhang W, Ding Q (2019) Deep learning-based remaining useful life estimation of bearings using multi-scale feature extraction. Reliab Eng Syst Saf 182:208–218
10. Rumelhart D, Hinton G, Williams R (1986) Learning representations by back-propagating errors. Nature 323:533–536
11. Hsu CS, Jiang JR (2018) Remaining useful life estimation using long short-term memory deep learning. In: 2018 IEEE international conference on applied system invention (ICASI). IEEE, pp 58–61
12. Yuan M, Wu Y, Lin L (2016) Fault diagnosis and remaining useful life estimation of aero engine using LSTM neural network. In: 2016 IEEE international conference on aircraft utility systems (AUS). IEEE, pp 135–140

13. Heimes FO (2008) Recurrent neural networks for remaining useful life estimation. In: 2008 IEEE International Conference on Prognostics and Health Management. IEEE, pp 1–6
14. Babu GS, Zhao P, Li XL (2016) Deep convolutional neural network based regression approach for estimation of remaining useful life. In: International conference on database systems for advanced applications. Springer, Cham, pp 214–228

Evaluation of Two Feature Extraction Techniques for Age-Invariant Face Recognition

Ashutosh Dhamija and R. B. Dubey

Abstract Huge variation in facial appearances of the same individual makes Age-Invariance Face Recognition (AIFR) task suffer from the misclassification of faces. However, some Age-Invariant Feature Extraction Techniques (AI-FET) for AIFR are emerging to achieve good recognition results. The performance results of these AI-FETs need to be further investigated statistically to avoid being misled. Here, the means between the quantitative results of Principal Component Analysis–Linear Discriminant Analysis (PCA-LDA) and Histogram of Gradient (HoG) are compared using one-way Analysis of Variance (ANOVA). The ANOVA results obtained at 0.05 critical significance level indicate that the results of the HoG and PCA-LDA techniques are statistically well in line because the F-critical value was found to be greater than the value of the calculated F-statistics in all the calculations.

Keywords AIFR · Statistical evaluation · Feature extraction techniques · ANOVA

1 Introduction

The great deal of research on human face reorganization (FR) has been reported in the previous three decades. Different FR calculations that can manage faces under various outward appearances, lighting conditions, and postures have been proposed and can accomplish agreeable exhibitions. In any case, adjustments in face appearance brought about by age movement have gotten constrained regard for date; this impact significantly affects FR calculations. There are two distinct methodologies for AIFR. First is the generative methodology. In this methodology, face pictures of different ages will be produced before FR is performed. For this methodology, age of face picture should be evaluated, and adequate preparing tests are vital for learning

A. Dhamija (✉) · R. B. Dubey
SRM University, Delhi NCR, Sonepat, Haryana, India
e-mail: dhamija.ashutosh@gmail.com

R. B. Dubey
e-mail: rbdubey@srmuniversity.ac.in

V. Singh et al. (eds.), *Computational Methods and Data Engineering*, Advances in Intelligent Systems and Computing 1227,
https://doi.org/10.1007/978-981-15-6876-3_15

connection between face at two unique ages. Subsequent methodology depends on discriminative models, which utilize facial highlights that are inhumane toward age movement to accomplish AIFR. Since facial maturing is for the most part found in more youthful age gatherings and it is likewise spoken to by enormous surface changes and minor shape changes because of difference in weight, nearness of wrinkles, and firmness of the skin in age over 18 years. Performance results of AI-FETs need to be investigated statistically to avoid misclassifications. In this paper, means of quantitative results of AI-FET HoG and PCA-LDA are compared to determine if statistically knowingly diverse from each other using one-way Analysis of Variance (ANOVA) [1].

AIFR is a difficult issue on FR look into in light of the fact that one individual can display generously unique appearance at changed ages which essentially increment acknowledgment trouble. Also, it is winding up progressively significant and has wide application, for example, finding missing youngsters, distinguishing culprits, and international ID confirmation. It is especially reasonable for applications where different biometrics systems are not accessible. Customary strategies depend on generative models emphatically rely upon parameters suspicions, precise age marks, and generally clean preparing information, so they don't function admirably in certifiable FR. To address this issue, some discriminative strategies [2–6] are proposed which is nonlinear factor examination strategy to isolate personality including from face portrayal [2–8].

FR can be performed utilizing 2D facial pictures, 3D facial sweeps, or their blend. 2D FR has been widely explored during the previous couple of decades. What's more, it is as yet confronting tested by various components including brightening varieties, scale contrasts, present changes, outward appearances, and cosmetics. With the fast improvement of 3D scanners, 3D information obtaining is winding up progressively less expensive and non-meddling. Furthermore, 3D facial outputs are increasingly hearty to lighting conditions, present varieties, and facial cosmetics. 3D geometry spoke by facial sweep additionally gives another sign to exact FR. 3D FR is in this way accepted can possibly beat numerous confinements experienced by 2D FR, and has been considered as an option or corresponding answer for regular 2D FR approaches. Existing 3D FR calculations can be sorted into two classes: all-encompassing and neighborhood highlight based calculations (LFBA). All-encompassing calculations utilize data of whole face or enormous districts of 3D face to perform FR. Noteworthy confinement of comprehensive calculations are required that they require exact standardization of 3D appearances, and they are generally progressively touchy to outward appearances and impediments. Conversely, LFBA distinguishes and coordinate a lot of nearby highlights (e.g., milestones, bends, patches) to perform 3D FR. Contrasted with all-encompassing calculations, LFBA is increasingly strong to different disturbances including outward appearances and impediment [9–13].

For most part, FR calculations yield agreeable outcome for frontal appearances. Anyway coordinating non-frontal faces straightforwardly is a troublesome errand. Posture invariant FR calculations are fundamentally arranged into three classifications, for example, invariant component extraction based techniques, multi-view based strategies, and present standardization based strategies. Definitive thought of posture standardization is by producing novel posture of either test picture as like that of exhibition picture or by the turn around dependent on 3D model. Another thought of posture standardization is by integrating frontal perspective on display and test picture which is also called frontal face recreation. It has already been reported that FR exactness recognition accuracy (RA) is useful for frontal appearances. Anyway in continuous situation face pictures caught isn't constantly frontal and have subjective posture varieties involving every single imaginable course. Consequently, it is popular for FR techniques that can ready to deal with these types of faces and is implemented to recreate frontal face from non-frontal face for better RA [14].

The rest of the paper is organized as follows. In Sect. 2, describes the architectures of methods. Section 3, feature extraction techniques are explained. Section 4 describes the methodology and its implementation. Results and discussion are given in Sect. 5. The conclusions and future directions are drawn in Sect. 6.

2 Architectures of Methods

2.1 *Architect of Histogram of the Oriented Gradient*

The HOG feature brings the distribution of observations in the image area which are useful for textured identification of objects having deformable shapes. The original HOG feature captured is a suitable factor of an image likewise used in scale-invariant feature transformation technique. Overall collected histogram features present an image [1]. Orientation may be indicated as a single angle/double angle [15]. The single angle presentation shows far better results than the double angle presentation. An image window can be defined by

$$I = \bigcup_{t=1}^{N} C_t \tag{1}$$

Let I is evenly separated into N cells, where I the image window of a key point, and C_t is the set of overall pixels of the tth cell. In any pixel $p(x, y)$ of I, the contrast is defined as

$$g_p = g(x, y) = \sqrt{\Delta x^2 + \Delta y^2} \tag{2}$$

and the gradient direction is given by

$$\theta_p = \theta(x, y) = \arctan \frac{\Delta y}{\Delta x} \tag{3}$$

Let the histogram vector length is H for every cell and the inclination is equally separated into H bins. Now, the histogram vector defined as

$$b_t^i = \frac{\sum \{g_p | p \in C_t, \theta_p \in [i\theta_0 - \theta_0/2, i\theta_0 + \theta_0/2]\}}{|C_t|} \quad (4)$$

$$v_t = \{b_t^0, b_t^1, b_t^2, \cdots, b_t^{H-1}\} \quad (5)$$

$|C_t|$ represents the physical size of C_t.

For good invariance to illumination and noise, four different normalization steps namely; L2-norm, L2-Hys, L1-sqrt, and L1-norm are suggested [16]. We applied the L2-norm step for its good result [16]:

$$\mathbf{v}_t' = \mathbf{v}_t \Big/ \sqrt{\|\mathbf{v}_t\|_2^2 + \varepsilon^2} \quad (6)$$

where ε having a small positive value.

A fast computation of histogram bin weights is done using Fig. 1 [16, 17].

$$g_n = \frac{\sin(n+1)\theta_0}{\sin\theta_0} g_x - \frac{\cos(n+1)\theta_0}{\sin\theta_0} g_y \quad (7)$$

$$g_{n+1} = -\frac{\sin n\theta_0}{\sin\theta_0} g_x + \frac{\cos n\theta_0}{\sin\theta_0} g_y \quad (8)$$

$$b_t^i = \sum_{p_j \in C_t} g_{i,j} \quad (9)$$

For matching two facial images, during varying light conditions/motion blurring, the accuracy of the eye orientation is utmost care. The histogram gives some solution

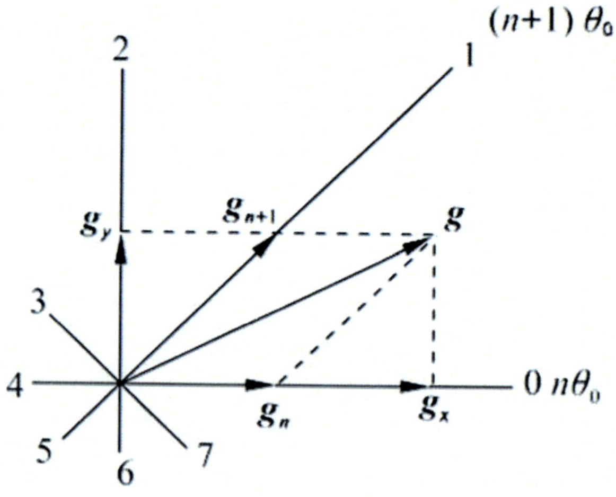

Fig. 1 Determination of projection of gradient magnitude [18]

to balance to this limitation, however, it is not sufficient. To overcome and compensate for this problem, the overlapped HOG feature was suggested [16].

2.2 *Principle Component Analysis (PCA)*

PCA is mainly employed for image identification and compression purposes. It converts the large 1D vector from 2D image into the eigenspace representation which in turn is determined by the eigenvectors of the covariance matrix of images by choosing a suitable threshold [19]. It involves the following steps:

(i) At each step, 2-D data set is utilized.
(ii) To subtract the mean from every data dimensions to get the average.
(iii) Compute the covariance matrix.
(iv) Computation of the eigenvectors and values of the covariance matrix. These are representations of unit eigenvectors.
(v) The components are chosen to form a feature vector. The eigenvalues are arranged, highest to lowest order.
(vi) Here, a new data set is derived. After selection of eigenvectors, we took the inverse of the vector which is multiplied with the left of actual database. This is the actual data purely in vectors form [19].

Each image is treated as a vector. Let image components $= w * h$, where w and h are width and height, respectively. The optimal space vectors are principle components. The linear algebra is applied to determining eigenvectors of the covariance matrix images in a set. The number of eigenfaces are equivalent to face images in the trained database. But faces are further estimated by prime eigenfaces of largest values [19–21]. The faces images are changed to binary using Sobel algorithm. The similarity between the two points sets is calculated by the Hausdorff distance using Eq. (10)

$$h(A, B) = \frac{1}{N_a} \sum_{a \in A} \min_{b \in B} \|a - b\| \tag{10}$$

The Line Segment Hausdroff Distance considers the different structures of line orientation, line-point conjunction and therefore has a superior different power than line edge map [22].

2.3 *Linear Discriminant Analysis (LDA)*

It gives directions along which the classes are best classified. The main purpose of PCA is reduction of dimensionality and elimination the empty spaces of the two scatter matrices. The direct LDA methods are used for further improvement [23]. Following are the main steps:

(i) Computation of within-class scatter matrix S_w

$$S_W = \sum_{i=1}^{c} \sum_{x_k \in x_i} (x_k - \mu_i) \cdot (x_k - \mu_i)^T \tag{11}$$

where x_k is the ith sample of class i,μ_i is the mean of class i, and C is the number of classes.

(ii) Computation of between-class scatter matrix, b:

$$S_W = \sum_{i=1}^{c} \sum_{x_k \in x_i} (x_k - \mu_i) \cdot (x_k - \mu_i)^{\mathrm{T}} \tag{12}$$

where μ is the mean of all classes.

(iii) Computation of the eigenvectors of projection matrix

$$W = \mathrm{eig}\left(S_W^{-1} \cdot S_B\right) \tag{13}$$

The test images estimate matrix, and the estimate matrix of every training image is estimated employing similarity measure. The resulting is the training image and is nearest to the test image.

For high-dimensional data, LDA measures an optimal transformation to maximize the ratio.

$$J_{\mathrm{LDA}}(W) = \arg_W \max \frac{\left|w^T S_B w\right|}{\left|w^T s_w w\right|} \tag{14}$$

where $A = \pi r^2$ is the between-class and S_w is the within-class scatter matrices, respectively [23].

3 Feature Extraction Techniques

3.1 Histogram of Gradient (HoG)

A new method HoG was introduced [24] based on Hidden Factor Analysis (HFA). It is based on the fact that the facial picture of an individual is composed of two segments: character explicit part that is steady over the maturing procedure, and other segments that mirror maturing impact. Instinctively, every individual is related with unmistakable personality factors, which is to a great extent invariant over maturing procedure and subsequently can be utilized as a steady element for FR; while age factor changes as individual develop.

In testing, given a couple of face pictures with obscure ages, a suitable score among them was based on the back mean of their character parameters. Each face picture is separated into various patches for the faithful implementation of HoG. Before applying HoG face pictures were pre-handled through accompanying advances:

(i) Rotate images to adjust in vertical direction;
(ii) Scaled images to keep separations between two eyes;
(iii) Crop images to eliminate background and hair area;
(iv) Impose histogram equalization on cropped image for standardization.

During training, preparation countenances were first gathered by their personalities and ages, trailed by highlight extraction on each picture. With each preparation face spoken to by HoG highlight, the element of these highlights was decreased with cutting utilizing PCA and LDA. Lastly, HFA models have adjusted autonomously on every one of cut highlights of dataset, getting a lot of model parameters for each cut. Attesting stage, coordinating score of given face pair was processed by first experiencing highlight extraction and measurement decrease steps equivalent to preparing, at that point assessing character dormant factors for each cut of two face highlights. Last coordinating score was given by cosine separation of connected personality highlights [24].

3.2 *Principal Component Analysis–Linear Discriminant Analysis (PCA-LDA)*

Comprehensive methodologies dependent on PCA and LDA experience ill effects of high levels of dimensionality [25]. Here, the required time for calculation develops exponentially, rendering calculation unmanageable in very high-dimensional issues. Endeavor was made to build up increasingly hearty AI-FET, PCA, and subspace LDA techniques for highlight extraction of face pictures. PCA ventures pictures into subspace with the end goal that primary symmetrical element of this subspace catches the best measure of difference among pictures and last component of this subspace catches minimal measure of fluctuation among pictures. In this regard, eigenvectors of covariance grid are registered which relate to headings of primary segments of first information and their factual hugeness is given by their comparing eigen esteems. PCA was utilized with the end goal of measurement decrease by summing up information while Quadratic Support Vector Machine (QSVM) was utilized for the last order [26].

4 Methodologies Implementation

In this section, the implementation of AI-FET and statistical significance tests are discussed. SRMUH aging database is used here and is composed of 400 images of 30 subjects (6–32 images per subject) in the age group 6–70 years. What's more, data is accessible for every one of the pictures in the dataset to be specific: picture size, age, sexual orientation, exhibitions, cap, mustache, whiskers, flat posture, and vertical posture. Since pictures were recovered from genuine collections of various subjects, perspectives, for example, light, head posture, and outward appearances are uncontrolled in this dataset. Table 1 shows a few examples of images from SRMUH database.

The evaluation parameters used to evaluate FETs are False Accept (FA), False Reject (FR), Recognition Accuracy (RA), and Recognition Time (RT).

(i) *FAR*: This is the level of tests framework dishonestly acknowledges despite the fact that their asserted characters are inaccurate [27].

$$\text{FAR} = \frac{\text{Number of false accepts}}{\text{Number of impostor scores}} \tag{15}$$

(ii) *FRR*: This is level of tests framework dishonestly rejects regardless of the way that their asserted characters are right. A false acknowledges happens when acknowledgment framework chooses the bogus case is valid and bogus reject happens when framework chooses genuine case is false [27].

$$\text{FRR} = \frac{\text{Number of false rejects}}{\text{Number of genuine scores}} \tag{16}$$

(iii) *RA*: This is principle estimation to portray exactness of acknowledgment framework. It speaks to quantity of appearances that are accurately perceived from all outnumber of countenances tried [27].

$$\text{RA} = \frac{\text{Number of correctly recognized persons}}{\text{Total number of person tested}} \times 100\% \tag{17}$$

(iv) *RT*: It is time to process and perceive all countenances in testing set.

The means between the quantitative results of HoG and PCA-LDA were compared using one-way ANOVA. ANOVA results obtained at 0.05 critical significance level indicate that the results of the HoG and PCA-LDA techniques are statistically well in line because the F-critical value was found to be greater than the value of the calculated F-statistics in all the calculations. ANOVA is a statistical approach for testing differences between two or more means. It is determined by:

$$H_o : \mu_1 = \mu_2 = \mu_3 = \cdots = \mu_k \tag{18}$$

Table 1 SRMUH database

(continued)

Table 1 (continued)

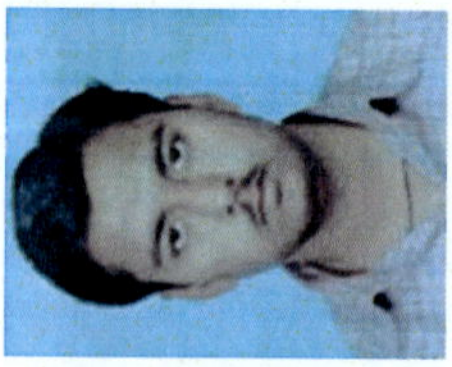		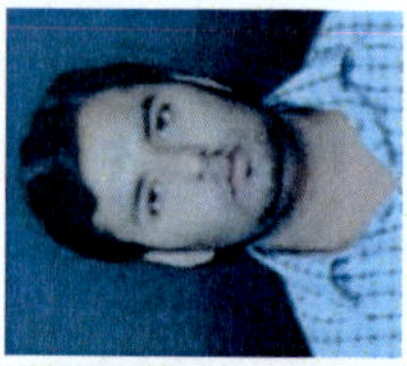	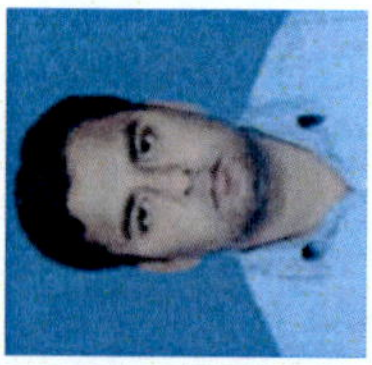	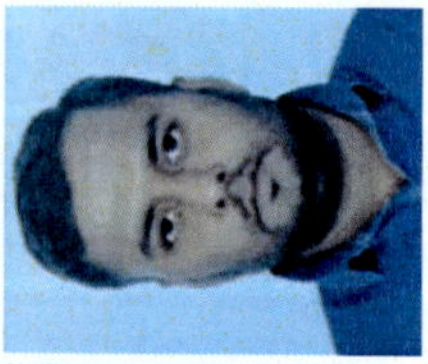	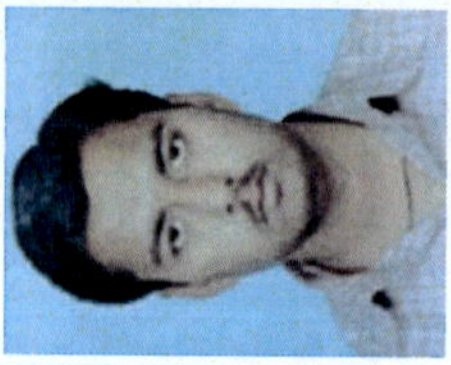

is tested where μ = group mean and k = number of groups. ANOVA performs F-test to check the variation between group means is greater than the variation of the experiments within the groups. Fisher-statistics is a ratio based on mean squares and used to assess the equality of variances. To determine which specific approach differs, a Least Significance Difference (LSD) Post Hoc test is conducted. LSD Post Hoc test is conducted in situations the results are found to be statistically quite similar [27].

$$\text{LSD Post Hoc Test} = t\sqrt{\text{MSW}}\left(\frac{1}{N_1} + \frac{1}{N_2}\right) \tag{19}$$

where t = critical value of the tail, N is sample size of each method and MSW is the Mean Square Within.

5 Results and Discussion

PCA-LDA yielded FA 22, FR 30, RA of 98.8%, and RT 80 s while, HoG produced FA 20, FR of 29, RA of 87%, and RT of 126 s. The summary of the results is given in Table 2.

The one-way ANOVA results are quite similar to the quantitative results of HoG and PCA-LDA techniques. Because the F-statistic and F-critical values are 0.125 and 9.56, respectively, for FAR values. Similarly, F-statistic and F-critical value for these techniques is 0.224 and 9.589, respectively, for RT values. While analyzing RA values, 0.123 and 9.467 are obtained as the F-statistic and F-critical value for these techniques, respectively.

Furthermore, F-statistic and F-critical value for HoG and PCA-LDA techniques obtained using FRR values are 0.934 and 9.126, respectively. In all the statistical evaluations conducted at 0.05 critical significance level, the F-critical values were found to be greater than the value of the calculated F-statistics. Hence, since one-way ANOVA did not return a statistically significant result, an alternative hypothesis with least two group means that are statistically knowingly diverse from each other is rejected. This implies that the results of HoG and PCA-LDA techniques are not statistically different.

Table 2 Results of AI-FETs

FET	FA	FR	RA (%)	RT (s)
HoG	20	29	87	126
PCA-LDA	22	30	98.8	80

6 Conclusions

In this paper, we have presented a statistical evaluation of HoG and PCA-LDA using one-way ANOVA. The ANOVA results obtained at 0.05 critical significance level indicate that the quantitative results of the HoG and PCA-LDA techniques are statistically well in line because the F-critical value was found to be greater than the value of the calculated F-statistics in all the calculations. Further work is in progress to test more emerging AI-FETs on different datasets to improve the accuracy of recognition.

References

1. Shu C, Ding X, Fang C (2011) Histogram of oriented gradient (HOG) of the oriented gradient for face recognition. Tsinghai Sci Technol 16(2):216–224
2. Zhifeng L, Dihong G, Xuelong L, Dacheng T (2016) Aging face recognition: a hierarchical learning model based on local patterns selection. IEEE Trans Image Process 25(5):2146–2154
3. Dihong G, Zhifeng L, Dahua L, Jianzhuang L, Xiaoou T (2013) Hidden factor analysis for age invariant Face Recognition. In: Proceedings of the IEEE international conference on computer vision, pp 2872–2879
4. Dihong G, Zhifeng L, Dacheng T, Jianzhuang L, Xuelong L (2015) A maximum entropy feature descriptor for age-invariant Face Recognition. In: Proceedings of the IEEE conference on computer vision and pattern recognition, pp 5289–5297
5. Zhifeng L, Unsang P, Anil KJ (2011) A discriminative model for age invariant face recognition. IEEE Trans Inf Forensics Secur 6(3):1028–1037
6. Haibin L, Stefano S, Narayanan R, David WJ (2010) Face verification across age progression using discriminative methods. IEEE Trans Inf Forensics Secur 5(1):82–91
7. Chenfei X, Qihe L, Mao Y (2017) Age invariant FR and retrieval by coupled auto-encoder networks. Neurocomputing 222:62–71
8. Di H, Mohsen A, Yunhong W, Liming C (2012) 3-D FR using e LBP-based facial description and local feature hybrid matching. IEEE Trans Inf Forensics Secur 7(5):1551–1565
9. Yulan G, Yinjie L, Li L, Yan W, Mohammed B, Ferdous S (2016) EI3D: Expression-invariant 3D FR based on feature and shape matching. Pattern Recogn Lett 83:403–412
10. Stefano B, Naoufel W, Albertodel B, Pietro P (2013) Matching 3D face scans using interest points and local histogram descriptors. Comput Graph 37(5):509–525
11. Stefano B, Naoufel W, Alberto B, Pietro P (2014) Selecting stable key points and local descriptors for person identification using 3D face scans. Vis Comput 30(11):1275–1292
12. Alexander MB, Michael MB, Ron K (2007) Expression-invariant representations of faces. IEEE Trans Image Process 16(1):188–197
13. Di H, Caifeng S, Mohsen A, Yunhong W, Liming C (2011) Local binary patterns and its application to facial image analysis. IEEE Trans Syst Man Cybern Part C Appl Rev 41(6):765–781
14. Kavitha J, Mirnalinee TT (2016) Automatic frontal face reconstruction approach for pose invariant face recognition. Procedia Comput Sci 87:300–305
15. Goesta HG (1978) In search of a general picture processing operator. Comput Graph Image Process 8(2):155–173
16. Navneet D, Bill T (2005) Histograms of oriented gradients for human detection. In: IEEE conference on computer vision and pattern recognition (CVPR). San Diego, CA, USA, pp 886–893

17. Liu CL, Nakashima K, Sako H (2004) Handwritten digit recognition: investigation of normalization and feature extraction techniques. Pattern Recogn 37(2):265–279
18. Liu H (2006) Offline handwritten character recognition based on descriptive model and discriminative learning [Dissertation]. Tsinghua University, Beijing, China
19. Lawrence S, Kirby M (1987) A low dimensional procedure for the characterization of human face. JOSA 4(3):519–524
20. Peter NB, Joao PH, David JK (1977) Eigen faces vs. fisher faces: recognition using class specific linear projection. IEEE Trans Patt Anal Mach Intell 9(7):711–720
21. Ravi S, Nayeem S (2013) A study on face recognition technique based on eigen face. Int J Appl Inf Syst 5(4):57–62
22. Sakai T, Nagao M, Fujibayashi S (1969) Line extraction and pattern recognition in a photograph. Pattern Recogn 1:233–248
23. Nikolaos G, Vasileios M, Ioannis K, Tania S (2013) Mixture subclass discriminant analysis link to restricted Gausian model and other generalizations. IEEE Trans Neur Netw Learn Syst 24(1):8–21
24. Dihong G, Zhifeng L, Dahua L, Jianzhuang L, Xiaoou T (2013) Hidden factor analysis for age invariant face recognition. In: IEEE international conference on computer vision, pp 2872–2879
25. Priti VS, Bl G (2012) particle swarm optimization—best feature selection method for face images. Int J Sci Eng Res 3(8):1–5
26. Issam D (2008) Quadratic kernel-free non-linear support vector machine. Springer J Glob Optim 41(1):15–30
27. Ayodele O, Temitayo MF, Stephen O, Elijah O, John O (2017) Statistical evaluation of emerging feature extraction techniques for aging-invariant face recognition systems. FUOYE J Eng Technol 2(1):129–134

XGBoost: 2D-Object Recognition Using Shape Descriptors and Extreme Gradient Boosting Classifier

Monika, Munish Kumar, and Manish Kumar

Abstract In this chapter, the performance of eXtreme Gradient Boosting Classifier (XGBClassifier) is compared with other classifiers for 2D object recognition. A fusion of several feature detector and descriptors (SIFT, SURF, ORB, and Shi Tomasi corner detector algorithm) is taken into consideration to achieve the better object recognition results. Various classifiers are experimented with these feature descriptors separately and various combinations of these feature descriptors. The authors have presented the experimental results of public datasets, namely Caltech-101 which is a very challenging image dataset. Various performance measures, i.e., accuracy, precision, recall, F1-score, false positive rate, area under curve, and root mean square error, are evaluated on this multiclass Caltech-101 dataset. A comparison among four modern well-known classifiers, namely Gaussian Naïve Bayes, decision tree, random forest, and XGBClassifier, is made in terms of performance evaluation measures. The chapter demonstrates that XGBClassifier outperforms rather than other classifiers as it achieves high accuracy (88.36%), precision (88.24%), recall (88.36%), F1-score (87.94%), and area under curve (94.07%) when experimented with the fusion of various feature detectors and descriptors (SIFT, SURF, ORB, and Shi Tomasi corner detector).

Keywords Object recognition · Feature extraction · Gradient boosting · XGBoost

---

Monika
Department of Computer Science, Punjabi University, Patiala, India

M. Kumar (✉)
Department of Computational Sciences, Maharaja Ranjit Singh Punjab Technical University, Bathinda, Punjab, India
e-mail: munishcse@gmail.com

M. Kumar
Department of Computer Science, Baba Farid College, Bathinda, Punjab, India

V. Singh et al. (eds.), *Computational Methods and Data Engineering*,
Advances in Intelligent Systems and Computing 1227,
https://doi.org/10.1007/978-981-15-6876-3_16

1 Introduction

Image classification problem is a key research area in computer vision. Image classification and object recognition are used interchangeably. The function of image classification is to classify similar images/objects under the same label. For this, the system extracts the features of the images/objects and groups the images/objects having similar features under one class. Feature extraction plays a very important role in the performance of a recognition system. An object can be recognized with its color, shape, texture, or some other features. Shape is an essential feature of an object that makes it easily identifiable. For example, a bench and a chair can easily be differentiated by their shape. There are so many objects present in the real world which are identified by their shape. In this chapter, the authors have used four shape feature detectors and descriptors, namely SIFT [1], SURF [2], ORB [3], and Shi Tomasi corner detector algorithm [4] for feature extraction. A hybrid of these feature descriptors is taken for experimental work, as individuality of these feature descriptors is not up to mark for providing acceptable recognition results. On the other hand, classification is also an important tool used in the object recognition problems. There are various classification algorithms available. They all have different ways of classification and perform differently on different datasets. The objective of this research is to explore the efficiency of XGBClassifier [5]. In modern research work, XGBClassifier is performing better than other existing classifiers in the field of image processing and pattern recognition. This algorithm is based on a gradient boosting algorithm. Gradient boosting algorithm boosts weaker classifiers and trains the data in an additive manner. The algorithm produces a predicted model in the form of an ensemble of weaker classifiers. Generally, it uses one decision tree as a weak classifier at a time. So, it takes more time and space for classification. This chapter describes the comparison of the performance of XGBClassifier with some well-known modern classification methods—Gaussian Naïve Bayes, decision tree [6], and random forest [7]. Caltech-101 dataset is chosen for the experimental work as this dataset contains many classes and images as it has 101 classes with a total of 8678 images [8]. Each class in the Caltech-101 dataset contains numerous images in the range of 40–800 images. So, there is an unbalance count of images in each class. Considering this unbalancing of the classes in the dataset, an averaging of the performance of overall dataset is evaluated. Classification on the dataset can be made based on two methods—dataset partitioning method or cross-validation method. The authors have selected dataset partitioning methodology where 80% of images of each class are taken as training data and the remaining 20% images of each class are taken as testing data. Seven performance evaluation measures are evaluated in the experiment to examine the efficiency of all these classifiers. The measures are balanced accuracy, precision, recall, F1-score, area under curve, false positive rate, and root mean square error. These computed measures are based on multiclass classification for which an average of the performance of all classes is taken to measure the overall efficiency. The paper presents a comparative view of all the four classification methods and the results depict that the XGBoost classifier outperforms other classifiers.

The rest of this paper is organized in various sections as follows: Sect. 2 presents a survey on XGBClassifier and measures used for unbalanced dataset. Section 3 describes about a brief detail on shape-based feature descriptors—SIFT, SURF, ORB, and Shi Tomasi corner detector. Section 4 gives a discussion on various classifiers used in the experiment. Section 5 explains the XGBClassifier in detail with the parameter tuning. In Sect. 6, a detailed study of all the performance measures is discussed. In Sect. 7, the authors have mentioned about the dataset and the tools used for experiment. Section 8 reports the experimental results evaluated on different models. A comparative view on various feature extraction algorithms and classification methods is presented in tabular form. Finally, a conclusion is drawn in Sect. 9.

2 Related Work

Ren et al. [9] experimented the performance of CNN and XGBoost on MNIST and CIFAR-10 dataset. The authors proposed the combination of CNN and XGBoost for classification. The paper also exhibited the comparison of this combined classifier with other state-of-the-art classifiers and the proposed system outperformed other classifiers with 99.22% accuracy on MNIST dataset and 80.77% accuracy on the CIFAR-10 dataset. Santhnam et al. [10] presented a comparison of the performance of the XGBoost algorithm with the gradient boosting algorithm on different datasets in terms of accuracy and speed. The results are derived on four datasets—Pima Indians Diabetes, Airfoil Self-Noise, Banknote Authentication, and National Institute of Wind Energy (NIWE) datasets. The authors concluded that the accuracy computed on these datasets is not always high using XGBoost methodology. They adopted both training/testing model and 10-fold cross-validation model for regression and classification. Bansal et al. [11] observed the performance of XGBoost on intrusion detection system. They compared the efficiency of XGBoost with AdaBoost, Naïve Bayes, multilayer perceptron (MLP), and K-nearest neighbor classification methods. The results proved that XGBoost classifier is achieving more efficiency than other classifiers. The performance of the model is also measured based on binary and multiclass classification. Two new evaluation measures—average class error and overall error based on multiclass classification—are considered in the paper. Vo et al. [12] proposed a hybrid deep learning model for smile detection on both balanced and imbalanced datasets and achieved high efficiency as compared to other state-of-the-art methods. Features are extracted using convolutional neural network. The authors used extreme gradient boosting to train the extracted features for imbalance dataset. The performance of the model is derived based on accuracy and area under curve.

3 Shape Descriptors

The object recognition system performs a few steps for identifying the object. The system starts with preprocessing of image, feature extraction, feature selection, dimensionality reduction, and then finally image classification. But the performance of an object recognition system basically depends on two tasks—feature extraction and image classification technique. In this section, a description of various shape feature extraction algorithms is presented. Scale Invariant Feature Transform (SIFT), Speed Up Robust Feature (SURF), Oriented FAST and Rotated BRIEF (ORB), and Shi Tomasi corner detectors are used in the experiment for feature extraction. SIFT, SURF, and ORB are shaped feature detectors and descriptors. Shi Tomasi corner detector algorithm extracts the corner of the objects that helps to find the shape of the object. Further, k-means clustering and Locality-Preserving Projection (LPP) are applied on these extracted features for feature selection and dimensionality reduction. As Shi Tomasi corner detector algorithm is not able to detect the corners of blurry image so a saliency map is used on the images before applying this algorithm to the image. Saliency map improves the quality of the image that helps in corner detection. The following is the description of these feature descriptor methods.

3.1 Scale Invariant Feature Transform (SIFT)

Lowe [1] proposed a powerful frame Scale Invariant Feature Transform (SIFT) for recognizing the objects. The algorithm produces distinct key descriptors of a 128-dimensional vector size of an image. SIFT works in four stages. First, locations are detected using Difference-of-Gaussian (DoG) algorithm on the image. These locations are invariant to scale. In the second stage, the detected keypoints are localized to improve the accuracy of the model and only selected keypoints are considered. Third stage computes the directions of gradients around the keypoint descriptors for the orientation assignment. This makes SIFT invariant to rotation. Finally, in the fourth stage, the keypoints computed are transformed into a feature vector of size 128 dimensions.

3.2 Speed Up Robust Feature (SURF)

Speed Up Robust Feature (SURF) is a feature extraction method that is proposed by Bay et al. [2]. It is a variant of SIFT algorithm. Like SIFT, SURF also uses four stages to extract the features from an image. The difference lies in first stage, where an image convolution of Gaussian derivatives is created using Box filter. And the results are represented in Hessian matrix. SURF is also invariant to scale, rotation, and translations. SURF produces feature vector of 64 or 128 dimensions.

3.3 Oriented FAST and Rotated BRIEF (ORB)

Oriented FAST and Rotated BRIEF (ORB) is proposed by Rublee et al. [3]. The authors developed this speedy and efficient algorithm over SIFT and SURF algorithm. ORB creates a feature vector of only 32 dimensions. ORB used Features from Accelerated Segment Test (FAST) and binary BRIEF algorithm for feature detection. The Harris corner detector is also used to determine the corners of the image. The features extracted through ORB are invariant to scale, rotation, translation, and less sensitive to noise.

3.4 Shi Tomasi Corner Detector

Shi et al. [4] proposed this corner detector algorithm. This is the best algorithm for corner detection. Shi Tomasi algorithm is entirely based on the Harris corner detector with a change in the selection criteria proposed by Shi and Tomasi [13]. This criterion improves the accuracy of corner detection of an image. The score criterion is as follows:

$$R = \min(\lambda_1, \lambda_2) \tag{1}$$

where R represents score criteria. If R is greater than a certain predefined value, then this region is accepted as corner.

4 Classification Techniques

There are many machine learning algorithms used for image classification. In this chapter, four renowned multiclass classification algorithms are used with extracted features for 2D object recognition. These four methods are Gaussian Naïve Bayes, decision tree, random forest, and XGBClassifier. All these classifiers are performing well in the field of image processing and pattern recognition. The authors have chosen them to show a comparison between their performance on multiclass parameter and to acknowledge other researchers about the efficiency of XGBClassifier over other conventional classifiers.

4.1 Gaussian Naïve Bayes

Gaussian Naïve Bayes is an extension of the Naïve Bayes algorithm. It deals with distribution of the continuous data associated with each class according to the normal (or Gaussian) distribution. Gaussian Naïve Bayes is a probabilistic approach that is used for classification by substituting the predicted values x_i of class c in Eq. 2 for normal distribution. The probability is formulated as:

$$P(x_i|c) = \frac{1}{\sqrt{2 * \pi * \sigma_c^2}} e^{-\frac{(x_i-\mu_c)^2}{2\sigma_c^2}} \tag{2}$$

Here π be the mean of the values in x_i of class c and σ_c be the standard deviation of the values in x_i of class c. Gaussian Naïve Bayes is a very simple method of classification.

4.2 Decision Tree

The decision tree classifier was proposed by Quinlan in 1986 [6]. This classifier uses a top-down approach to recursively break the decision-making problem into a few simpler decision problems that are easy to interpret. By splitting a large dataset into individual classes, this model makes a tree like structure where internal nodes represent the features, branches represent the decision rule and leaf nodes represent the outcome, e.g., the label for the object in an object recognition system. Decision tree classifier utilizes the process of breaking up the data till further division is not possible. The root of the tree contains all the labels. Decision tree takes less time for classification and gives more accuracy. But this algorithm has some drawbacks as it leads to poor performance on unseen data. The algorithms have a problem of over-fitting of the data.

4.3 Random Forest

Random forest classifier is an ensemble classifier that consists of many decision trees from a randomly selected subset of training data. This classifier is developed by Kleinberg in 1996. Random forest is a categorical classifier, so it is best suitable for object recognition problem. It takes more time for classification than a decision tree algorithm, but it gives more accurate results than decision tree. This algorithm selects the class for an object by aggregating the votes of all the decision trees. Random forest gives more accurate results than decision tree, but it takes a lot of time to classify the large data. It overcomes the problem of over-fitting of data. It assumes to produce stronger classifiers from a collection of weaker classifiers.

4.4 *XGBClassifier*

eXtreme Gradient Boosting Classifier (XGBClassifier) is proposed by Chen et al. [5]. It is an improvement over gradient boosting classifier. A detailed discussion of this model is presented in Sect. 5.

5 XGBClassifier Model

XGBClassifier is a scalable end-to-end tree boosting algorithm that achieves state-of-the-art results on many machine learning algorithms. The model is proposed by Chen et al. [5]. It is a tree ensemble model that had made many weaker models into stronger model by iteratively predicting each tree. Nowadays, XGBClassifier is producing efficient results as compared to other machine learning algorithms and solves the problem of over-fitting. The main advantage of XGBClassifier is its limited use of resources. This model takes very less time and space for classification [14]. XGBClassifier is summarized as follows.

5.1 *Regularized Learning Objective*

XGBClassifier is a tree ensemble model that uses both classification and regression trees (CART). Mathematically, the model is written as follows:

$$\hat{y}_i = \sum_{k=1}^{K} f_k(x_i), \quad f_k \in F \tag{3}$$

where K is the number of trees and f is the function that works on a set of all possible CARTs. Further, a training objective is formulated to optimize the learning using eq.

$$£(\Phi) = \sum_i l(\hat{y}_i, y_i) + \sum_k \Omega(f_k) \tag{4}$$

where ϕ represents the parameters of the model £. l is the loss function that evaluates the difference between actual label (y_i) and predicted label ($\hat{y}_i$). Ω is a regularization of the model that is used to measure the complexity of the model in order to avoid the over-fitting problem. It is mathematically written as:

$$\Omega(f) = \gamma T + \frac{1}{2}\lambda||w||^2 \tag{5}$$

5.2 Gradient Tree Boosting

It used the additive method to train the data so the function $£(\Phi)$ is modified as

$$£^t = \sum_{i=1}^{n} l(\hat{y}_i^{t-1}, y_i + f_t(x_i)) + \sum_{i=1}^{t} \Omega(f_i) \tag{6}$$

where t represents the iteration and f_t is added to minimize the objective. Then a second-order Taylor expansion is used to remove the constant terms to approximate the objective at step t as follows:

$$\tilde{£}^t = \sum_{n=1}^{N} \left[g_i f_i(x_i) + \frac{1}{2} h_i f_i^2(x_i) \right] + \Omega(f_t) \tag{7}$$

where $g_i = \partial_{\hat{y}_i^{t-1}} l(\hat{y}_i^t, y_i)$ and $h_i = \partial^2_{\hat{y}_i^{t-1}} l(\hat{y}_i^t, y_i)$ are first- and second-order gradient statistics on the loss function.

5.3 Tree Structure

Finally, using Eqs. 5 and 7, Gain is generated to measure the efficiency of the model.

$$\text{Gain} = \frac{1}{2} \left[\frac{G_L^2}{H_L + \lambda} + \frac{G_R^2}{H_R + \lambda} - \frac{(G_L + G_R)^2}{H_L + H_R + \lambda} \right] - \gamma \tag{8}$$

5.4 Parameter Tuning

For multi-classification, few parameters are selected in XGBClassifier. In the experiment, the authors choose objective parameter as multi: softmax and num_class of 101 values for multi-classification. Num_class occupied the value of total classes used for classification. Random_state has value 10. Maximum tree depth is set to 5.

6 Evaluation Measures

Various performance evaluation measures are considered during experiments on the proposed task. These measures are evaluated using multiclass classification parameters. Multiclass classification works on the dataset in which all the classes are mutually exclusive. The Caltech-101 dataset contains 101 classes where all the instances/images are assigned to one and only one class. For example, a flower can be either a lotus or a sunflower but not both at the same time. In multiclass classifier, the evaluation measures of individual classes are averaged out to determine the performance on overall system across the sets of data. There are three methods referring to average results: micro-averaged, macro-averaged, and weighted. Here, the authors have adopted macro-averaging as macro-averaging estimates the performance by averaging the predictive results of each class rather than averaging the predictive results over the whole dataset. Macro-averaging results [15] are computed as:

$$B_{\text{macro}} = \frac{1}{n} \sum_{\lambda=1}^{n} B(\text{TP}_\lambda, \text{FP}_\lambda, \text{TN}_\lambda, \text{FN}_\lambda) \tag{9}$$

where $L = \{\lambda_j : j = 1 \ldots n\}$ is the set of n labels. TP_λ, FP_λ, TN_λ, and FN_λ are the counts for true positives, false positives, true negatives, and false negatives, respectively for a label λ. TP_λ, FP_λ, TN_λ, FN_λ counts are obtained from confusion matrix. The confusion matrix presents the results of the classification of images. It gives a better idea of what the classifier is predicting right and where is it making an error. Confusion matrix helps to compute the accuracy, precision, recall (true positive rate), false positive rate, F1-score, area under curve, root mean squared error of the classification model. These measures are evaluated by substituting true positives (TP), true negatives (TN), false positives (FP), and false negatives (FN) of all the classes (Fig. 1). Each row in the confusion matrix represents the actual class label and each column represents the predicted class label. And each cell in the matrix represents the predicted instances for the given actual class.

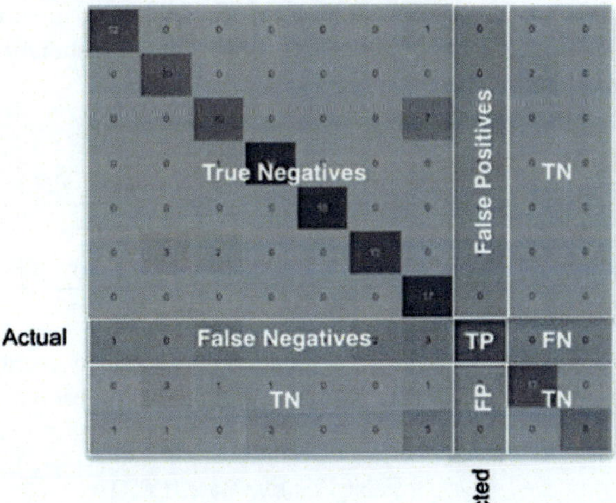

Fig. 1 Confusion matrix for multi-class classifier

True positive (TP): represents the number of correct predicted instances of the positive class.
False positive (FP): represents the number of incorrect predicted instances of the positive class.
True negative (TN): represents the number of correct predicted instances of the negative class.
False negative (FN): represents the number of incorrect predicted instances of the negative class.

Seven commonly used measures, i.e., accuracy, precision, recall (true positive rate), F1-score, area under curve, false positive rate, and root mean square error, are evaluated on the Caltech-101 dataset. Mathematically, they are defined as follows: Balanced accuracy is used as an average of recall obtained in each class [16]. When multiclass classifier is used, balanced accuracy (ACC) gives more accurate results than accuracy measure.

$$\text{Accuracy (ACC)} = \frac{1}{n}\sum_{i=1}^{n}\frac{\text{TP}_i}{\text{TP}_i + \text{FN}_i} \tag{10}$$

Precision is the proportion of correct positive identifications over all positive instances. It is computed as average of true positive instances (TP) over all the predicted instances of each class in multiclass classification [17].

$$\text{Precision } (P) = \frac{1}{n}\sum_{i=1}^{n}\frac{\text{TP}_i}{\text{TP}_i + \text{FP}_i} \tag{11}$$

Recall is the proportion of actual correct positive identifications over actual instances. For multiclass classification, it is used as an average of the true positive rate (TPR) of each class where the true positive rate (TPR) is computed as true positive (TP) matches out of all the actual instances (TP + FN) of the given class [17].

$$\text{TPR} = \frac{\text{TP}}{\text{TP} + \text{FN}} \tag{12}$$

$$\text{Recall (R)} = \frac{1}{n}\sum_{i=1}^{n}\text{TPR}_i \tag{13}$$

False positive rate (FPR) is computed as an average of false positives (FP) over all the instances except for the given label on each class.

$$\text{False Positive Rate (FPR)} = \frac{1}{n}\sum_{i}^{n}\frac{\text{FP}_i}{\text{FP}_i + \text{TN}_i} \tag{14}$$

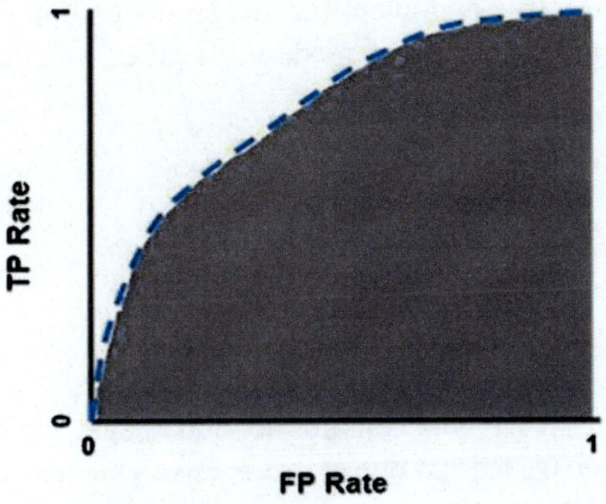

Fig. 2 Diagram representing area under curve (AUC)

F1-score is determined as an average of the harmonic mean of the precision and recall computed on each class [17].

$$\text{F1 - score} = \frac{1}{n}\sum_{i=1}^{n}\frac{2 \times \text{Precision}_i \times \text{Recall}_i}{\text{Precision}_i + \text{Recall}_i} \tag{15}$$

Area under curve (AUC) is used as probability estimates of the performance of the classification model. The value of AUC lies between 0 and 1. Higher the AUC, better the model is predicted. AUC is computed using ROC curve by plotting on TPR against FPR where TPR is taken on y-axis and FPR is on the x-axis of the graph (Fig. 2).

Root mean square error (RMSE) is the standard deviation of the predicted errors. It is used to verify the experimental results of the classifier model.

$$\text{RMSE} = \sqrt{\frac{\sum_{i=1}^{n}(f_i - o_i)^2}{n}} \tag{16}$$

where f is the predicted results and o is the actual results.

7 Data and Methods

7.1 Classification Algorithms

Gaussian Naïve Bayes, decision tree, random forest, and XGBClassifier are used as classification methods in the experiment to determine the comparison among these classifiers. These classifiers are essentially selected because these are basically well-known modern classifiers and able to better optimize the results as compared to other classifiers for image recognition task. The experiment aimed to represent the efficiency of modern classifier, i.e., XGBClassifier over other state-of-the-art classifiers.

7.2 Datasets

The study of multiclass classification measures is carried out on various classifiers on the Caltech-101 image dataset. Caltech-101 dataset consists of 101 categories and each category contains 40–800 images. Caltech-101 is a very challenging dataset for image recognition as it contains a total of 8678 images. All the images of Caltech-101 dataset managed in different categories are mutually exclusive. Partitioning strategy is adopted for the classification task in the experiment where 80% of data of each class is used as training and the remaining 20% of each class is taken as testing data. The overall performance of the system depends on the selection of size of training data. A system with more training data performs more accurately as compared with a system having less training data.

7.3 Software and Code

Feature extraction for the object recognition task is implemented using Open Source packages of Python and OpenCV. The experiment results were calculated by the means of a classification toolbox of Scikit-learn. The classification task is implemented on Online Open Source framework—Jupyter.

8 Experimental Analysis

The proposed system implemented experiments on Caltech-101 image dataset and the report results of a comparison among four state-of-the-art multiclass classification methods. The performance of XGBClassifier is compared with Gaussian Naïve Bayes, decision tree, random forest multiclass classification algorithms. All evaluation measures discussed in the paper are observed on individual classes and the aggregate value is averaged over all the classes. All the experiments are performed on a machine with Microsoft Windows 10 Operating System (original), Intel Core i5 processor with 4 GB RAM. The comparison among the classifier is shown in different tables using performance evaluation measure. Tables 1 and 3 show the comparative view of balanced accuracy and recall, respectively, computed by each multiclass classification method. The results prove that XGBClassifier is 3% better than Gaussian Naïve Bayes classifier and 2% better than random forest and decision tree (Table 2).

Table 1 Quantitative comparison of different classifiers with shape descriptors for object recognition (classifier wise recognition accuracy)

Shape descriptor	Gaussian Naïve Bayes (%)	Decision tree (%)	Random forest (%)	XGBoost (%)
SIFT (I)	53.51	55.02	56.22	64.78
SURF (II)	48.79	48.15	50.07	59.58
ORB (III)	57.03	57.96	60.89	72.01
Shi_Tomasi (IV)	50.27	53.39	57.84	64.84
I + II + III + IV	85.69	86.87	86.74	88.36

Table 2 Quantitative comparison of different classifiers with shape descriptors for object recognition (classifier wise precision)

Shape descriptor	Gaussian Naïve Bayes (%)	Decision tree (%)	Random forest (%)	XGBoost (%)
SIFT (I)	51.27	54.38	55.11	63.14
SURF (II)	45.74	46.17	47.91	57.66
ORB (III)	54.83	58.19	60.71	70.68
Shi_Tomasi (IV)	47.53	53.17	57.93	63.63
I + II + III + IV	85.98	86.48	86.47	88.24

Table 3 Quantitative comparison of different classifiers with shape descriptors for object recognition (classifier wise recall)

Shape descriptor	Gaussian Naïve Bayes (%)	Decision tree (%)	Random forest (%)	XGBoost (%)
SIFT (I)	53.51	55.02	56.22	64.78
SURF (II)	48.79	48.15	50.07	59.58
ORB (III)	57.03	57.96	60.89	72.01
Shi_Tomasi (IV)	50.27	53.39	57.84	64.84
I + II + III + IV	85.69	86.87	86.74	88.36

Table 2 shows the comparison among the precision computed by each multi-class classification method. Table 4 shows the comparison among F1-score. The comparison based on false positive rate shown in Table 5 depicts that XGBClassifier is achieving 0.22%, which is very less as comparative to other classifiers. Table 6 represents the area under curve (AUC) obtained by all the classifiers and XGBClassifier achieved higher results. Table 7 shows the results of root square mean error (RMSE) computed by each classification model.

Table 4 Quantitative comparison of different classifiers with shape descriptors for object recognition (classifier wise F1-score)

Shape Descriptor	Gaussian Naïve Bayes (%)	Decision tree (%)	Random forest (%)	XGBoost (%)
SIFT (I)	51.58	54.17	54.68	63.44
SURF (II)	45.91	46.37	47.61	57.59
ORB (III)	54.84	57.19	59.77	70.72
Shi_Tomasi (IV)	47.75	52.66	56.63	63.35
I + II + III + IV	85.53	86.29	85.78	87.94

Table 5 Quantitative comparison of different classifiers with shape descriptors for object recognition (classifier wise false positive rate)

Shape descriptor	Gaussian Naïve Bayes (%)	Decision tree (%)	Random forest (%)	XGBoost (%)
SIFT (I)	0.57	0.56	0.54	0.51
SURF (II)	0.57	0.61	0.56	0.53
ORB (III)	0.48	0.45	0.41	0.38
Shi_Tomasi (IV)	0.53	0.52	0.47	0.45
I + II + III + IV	0.24	0.24	0.23	0.22

Table 6 Quantitative comparison of different classifiers with shape descriptors for object recognition (classifier wise area under curve)

Shape descriptor	Gaussian Naïve Bayes (%)	Decision tree (%)	Random forest (%)	XGBoost (%)
SIFT (I)	51.93	52.07	52.30	82.14
SURF (II)	50.13	50.17	50.32	79.53
ORB (III)	53.56	53.55	53.70	85.82
Shi_Tomasi (IV)	52.72	52.65	52.65	82.20
I + II + III + IV	92.73	93.32	93.26	94.07

Table 7 Quantitative comparison of different classifiers with shape descriptors for object recognition (classifier wise root square mean error)

Shape descriptor	Gaussian Naïve Bayes (%)	Decision tree (%)	Random forest (%)	XGBoost (%)
SIFT (I)	29.21	29.83	29.28	28.68
SURF (II)	33.56	34.21	32.90	31.20
ORB (III)	30.48	29.20	28.08	26.14
Shi_Tomasi (IV)	30.99	30.48	30.28	28.72
I + II + III + IV	23.52	23.09	22.22	22.59

9 Conclusion

This chapter presents the efficiency of XGBClassifier for 2D-object recognition. Various handcrafted methods of shape feature extraction are used in the experimental work. The methods used are SIFT, SURF, ORB, and Shi Tomasi corner detector. A comparative analysis is made among these feature extraction algorithms and a hybrid of all these algorithms is considered. The experimental results are reported on all these features using various classifiers. The classification methods used are Gaussian Naïve Bayes, decision tree, random forest, and XGBClassifier. Various performance evaluation measures on multiclass classification are described in this chapter. Dataset partitioning method on multiclass dataset Caltech-101 is adopted for the classification of the dataset in the ratio of 4:1 for training and testing dataset. The experiment proved that XGBClassifier is the best among other existing classifiers. In future work, other handcrafted and deep learning methods of feature extraction will be developed to improve the efficiency of the model.

References

1. Lowe DG (2004) Distinctive image features from scale-invariant keypoints. Int J Comput Vision 60(2):91–110
2. Bay H, Tuytelaars T, Van-Gool, L (2006) Surf: speeded up robust features. In: Proceedings of the European conference on computer vision, pp 404–417
3. Rublee E, Rabaud V, Konolige K, Bradski GR (2011) ORB: an efficient alternative to SIFT or SURF. In: International conference computer vision, vol 11, no 1, p 2
4. Shi J, Tomasi (1994) Good features to track. In: Proceedings of IEEE conference on computer vision and pattern recognition, pp 593–600
5. Chen T, Guestrin C (2016) XGBoost: a scalable tree boosting system. In: Proceedings of the 22nd ACM SIGKDD international conference on knowledge discovery and data mining, pp 785–794
6. Quinlan JR (1986) Induction of Decision Trees. Mach Learn 1(1):81–106
7. Kleinberg EM (1996) An overtraining-resistant stochastic modeling method for pattern recognition. Ann Stat 24(6):2319–2349
8. Fei-Fei L, Fergus R, Perona P (2004) Learning generative visual models from few training examples: an incremental Bayesian approach tested on 101 object categories. In: Proceedings of the conference on computer vision and pattern recognition workshop, pp 178–178
9. Ren X, Guo H, Li S, Wang S, Li J (2017) A novel image classification method with CNN-XGBoost model. Lecture Notes in Computer Science, pp 378–390
10. Santhanam, R., Uzir, N., Raman, S. and Banerjee, S. (2017) Experimenting XGBoost algorithm for prediction and classification of different datasets. In: Proceedings of the national conference on recent innovations in software engineering and computer technologies (NCRISECT), Chennai
11. Bansal A, Kaur S (2018) Extreme gradient boosting based tuning for classification in intrusion detection systems. In: Proceedings of the international conference on advances in computing and data sciences, pp 372–380
12. Vo T, Nguyen T, Le CT, A hybrid framework for smile detection in class imbalance scenarios. Neur Comput Appl, 1–10
13. Harris C, Stephens M (1988) A combined corner and edge detector. In: Proceedings of the fourth Alvey vision conference, pp 147–151

14. Song R, Chen S, Deng B, Li L (2016) eXtreme gradient boosting for identifying individual users across different digital devices. Lecture Notes in Computer Science, pp 43–54
15. Asch VV (2013) Macro-and micro-averaged evaluation measures. CLiPS, Belgium, pp 1–27
16. Brodersen KH, Ong CS, Stephan KE, Buhmann JM (2010) The balanced accuracy and its posterior distribution. In: Proceedings of the 20th international conference on pattern recognition, pp 3121–3124
17. Godbole S, Sarawagi S (2004) Discriminative methods for multi-labeled classification. In: Proceedings of the Pacific-Asia conference on knowledge discovery and data mining, pp 22–30

Comparison of Principle Component Analysis and Stacked Autoencoder on NSL-KDD Dataset

Kuldeep Singh, Lakhwinder Kaur, and Raman Maini

Abstract In the traditional era, there was no concern of time and the memory space, the processing was the main issues to solve any problem. But in the modern era, large space and high processing are available. The main concern is to reduce the time to solve any problem. In computer networks, malicious activities are increasing rapidly due to the exponential growth of the number of users on the Internet. Many classification models have been developed that classify the malicious user from benign, but all requires large amount of training data. The main challenge of this field is to reduce the volume and dimension of the data used for training that will speed up the detection process. In this work, the two dimensionality reduction techniques principal component analysis (PCA) and autoencoders are compared on standard NSL-KDD dataset using 10% data for training the classifiers. The results of these techniques are tested on different machine learning classifiers like tree-based, SVM, KNN and ensemble learning. Most of the intrusion detection technique tested on benchmark NSL-KDD dataset. But the standard NSL-KDD dataset is not balanced, i.e., for some classes, this dataset has an insufficient number of records that are difficult to train and test the model for multiclass classification. The imbalance problem of the dataset is solved by creating extended NSL-KDD dataset by merging the standard NSL-KDD train and test set. From the experiments, it is evident that autoencoders extract the better deep features than PCA on binary class and multiclass classification. The achieved accuracy by autoencoders on 2-class (95.42%), 5-class (95.71%), 22-class (97.63%) and F-score on 2 class (95.49%), 5-class (74.79%), 22-class (79.18%), which is significant more than other compared classifiers that are trained using extracted features by PCA.

K. Singh (✉) · L. Kaur · R. Maini
Department of Computer Science and Engineering, Punjabi University Patiala, Patiala, India
e-mail: sidhu.kuldeep89@gmail.com

L. Kaur
e-mail: mahal2k8@gmail.com

R. Maini
e-mail: research.raman@gmail.com

V. Singh et al. (eds.), *Computational Methods and Data Engineering*,
Advances in Intelligent Systems and Computing 1227,
https://doi.org/10.1007/978-981-15-6876-3_17

Keywords Intrusion detection system · Deep learning · Dimensionality reduction

1 Introduction

Due to rapidly increase in size of the Internet, the security-related services like authentication, availability, confidentiality, non-repudiation, etc., are difficult to ensure for all users. Malicious activity like infected source code, denial of services, probe, virus and worms is increasing day by day. These types of events are difficult to identify because, in today's technology, they are using self-modifying structure after a while. That is why most of the intrusion detection models become obsolete after some time. The second major challenge for the current intrusion detection system (IDS) is to monitor a large amount of data present on the Internet to identify multiple attacks without degrading the performance of the network.

Deep learning changes the way we interpret data in various domains like network security, speech recognition, image processing, hyperspectral theory, etc. [1–3]. It works efficiently on large datasets in which other conventional methods are not well suited. Deep learning is a subfield of machine learning that is based on the learning algorithm which represents the complex relationship among data. It processes the information by using forward and backwards learning and describes the higher level concepts from lower level concepts [2]. Deep learning also used to represent high-dimensional information to low-dimensional information that makes easy to process the enormous amount of data present on the network.

Artificial neural network-based intrusion detection models that used machine learning and deep learning require period training to update their definitions. To classify the user present on the network whether it is abnormal or benign, the large number of variables is required that are known as features of the networks. If all of the features are selected to train and test the IDS model, then model takes a huge amount of time and if a smaller number of features are selected for training and testing, the performance of the model to classify the malicious activities degrades. Here is the place where dimensionality reduction plays a vital role to reduce the IDS training and testing time. The dimensionality reduction process either selects the high ranked features or information of a large number of features are represented by a low number of features that are known as feature extraction. Many linear and nonlinear-based dimensionality reduction algorithms like PCA, Isomap, locality preserving projections (LPP), locally linear embedding (LLE), linear discriminant analysis (LDA), autoencoders, etc., are used in the literature [4]. This work compares the linear dimensionality reduction (DR) algorithm PCA with nonlinear-based DR autoencoder on NSL-KDD extended dataset. The experiment results show that DR using nonlinear autoencoders achieves higher accuracy than linear-based PCA algorithm.

2 Related Work

Almotiri et al. [5] compared the principal component analysis (PCA) and deep learning autoencoder on Mixed National Institute of Standard and technology (MNIST) of handwritten character recognition dataset. The authors demonstrate that the autoencoder has better dimensionality reduction that has 98.1% accuracy than PCA which has 97.2% accuracy on the considered dataset. Sakurada and Yairi [6] proposed the anomaly detection technique based on autoencoder. The authors also compare the dimensionality reduction using autoencoder with linear and kernel PCA using artificial data generated from the Lorenz system and the real data (The spacecraft's data). The better results are shown by autoencoder than PCA.

Wang et al. [7] demonstrate the dimensionality reduction ability of autoencoder. The authors used the MNIST handwritten character recognition and Olivetti face detection datasets. The work is compared with state-of-the-art dimensionality reduction techniques; PCA, LDA, LLE and Isomap and experimentally showed that autoencoder performs better denasality reduction than compared techniques. Lakhina et al. [8] used the PCA algorithm for dimensionality reduction on NSL-KDD dataset. The authors reduce the training time by reducing the features of the dataset from 41 to 8 and achieved the same detection rate as with whole 41 features achieved. But the authors used 100% NSL-KDD train dataset to train the ANN which can be further reduced. Mukherjeea and Sharmaa [9] proposed the feature vitality-based reduction method (FVBRM) of feature selection on NSL-KDD dataset. In this work, the 24 features are selected out of total 41 features, and authors compared the proposed method with other feature selection methods, information gain (IG), gain ratio (GR) and correlation-based feature selection (CFS). The proposed method achieved the 97.8% overall accuracy which is more than the other compared techniques. Salo et al. [10] proposed the information gain—PCA (IG-PCA) hybrid technique of dimensionality reduction for IDS. The authors tested the proposed technique by ensemble classifier that is based on SVM, IBK and MLP on three datasets, NSL-KDD, Kyoto-2006 + and ISCx-2012. The selected features in NSL-KDD (13), Kyoto-2006 + (10) and ISCX-2012(9) achieved accuracy 98.24%, 98.95%, 99.01%, respectively. Panda et al. [11] proposed a discriminative multinomial Naïve Bayes (DMNB) technique for network intrusion detection. PCA is used for dimensionality reduction. The results are evaluated on NSL-KDD dataset with 96.50% accuracy. Singh et al. [12] proposed the online sequential extreme learning machine (OS-ELM) technique for intrusion detection. The time complexity is reduced by alpha profiling technique and hybrid technique of three techniques; hybrid, correlation and consistency-based are used for feature selection. The author used NSL-KDD dataset for result evaluation and achieved 98.66% accuracy for binary class and 97.67% for multiclass classification. The proposed method has also tested on Kyoto dataset with acquired accuracy 96.37%. DelaHoz et al. [13] presented a hybrid network anomaly classification technique that is based on statistical techniques and self-organizing maps (SOMS). PCA and fisher discriminant radio (FDR) have been used for feature selection. The presented technique is tested on NSL-KDD dataset with accuracy (90%), sensitivity (97%) and specificity (93%). Osanaiye et al. [14] proposed an ensemble base multi filtered feature selection method that is information gain, gain ratio, chi-square and relief techniques. The results are tested on NSL-KDD dataset. The final 13 reduced features from each technique have been selected by vote majority. The authors used J48 classifier and achieved 99.67% accuracy.

3 Dimensionality Reduction, Autoencoder and PCA

This section explains the dimensionality reduction process and compared DR algorithms autoencoders and PCA.

3.1 Dimensionality Reduction

In intrusion detection process, there are too many factors on the basis of which the classification of normal and abnormal user is performed. These factors are known as features of the network. Higher the number of features, the more time it takes to train and test the classification models. Moreover, some features are correlated in some way with other features. Hence, more duplicate information reduces the performance of the classification models. This is the situation where the dimensionality reduction algorithms play important role.

Dimensionality reduction is the process of reduction of high-dimensional feature space into low-dimensional feature space. It is mainly of two types (a) feature extraction and (b) feature selection. In feature extraction process, the high-dimensional feature space is mapped to low-dimensional feature space [4]. In feature selection process only, the high rank features are selected by filtration, wrapping and embedding methods. Many dimensionality reduction techniques have been used in the literature. Some of them are principle components analysis (PCA) [15], locality preserving projections (LPP) [16], Isomap [17], deep learning autoencoders [7], etc. (Fig. 1).

Fig. 1 Visualization of dimensionality reduction process

3.2 Autoencoders

AEs are unsupervised deep learning [18, 19] neural networks which has backpropagation [20, 21] algorithm for learning. AEs are used to represent the high order input vector space to intermediate low order vector space, and later, it reconstructs the output equivalent to given input from intermediate low order representation. This represents the dimensionality reduction characteristics like PCA. But PCA works only for linear transformation, and AEs work for both linear and nonlinear transformation of data.

AEs have three layers; input layer represented by X, hidden layer also known as bottleneck, represented by H and output layer which is represented by X' as shown in Fig. 2.

Single layer autoencoder has three layers as shown in Fig. 3. Hence, f is the activation function, W_1 is the input weight factor, and b_1 is the biased amount for input layer. Similarly, W_2 is weight, and $b2$ is bias amount for hidden layer. The following two steps define how the intermediate representation from the input layer and reconstruction from the hidden layer is performed.

$$h(X) = f(W_1 x_i + b_1) \tag{1}$$

$$X' = f(W_2 h(X) + b_2) \tag{2}$$

The following optimization function is used to minimize the error between input vector space and reconstructed vector space from the hidden layer.

$$\underset{W_1, b_1, W_2, b_2}{\arg\min}\ [K] = \underset{W_1, b_1, W_2, b_2}{\arg\min} \left[(1/2) \sum_{i=1}^{d} \| x_i - x_i' \| + K_1 + K_2 \right] \tag{3}$$

where K is the squared reconstructed error, K_1 and K_2 are weight decay and sparse penalty terms, x_i is the ith value of input vector and x_i' corresponding reconstructed vector.

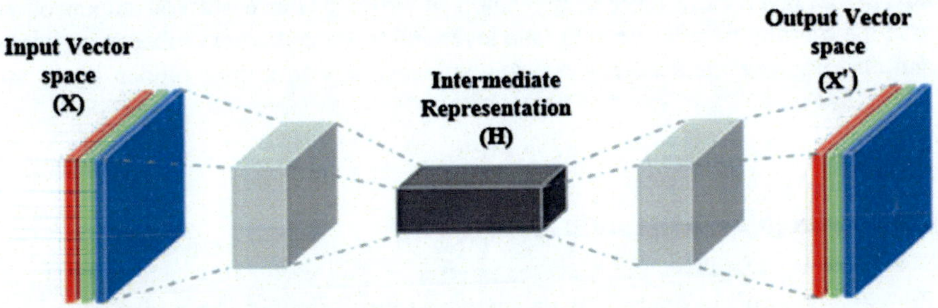

Fig. 2 Visualization of dimensionality reduction process

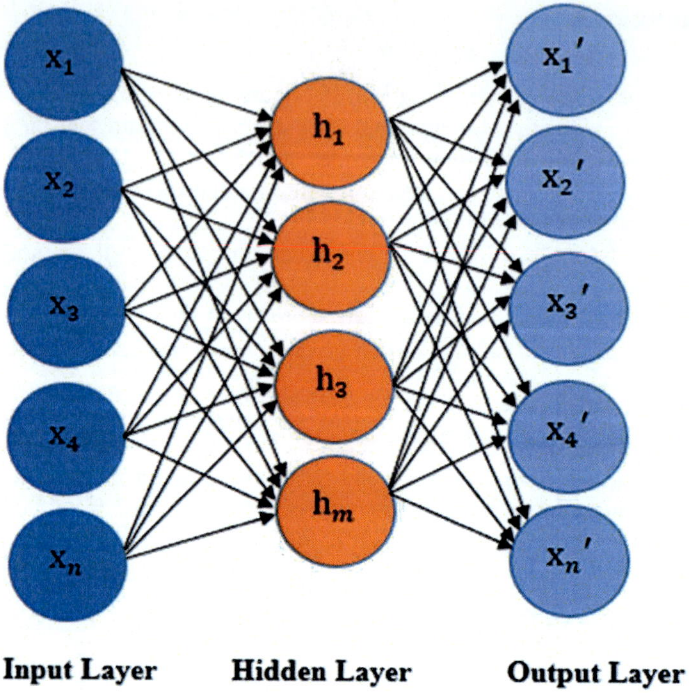

Fig. 3 Single layer autoencoder

4 Stacked Autoencoders

Stacked autoencoders are the extension of the simple autoencoder in which multiple hidden layers clubbed together to learn the deep features from the given input data [22]. The output of the one layer is given as an input to the next layer. Hence, the first hidden layer of stacked autoencoder learns the first order deep features from the input raw data. The second layer of stacked autoencoder learns the second order deep features corresponding to the features learn by the first layer, and similarly, the next higher layer learns more deep features of the data. Hence, stacked autoencoders save the training time by freezing one layer and training the next subsequent layers and also improve the accuracy. Stacked autoencoders with three hidden layers are shown in Fig. 4.

4.1 Principle Component Analysis

Principle component analysis is unsupervised dimensionality reduction algorithm that is used to transform the high-dimensional vector space to low-dimensional vector.

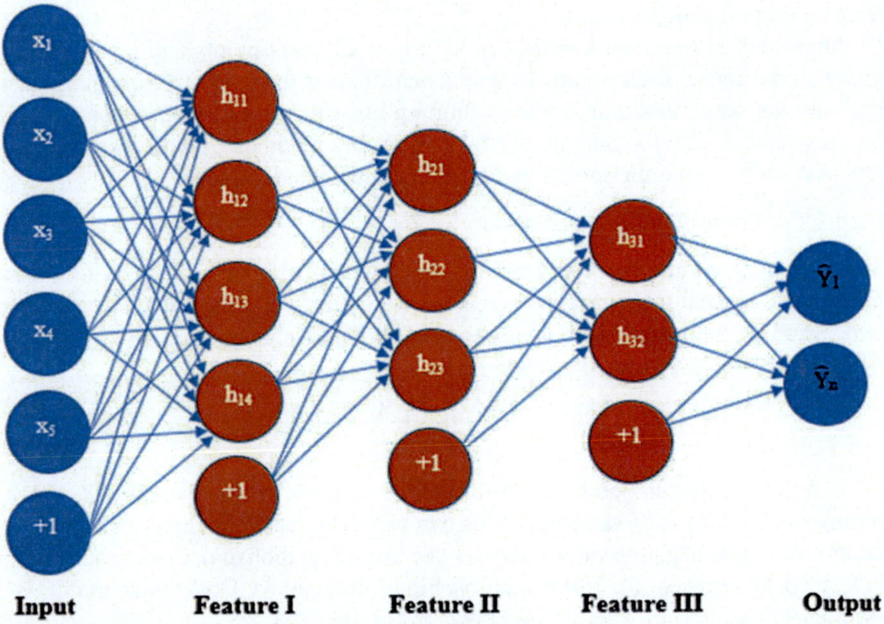

Fig. 4 Stacked autoencoders with three hidden layers

It is also used to visualize the data in low dimensions space, for noise reduction and for outlier detection. Among the 'N' features in dataset, PCA preserves the 'd' features with maximum covariance where $d \ll$ N. These features are orthogonal to each other, and these are known as principal components as shown in Fig. 5. Two methods are used to calculate the principal component: Covariance matrix and singular value decomposition.

Steps in PCA algorithm

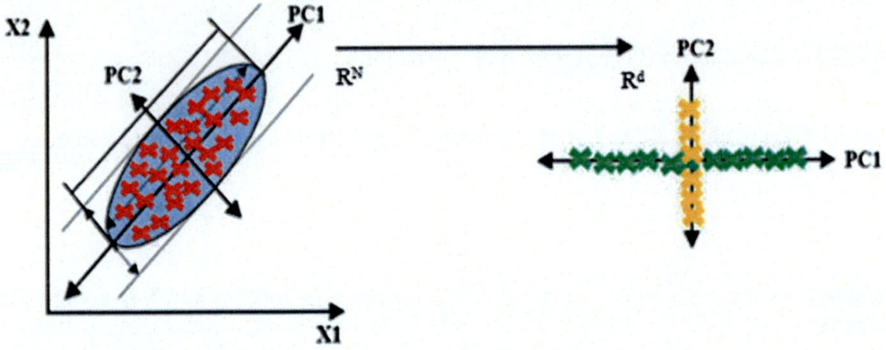

Fig. 5 Principal components of data across the maximum variance

Step-1: Preprocessing of data

Suppose X is a dataset having $\{x_1, x_2, x_3 \ldots x_n\}$ data points and having R^N, N dimensional space. Data preprocessing is mainly used to remove the errors in data, scale up the data, removing outlier, filling up missing values and transforming the un-computed values to some units of values. Many methods of data preprocessing are used such as column normalization and column standardization.

Step-2 Covariance matrix calculation

Variance of any variable represents the deviation of that variable from its mean. Covariance is used to represent the relation between two variables. If X and Y are two variables, then covariance is represented by C_{xy} as shown in Eq. (4)

$$C_{xy} = \frac{\sum(X_i - \mu_x)\left(Y_i - \mu_y\right)}{N} \tag{4}$$

X_i represents the all points in variable X, and μ_x is the mean of variable X, and μ_y represents the means of variable Y. Positive value indicates the direct or increasing relationship, and negative value indicates decreasing relationship. Covariance matrix is formed to represent the linear relationship of data points. Covariance matrix is a symmetric matrix (i.e., $X = X^{\mathrm{T}}$) as shown in Eq. (5)

$$\begin{pmatrix} \mathrm{Var}(x_1, x_1) & \mathrm{Cov}(x_1, x_2) & \mathrm{Cov}(x_1, x_3) & \cdots & \mathrm{Cov}(x_1, x_M) \\ \mathrm{Cov}(x_2, x_1) & \mathrm{Var}(x_2, x_2) & \mathrm{Cov}(x_2, x_3) & \cdots & \mathrm{Cov}(x_2, x_M) \\ \mathrm{Cov}(x_3, x_1) & \mathrm{Cov}(x_3, x_2) & \mathrm{Var}(x_3, x_3) & \cdots & \mathrm{Cov}(x_3, x_M) \\ \vdots & \vdots & \vdots & \vdots & \vdots \\ \mathrm{Cov}(x_3, x_M) & \mathrm{Cov}(x_M, x_2) & \mathrm{Cov}(x_M, x_3) & \cdots & \mathrm{Var}(x_M, x_M) \end{pmatrix} \tag{5}$$

Covariance matrix is formed to represent the linear relationship of data points as shown is Eq. (6) represented by E

$$E = XX^{\mathrm{T}} \tag{6}$$

Step-3 Calculation of Eigenvector and eigenvalue

Eigenvector is non-zero vectors that are used to represent the direction of the data points. Eigenvalues are scalar values which represent the magnitude or spreadness of data around that particular eigenvector.

$$EV = \lambda V \tag{7}$$

where E is the covariance matrix, V is the eigenvector matrix, and λ is eigenvalue matrix.

Step-4 Construction of lower-dimensional space

Eigenvalues and eigenvector are used to construct the lower-dimensional space. Select the d number of eigenvectors corresponding to d number of largest eigenvalues. $\lambda_1, \lambda_2, \lambda_3, \ldots \lambda_d$ are the first d largest eigenvalues corresponding to eigenvectors V_1, $V_2, V_3, \ldots V_d$.

$$F_{N*d} = X_{N*M} * V_{M*d} \quad (8)$$

X is the original data matrix having N rows and M features, and V is the eigenvector matrix having M rows and d number of features. F is the data matrix formed by PCA after applying transformations.

Methodology

This section explains the extended NSL-KDD dataset that uses to test the performance to two DR algorithm, normalization used to preprocess the dataset, various types of the attacks present in the dataset and different types the metrics used to evaluate the performance of the DR algorithms.

4.2 Extended NSL-KDD Dataset

In this work, the standard benchmark NSL-KDD dataset has been used which was initially collected by Cyber System and Technology Group MIT Laboratory as a KDD'99 dataset. But original dataset had many duplicates records which was removed by Tavallaee et al. [23] and proposed new dataset known as NSL-KDD.

NSL-KDD dataset contains two sets, first set train data having a total of 125,917 records and the second set is test data which have 22,544 records. Due to insufficient number of records in some classes, the dataset creates a problem to test the efficacy of the designed IDS models. This problem is solved by combining both train and test set of NSL-KDD dataset named as extended NSL-KDD, which has total 148,517 records (as shown in Table 1) and 41 types of features with one class label. For binary class, the label has two values, normal and anomalous connections, and for multiclass, the labels are classified as normal and attack group which are categorized mainly in four types, denial of services (DoS), probe, user to root (U2R), root to local (R2L).

All 41 features are mainly under three data types, nominal, binary and numeric. Features 2, 3, 4 are nominal type, 7, 12, 14, 15, 21, 22 are binary, and all remaining features are a numeric type. To perform the experiment, the nominal features are

Table 1 Composition of NSL-KDD train and test data in totality

NSL-KDD	Records	Normal	DoS	Probe	R2L	U2R
Train + Test	148,517	77,054	53,387	14,077	3880	119
	%	51.88	35.94	9.47	2.6	0.08

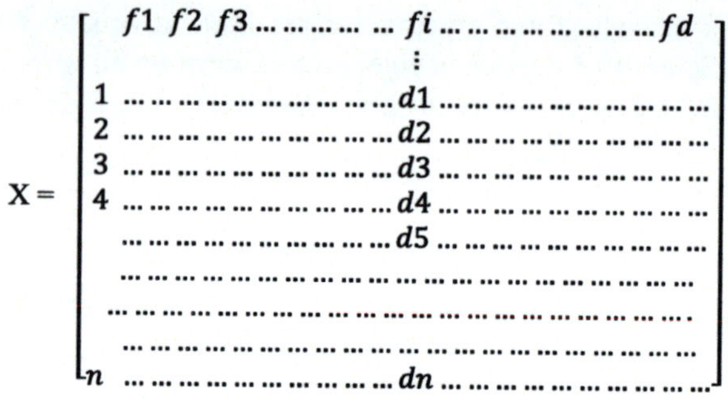

Fig. 6 Visualization representation of dataset

converted into numeric features by assigning numbers (like tcp = 1, udp = 2, …). One feature contains all records having zero values, so it is eliminated as it does not have any effect on the experiment.

4.3 Normalization

Normalization is the process of transforming the features on a common scale and changing the statistics like mean and standard deviation to speed up the calculations used in training and testing of the dataset. In this paper, two types of data normalization have been used: column normalization and column standardization.

Let $f_1, f_2, f_3, \ldots, f_d$ are the total features and n is the total number of records in dataset as shown in Fig. 6. $f_i = \{d_1, d_2, d_3, \ldots d_n\}$ are the data points in each feature.

4.4 Column Standardization

This method, mean of each feature in the data is shifted to origin and standard deviation of all the features transformed to unity.

$$d_i' = \frac{d_i - \bar{d}}{\sigma} \tag{9}$$

Column standardization technique transformed data $d, d_2, d_3 \ldots d_n$ in each feature into points into standard values, i.e., $d_1', d_2', \ldots d_n'$ by setting the mean of the transformed data to zero and standard deviation σ to 1.

$$\bar{d} = \sum_{i=1}^{n} d_i' = 0 \tag{10}$$

$$\sigma = \sqrt{\frac{\sum_{i=1}^{n} \left(d_i' - \bar{d}\right)^2}{n}} = 1 \tag{11}$$

where $\bar{d}$ is the sample mean of data and σ is the sample standard deviation of column standardized data.

4.5 Performance Metrics

All the techniques have been evaluated using the following metrics [24].

- **Precision**: It is the measure of the number of abnormal users or events present in the network rightly classified as abnormal to the total users predicted as abnormal, i.e., true positive and false positive.

$$\textbf{Precision}: \frac{\text{TP}}{\text{TP} + \text{FP}} \tag{12}$$

- **True Negative (TN) rate**: TN is also known as specificity. It is the measure of the number of normal users or events present in the network rightly classified as normal.

$$\textbf{TN (Specificity)}: \frac{\text{TN}}{\text{TN} + \text{FP}} \tag{13}$$

- **True Positive (TP) rate**: TP is also known as recall or probability of detection or sensitivity. It is the measure of the number of abnormal users or events present in the network rightly classified as abnormal.

$$\textbf{TP (Sensitivity)}: \frac{\text{TP}}{\text{TP} + \text{FN}} \tag{14}$$

- **False Negative (FN) rate**: It is the measure of the number of abnormal users or events present in the network misclassified as normal.

$$\textbf{FN (Miss Rate)}: \frac{\text{FN}}{\text{FN} + \text{TP}} = 1 - \text{Sensitivity} \tag{15}$$

- **False Positive (FP) rate**: It is the measure of the number of normal users or events present in the network misclassified as abnormal.

$$\textbf{FP (Fallout)}: \frac{\text{FP}}{\text{FP} + \text{TN}} = 1 - \text{Specificity} \tag{16}$$

- **Accuracy**: It is the measure of the number of users or events present in the network correctly classified by the total number of users.

$$\textbf{Accuracy}: \frac{\mathrm{TP}+\mathrm{TN}}{\mathrm{TN}+\mathrm{TP}+\mathrm{FN}+\mathrm{FP}} \tag{17}$$

- **F-Score**: It is the measure of the harmonic mean of precision and recall which represents the predictive power of classification model.

$$\textbf{F–Score}: 2 \cdot \frac{\mathrm{Precision} \cdot \mathrm{Recall}}{\mathrm{Precision}+\mathrm{Recall}} \tag{18}$$

All these metrics are calculated on individual class, and overall accuracy is calculated for all classes.

5 Experimental Result and Analysis

This work compares the two DR algorithms on extended NSL-KDD dataset which is benchmark for intrusion detection. The experiment starts with preprocessing of dataset in which all non-numeric fields are converted to numeric fields, column standardization normalization is used which transformed the mean of all data items to zero and standard deviation to unity. Then, 10% of the total extended NSL-KDD dataset is randomly selected and applied SAEs to extract the deep features, and SoftMax layer of SAEs is trained using these deep features, and finally, trained model is tested using remaining 90% data. Similar work is done for PCA algorithm, and the reduced dimensions obtained by PCA are used to train different ML classifiers, and then, all trained classifiers are tested using remaining 90% data. The flowchart of the performed work is shown in Fig. 7.

In this research work, the simulation is carried out with 10% training sample of extended NSL-KDD dataset. Classification accuracy of stacked autoencoder is compared with 21 machine learning classifiers by using PCA algorithm. All 40 features of extended NSL-KDD dataset are given as input to two-layer SAEs. Out of 40 features, the 35 deep features are extracted by first layer, and these 35 features are given as input to second layer that further extract the 30 deep features. Then, softmax regression layer is applied to classify the labels of the data.

Similarly, 40 input features are given as an input to PCA, and most promising 30 principle components are selected for the comparison with SAEs. These selected components are used to train the 21 ML classifiers. All trained classifiers are tested using remaining 90% data, and values of various performance metrics like precision, recall, FN, specificity, FP, class wise accuracy, overall accuracy and F-Score are shown in Table 2 and Figs. 8, 9, 10 and 11.

The presented result in Table 2 and Figs. 8, 9, 10 and 11 shows that the deep features extracted by SAEs are more significant than the similar features extracted

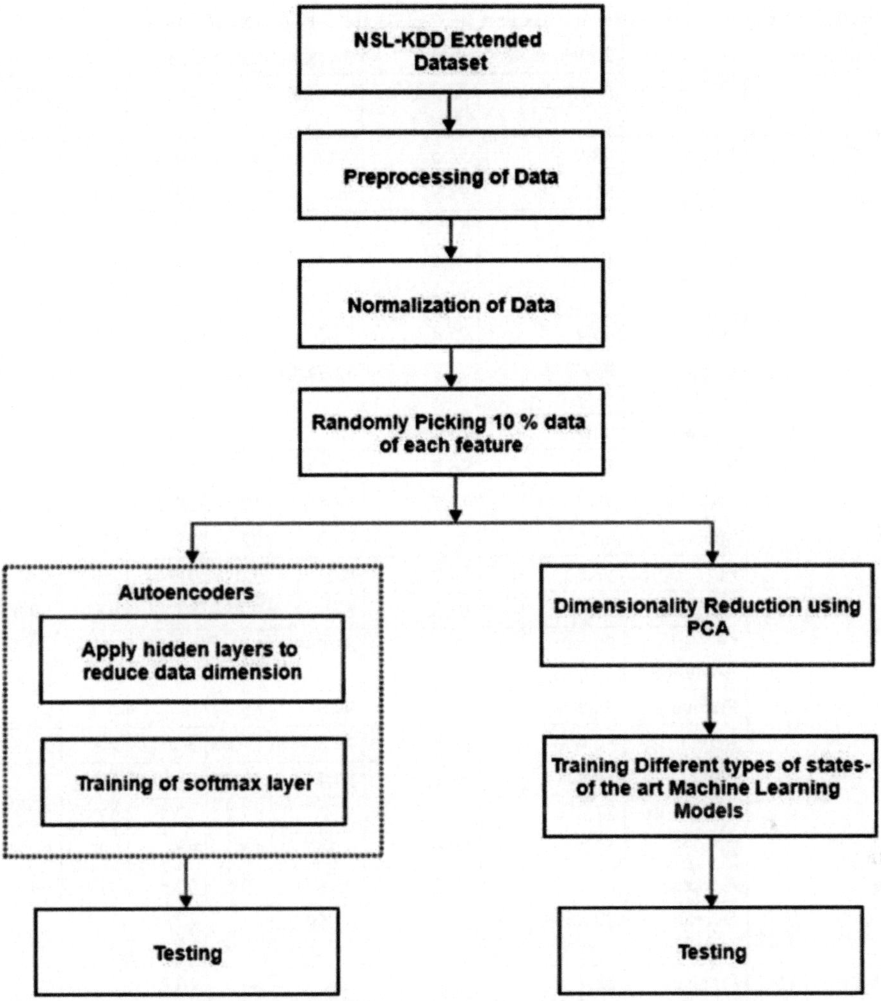

Fig. 7 Flowchart of compared DR algorithms autoencoders and PCA

by PCA. The obtained accuracy by using the extracted feature from SAEs on 2-class is 95.42%, on 5-class is 95.71% and on 22-class is 97.63%, whereas the accuracy of models trained by using extracted feature with PCA is maximum on 2-class is 85.99%, on 5-class is 83.09% and on 22-class is 83.97%. Experiment shows that the SAEs perform better dimensionality reduction than PCA on intrusion detection dataset.

Table 2 Performance of different classifiers for 2-class and 5-class classification

Techniques	Metrics (%)	2-class classification		5-class classification				
		Normal class	Abnormal class	Normal	Dos	Probe	U2R	R2L
SAEs	Precision	*92.9*	*98.5*	*94.1*	*99.1*	*94.2*	*81.9*	*52.8*
	Recall	*98.7*	*91.9*	*99*	*96.6*	*93.1*	*31*	*16*
	FN	*1.3*	*8.1*	*1*	*3.4*	*6.9*	*69*	*84*
	Specificity	*91.9*	*98.7*	*93.3*	*99.5*	*99.4*	*99.8*	*100*
	FP	*8.1*	*1.3*	*6.7*	*0.5*	*0.6*	*0.2*	*0*
	Accuracy	*95.4*	*95.4*	*96.2*	*98.5*	*98.8*	*98*	*99.9*
	Overall Accuracy	**95.42%**		**95.71%**				
Medium tree	Precision	88.7	78	97	86.2	18.7	0.2	0
	Recall	81.3	86.5	81.5	87.9	60.4	7.2	X
	FN	18.7	13.5	18.5	12.1	39.6	92.8	X
	Specificity	86.5	81.3	95.9	92.3	92.1	97.4	99.9
	FP	13.5	18.7	4.1	7.7	7.9	2.6	0.1
	Accuracy	83.5	83.5	87	90.8	91.1	97.3	99.9
	Overall Accuracy	83.54%		83.09%				
Coarse Tree	Precision	83.5	78	99.8	85.7	0	0	0
	Recall	80.4	81.4	79.8	87.8	5.1	X	X
	FN	19.6	18.6	20.2	12.2	94.9	X	X
	Specificity	81.4	80.4	99.7	92.1	90.5	97.4	99.9
	FP	18.6	19.6	0.3	7.9	9.5	2.6	0.1
	Accuracy	80.9	80.9	86.8	90.6	90.5	97.4	99.9
	Overall Accuracy	80.88%		82.59%				
Linear SVM	Precision	88.7	76.7	86.6	64.9	16.6	0.3	0
	Recall	80.4	86.3	78.9	71.3	15.4	12.7	0
	FN	19.6	13.7	21.1	28.7	84.6	87.3	100
	Specificity	86.3	80.4	83.9	81.3	91.2	97.4	99.9
	FP	13.7	19.6	16.1	18.7	8.8	2.6	0.1
	Accuracy	82.9	82.9	81	78	83.5	97.3	99.9
	Overall Accuracy	82.93%		69.85%				
Coarse Gaussian SVM	Precision	91.8	74.5	96.5	65.8	15	0	0
	Recall	79.5	89.4	77.7	87.8	16.6	0	X
	FN	20.5	10.6	22.3	12.2	83.4	100	X

(continued)

Table 2 (continued)

Techniques	Metrics (%)	2-class classification		5-class classification				
		Normal class	Abnormal class	Normal	Dos	Probe	U2R	R2L
	Specificity	89.4	79.5	94.8	83.2	91.2	97.4	99.9
	FP	10.6	20.5	5.2	16.8	8.8	2.6	0.1
	Accuracy	83.5	83.5	83.8	84.4	84.8	97.4	99.9
	Overall Accuracy	83.47%		75.13%				
Fine KNN	Precision	88.5	75.7	84.7	83	13.3	0.4	0
	Recall	79.7	85.9	78.4	73.7	36.7	9.6	0
	FN	20.3	14.1	21.6	26.3	63.3	90.4	100
	Specificity	85.9	79.7	81.9	89.7	91.5	97.4	99.9
	FP	14.1	20.3	18.1	10.3	8.5	2.6	0.1
	Accuracy	82.3	82.3	80	83.3	89.6	97.3	99.9
	Overall Accuracy	82.32%		75.03%				
Weighted KNN	Precision	87.2	76.9	83.9	85.7	7.7	0.1	0
	Recall	80.3	84.8	79.4	72.1	30.9	1.4	0
	FN	19.7	15.2	20.6	27.9	69.1	98.6	100
	Specificity	84.8	80.3	81.6	91	91	97.4	99.9
	FP	15.2	19.7	18.4	9	9	2.6	0.1
	Accuracy	82.2	82.2	80.4	83	89.6	97.3	99.9
	Overall Accuracy	82.23%		75.09%				
Boosted Tree	Precision	92.2	77.5	98.7	79.5	9.6	0.1	0
	Recall	81.5	90.2	78.2	89	38.7	2.1	X
	FN	18.5	9.8	21.8	11	61.3	97.9	X
	Specificity	90.2	81.5	98	89.2	91.2	97.4	99.9
	FP	9.8	18.5	2	10.8	8.8	2.6	0.1
	Accuracy	85.1	85.1	85	89.1	90	97.3	99.9
	Overall Accuracy	85.13%		80.67%				
RUSBoosted Tree	Precision	88.6	83.2	87.3	82	12.8	23.7	61.3
	Recall	85	87.1	82.9	86.4	23.6	22.5	1.5
	FN	15	12.9	17.1	13.6	76.4	77.5	98.5
	Specificity	87.1	85	85.5	90.2	91.3	97.9	100
	FP	12.9	15	14.5	9.8	8.7	2.1	0
	Accuracy	86	86	84	88.9	87.8	95.9	96.7

(continued)

Table 2 (continued)

Techniques	Metrics (%)	2-class classification		5-class classification				
		Normal class	Abnormal class	Normal	Dos	Probe	U2R	R2L
	Overall Accuracy	85.99%		76.65%				

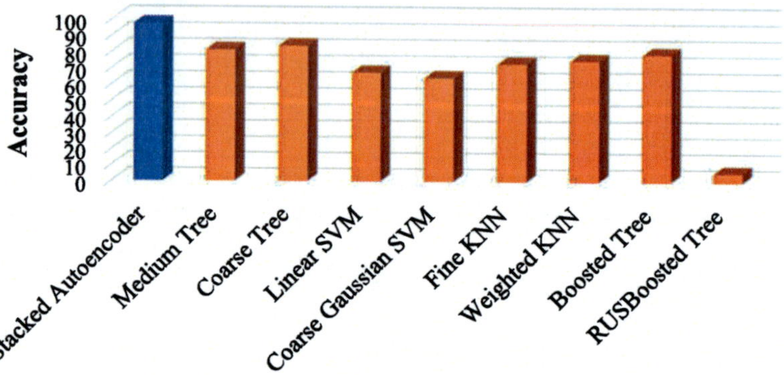

Fig. 8 Performance of different classifiers for 22-class classification

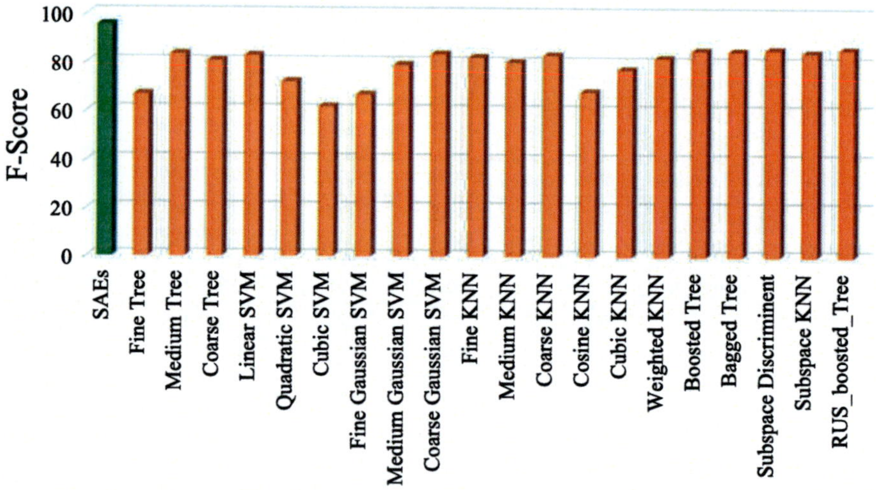

Fig. 9 F-score value of classifiers in binary class classification

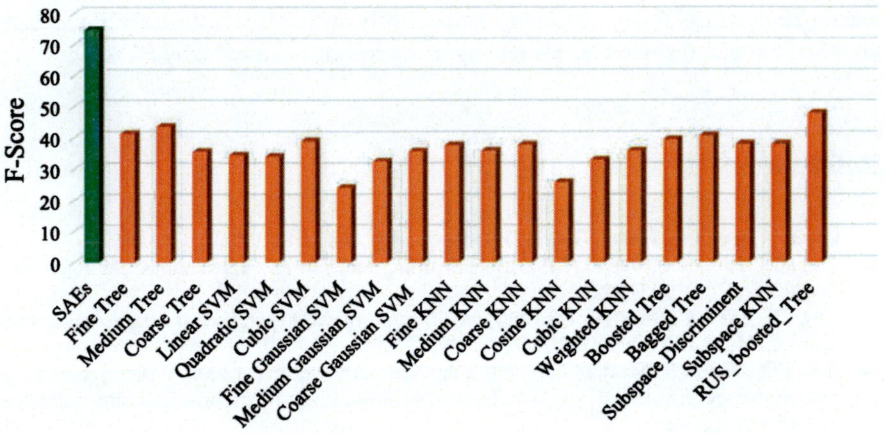

Fig. 10 F-score value of classifiers in 5-class classification

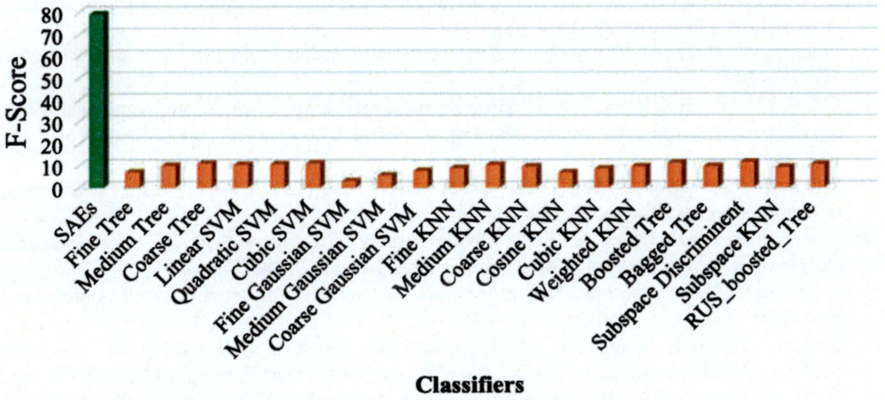

Fig. 11 F-score value of classifiers in 22-class classification

6 Conclusion

This work compares the neural network-based linear dimensionality reduction technique PCA and nonlinear dimensionality reduction technique autoencoders. The standard NSL-KDD extended dataset is used to test the efficacy of both techniques. Stacked autoencoder and different ML-based classifiers are trained using 10% dataset, obtained after selecting 30 deep features from all 41 features. Experimentally, it is observed that deep features extracted by autoencoders are more useful to train the classifiers for intrusion detection which increase accuracy and F-score of the classifier as compared to feature extracted by PCA technique. The achieved accuracy and F-score on 2-class (95.42%), (95.49%), on 5-class (95.71%), (74.79%)

and on 22-class (97.63%), (79.18%), respectively, which is significantly higher than all other compared classifiers trained by using features extracted by PCA.

References

1. Lecun Y, Bengio Y, Hinton G (2015) Deep learning. Nature 521(7553):436–444
2. Ahmad J, Farman H, Jan Z (2019) Deep learning methods and applications. SpringerBriefs Comput Sci 7(2013):31–42
3. Singh S, Kasana SS (2018) Efficient classification of the hyperspectral images using deep learning. Multimed Tools Appl 77(20):27061–27074
4. Box PO, Van Der Maaten L, Postma E, Van Den Herik J (2009) Tilburg centre for creative computing dimensionality reduction: a comparative review dimensionality reduction: a comparative review
5. Almotiri J, Elleithy K, Elleithy A (2017) Comparison of autoencoder and principal component analysis followed by neural network for e-learning using handwritten recognition. In: 2017 IEEE long island Systems, applications and technology conference (LISAT 2017), no October
6. U. S. A. C. of Engineers (1994) Distribution restriction statement approved for public release ; distribution is. U.S. Army Corps Eng.
7. Wang Y, Yao H, Zhao S (2016) Auto-encoder based dimensionality reduction. Neurocomputing 184:232–242
8. Lakhina S, Joseph S, Verma B (2010) Feature reduction using principal component analysis for effective anomaly-based intrusion detection on NSL-KDD. Int J Eng Sci Technol 2(6):1790–1799
9. Mukherjee S, Sharma N (2012) Intrusion detection using Naive Bayes classifier with feature reduction. Procedia Technol 4:119–128
10. Salo F, Nassif AB, Essex A (2019) Dimensionality reduction with IG-PCA and ensemble classifier for network intrusion detection. Comput Netw 148:164–175
11. Abraham A (2010) Discriminative multinomial Naïve Bayes for network Intrusion detection, pp 5–10
12. Singh R, Kumar H, Singla RK (2015) An intrusion detection system using network traffic profiling and online sequential extreme learning machine. Expert Syst Appl 42(22):8609–8624
13. De la Hoz E, De La Hoz E, Ortiz A, Ortega J, Prieto B (2015) PCA filtering and probabilistic SOM for network intrusion detection. Neurocomputing 164:71–81
14. Osanaiye O, Cai H, Choo KKR, Dehghantanha A, Xu Z, Dlodlo M (2016) Ensemble-based multi-filter feature selection method for DDoS detection in cloud computing. Eurasip J Wirel Commun Netw 1:2016
15. Eid HF, Darwish A, Ella Hassanien A, Abraham A (2010) Principle components analysis and support vector machine based intrusion detection system. In: Proceedings of the 2010 10th international conference on intelligent systems design and applications (ISDA'10), pp 363–367
16. N. Info and N. Info (1998) Sam T. Roweis 1 and Lawrence K. Saul 2, vol 2, no 1994
17. de Silva V, Tenenbaum JB (2003) Global versus local methods in nonlinear dimensionality reduction. Adv Neural Inf Process Syst 15:705–712
18. Chuan-long Y, Yue-fei Z, Jin-long F, Xin-zheng H (2017) A deep learning approach for intrusion detection using recurrent neural networks, vol 3536, no c
19. Shone N, Ngoc TN, Phai VD, Shi Q (2018) A deep learning approach to network intrusion detection, vol 2, no 1, pp 41–50
20. Farahnakian F, Heikkonen J (2018) A deep auto-encoder based approach for intrusion detection system
21. Lee B, Green C (2018) Comparative study of deep learning models for network intrusion detection, vol 1, no 1

22. Singh S, Kasana SS, Efficient classification of the hyperspectral images using deep learning, pp 1–19
23. Tavallaee M, Bagheri E, Lu W, Ghorbani AA (2009) A detailed analysis of the KDD CUP 99 data set. In: Computational intelligence for security and defense applications, no. Cisda, pp 1–6
24. Hodo E, Bellekens X, Hamilton A, Tachtatzis C, Shallow and deep networks intrusion detection system : a taxonomy and survey, pp 1–43

Maintainability Configuration for Component-Based Systems Using Fuzzy Approach

Kiran Narang, Puneet Goswami, and K. Ram Kumar

Abstract Maintenance is one of the extremely important and tricky missions in the area of component-based software. Numerous maintainability models are proposed by the scientist and researchers, to reduce the cost of maintenance, for improving the excellence and life period of a component-based system. Various quality models have been discussed briefly to show importance of maintainability. This research will facilitate the software designer to assemble maintainable component-based softwares. The proposed configuration confers a fuzzy-based maintainability model that chooses four fundamental features that enormously influence maintainability of component-based software system, i.e., Document Quality, Testability, Coupling, and Modifiability (DTMC). MATLAB's fuzzy logic toolbox is utilized to implement this configuration and output values are confirmed using center of gravity formula, as we have taken centroid defuzzification method. For a particular set of input, output provided by the model is 0.497 and output value from center of gravity formula comes up to be 0.467 which is around the value specified by the model.

Keywords Boehm's quality model · Component-based system · Coupling · Document quality · ISO 9126 · Maintainability · MATLAB fuzzy logic · McCall's quality models · Modifiability · Reusability · Testability

---

K. Narang (✉) · P. Goswami · K. Ram Kumar
SRM University, Sonepat, Haryana, India
e-mail: kkirannarang@gmail.com

P. Goswami
e-mail: goswamipuneet@gmail.com

K. Ram Kumar
e-mail: ramkumar@srmuniversity.ac.in

V. Singh et al. (eds.), *Computational Methods and Data Engineering*, Advances in Intelligent Systems and Computing 1227,
https://doi.org/10.1007/978-981-15-6876-3_18

1 Introduction

This component-based software engineering (CBSE) is a technique to design and develop software by reusing already built software components [1]. The principal objective of component-based development is 'Buy—Don't Build.' In modern era, CBSE has acquired esteem recognition because of growing demand of complex and up to date software [2]. The major advantages of CBSE include cost-effectiveness, express, and modular method of developing complex software with compact release time [3–5].

Maintainability of CBSE is defined as, lenience to modify the software product after delivery, so as to make it more efficient and to adopt new technology [6]. Maintenance became necessary because of business growth; bug fixing, update access, user adoption, and reengineering [7, 8]. The expenditure on maintenance can be as high as 65% of the total expenditure of the software. Generally, the development stage of software is merely three to five years while maintenance stage may possibly be of twenty years or more [9–11]. Software maintenance is categorized into the following four branches [12]:

1.1 Corrective Maintenance

Corrective maintenance is required to correct or fix problems, which are observed by the user at the time of system being used.

1.2 Adaptive Maintenance

Adaptive maintenance is required by the software to keep it up to date according to the latest technology available in the market.

1.3 Perfective Maintenance

Perfective maintenance is required in order to keep the software functional over an elongated period of time.

1.4 Preventive Maintenance

Preventive maintenance is performed regularly on working software with the motivation to deal with forthcoming problems and unexpected failure [13–15].

The proposed model is able to predict the maintainability and to reduce the maintenance effort of component-based system (CBS), by selecting only maintainable components as the part of system and rejecting the other ones [16]. The proposed DTMC model will help the modern clients of today's digital world, who does not want their software to be down for even a microsecond. Further the cost to maintain the software after it goes down and time to maintain it can also be saved by the use of this configuration.

Maintainability of CBS cannot be determined directly, we require some factors to evaluate it. Several factors that influence the maintainability such as complexity, modularity, understandability, usability, analyzability, coupling, and cohesion, etc. are discussed in Table 1 [17–19]. We have proposed a fuzzy approach to determine maintainability of CBS with the features supported by numerous investigators, i.e., Document Quality, Testability, Coupling, and Modifiability (DTCM).

This paper is organized in five sections. Second section describes significance of maintainability in quality models. Third section discusses literature survey. Fourth section discusses proposed approach for calculating maintainability. Fifth part shows results and comparative analysis of the proposed research. Sixth section discusses conclusion and future scope of the research. Last section shows the list of references.

Table 1 A brief description of few factors that influence maintainability

Documentation quality	Complete and concise documentation along with the software makes it easy to operate and use
Cohesiveness	It is the degree of belongingness inside the modules of a software component
Coupling	Frequency of the communication among the assorted components in a system is referred to as its coupling
Modularity	Modularity is the extent to which a system or a component may be separated (break down), recombined, for the purpose of flexibility and the diversity to utilize it
Understandability	Property with which a system can be understood effortlessly
Extensibility	Extensibility is the convenience enjoyed while adding innovative functionalities and features to the existing software
Modifiability	Modifiability is the extent of easiness with which amendments can be made in a system
Granularity	It refers to flouting down of larger tasks into smaller and new light tasks
Testability	It is the measure to which a system assist the establishment of analysis and tests [20]

2 Importance of Maintainability in Quality Models

Various quality models in the literature discuss the characteristics and sub-characteristics that influence eminence of the software product. Out of all these models, it is tricky to find the best model; however, all of these models have given place to maintainability as an important characteristic to attain a good quality product [21, 22].

2.1 *ISO 9126 Quality Model*

ISO 9126 quality model is an element of ISO 9000 which was launched to authenticate excellence of a software package [23]. Basic quality characteristics according to this model are:

- Functionality
- Reliability
- Usability
- Efficiency
- Maintainability
- Portability.

2.2 *Mccall's Quality Model*

McCall classified quality features into three components:

- Product Revision
- Product Transition
- Product Operation.

Further these components have eleven quality characteristics within them. Maintainability comes under the Product Revision.

2.3 *Boehm's Quality Model*

Boehm symbolized a model with three levels of hierarchy. Factors that belong to upper level in hierarchy have greater impact to quality of software as compared to the factors at the lower level.

- High-level characteristics
- Intermediate level characteristics
- Primitive characteristics.

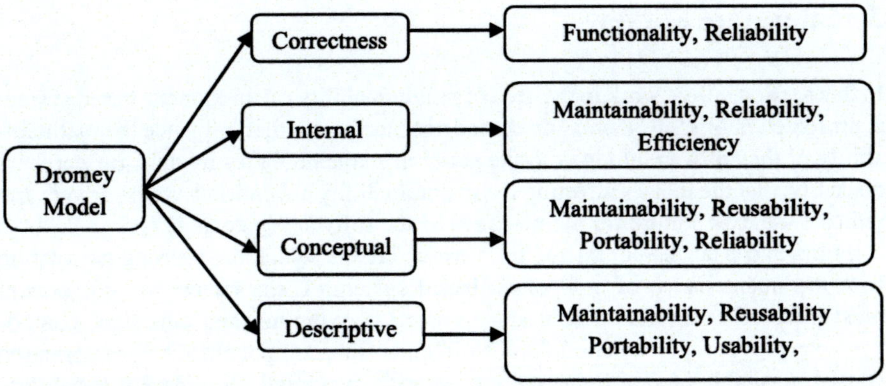

Fig. 1 Dromey model

High level Characteristics of Boehm's Quality model includes Maintainability As-Is Utility and Portability [24, 25].

2.4 Dromey Model

Dromey discussed the correlation among the quality features and their sub-attributes. This model tried to hook up the software package attributes with its quality attributes. It is a quality model based on product representation and it distinguishes that quality assessment procedure differs from product to product [26]. Figure 1 shows Dromey model which is based on the perspective of product quality.

2.5 Functionality Usability Reliability Performance Supportability (FURPS)

The following specialties are considered by FURPS model:

- Functionality
- Usability
- Reliability
- Performance
- Supportability.

Maintainability is included in Supportability as its sub-characteristics.

3 Literature Survey

In this section, allied work in the area of maintainability of component-based system is presented in brief. B. Kumar discussed various factors that influence the maintainability of the software along with the study of maintainability models. Researchers discussed that the factors affecting the maintainability are entitled as cost drivers and indicate the expenditure for maintenance of the software systems [27].

Punia and Kaur had extracted the various factors which have strong correlation with the maintenance of component-based system. These factors are component reusability (CR), available outgoing interaction, coupling and cohesion, component interface complexity, modifiability, integrability, component average interaction density, available incoming interaction, average cyclomatic complexity, testability, granularity [19, 28]. Kumar and Dhanda discussed that maintainability of a system design is influenced by numerous factors, including Extendibility and Flexibility as high impact factors [9]. Sharma and Baliyan compared maintainability attributes of quality models; McCall, Boehm's and ISO 9126 for component-based systems. Novel features tailorability, tractability, reusability, and scalability that affect the maintainability of CBS are introduced [13].

Aggarwal et al. proposed a fuzzy maintainability system that incorporates four factors, i.e., Live Variables (LV), Comment Ratio (CR), Average Cyclomatic Complexity (ACC) and average Life Span (LS) of variables [29]. Different analyst disclosed diverse attributes that affect the maintainability of component-based software (CBS). Table 1 reveals few of these aspects briefly.

4 Proposed DTCM Methodology Using Fuzzy Approach

The research proposes a fuzzy maintainability configuration for component-based system. Complicated problems can be effortlessly solved by using fuzzy logic which in turn comprises fuzzy set theory. One main attribute of fuzzy approach is that it can utilize English words as input and output instead of numerical data, which are named as Linguistic Variables (LV) [30]. Figure 2 conceptualizes functionality of

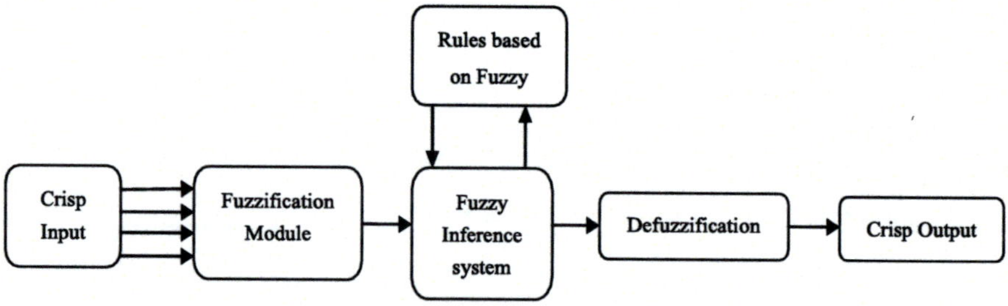

Fig. 2 Fuzzy logic-based DTMC model

fuzzy DTMC model for maintainability.

4.1 Crisp Input

For the purpose of input, this research chooses four factors out of several factors described in Table 1, on the basis of their deep correlation with the maintainability. These factors are **D**ocumentation Quality, **T**estability, **C**oupling, and **M**odifiability (DTCM) and output maintainability. Figure 3 shows the inputs and output of the fuzzy DTMC model.

Documentations Quality: The documentation clarifies the way to operate and use software and also it may be utilized by different people for different purposes. Reliable and good quality documentation of a component is the sole way to judge its applicability and to bring confidence among client and collaborators. A component with high-quality documentation is said to be more maintainable as compared to poorly documented component. So maintainability is directly proportional to Documentation Quality [31, 32].

Testability: It is the measure to which a system assists the establishment of analysis and tests. Higher the testability of a software component, easy will be the faultfinding process and hence easy to maintain it [2, 8, 28].

Coupling: It is the degree of closeness or relationship of various components or modules [33]. Low coupling is an indication of a good design and supports the universal aim of high maintainability, reduced maintenance, and modification costs [7, 34]. So maintainability is inversely proportional to coupling.

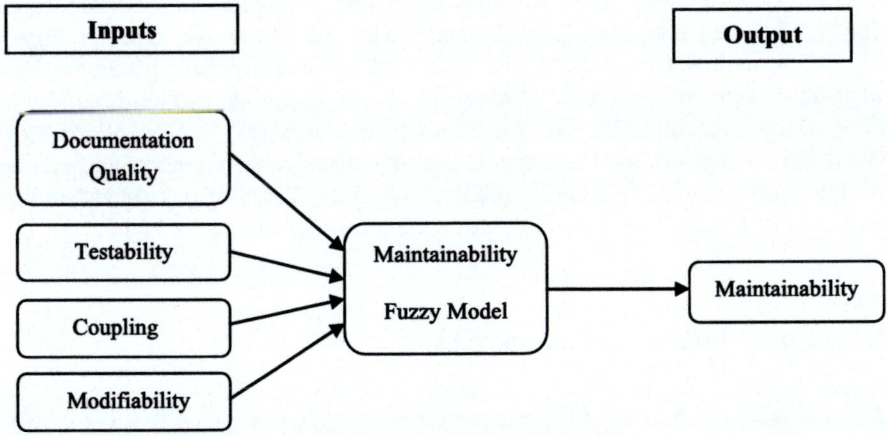

Fig. 3 Inputs and outputs of proposed fuzzy maintainability model

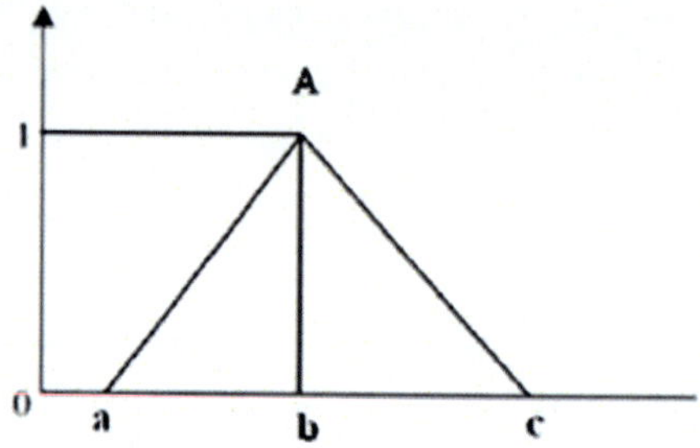

$$\mathrm{triangle}(x; a, b, c) = \begin{cases} 0, & x \leq a. \\ \frac{x-a}{b-a}, & a \leq x \leq b. \\ \frac{c-x}{c-b}, & b \leq x \leq c. \\ 0, & c \leq x. \end{cases}$$

Fig. 4 Triangular membership function (TFN)

Modifiability: It is the extent of easiness with which amendments can be made in a system, and the system adapts these changes such as new environment, requirements, and functional specification. Higher the modifiability parameter, easy will be the maintenance; hence, maintainability is directly proportional to modifiability [8, 12].

4.2 *Fuzzification Module*

Fuzzification module converts these inputs into their corresponding fuzzy value. Fuzzification is the method of converting a true input value into a fuzzy value. Various fuzzifiers available in MATLAB fuzzy toolbox to perform fuzzification are Trapezoidal, Gaussian, S Function, Singleton, and Triangular fuzzy number (TFN). We have utilized TFN in proposed model, due to its simplicity. Figure 4 demonstrates TFN which have a lower bound, center bound, upper bound; 'a,' 'b,' and 'c,' respectively.

For DTCM model's input, we have taken three triangular membership functions (TFN), i.e., minimum, average, and maximum. Maximum indicates higher value of Testability, Coupling, Modifiability, and Document Quality. Minimum indicates lower value of Document Quality, Testability, Coupling, and Modifiability in fuzzy. Output maintainability has five membership functions (TFN), i.e., Worst, Bad, Good, Fair, and Excellent. Figures 5, 6, 7, and 8 visualize the membership function for the input variables Document Quality, Testability, Coupling, and Modifiability, respectively. Figure 9 describes TFN for maintainability (output variable).

4.3 *Fuzzy Inference System (FIS)*

A fuzzy inference system (FIS) is a method of mapping an input to an output by means of fuzzy logic. It is the most important module in the proposed maintainability model, as the whole model relies upon the decision-making capability of FIS. FIS perform decision-making with the help of the rules that are inserted into the rule

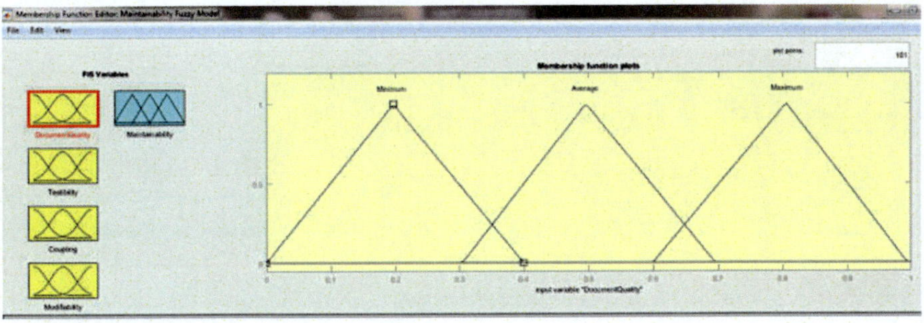

Fig. 5 TFN for input value Document Quality

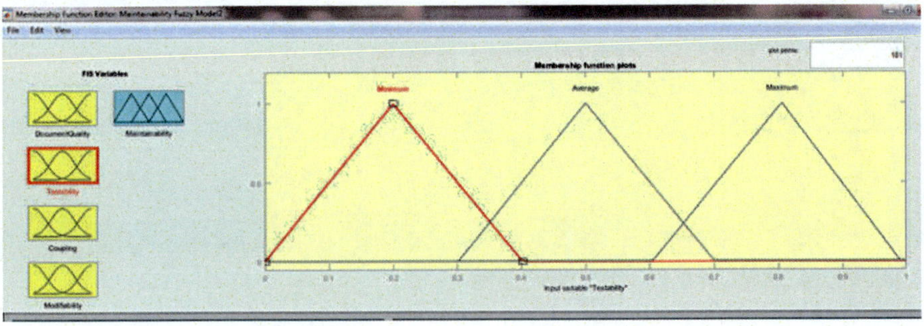

Fig. 6 TFN for input value testability

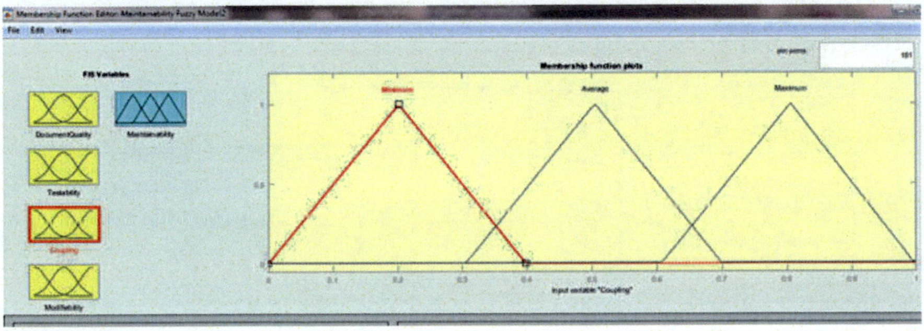

Fig. 7 TFN for input value coupling

editor. MATLAB fuzzy logic toolbox has two kinds of fuzzy inference systems, i.e., Mamdani-type and Sugeno-type. Figure 10 conceptualizes the Mamdani-type FIS of the DTMC model.

Formula for the calculation of total number of rules is given by the following equation:

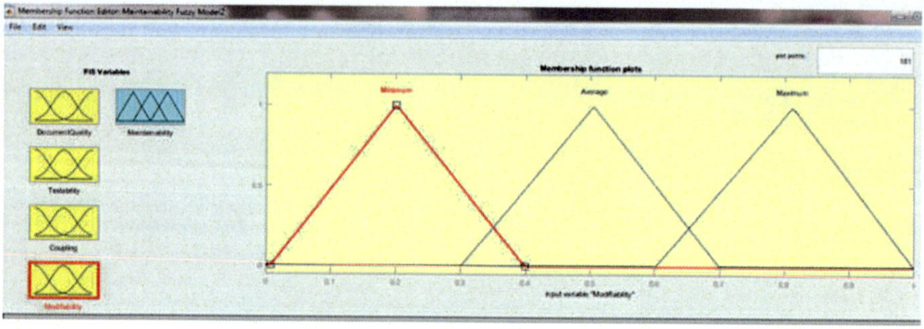

Fig. 8 TFN for input value modifiability

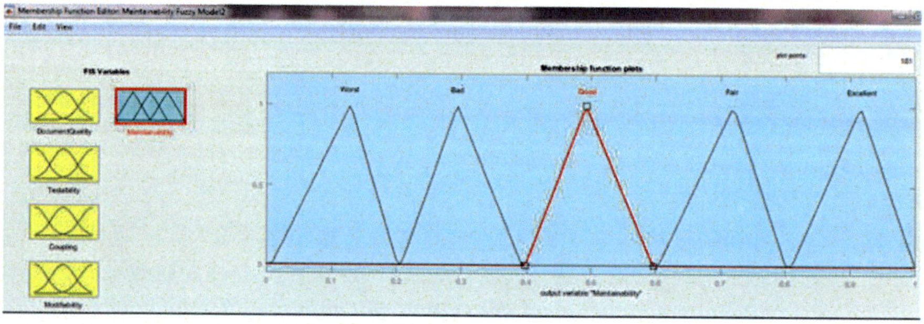

Fig. 9 TFN for the output variable maintainability

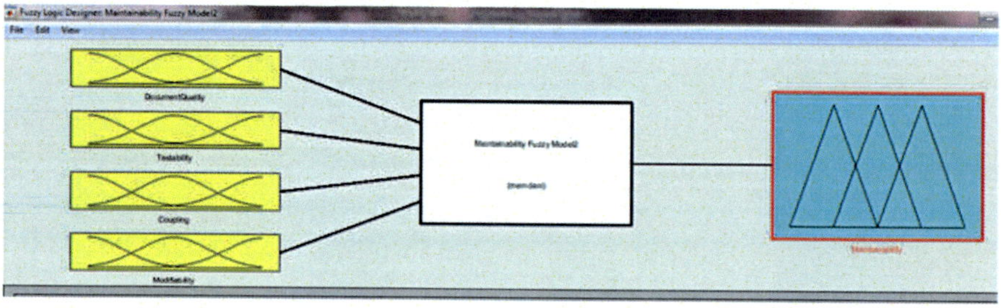

Fig. 10 FIS of the DTMC model

$$(\text{Number of the membership functions})^{\text{Number of inputs}}$$

In the proposed DTMC model, numbers of membership functions are three, i.e., maximum, average, minimum, and number of inputs are four, i.e., Document Quality, Testability, Coupling, and Modifiability. So the rules formed according to the equation will be $3^4 = 81$. Some of rules for the DTMC model are illustrated below:

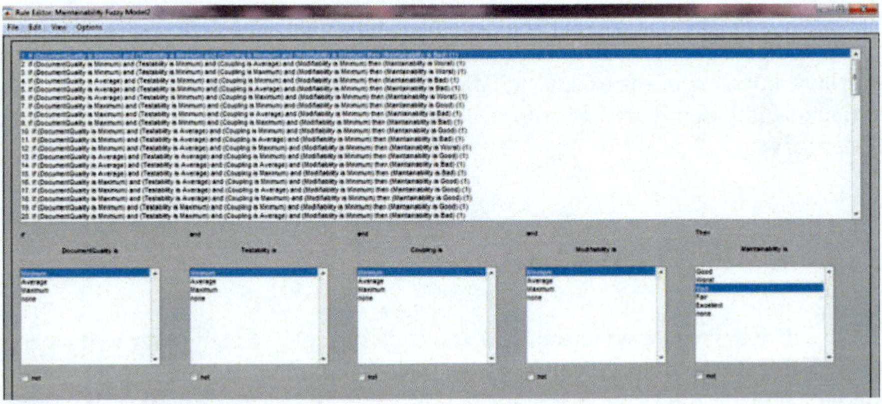

Fig. 11 Rule editor for DTMC configuration

- Document Quality—Minimum, Testability—Minimum, Coupling—Maximum, Modifiability—Minimum.
  Maintainability—Worst.
- Document Quality—Minimum, Testability—Minimum, Coupling—Average, Modifiability—Minimum.
  Maintainability—Bad.
- Document Quality—Maximum, Testability—Maximum, Coupling—Minimum, Modifiability—Maximum.
  Maintainability—Excellent.
- Document Quality—Maximum, Testability—Maximum, Coupling—Average, Modifiability—Maximum.
- Maintainability—Fair Document Quality—Maximum, Testability—Average, Coupling—Minimum Modifiability—Minimum.
  Maintainability—Good.

All rules are created by this method and pop into the rule editor, to form a rule base for the fuzzy DTMC model. Depending on the accurate information supplied by experts, the rules are fired to get the values for output maintainability and related graphs are plotted. Figure 11 shows the rule editor for DTMC configuration.

4.4 *Aggregation, Defuzzification, Crisp Output*

Result calculation process for a particular input uses aggregation which is done in the defuzzification module. It means that output for a particular input is calculated by testing and combining certain rules into a single one. Aggregation combines the output of all the rules that satisfies the given input.

Defuzzifier converts the aggregated fuzzy output value into crisp value. MATLAB fuzzy toolbox supports five built in defuzzification schemes, i.e., smallest of maximum, largest of maximum, middle of maximum, bisector, and centroid. This maintainability configuration utilizes centroid method that finds the center of area under curve.

5 Results of DTCM Model

To visualize the result, go to the view and click on 'rule.' Rule viewer will emerge to demonstrate the defuzzification of the input values. For determining the output, we will provide the input at the left bottom input box of the rule viewer. Output will be displayed on the top right corner. Table 2 shows the result for certain input values; Document Quality 0.5, Testability 0.75, Coupling 0.7, Modifiability 0.5. Maintainability model gives the outcome to be 0.497. MATLAB rule viewer shows the output for the same input in Fig. 12. For verification of the output, we have employed center of gravity formula given in Eq. 1 and it comes out to be 0.476, and it is roughly similar to the output specified by the proposed model.

Table 2 CBSE maintainability results for a particular input set

Input values				Output value
Document Quality	Testability	Coupling	Modifiability	Maintainability
0.5	0.75	0.7	0.5	0.497

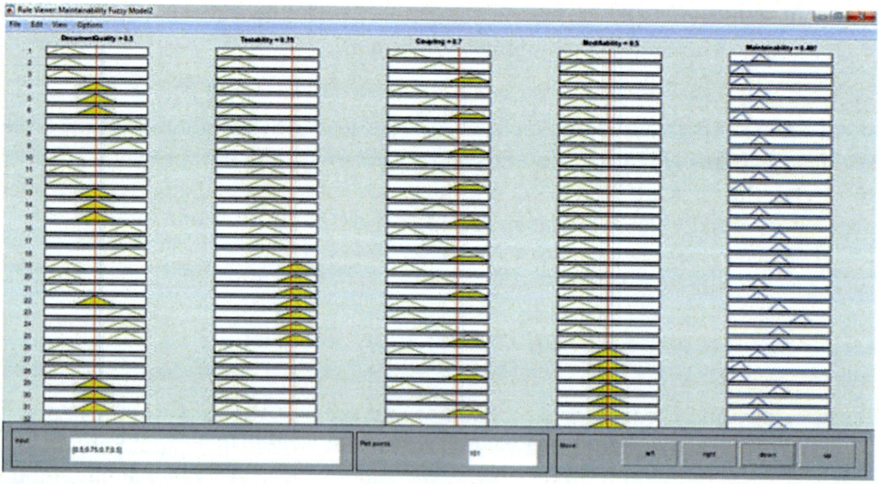

Fig. 12 Rule viewer

$$\text{Centre of gravity} = \frac{\int yx\mathrm{d}x + \int yx\mathrm{d}x}{\int y\mathrm{d}x + \int y\mathrm{d}x} = 0.476 \quad (1)$$

Figure 13 represents the surface viewer for 3D view of Document Quality, Testability, and Maintainability. Figure 14 represents the surface viewer for 3D view of Document Quality, Coupling, and Maintainability. Figure 15 represents the 3D view of Document Quality, Modifiability, and Maintainability. Figure 16 shows evident output for the same input data set.

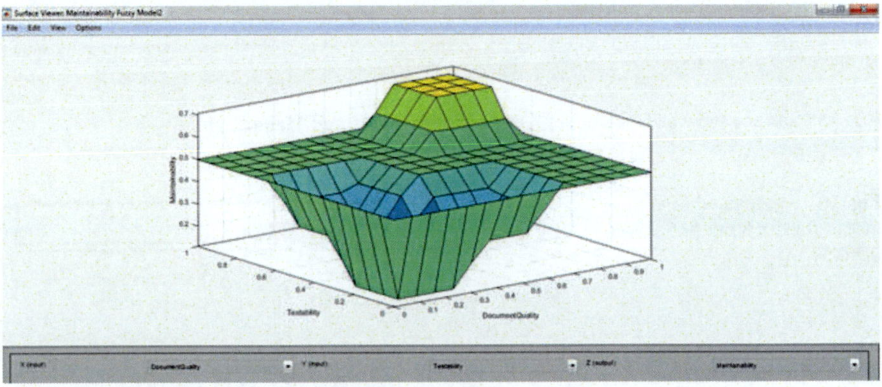

Fig. 13 Surface view of Document Quality, Testability, and Maintainability

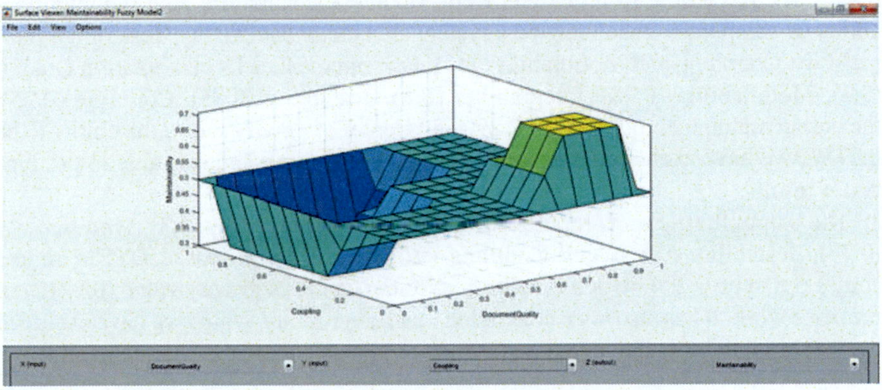

Fig. 14 Surface view of Document Quality, Coupling, and Maintainability

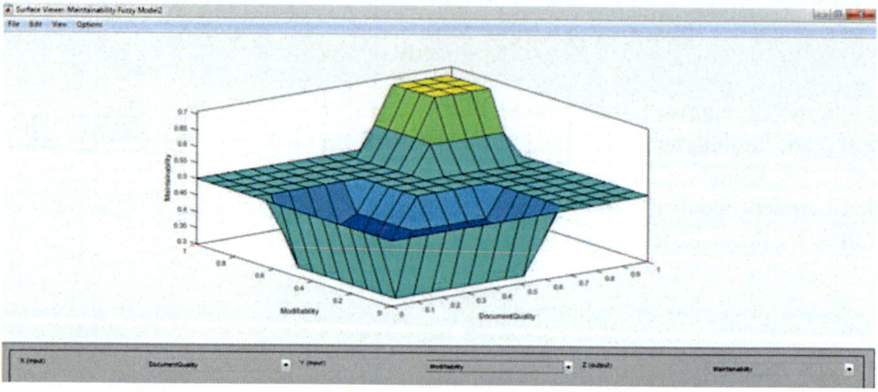

Fig. 15 Surface view of Document Quality, Modifiability, and Maintainability

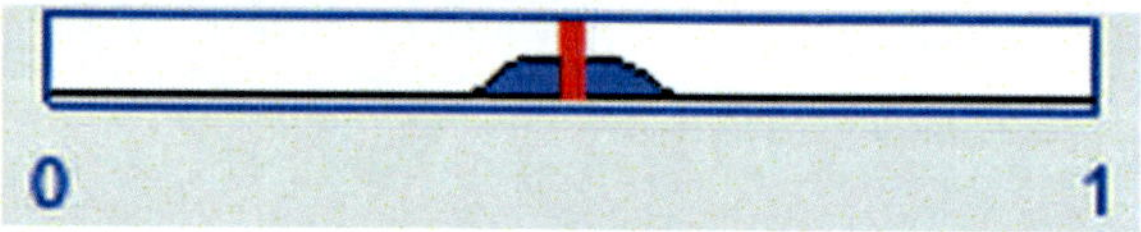

Fig. 16 Aggregated final output for the input [0.5 0.75 0.7 0.5]

5.1 Comparative Analysis of the Proposed DTCM Model

Now, we compare our DTCM fuzzy model with the research published by other researcher. According to Punia and Kaur, immense factors affecting the maintainability of component-based software system are Document Quality, Testability, Integrability, Coupling, and Modifiability [19]. For input values Documentation Quality (0.3), Modifiability (0.78), Integrability (0.8), Testability (0.85), Coupling (0.25), the output maintainability arise to 0.299 and Center of gravity shows the output to be 0.2712. We have excluded Integrability from our research and still able to produce better results.

The output for the same input values, i.e., Document Quality (0.3), Modifiability (0.78), Testability (0.85), and Coupling (0.25) from the proposed DTCM model comes out to be 0.294, which is more near to the center of gravity value (0.2712) of previous work. It has also been concluded that Integrability, which we have excluded in DTCM model was having a least impact on the calculation of maintainability.

6 Conclusions and Future Scope

Maintenance in component-based system is necessary for amendments and to enhance the adaptability of software in changing environment. The quality attribute maintenance plays central role in all varieties of software developments, for example, iterative development and agile technology. In our research, we have proposed a fuzzy logic-based method to robotically forecast component-based system's maintainability ranks, i.e., Worst, Bad, Good, Fair, and Excellent. We have concluded that the four factors described above, have immense influence on the maintainability though it is influenced by numerous attributes.

In this modern era, more significant factors are desirable to be explored to determine the maintainability. Early stage maintainability determination results in highly maintainable software and thereby reducing the maintenance efforts greatly.

MATLAB's fuzzy toolbox is used here to authenticate the same and demonstrates high correlation with maintainability. In case we increase the number of attributes influencing maintainability from four to five so as to improve precision, the complexity of the model became very high and number of rules to be inserted into the rule editor became 3^5 instead of 3^4 and we have to fetch out the values for 5 features. For future research, a comparison of DTMC model with other models can be done to find out its precision, usefulness, and accuracy. The improvement in the proposed technique can be comprehended by making use of neuro-fuzzy technique which will develop the wisdom ability and interpretability of the model.

References

1. Anda B (2007) Assessing software system maintainability using structural measures and expert assessments. IEEE Int Conf Softw Maintenance 8(4):204–213
2. Vale T, Crnkovice I, Santanade E, Neto PADMS, Cavalcantic YC, Meirad SRL, Meira SRDL (2016) Twenty-eight years of component-based software engineering. J Syst Softw 128–148
3. Lakshmi V (2009) Evaluation of a suite of metrics for component based software engineering. Issues Inf Sci Inf Technol 6:731–740
4. Pressman R (2002) Software engineering tata. Mc Graw Hills, pp 31–36
5. ISO/IEC TR 9126 (2003) Software engineering—product quality—part 3. Internal metrics, Geneva, Switzerland, pp 5–29
6. Grady RB (1992) Practical software metrics for project management and process improvement. Prentice Hall, vol 32
7. Siddhi P, Rajpoot VK (2012) A cost estimation of maintenance phase for component based Software. IOSR J Comput Sci 1(3):1–8
8. Freedman RS (1991) Testability of software components. IEEE Trans Softw Eng 17(6):553–564
9. Kumar R, Dhanda N (2015) Maintainability measurement model for object-oriented design. Int J Adv Res Comput Commun Eng 4(5):68–71. ISSN (Online) 2278-1021, ISSN (Print) 2319-5940
10. Mari M, Eila N (2003) The impact of maintainability on component-based software systems. In: Proceedings of the 29th EUROMICRO conference new waves in system architecture (EUROMICRO'03)
11. Malviya AK, Maurya LS (2012) Some observation on maintainability metrics and models for web based software system. J Global Res Comput Sci 3(5):22–29
12. Abdullah D, Srivastava R, Khan MH (2014) Modifiability: a key factor to testability. Int J Adv Inf Sci Technol 26(26):62–71
13. Sharma V, Baliyan P (2011) Maintainability analysis of component based systems. Int J Softw Eng Its Appl 5(3):107–117

14. Chen C, Alfayez R, Srisopha S, Boehm B, Shi L (2017) Why is it important to measure maintainability, and what are the best ways to do it? IEEE/ACM 39th IEEE international conference on software engineering companion, pp 377–378
15. Jain D, Jain A, Pandey AK (2018) Quantification of dynamic metrics for software maintainability prediction. Int J Recent Res Aspects 5(1):164–168. ISSN: 2349-7688
16. Saini R, Kumar S, Dubey and Rana A (2011) Aanalytical study of maintainability models for quality evaluation. Ind J Comput Sci Eng (IJCSE) 2(3):449–454. ISSN: 0976-5166
17. Muthanna S, Kontogiannis K, Ponnambalaml K, Stacey BA (2000) Maintainability model for industrial software systems using design level metrics, pp 248–256
18. Olatunji SO, Rasheed Z, Sattar KA, Mana AM, Alshayeb M, Sebakhy EA (2010) Extreme learning machine as maintainability prediction model for object-oriented software systems. J Comput 2(8):49–56
19. Punia M, Kaur A (2014) Software maintainability prediction using soft computing techniques. IJISET 1(9):431–442
20. Narang K, Goswami P (2018) Comparative analysis of component based software engineering metrics. In: 8th international conference on cloud computing, data science & engineering (Confluence), IEEE, pp 1–6
21. McCall J, Walters G (1997) Factors in software quality. the national technical information service (NTIS). Springfield, VA, USA, pp 1–168
22. Mittal H, Bhatia P (2007) Optimization criterion for effort estimation using fuzzy technique. CLEI EJ 10(1):2–8
23. Koscianski A, Candido B, Costa J (1999) Combining analytical hierarchical analysis with ISO/IEC 9126 for a complete quality evaluation framework. international symposium and forum on software engineering standards, pp 218–226
24. Boehm B (19996) Identifying quality-requirement conflicts. IEEE Softw 13:25–35
25. Boehm BW, Brown JR, Kaspar H, Lipow M, McLeod G, Merritt M (1978) Characteristics of software quality. North Holland Publishing, Amsterdam, The Netherlands
26. Dromey RG (1995) A model for software product quality. IEEE transactions on software engineering, pp 146–162
27. Kumar B (2012) A survey of key factors affecting software maintainability. international conference on computing sciences, pp 263–266
28. Oquendo F, Leite J, Batista T (2016) Designing modifiability in software architectures in action. Undergraduate Topics in Computer Science, Springer, Cham
29. Aggarwal KK, Singh Y, Chandra P, Puri M (2005) Measurement of software maintainability using a fuzzy model. J Comput Sci 1(4):538–542
30. https://in.mathworks.com/help/fuzzy/fuzzy-inference-process.html
31. Lenarduzzi V, Sillitti A, Taibi D (2017) Analyzing forty years of software maintenance models. In: IEEE/ACM 39th IEEE international conference on software engineering companion, pp 146–148
32. Narang K, Goswami P (2019) DRCE maintainability model for component based systems using soft computing techniques. Int J Innovat Technol Exploring Eng 8(9):2552–2560
33. Perepletchikov M, Ryan C, Frampton K, Tari Z (2007) Coupling metrics for predicting maintainability in service-oriented designs. In: Proceeding of Australian software engineering conference (ASWEC'07). Melbourne, Australia, pp 329–340
34. Rizvi SWA, Khan RA (2010) Maintainability estimation model for object- oriented software in design phase (MEMOOD). J Comput 2(4):26–32. ISSN2151-9617

Development of Petri Net-Based Design Model for Energy Efficiency in Wireless Sensor Networks

Sonal Dahiya, Ved Prakash, Sunita Kumawat, and Priti Singh

Abstract Wireless networks mainly wireless sensor networks have an abundant application in area of science and technology and energy is one of the chief design limitations for these types of networks. Energy conservation is a very prominent way to improve energy efficiency specially in communication. It is evident from the research in recent past that major part of energy is consumed in inter-node data transmission. This chapter is dedicated to design and development of antenna array design process modeling using Petri Net for energy efficient WSN. We worked on the model, which will study the dynamic nature of design process and evaluate for the deadlock conditions. On the basis of proposed model, a single band antenna resonating at a frequency of 2.4 GHz (Wireless LAN band) and a linear (2 × 1) antenna array for the same frequency is designed and simulated. The antenna array has improved gain as compared to single element and it can be utilized to improvise the total energy consumption inside the network.

Keywords WSN · Petri Net · Antenna array · Design procedure modeling

1 Introduction

Wireless sensor networks (WSN) have prominent applications in communication and information technology industry as well as scientific communities for monitoring surrounding environments. These have been used in each area of day-to-day,

S. Dahiya (✉) · V. Prakash · S. Kumawat · P. Singh
Amity University Gurugram, Haryana 122413, India
e-mail: sdahiya@ggn.amity.edu

V. Prakash
e-mail: vprakash@ggn.amity.edu

S. Kumawat
e-mail: skumavat@ggn.amity.edu

P. Singh
e-mail: priti.panwar2004@gmail.com

V. Singh et al. (eds.), *Computational Methods and Data Engineering*,
Advances in Intelligent Systems and Computing 1227,
https://doi.org/10.1007/978-981-15-6876-3_19

for example, logistics and transportation, agriculture, wildlife tracking, monitoring of environmental changes, structure monitoring, terrain tracking, entertainment industry, security, surveillance, healthcare monitoring, energy control applications, industrial societies, etc. [1].

These networks contain self-sufficient and multi-functional motes, i.e., sensor nodes that are distributed over the area of interest for either monitoring an event or for recording physical parameters of interest. This recorded data from the field of application is then transmitted to remote base station for further processing [2, 3].

In WSNs each and every node incorporates the capability for sensing, investigating, and transferring the collected data from its surrounding environment. The larger is the area, more is the number of sensor nodes required for investigation and information collection. While energy is consumed in every process occurring inside a node, a major portion is spent in the process of communicating data either among nodes or between nodes and base station [4]. Therefore, energy consumption in communication systems of a network has to be minimized by using efficient communication elements like antenna subsystem.

The sensor nodes in a network are usually power-driven by small batteries which are generally non-replaceable. Therefore, to increase the life of the network energy has to be conserved by efficient usage [5, 6].

Petri Nets have myriad applications in the field of wireless sensor networks. They are evidently used for designing, analyzing, and evaluation of these networks. PN is efficient for modeling discrete event systems, concurrent systems, etc. [7]. It is efficient for modeling of a network, a node in a network and even processor inside a node and its performance is better than the simulation based and formal methods of modeling a network [8]. A WSN model for energy budget evaluation with an insight to packet loss is also used for maximizing the life of a network [9].

In this paper, a multi-node sensor network is presented in which the design process model of antenna as shown in Fig. 1, is developed by using Petri Nets. The developed model is then simulated and analyzed in MATLAB. Also, based on this design process model, a single band radiator working at 2.4 GHz (Wireless LAN) frequency is designed and analyzed on ANSYS HFSS, i.e., High Frequency Structure Simulator Software. Owing to its small gain and other antenna parameters, antenna array has also been designed. Researchers in the recent decades have proposed various array designs, structures, multi-feed, and their applications in the IEEE standard bands. An array with tooth like patches, filtering array, array using butler matrix for indoor wireless environment, 16-element corporate fed array with enhanced gain and reduced sidelobe level have been proposed [10–15]. High gain series fed 2 × 1 and 4 × 1 array using metamaterials, Teflon substrate, and multi-layered substrates had been simulated and designed by certain researchers [16–20]. Arrays for various applications like biomedical, RF harvesting, mm-waves wireless applications of broadband mobile communication [18–21]. These antenna arrays can be used alongside the sensor nodes of the network for augmentation of energy, efficiency, and some other network parameters as well. Therefore, a linear (2 × 1) antenna array for the same frequency is designed and simulated for enhancing antenna parameters viz. gain and overall energy efficiency in a network [22].

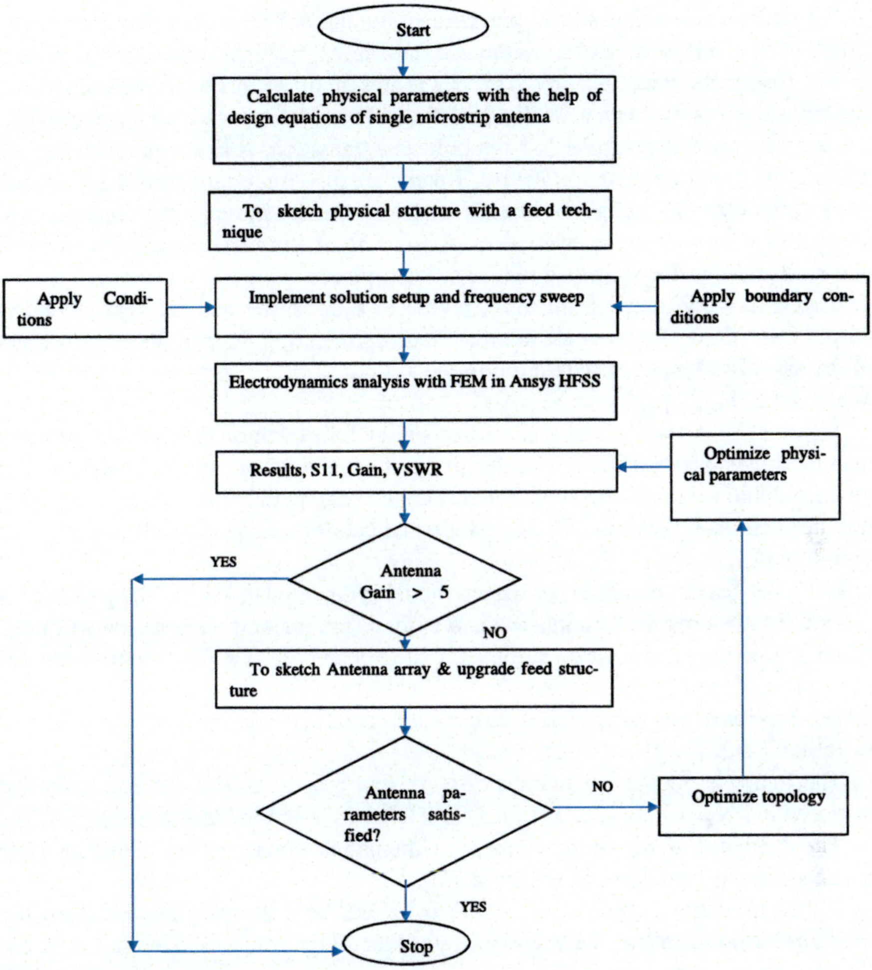

Fig. 1 Antenna array design model flowchart

2 Development of Antenna Design Process Model Based on PN

A PN is a multigraph which is weighted as well as directed and it is efficiently used to define and analyze a system. Like graphical modeling tool, it shows efficacies of flowchart and block diagrams while similar to a mathematical and formal tool it allows the user to develop state equations for representing the systems. Petri Nets were first presented by Carl Adam Petri for representing chemical equations in year 1962 [7].

A Petri Net is considered as a place/transition net or P/T net comprises of places which symbolize the state of a system and transitions. It symbolizes activities necessary to change the states and arcs. The arcs represent interconnection between places and transitions or vice versa. While depicting a Petri Net system a circle symbolizes a place whereas a rectangular bar symbolizes a transition. A line symbolizes an arc making the system directed in nature. Tokens are also important part of PN model. They symbolize the change in the state of the system by moving from one place to another and are denoted by black dots. Also firing of a transition indicates an event occurred and it is dependent on tokens at input places.

Tokens are consumed from input places in a transition and are reproduced in output transitions. Token consumptions and reproduction depend upon the weight of the associated input or output arc in the system [23]. The process of Petri Net is explained in Fig. 2 [22].

We must have a well-defined execution policy for execution of Petri Nets as more than one transition can be enabled for firing at same moment of time. PNs are very much suitable for modeling synchronous, concurrent, parallel, distributed, and even non-deterministic systems [23, 24]. Mathematical definitions are well explained in literature [25].

Petri Net-based antenna array design model is depicted in Fig. 3. The position p4 and p5 signifies required conditions and is therefore marked with tokens while the token at position p8 represents sufficiency of gain and token at position p10 denotes fulfillment of antenna parameters.

Description of positions as well as transitions used in the design model is explained in Tables 1 and 2.

The Petri Net model for antenna array design process model can be drawn and explored in PN Toolbox [26, 27] with MATLAB as described in Fig. 4.

The incidence matrix for developing mathematical equations for further analysis of the system is calculated as shown in Fig. 5.

All the transitions used in the model can be fired at least once, and therefore, the model is found to be live. This demonstrates that all the states used in the model are significant, as can be seen in Fig. 6.

The cover-ability tree explains the inter-state movement which can be represented in graphic mode with the help of PN Toolbox as presented in Fig. 7.

It validates that each and every state in the model is finite, feasible and denotes absence of deadlock or undefined situations.

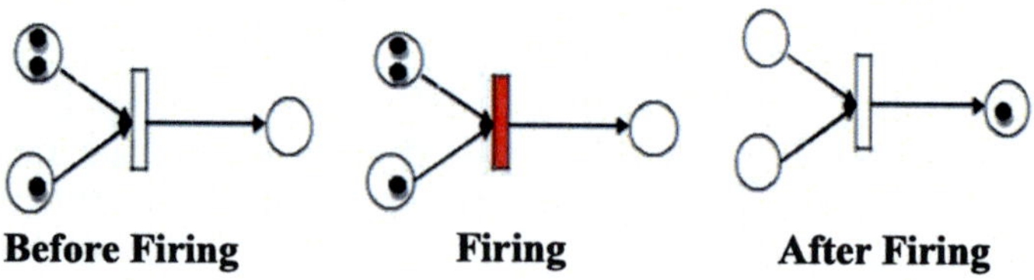

Fig. 2 Basic Petri Net model working

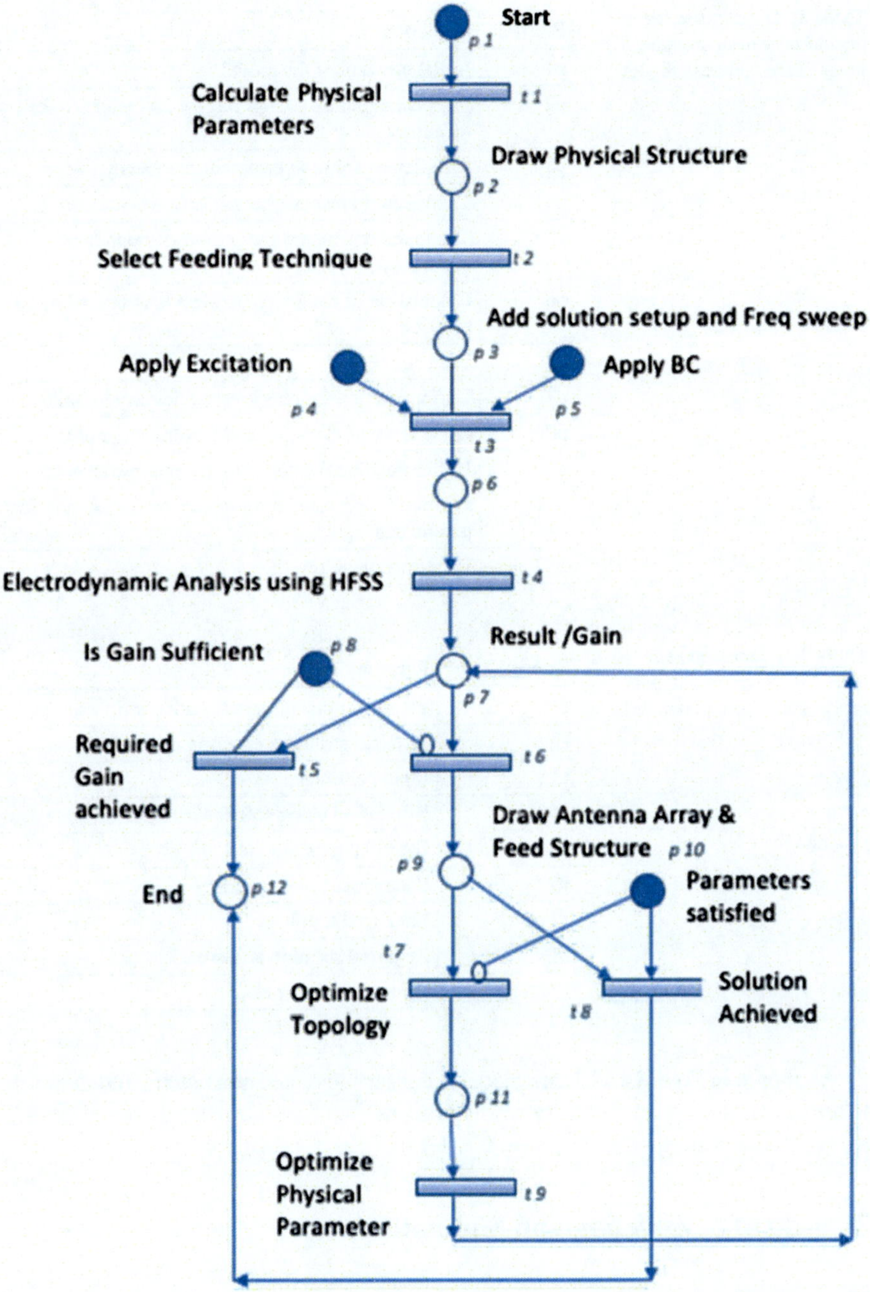

Fig. 3 Petri Net model of array

Table 1 Description for positions used in antenna array design process model

Position	Depiction
p1	Initial position of the model
p2	Physical structure design stage with insight to feed technique
p3	Addition of solution setup and frequency sweep
p4	The token represents that excitation is applied
p5	The token represents application of boundary conditions
p6	Calculation of design parameters is indicated by this position
p7	Gain and results calculated
p8	Buffer position for checking sufficiency of gain
p9	Draw array of elements and feeding structure
p10	Buffer position for checking antenna parameters
p11	Position to indicate requirements to modify physical parameters
p12	Final state position

Table 2 Description for transitions used in antenna array design process model

Transition	Purpose
t1	Parameter calculation for microstrip antenna
t2	Physical structure assessment
t3	Input parameters
t4	Evaluation using softwares
t5	Essential gain attained
t6	Parametric investigation
t7	Enhanced topology
t8	Required solution achieved
t9	Parameters optimized

As shown in Figs. 8 and 9 this model is conservative and consistent, and therefore, tokens are not consumed during whole process.

3 Antenna Modeling and Simulation

The designing and simulation of antenna element and array are discussed in this section. An inset feed antenna is described resonating at 2.4 GHz. Figure 10 shows the simulated design of single patch element whose dimensions are calculated from the standard design equations [10].

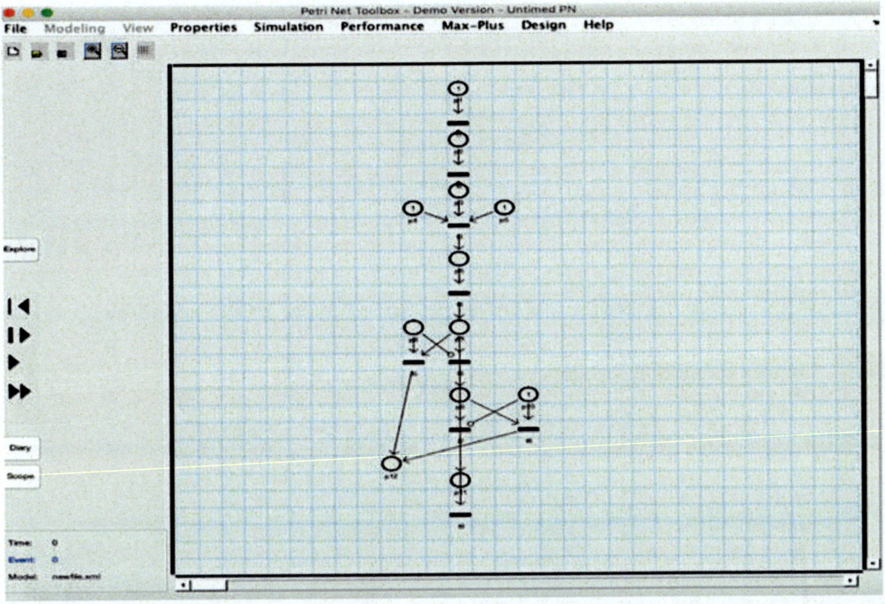

Fig. 4 PN model for developed model

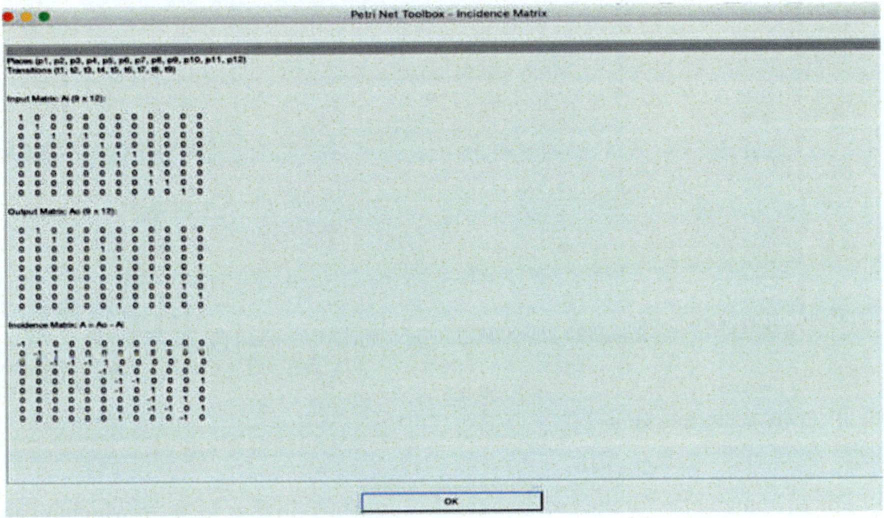

Fig. 5 Incidence matrix for developed model

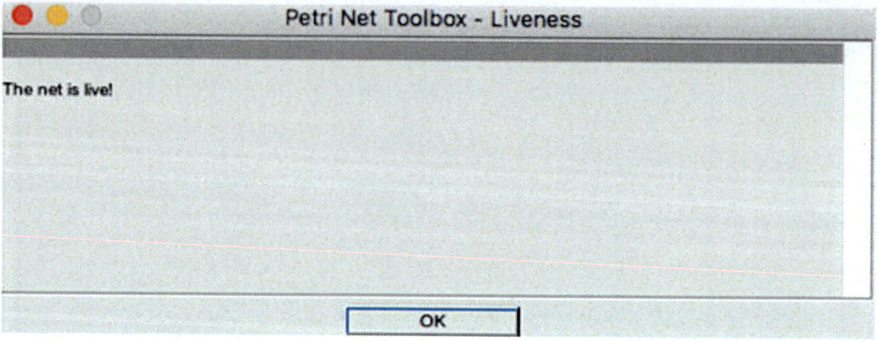

Fig. 6 Liveness result for the model

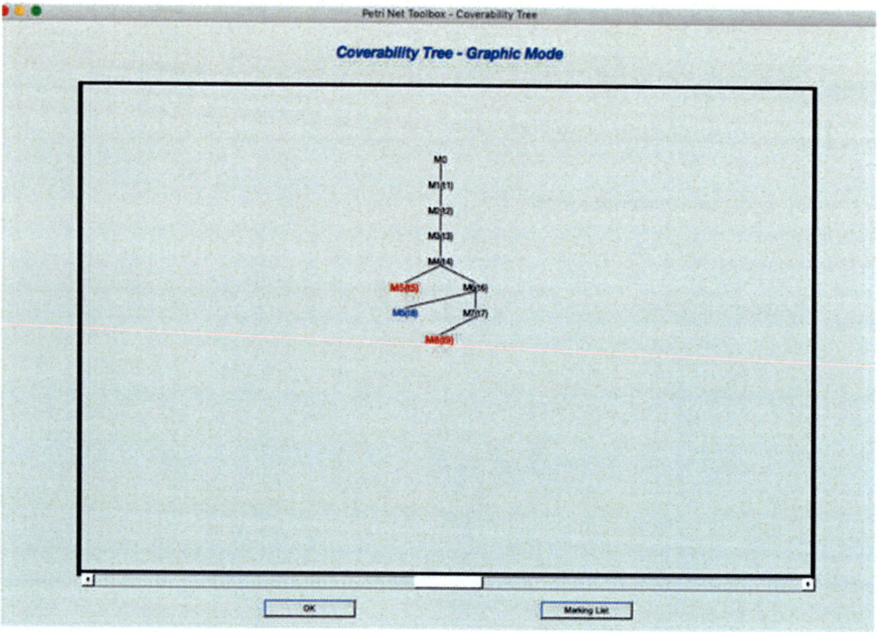

Fig. 7 Cover-ability tree for model

The S_{11} parameter for the radiator element is shown in Fig. 11 from which it can be predicted that the antenna resonates at 2.45 GHz will a S_{11} parameter well below -10 dB. This means that it is able to transmit more than 90% of the input power.

Figure 12 shows the value of VSWR which is well below 2 in antenna element.

The radiation pattern of the radiator is depicted in Fig. 13 which reveals that it radiates in the upper half of the space.

When certain antenna elements are placed in a predefined pattern either along a line or in a plane so that constructive interference of the electric field takes place,

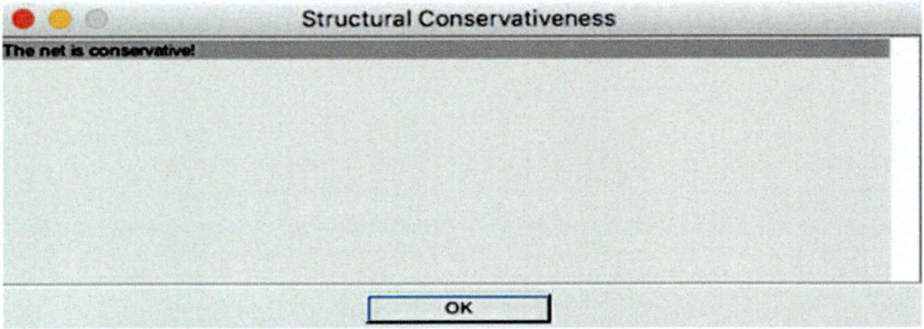

Fig. 8 Conservativeness of antenna array design model presented in Fig. 4

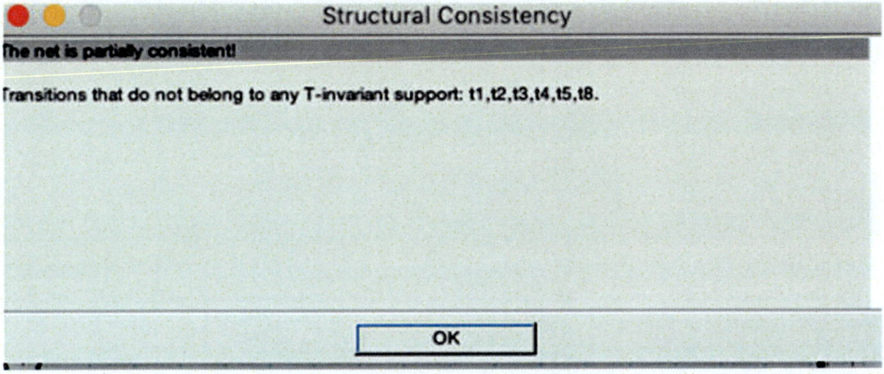

Fig. 9 Consistency of antenna array design model shown in Fig. 4

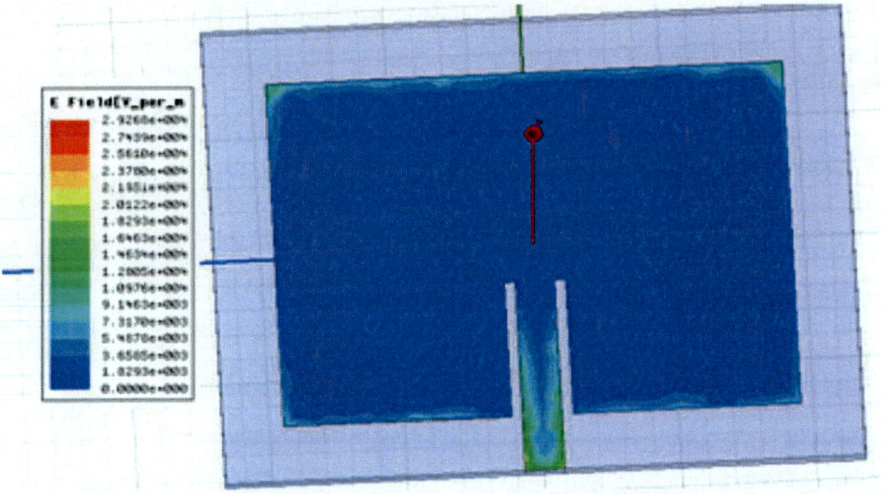

Fig. 10 Antenna element prototype

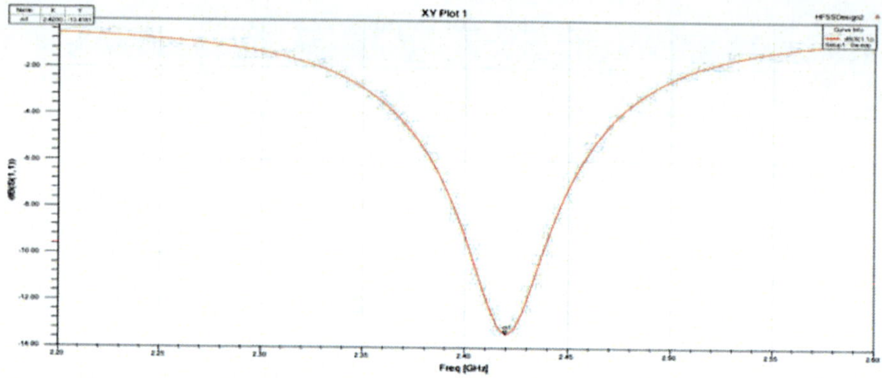

Fig. 11 S-11 parameter of antenna element

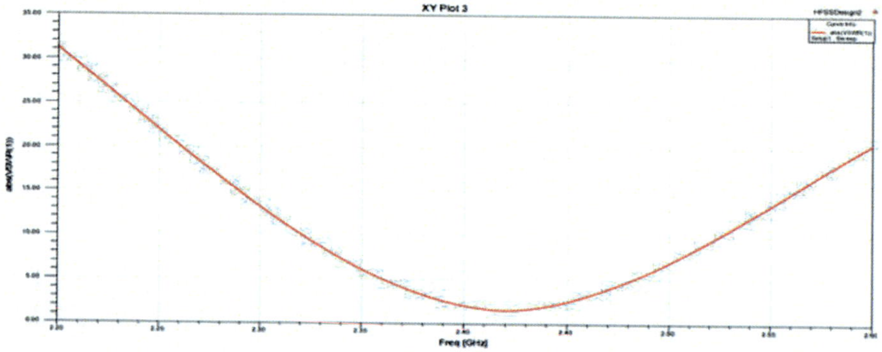

Fig. 12 VSWR of the element

Fig. 13 Radiation pattern of the element

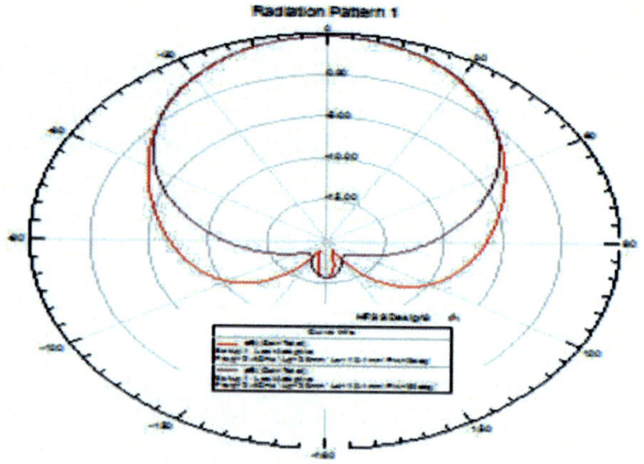

then an array is said to be formed. An array can be classified into linear and planar depending on the geometrical configuration. Elements in a linear array spread out along a single line whereas in a planar they are placed into a plane. The key parameters which play a vital role in deciding the network antenna are the geometrical alignment of patch elements, their inter-element spacing, and its excitation amplitude and phase.

The net electric field for array is estimated by the vector sum of individual electric fields is given by:

$$E(\text{total}) = \left[E(\text{single element at reference point})\right] \times \left[\text{Array factor}\right] \quad (1)$$

where array factor AF is given by

$$AF = 2\cos\left[\frac{1}{2}(kd\cos\theta + \beta)\right] \quad (2)$$

The array factor for N elements can be written as

$$AF = \frac{\sin\frac{N\psi}{2}}{\sin\frac{\psi}{2}} \quad (3)$$

where $\psi = kd\cos\theta + \beta$ and the gain of the array is given by

$$D = 2N(\mathrm{d}\lambda) \quad (4)$$

Figure 14 shows linear array of 2 **x** 1 element, simulated on HFSS software.

Figure 15 shows the S-parameter plot of the array where it can be easily seen the return loss is below −20 db.

Figure 16 shows the gain of the array is 5.3441 db.

The simulation results can be summarized as shown in Table 3.

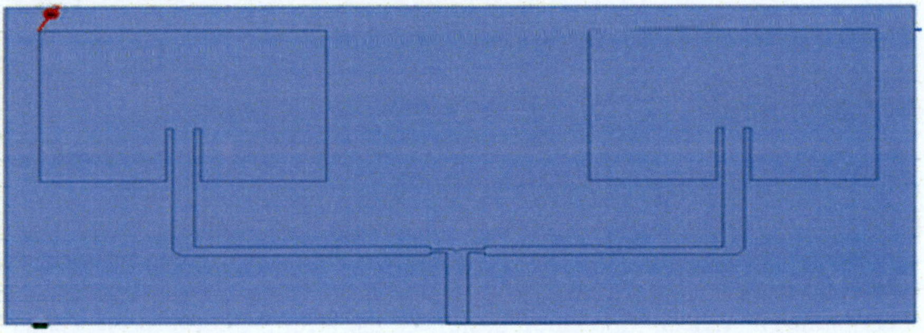

Fig. 14 2 × 1 antenna array

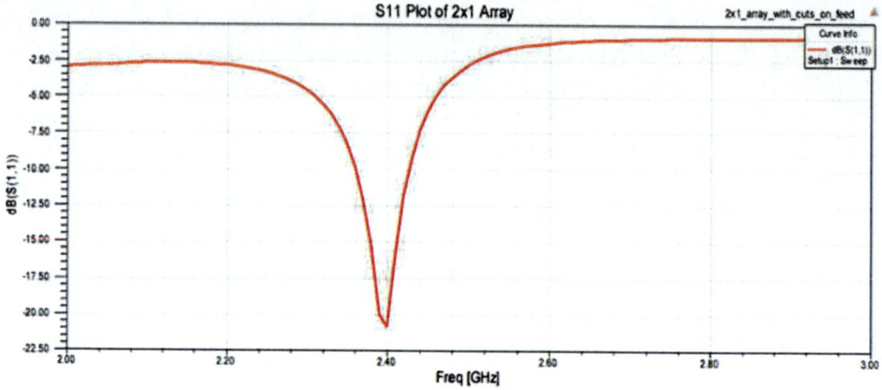

Fig. 15 S11 parameter of 2 × 1 array

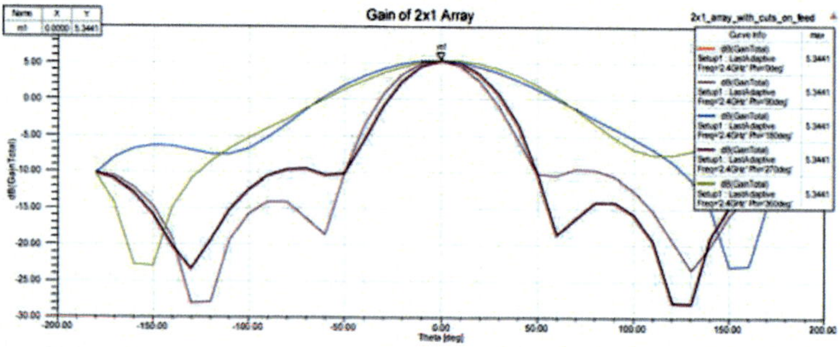

Fig. 16 Gain of 2 × 1 array

Table 3 Simulation results

Type of antenna	VSWR	GAIN	S_{11} (dB)
Single element	1.4	2.62	−13.48
2 × 1 array with power divider	1.5	5.34	−20.71

4 Conclusion

Petri Nets have applications in every field of engineering and technology, especially in the communication models and process modeling. To garner benefits of modeling and analysis of systems, Petri Nets are used very frequently in current scenarios. Petri Net theory is used for analyzing antenna array design process modeling for energy efficient WSNs. This model facilitates the investigation of design process dynamics and assesses the design process for existence of any deadlock and uncertain conditions in the system. Property analysis of this model demonstrates that the developed model is finite and feasible for every state and there is no deadlock or uncertain condition. On the basis of this model, a single band antenna and a linear

(2 × 1) antenna array resonating at a frequency of 2.4 GHz (Wireless LAN band) have been designed. It is observed that array gain is enhanced to 5 db as compared to 2.62 db as in case of single element. Also, the voltage standing wave ratio (VSWR) of the array is measured to be 1.6. Therefore, energy efficiency of network can be increased by using an antenna array in spite of single element. The developed model and antenna array can be utilized in energy efficient WSNs.

References

1. Rashid B, Rehma MH (2016) Applications of wireless sensor networks for Urban areas: a survey. J Network Comput Appl 60:192–219
2. Martino CD (2009) Resiliency assessment of wireless sensor networks: a holistic approach. PhD Thesis, Federico II, University of Naples, Itly
3. Yahya B, Ben-Othman J, Mokdad L, Diagne S (2010) Performance evaluation of a medium access control protocol for wireless sensor networks using Petri Nets. In: HET-NET's 2010, 335–354
4. Akyidiz IF, Su W, Sankarasubramaniam Y, Cayirci E (2002) Wireless sensor networks: a survey. Comput Networks 38(4):392–422
5. Anastasi G, Counti M, Francesco MD, Pasrella A (2009) Energy conservation in wireless sensor networks: a survey. Adhoc Networks 7(3):537–568
6. Francomme J, Godary K, Val T (2009) Validation formelle d'un mechanism de synchrinisation pour reseaux sans fil. CFIP'2009
7. Murata T (19982) Petri nets: properties, analysis and applications. In: Proceedings of the IEEE, vol 77, pp 541–580
8. Shareef A, Zhu Y (2012) Effective stochastic modelling of energy constrained wireless sensor networks. J Comput Network Commun
9. Berrachedi A, Boukala-Ioualalen M (2016) Evaluation of energy consumption and the packet loss in WSNs using deterministic stochastic petri nets. In: 30th international conference on advanced information networking and applications workshop
10. Wong K-L (2002) Compact and broadband microstrip antennas. Wiley Publications
11. Secmen M (2011) Active impedance calculation in uniform microstrip patch antenna arrays with simulated data. In: EURCAAP
12. Wang H, Huang XB, Fang DG, Han GB (2007) A microstrip antenna array formed by microstrip linefed tooth-like-slot patches. In: IEEE transactions on antennas and propagation 55(4)
13. Lin C-K, Chung S-J (2011) A filtering microstrip antenna array. In: IEEE transactions on microwave theory and techniques 59(11)
14. Elhefnawy M, Ismail W (2009) A microstrip antenna array for indoor wireless dynamic environments. In: IEEE Trans Antennas Propag 57(12)
15. Ali MT, Rahman TA, Kamarudin MR, Md Tan MN (2009) A planar antenna array with separated feed line for higher gain and sidelobe reduction. Progress in Electromagnet Res 8:69–82
16. Gupta V (2013) Design of a microstrip patch antenna with an array of rectangular SRR using left-handed metamaterial. CREST J 1
17. Yahya SH (2012) Khraisat.: design of 4 elements rectangular microstrip patch antenna with high gain for 2.4 GHz Applications. Modern Appl Sci
18. Hamsagayathri P, Sampath P, Gunavathi M, Kavitha D (2016) Design of slotted rectangular patch array antenna for biomedical applications. IJRET 3
19. Tawk Y, Ayoub F, Christodoulou CG, Costantine J (2015) An array of inverted-F antennas for RF energy harvesting. In: IEEE AP-S, pp 278–1279
20. Santos RA, Penchel RA, Bontempo MM, Arismar Cerqueira S Jr (2016) Reconfigurable printed antenna arrays for mm-wave applications. In: EuCAP

21. Prakash V, Kumawat S, Singh P (2016) Circuital analysis of coaxial fed rectangular and U-slot patch antenna. In: ICCCA 2016, pp 1348–1351. IEEE, Noida
22. Dahiya S, Kumawat S, Singh P, Sekhon KK (2019) Modeling and analysis of communication subsystem design process for wireless sensor networks based on petri net. Int J Recent Technol Eng 8(3):10124–10128
23. Kumawat S (2013) Weighted directed graph: a petri net based method of extraction of closed weighted directed euler trail. Int J Serv Econom Manage 4(3):252–264
24. Khomenko V, Roux OH (2018) Application and theory of petri net and concurrency. In: Proceedings of 39th international conference, PETRI NETS 2018, Bratislava, Slovakia
25. Dahiya S, Kumawat S, Singh P (2019) Petri net based modeling and property analysis of distributed discrete event system. Int J Innov Technol Explor Eng 8(12):3887–3891
26. Jie TW, Ameedeen MAB (2014) A survey of petri net tools. ARPN J Eng Appl Sci 9(8):1209–1214
27. Mortensen KH (2003) Petri nets tools and software. http://www.daimi.au.dk/PetriNets/tools

Lifting Wavelet and Discrete Cosine Transform-Based Super-Resolution for Satellite Image Fusion

Anju Asokan and J. Anitha

Abstract Super-resolution creates a high-resolution image from an input low-resolution image. The availability of low-resolution images for analysis has degraded the quality of image processing. We propose a lifting wavelet and discrete cosine transform-based super-resolution technique for satellite image enhancement. Here, the low-resolution images are decomposed using Lifting Wavelet Transform (LWT) and Discrete Cosine Transform (DCT). The high-frequency components and the source image are interpolated and all these images are combined to generate the high-resolution image using Inverse Lifting Wavelet Transform (ILWT). The enhanced source images are further fused using curvelet transform. The proposed work is assessed on a set of multispectral images and the results indicate that the proposed framework generates better quality high-resolution satellite images and further enhances the image fusion results compared to the traditional wavelet-based transforms and spatial domain interpolation schemes.

Keywords Super-resolution · Satellite image · Lifting Wavelet Transform · Curvelet transform · Lifting scheme · Multispectral · Image fusion

1 Introduction

Super-resolution image reconstruction is a very promising research domain as it can overcome some of the existing resolution related limitations of the imaging sensors. High-resolution images are required in most digital imaging applications for proper analysis. These high-resolution images play a crucial role in areas such as defense, biomedical analysis, criminology, surveillance, etc.

A. Asokan (✉) · J. Anitha
Department of Electronics
and Communication Engineering, Karunya Institute of Technology and Sciences, Coimbatore 641114, India
e-mail: anju.asok@yahoo.com

J. Anitha
e-mail: anithaj@karunya.edu

V. Singh et al. (eds.), *Computational Methods and Data Engineering*,
Advances in Intelligent Systems and Computing 1227,
https://doi.org/10.1007/978-981-15-6876-3_20

Image resolution gives the information content in the image. Super-resolution involves constructing the high-resolution images from various low-resolution images. Imaging devices and systems such as sensors affect image resolution. Using optical elements to acquire high-resolution satellite images is very expensive and not feasible. Apart from the sensors, the image quality is also affected by the optics mainly due to lens blurs, diffractions in lens aperture, and blurring due to lens movement. In recent years, the need for high-resolution imagery is increasing. Many researches are carried out to get high-resolution image.

Traditional super-resolution techniques can be classified as: interpolation based, reconstruction based, and example based [1]. Interpolation-based techniques use pixel correlation to get an approximation of the high-resolution image. Though this technique is faster and simpler to implement, there is a loss of high-frequency data. Reconstruction-based techniques use information from a series of images to generate high-resolution image. Example-based techniques utilize machine learning so that by learning the co-occurrence relation between low-resolution and high-resolution image patches, high-frequency information lost in the low-resolution image is predictable. Different techniques to improve the image super-resolution have been developed. A super-resolution method using Discrete Cosine Transform (DCT) and Local Binary Pattern (LBP) is shown in [2, 3]. Gamma correction is applied to the low-frequency band to preserve the edge information. A Stationary Wavelet Transform (SWT) and DWT-based super-resolution method is presented in [4]. A dictionary pair-based learning method to partition the high-resolution patches and low-resolution patches is described in [5]. Here the high-resolution patches are linearly related to low-resolution patches [6]. This technique can utilize the contextual details in an image over a large area and effectively recover the image details. It can overcome the computational complexities associated with a deep Convolutional Neural Network (CNN)-based model. Fractional DWT (FDWT) and Fractional Fast Fourier Transform for image super-resolution are presented in [7]. Directional selectivity of FDWT is responsible for the high quality of the image. A discrete curvelet transform and discrete wavelet transform method for enhancing the image is described in [8]. Multiscale transforms like curvelet transforms can effectively reconstruct the image and can deal with edge dependent discontinuities. An image reconstruction using granular computing technique is presented in [9]. This method uses transformation from image space to granular space to get high-resolution image. A fractional calculus-based enhancement technique is proposed in [10]. A hybrid regularization technique for PET enhancement is presented in [11]. A Hopfield neural network based on the concept of fractal geometry for image super-resolution is described in [12].

Fusion adds complementary information from images of different modalities and gets the entire information in a single image. It is of extreme importance in areas like remote sensing, navigation, and medical imaging. A cascaded lifting wavelet and contourlet transform-based fusion scheme for multimodal medical images is proposed in [13]. A remote sensing fusion technique using shift-invariant Shearlet transform is described in [14]. This technique can reduce the spectral distortion to a great extent. An image fusion technique for fusing multimodal images using cartoon-texture decomposition and sparse representation is presented in [15].

Proposed work describes an LWT- and DCT-based super-resolution scheme on low-resolution multispectral images and fusion of the enhanced images using curvelet transform. The fusion results are compared using performance metrics such as PSNR, entropy, FSIM, and SSIM. The results are compared against fusion results for enhancement schemes like bicubic interpolation, SWT-based super-resolution, and LWT-based super-resolution. Source images used are LANDSAT 7 multitemporal images with dimensions 512 × 512.

The paper is arranged as: Sect. 2 describes the proposed satellite image super-resolution method. Section 3 gives the results and discussion and Sect. 4 presents the conclusion.

2 Methodology

The proposed method is executed in MATLAB 2018a on an Intel® Core™ i3-4005U CPU @1.70 GHz system on different sets of multispectral satellite images. Two multitemporal LANDSAT images are taken and subjected to LWT- and DCT-based image super-resolution. The enhanced source images are further fused using curvelet transform.

Figure 1 shows the framework of the proposed method.

The data used are LANDSAT images. A set of 50 images are available and 5 samples are used for analysis. The source images are low-resolution satellite images and fusion result of the low-resolution images does not give good quality images. Hence, a super-resolution scheme using LWT and DCT is used to construct high-resolution imagery from the low-resolution source images to improve the quality of fusion.

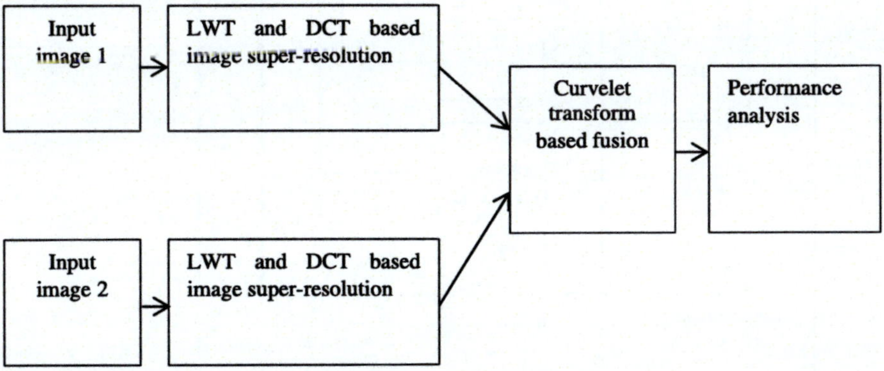

Fig. 1 Block diagram of the proposed method

2.1 LWT- and DCT-Based Super-Resolution

The lifting scheme is basically used to create wavelet transform. It generates a new wavelet with added properties by incorporating a new basis function. The frequency components of the source image are created by decomposing the image. Figure 2 shows the LWT- and DCT-based super-resolution framework.

The generated frequency components comprise of three high-frequency components and one low-frequency component: horizontal, vertical, and diagonal information of the input image form the high-frequency components. These components are interpolated with factor γ. Low pass filtering of the source image creates the low-frequency component. Since this component has image information, input image

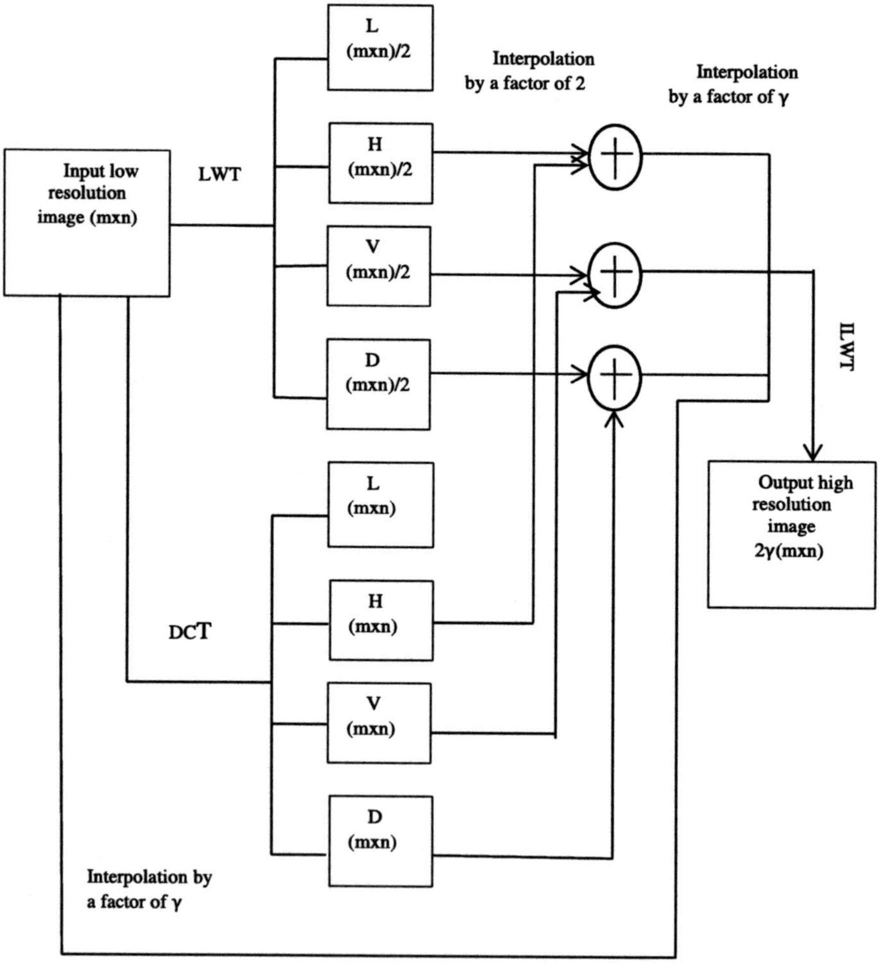

Fig. 2 DCT- and LWT-based super-resolution framework

is interpolated by factor γ using surface fitting in order to reconstruct the output satellite image. The interpolated high-frequency components and source image are considered as input to the Inverse Lifting Wavelet Transform (ILWT). All the input images interpolated by a factor 4. Source image resolution was 512 × 512. The images are interpolated to 2048 × 2048.

In DCT, biorthogonal filter creates the frequency components. They are interpolated by a factor of 2 using surface fitting. These components are modified by adding high-frequency components generated using LWT. High-resolution image is created from the two source images individually using the DCT- and LWT-based super-resolution scheme.

In LWT, initial interpolation of high-frequency components is necessary because it utilizes sampling to generate frequency components which are half the size of source image while DCT generates same size frequency components.

2.2 *Curvelet Transform-Based Image Fusion*

The enhanced source images are fused using curvelet transform. The main highlight of using this method is its ability to represent the edge details with minimum nonzero coefficients. Due to its anisotropic property, it can represent edges much better than wavelet and ridgelet transform. Here, the source images are divided into curvelet coefficients. These coefficients are then combined using the fusion rule. The final fusion result is created on applying the Inverse Curvelet Transform (ICT) on the transformed coefficients.

Curvelet coefficients differ for each scale and have different properties. The high-frequency components carry the important image features such as edges and contours and have more significance. Hence, the selection of high-frequency components is of utmost importance since it contains the salient image features. Local energy calculation is used for fusion here.

It is effective over single coefficient rule. This is so because single coefficient rule is decided by the absolute value of only single coefficient and presence of noise can affect the fusion result. But in local energy-based fusion, choosing single coefficient is decided by that particular coefficient along with its neighboring coefficients. Selecting coefficient using this method is effective in obtaining the edge details in the image. Noise has high absolute value and if noise is present in the image, it will be isolated and hence the neighboring coefficients affected by noise might have low absolute values. Therefore, the noise affected coefficients can be easily distinguished from the other coefficients.

Let C be the transformed coefficient matrix. A 3 × 3 window is considered and the local energy values are computed for all coefficients by moving the window throughout the image.

The local energy E for a particular coefficient $C(m, n)$ at pixel location (m, n) is computed using Eq. (1) as:

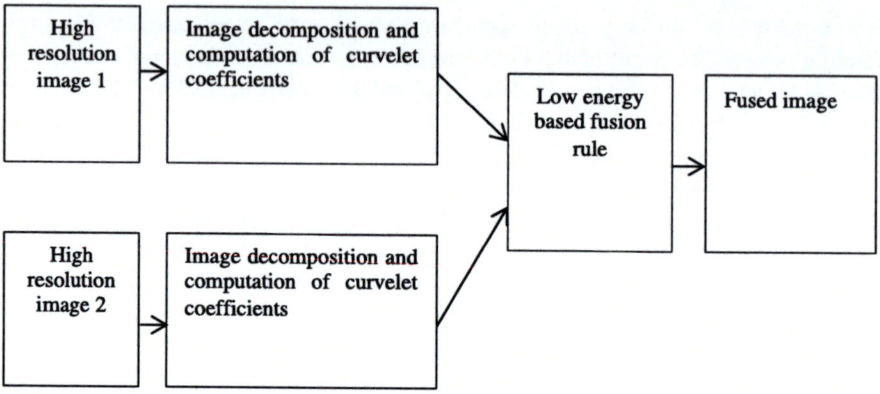

Fig. 3 Curvelet transform-based fusion

$$E_{m,n} = \sum_{m-1}^{m+1} \sum_{n-1}^{n+1} C(m, n)^2 \quad (1)$$

An edge corresponds to a high local energy value for the center coefficient. Once the local energy is computed, curvelet coefficients are compared depending on their local energy values and the coefficient with higher energy values are selected. Then, the coefficients of the fused image are found. Final image is formed by employing ICT. Figure 3 shows the curvelet transform-based fusion.

3 Results and Discussion

The source images are LANDSAT 7 images of size 512 × 512. The proposed super-resolution method enhances the source images and the enhanced images are fused using the curvelet transform. Super-resolution of the source images is further carried out using three existing techniques and fusion is done using the same transform. The fusion results of the images enhanced using different super-resolution techniques are compared in terms of Peak Signal-to-Noise Ratio (PSNR), entropy, Feature Similarity Index (FSIM), and Structural Similarity Index (SSIM). Table 1 shows the performance metric comparison for different super-resolution schemes.

The Peak Signal-to-Noise Ratio (PSNR) describes the accuracy of the output image and is dependent on the intensity values of the image. It is computed using Eq. (2) as:

$$\text{PSNR} = 10 \log_{10}\left(\frac{255 * 255}{\text{MSE}}\right) \quad (2)$$

where MSE is the mean square error.

Table 1 Comparison of the performance metrics for different super-resolution methods

Technique	Database	PSNR	Entropy	FSIM	SSIM
Bicubic interpolation	Sample 1	43.2119	4.8654	0.8312	0.8215
	Sample 2	40.6711	4.7025	0.8208	0.8129
	Sample 3	42.9133	4.8733	0.8367	0.8331
	Sample 4	44.2144	4.8024	0.8291	0.8267
	Sample 5	45.5616	4.9067	0.8203	0.8187
SWT-based super-resolution	Sample 1	48.9240	5.3322	0.8524	0.8462
	Sample 2	45.1067	5.2097	0.8448	0.8327
	Sample 3	47.1900	5.3424	0.8493	0.8409
	Sample 4	48.2388	5.4232	0.8417	0.8422
	Sample 5	47.1224	5.4099	0.8312	0.8312
LWT-based super-resolution	Sample 1	52.1899	5.6734	0.8824	0.8756
	Sample 2	49.2144	5.7209	0.8742	0.8522
	Sample 3	51.9056	5.6021	0.8767	0.8615
	Sample 4	55.1224	5.8024	0.8890	0.8702
	Sample 5	51.0933	5.5523	0.8654	0.8641
LWT–DCT-based super-resolution	Sample 1	62.1289	6.2033	0.9412	0.9556
	Sample 2	59.7223	5.9878	0.9378	0.9622
	Sample 3	60.4142	6.0211	0.9445	0.9412
	Sample 4	63.9011	6.3124	0.9477	0.9465
	Sample 5	58.3456	5.8977	0.9402	0.9337

It can be seen that the proposed super-resolution scheme-based images give better results in comparison with traditional super-resolution-based fusion. Bicubic interpolation and SWT-based fusion create high-resolution images in which the high-frequency components such as edges and corners are not preserved. But in LWT-based super-resolution scheme, the use of surface fitting enables the edges and curves in the image to be preserved.

The addition of the DCT-based decomposition adds another degree of resolution enhancement to sharpen the high-frequency details in the image. As a result, the blur effect on the edges and corners affect the PSNR values in the case of bicubic interpolation and SWT-based fusion results. The LWT-based fusion gives improved results due to the property of high-frequency edge preservation. The added DCT module to LWT improves the PSNR values since it can sharpen the high-frequency information in the image.

The entropy H is the information content in the image. From the table, it is observed that the information is better preserved in the fusion results obtained by the proposed scheme when compared to the traditional methods which suffer from the blurring of edges and corners.

FSIM describes the resemblance of input and final image features and SSIM describes the resemblance of input and final image structures. It is observed that the FSIM and SSIM values are high for the fusion results obtained by the proposed scheme when compared to the traditional methods. This is so because the features in the fused image resemble the source image features in the case of the proposed scheme due to the preserving of the high-frequency detail in the image. Figure 4 gives the fused image outputs of all the techniques.

From the table, it is concluded that the proposed super-resolution scheme-based image fusion produces improved outcome when compared to bicubic interpolation, SWT super-resolution-based fusion, and LWT super-resolution-based fusion results. The presence of DCT with LWT adds an additional level of sharpening of the image details thus improving PSNR, SSIM, FSIM, and entropy values.

4 Conclusion

An LWT and DCT super-resolution method for fusion of satellite images is proposed. This technique recovers the high-frequency image information. Individual source images are enhanced using LWT- and DCT-based super-resolution and are fused using curvelet transform. The results are compared for different traditional super-resolution schemes such as bicubic interpolation, SWT-based enhancement, and LWT-based enhancement. The effectiveness of the proposed technique is observed in the high-resolution fusion results. However, the presence of non-homogeneous textures in the satellite image can limit the accuracy of the proposed super-resolution scheme. Future work can be aimed at textural synthesis with the traditional super-resolution method for obtaining better image quality.

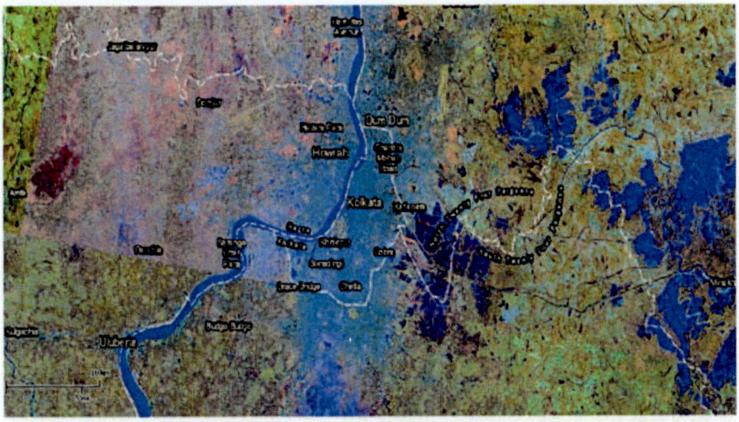

a.

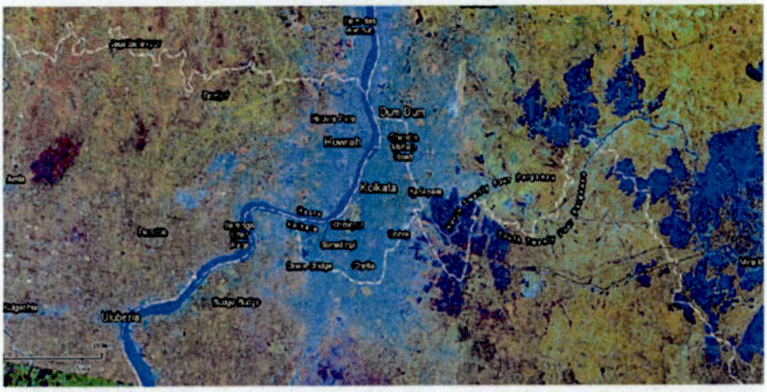

b.

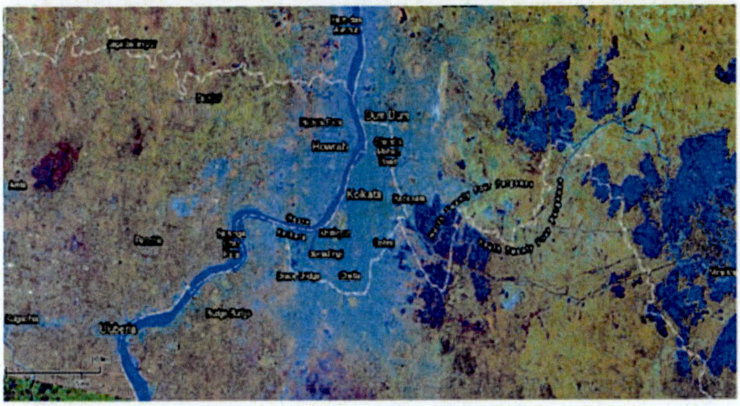

c.

Fig. 4 **a** Dataset 1 **b** Dataset 2 **c** Bicubic interpolation-based fusion **d** SWT super-resolution-based fusion **e** LWT super-resolution-based fusion **f** Proposed method

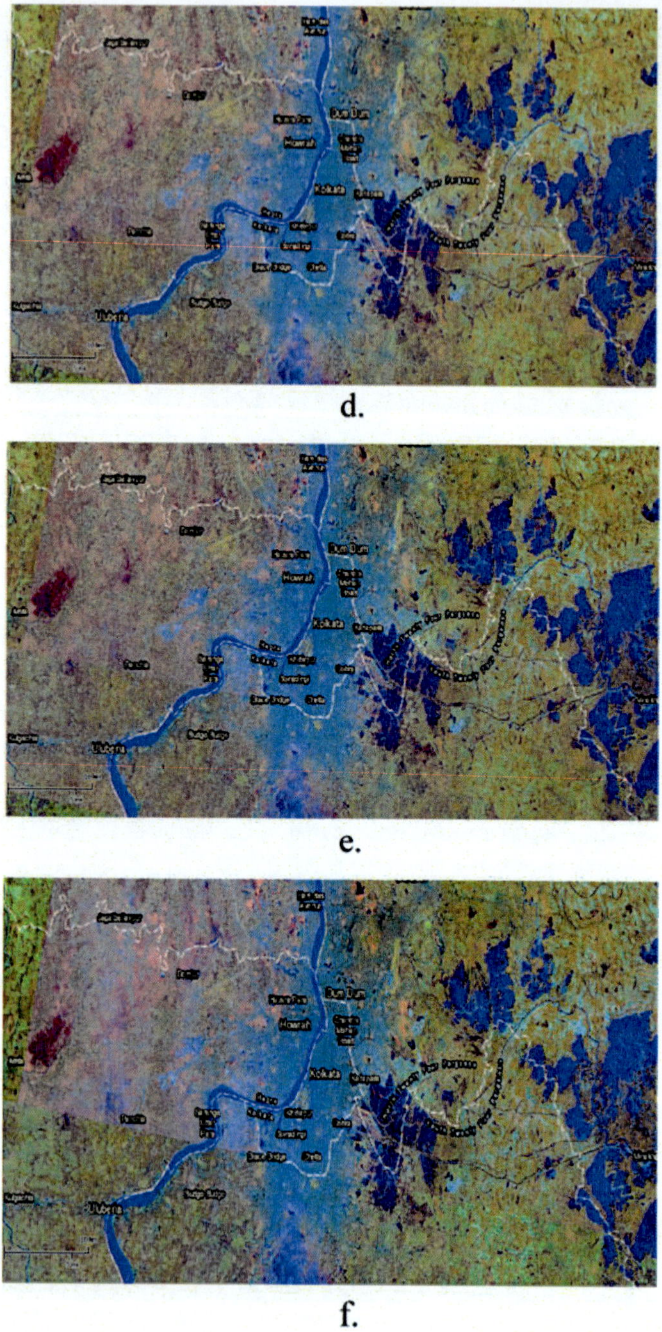

Fig. 4 (continued)

References

1. Yang X, Wu W, Liu K, Zhou K, Yan B (2016) Fast multisensor infrared image super-resolution scheme with multiple regression models. J Sys Archit 64:11–25. https://doi.org/10.1016/j.sysarc.2015.11.007
2. Doshi M, Gajjar P, Kothari A (2018) Zoom based image super-resolution using DCT with LBP as characteristic model. J King Saud Univ-Comput Inf Sci 1–14. https://doi.org/10.1016/j.jksuci.2018.10.005
3. Basha SA, Vijayakumar V (2018) Wavelet Transform based satellite image enhancement. J Engg and Appl Sci 13(4):854–856. https://doi.org/10.3923/jeasci.2018.854.856
4. Chopade PB, Patil PM (2015) Image super resolution scheme based on wavelet transform and its performance analysis. In: International conference on Computing. Communication and Automation (ICCCA), pp 1182–1186. https://doi.org/10.1109/ccaa.2015.7148555
5. Yang X, Wu W, Liu K, Kim PW, Sangaiah AK, Jeon G (2018) Multi-semi-couple super-resolution method for edge computing. Special Section on recent advances in computational intelligence paradigms for security and privacy for fog and mobile edge computing. IEEE Access 6:5511–5520 (2018). https://doi.org/10.1109/access.2019.2940302
6. Li F, Bai H, Zhao Y (2019) Detail-preserving image super-resolution via recursively dilated residual network. Neurocomputing 358:285–293. https://doi.org/10.1016/j.neucom.2019.05.042
7. Choudhary AR, Dahake VR (2018) Image super resolution using fractional discrete wavelet transform and fast fourier transform. In: 2nd international conference on I-SMAC (IoT in social, mobile, analytics and cloud) (I-SMAC). https://doi.org/10.1109/i-smac.2018.8653723
8. Shrirao AS, Zaveri R, Patil MS (2017) Image resolution enhancement using discrete curvelet transform and discrete wavelet transform. In: International Conference on Current Trends in Computer, Electrical, Electronics and Communication (CTCEEC) https://doi.org/10.1109/ctceec.2017.8455163
9. Liu H, Diao X, Guo H (2019) Image super-resolution reconstruction: a granular computing approach from the viewpoint of cognitive psychology. Sens Imaging 20:1–19. https://doi.org/10.1007/s11200-019-0241-3
10. Lei J, Zhang S, Luo L, Xiao J, Wang H (2018) Super-resolution enhancement of UAV images based on fractional calculus and POCS. Geo-spatial Inf Sci 21(1):56–66. https://doi.org/10.1080/10095020.2018.14
11. Mederos B, Sosa LA, Maynez LO (2017) Super resolution of PET images using hybrid regularization. Graphics and Sig Proc 1:1–9. https://doi.org/10.5815/ijigsp.2017.01.01
12. Su YF (2019) Integrating a scale-invariant feature of fractal geometry into the hopfield neural network for super-resolution mapping. Int J Remote Sens 40:8933–8954. https://doi.org/10.1080/01431161.2019.1624865
13. Bhardwaj J, Nayak A (2018) Cascaded lifting wavelet and contourlet framework based dual stage fusion scheme for multimodal medical images. J Electr Electron Sys 7:1–7. https://doi.org/10.4172/2332-0796.1000292
14. Luo X, Zhang Z, Wu X (2016) A novel algorithm of remote sensing image fusion based on shift-invariant Shearlet transform and regional selection. Int J Electron Commun 70:186–197. https://doi.org/10.1016/j.aeue.2015.11.004
15. Zhu Z, Yin H, Chai Y, Li Y, Qi G (2018) A novel multi-modality image fusion method based on image decomposition and sparse representation. Inf Sci 432:516–529. https://doi.org/10.1016/j.ins.2017.09.010

Biologically Inspired Intelligent Machine and Its Correlation to Free Will

Munesh Singh Chauhan

Abstract Human behavior is a complex combination of emotions, upbringing, experiences, genetics, and evolution. Attempts to fathom it have been a long sought-after human endeavor, and it still remains a mystery when it comes to actually interpreting or deriving it. One such trait, free will, or an ability to non-deterministically act without any external motivation has been one such instinct which has remained an enigma as so far as when it comes to fully understanding its genesis. Two schools of thoughts prevail, and both have attempted to understand this elusive quality. One school that has a long history has been exploring from the perspective of metaphysics, while the other one interprets it using rational science that includes biology, computing, and neuroscience. With the advent of artificial neural networks (ANN), a beginning has been made to computationally represent the biological neural structure. Despite the ANN technology in its infancy especially when it comes to actually mimic the human brain, major strides are self-evident in the field of object recognition, natural language processing, and other fields. On the other end of the spectrum, persistent efforts to understand let alone simulate the biologically derived unpredictability in thoughts and actions is still a far cry. This work aims to identify the subtle connections or hints between the biological derived unpredictability and the ANNs. Even an infinitesimal progress in this domain shall open the flood gates for more emotive human-like robots, better human–machine interface, and an overall spin-off in many other fields including neuroscience.

Keywords Free will · Artificial neural network · Consciousness

1 Unpredictability and Its Genesis

The main tenet of unpredictability in each of the animal's action or behavior is its evolutionary process [1, 2]. It is widely noticed in both the flora and the fauna that those species which were not able to change and adapt, dwindled, and few of them

M. S. Chauhan (✉)
School of Advanced Studies, University of Tyumen, Tyumen, Russia
e-mail: m.chauhan@utmn.ru

V. Singh et al. (eds.), *Computational Methods and Data Engineering*,
Advances in Intelligent Systems and Computing 1227,
https://doi.org/10.1007/978-981-15-6876-3_21

even became extinct. Hence, the unconscious or the conscious desire to be ad hoc and variable is inane to the very survival of a species. This has led to a surge in different studies on how species, both plants and animals maintain their competitive advantage to overcome extinction [3] at all odds. A very good discussion on how predators hoodwink their prey in their own game is described in the work by Bjorn Brembs [4]. The major deception that a prey deploys to protect itself from a fatal attack is to make its move unknown and uncertain to the predators. Another trait that can be considered as an extension of unpredictability is "*free will.*" *Free will* has been a contentious issue, of late, mostly been rebuffed by many neuroscientists, whereas on the other end of the spectrum, it is still being actively embraced in the world of metaphysics, as an ingenuity of a human to decide what can or cannot be done. The debate on determinism vs non-determinism has been ongoing for decades as most of the real-world scientific knowledge is based on deterministic outcome, but at the same time on the para-micro level (e.g., Heisenberg's uncertainty principle in quantum mechanics) or at large macro levels (black holes), non-determinism rules the roost. So, a categorical allegory cannot be postulated for all types of matter and behaviors. We, as humans, have to live and thrive between the two opposite poles of certainty and uncertainty.

Another notable factor about the predictability of animal behavior lies in its interaction with its immediate environment. The default behavior of a species always borders on its readiness to deal with any eventuality resulting from the sudden change in its habitat. Thus, it can be assumed that the default behavior of most of the species is quite unpredictable which is a necessary pre-condition for survival in the ever-changing world. Unpredictability can be visualized quite easily in humans. Psychiatric patients suffer from an extreme disability [5] while interacting with their immediate environments. They show a persistent stereotypical behavior toward all sorts of issues, thus entangling their thoughts further in the process. While in general, a normal healthy human being is able to wade through a world of uncertainty with relative ease and finesse.

A generalized assumption here can be made about an animal/human behavior is that *free will* as a trait is not completely random, but always carries with itself some trace of ambiguity. Brembs consider this dichotomy as a sort of nonlinearity in the behavior which can be represented possibly by a complex set of calculations with an accompanying variety of triggers or events.

2 Unpredictability and Machines

Representing "*free will*" as a combination of intelligently derived ambiguity interspersed with an ominous potential of predicated outcomes can become one of the pragmatic ways of bringing it closer to the artificial neural networks. Currently, neural networks bank on stochastic simulations of trainable data using a combination of nonlinearity and weights to derive functions that can make accurate prediction when fed with new data source. Scientifically, "*free will*" can also be assumed to follow the same lineage of stochastic decision making under controlled conditions but with a flair of uncertainty.

The fundamental question that arises is how "*free will*" can be replicated in machines [6]. Any entity having "*free will*" should possess this constant awareness in its backdrop in order to function in a manner that precludes any overt influence from the external world. Yes, environment acts as a big influencer in the development as well as exercise of conscious thought but the seed comes from the entity within and not exogenously. The work by Dehaene et al., points to a very primitive form of awareness, which he terms as unconsciousness or "C0 *consciousness*" [7]. All current neural nets exhibit C0 level of consciousness. The areas that lie in C0 fall in the domain of pattern recognition, natural language processing, and other typical tasks that involve training a machine to do a job based on a repository of trainable dataset. The current neural network technology falls under this C0 domain. The other two forms of consciousness according to Dehaene et al. are relatively higher forms which he terms as C1 and C2. The C1 form of consciousness is the ability to make the information available at all times also called broadcasting. This information availability can be called awareness, and it is normally but not always the precursor to the higher form called C2 consciousness. In C2 form, the organism is able to monitor and possibly act upon the available information. Figure 1 gives a general description of the alignments of C0, C1, and C2 level of consciousness with respect to a Turing machine [8].

The main limiting factor in artificial neural networks is that the system works on the premise that the data is supreme and everything can be derived stochastically from the inference of these datasets [9–11]. So far, the progress using this approach has been tremendous and immense. This type of probabilistic inferences is very limited [12, 13] when it comes to creating an ecosystem that can tap and measure or for that matter substitute consciousness, though in its most raw and primitive state. A possible path in sight that can facilitate consciousness simulation in machines is to go deeper into the working of human mind and correlate its functioning with the tasks that represent basic forms of awareness in a typical biological organism. Biological variability is an evolutionary trait [14] practiced by all life forms for their survival and progeny. Biological variability can be defined as something which is

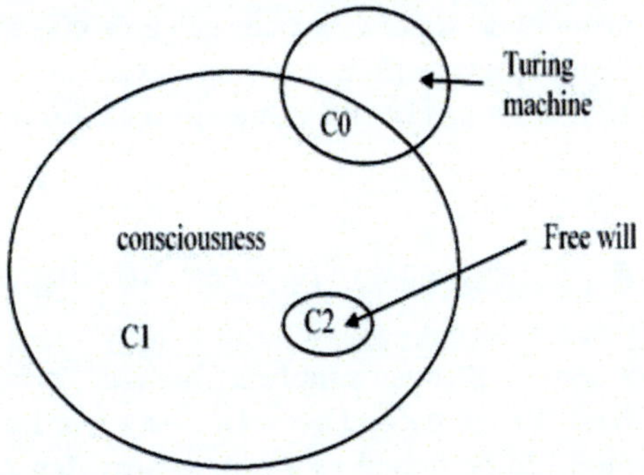

Fig. 1 States of consciousness (C0, C1, & C2) and how a Turing machine fits in

not exactly random but carries an immense dose of unpredictability. This aspect can be ascribed to the prevalence of quantum randomness in the brain [15], but how and when it is triggered is very difficult to predict. In sum, it can be said that the unpredictability has a connection with nonlinearity with an accompanied mix of quantum randomness. Quantum randomness has been studied in plants in the context of photosynthesis [16].

3 Limitations of Artificial Neural Networks

The current ANNs have reached their optimum potential, and according to many machine learning specialists, the neural networks have a black box approach [9] which makes it impossible to decipher how an output is derived based on the first principles (mathematically). Other issues which are quite restrictive are the necessity of large sets of training data, and the immediate consequence of this is the large computation time requirements coupled with ever dire need for highly parallel hardware.

Several real-life examples have been sighted in various researches that neural networks have in fact under-performed as compared to other traditional methods. Hence, even if we have enough data and computing power, it is not always the case for the neural networks to efficiently solve all problems. For example, it has been shown that traffic prediction [17] is not very amenable to neural network and the prediction using neural networks is on par with the linear regression method. Another example pertains to subpixel mapping [18] in remote sensing, where the best neural networks were employed for classification of subpixel mapping of land cover, and they carry an inherent limitation. A similar situation is cited in ICU patient database [19] research where ANNs were found wanting. There are many other examples cited by researchers that despite the presence of voluminous training data and the computation power, neural networks are not able to provide the required results. It can be argued that these examples are too specific and possibly need cross-verification before arriving to a final decision. Nevertheless, applying neural networks for predicting or simulating human-like conscious patterns is still beyond the present knowledge reach. As we progress toward large-scale automation in our economies, robotics is one such area that will become a prominent driver of growth. Building ever-efficient robots that are as close to human and can mimic vital human behaviors will be one of the most exciting of all human endeavors ever taken.

4 Biologically Inspired Intelligent System

A system that can simulate "*free will*" is logically and practically a distant dream at least for the current scenario. Any worthwhile step toward realization of this goal ought to be hailed as a progressive step albeit with a realization that an artificial system is completely different in terms of its genesis as compared to a biological one.

The present-day Turing machine is having a limited human-level ability and is still in many aspects' way behind even a 3-year-old child when it comes to basic cognitive tasks such as object detection. This limitation becomes even more pronounced when a computer has an advantage of tera to peta-bytes of data but still blunders in making correct recognition of images. Whereas on the contrary, this is an effortless exercise for a 3-year old who has just begun observing the world.

5 Free Will Model of a Machine

Transforming a machine to mimic "*free will*" type behavior requires a conglomeration of properties that can be ascribed to the study of stochasticity, indeterminism, and spontaneity. Factors related to the propensity of an organism to take action or recourse can be categorized under three broad situations. Firstly, as a "*planned action*"; an algorithmic, stepwise description of solving a particular problem. This behavior is simulated by the present-day computers, which can be programmed to deal with various conditions and events. The second situation can be termed as "*reflex.*" The reflex action behavior is very similar to the previously described "planned action" but with a major difference in terms of time required to accomplish the task. Now, the time becomes the limiting factor, immediate actions are warranted in a very short span of time triggered by some external event. Embedded systems and real-time systems fall under this *reflex* category. Finally, the third and the last situation is a "*free will*" type scenario in which the action is stochastically nonlinear, and generated from within the agent without any external form of trigger. Table 1 diagrammatically denotes the three proposed situations with examples.

Neural network limitations are key obstacles in realizing a free-will type behavior. The current neural network learns from data and is designed for an almost zero interaction with other computational modules that are in turn responsible for producing different sets of traits. On the contrary, the human brain carries multiple segmented

Table 1 Action/recourse options

	Planned action	Reflex action	Free will
Description	Algorithmic, stepwise, can be tested	Sudden, within very short time interval, sporadic	Inane, without any trigger from external source
Agents	Algorithms	Embedded systems, real-time systems	Humans
Examples	Shortest path algorithm	Press of buttons on a game console, control systems in a microwave	A fly doing indeterministic maneuvers to escape a predator

reasoning sites with dedicated neurons for specialized tasks. These segmented biological neural structures are interconnected using a type of gated logic with ascribed weights on interconnections. The sum effect is more advanced perception generated as a sum whole of the entire network. This unfortunately is absent in the current artificial systems. Hence, in order for a machine to simulate or mimic human *free-will* type trait is unrealizable at least for the time being. This work aims to identify the necessary ingredients needed to enable artificial networks becoming closer to a biologically conscious system.

The following is the list of factors that can be integrated in a nonlinear fashion to realize a machine behaving in a conscious mode:

1. Temporal awareness.
2. Space perception.
3. Genetic evolution (not possible in machines, but a series of generations of machine versions can add up to this evolution idea).
4. Environment (auditory, visual, etc., machines are capable of functioning well in this domain, especially the deep learning neural networks. These systems even surpass humans).
5. Memory or retention (machines have an upper edge in the retention mechanisms but do not have the intelligence to make sense out of this storage).
6. Energy potential (both biological organisms as well as machines use this potential difference to propagate/channelize energy in various forms).
7. Mutative (biologically structures are extremely mutative and adaptive to the surroundings, in fact this trait is a key factor in evolution. Machines too shall have this ability to adapt to different scenarios based on the availability of resources) (Fig. 2).

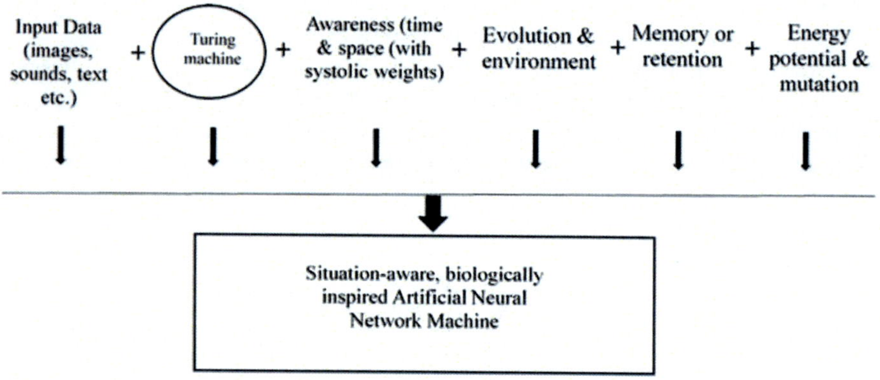

Fig. 2 Situation-aware, biologically inspired intelligent neural network prototype

6 Conclusion

The current development in neural networks is stupendous and has taken the world by storm. Almost all areas of human activity have been transformed with the applications of artificial intelligence. This has created over-expectations especially in the domain of robotics where human-level intelligence coupled with consciousness is desired and is starkly missing. The awareness akin to free-will can only be replicated if the current design of artificial neural networks is drastically modified to incorporate a fundamental change in how these machines simulate awareness and subtle consciousness.

Acknowledgements The author would like to sincerely thank the School of Advanced Studies, University of Tyumen, Russian Federation for funding the research on "Free Will."

References

1. Dercole F, Ferriere R, Rinaldi S (2010) Chaotic Red Queen coevolution in three-species food chains. Proc Royal Soc B: Biolog Sci 277(1692):2321–2330. https://doi.org/10.1098/rspb.2010.0209
2. Sole R, Bascompte J, Manrubia SC (1996) Extinction: bad genes or weak chaos? Proc Royal Soc London. Series B: Biolog Sci 263(1375):1407–1413. https://doi.org/10.1098/rspb.1996.0206
3. Scheffers BR, De Meester L, Bridge TCL, Hoffmann AA, Pandolfi JM, Corlett RT, Watson JEM (2016) The broad footprint of climate change from genes to biomes to people. Sci 354(6313):aaf7671. https://doi.org/10.1126/science.aaf7671
4. Brembs B (2010) Towards a scientific concept of free will as a biological trait: spontaneous actions and decision-making in invertebrates. Proc Royal Soc
5. Glynn LM, Stern H. S, Howland MA, Risbrough VB, Baker DG, Nievergelt CM, … Davis EP (2018) Measuring novel antecedents of mental illness: the questionnaire of unpredictability in childhood. Neuropsychopharmacology, 44(5):876–882. https://doi.org/10.1038/s41386-018-0280-9
6. Lin J, Jin X, Yang J (2004) A hybrid neural network model for consciousness. J Zhejiang Univ-Sci A 5(11):1440–1448. https://doi.org/10.1631/jzus.2004.1440
7. Dehaene S, Lau H, Kouider S (2017) What is consciousness, and could machines have it? Science 358:486–492
8. Petzold C (2008) The annotated turing: a guided tour through alan turing's historic paper on computability and the turing machine. Wiley, USA
9. Benítez JM, Castro JL, Requena I (1997) Are artificial neural networks black boxes? IEEE Trans Neural Networks 8(5):1156–1164
10. Braspenning PJ, Thuijsman, F, Weijters AJMM (1995) Artificial neural networks: an introduction to ANN theory and practice, vol 931. Springer Science & Business Media
11. Dreiseitl S, Ohno-Machado L (2002) Logistic regression and artificial neural network classification models: a methodology review. J Biomed Inform 35(5–6):352–359
12. Livingstone DJ, Manallack DT, Tetko IV (1997) Data modelling with neural networks: advantages and limitations. J Comput Aided Mol Des 11(2):135–142
13. Hush DR, Horne BG (1993) Progress in supervised neural networks. IEEE Signal Process Mag 10(1):8–39

14. Tawfik DS (2010) Messy biology and the origins of evolutionary innovations. Nat Chem Biol 6(11):692
15. Suarez A (2008) Quantum randomness can be controlled by free will-a consequence of the before-before experiment. ArXiv preprint arXiv:0804.0871
16. Sension RJ (2007) Biophysics: quantum path to photosynthesis. Nature 446(7137):740
17. Hall J, Mars P (1998) The limitations of artificial neural networks for traffic prediction. In: Proceedings third IEEE symposium on computers and communications. ISCC'98. (Cat. No.98EX166), Athens, Greece, pp 8–12
18. Nigussie D, Zurita-Milla R, Clevers JGPW (2011) Possibilities and limitations of artificial neural networks for subpixel mapping of land cover. Int J Remote Sens 32(22):7203–7226. https://doi.org/10.1080/01431161.2010.519740
19. Ennett CM, Frize M (1998) Investigation into the strengths and limitations of artificial neural networks: an application to an adult ICU patient database. Proc AMIA Symp 998

Weather Status Prediction of Dhaka City Using Machine Learning

Sadia Jamal, Tanvir Hossen Bappy, Roushanara Pervin, and AKM Shahariar Azad Rabby

Abstract Weather forecasting refers to understanding the weather condition for the days or moments ahead. It is one of the blessings of modern science to be able to make the weather predictions from previous quantitative data. Weather forecasts state the weather status and how the environment would behave for a chosen location in a specific time. Before inventing the machine learning techniques, people used different types of physical instruments like barometer and anemometer for predicting (forecasting) weather. But it took a lot of time for this phase and there were some issues like maintaining those instruments. Moreover, not every time their forecasts were as accurate. For these problems, people use machine learning techniques nowadays. The purpose of this work is to use machine learning techniques to forecast the weather for Dhaka City. Here, various types of algorithms are used to forecast Dhaka's environment, such as linear regression, logistic regression, and Naïve Bayes algorithm. The data are gathered from some websites [1] and a dataset was developed.

Keywords Barometer · Anemometer · Machine learning · Linear regression · Naïve Bayes classifier · Logistic regression

---

S. Jamal (✉) · T. H. Bappy · R. Pervin · A. S. A. Rabby
Department of CSE, Daffodil International University, Dhaka, Bangladesh
e-mail: sadia15-8084@diu.edu.bd

T. H. Bappy
e-mail: tanvir15-7706@diu.edu.bd

R. Pervin
e-mail: roushanara15-11666@diu.edu.bd

A. S. A. Rabby
e-mail: shahariar.cse@diu.edu.bd

V. Singh et al. (eds.), *Computational Methods and Data Engineering*, Advances in Intelligent Systems and Computing 1227,
https://doi.org/10.1007/978-981-15-6876-3_22

1 Introduction

The weather has an important effect on daily life. It is something that can change any time without any further notice. Because of some changes in the atmosphere, this instability happens. Weather prediction, however, is a vital part of daily life. So, weather forecasting accuracy is very critical. Nowadays, the weather data are interpreted by supercomputers. They obtain raw data from space-launched satellite. But the data collected in raw format do not contain any insightful information. So, cleaning of those data is needed for giving input in the mathematical model. This process is known as data mining. After cleaning the data, it is used to input into the mathematical model and predict the weather. In this research paper, the data of the previous months are collected and a dataset is created.

For avoiding complexity, the months are sorted in three seasons like summer, fall, and winter. Each of these seasons contains 4 months.

Then, algorithms like linear regression, logistic regression, and Gaussian Naïve Bayes are implemented on those datasets.

2 Background Study

Some researchers have already researched this topic. In this section, there will be some reference to previous works as well as the challenges. There are some works like using the time series modeling algorithm [2] here.

Holmstrom et al. [3] say that professional/traditional weather forecasting approaches perform well over the linear and practical algorithm regression. But in the long run the qualified approach efficiency decreased, and in that case, they suggested that machine learning should do well. By adding more data to the training model, the accuracy of linear regression can be increased.

Chauhan and Thakur [4], in their work, contrasted the K-meaning clustering and the tree of decision algorithm. They showed that at first the accuracy of the algorithm increases with the increase in size of the training set, but it starts to decline after some time.

Biswas et al. presented a weather forecast prediction using an integrated approach for analyzing and measuring weather data [5]. In their research, the authors used weather forecasting data mining techniques. They have developed a system, which predicts future data based on the present data. They used the chi-square algorithm and the Naïve Bayes algorithm for that method. This program takes user input and gives the output in the form of the expected result. They found that the observed values vary significantly from the predicted value by using the chi-square method. And Naïve Bayes also gave various values out of planned performance.

Wang et al. presented the weather forecast using data mining research based on cloud computing [6]. They demonstrated the data mining technology being applied along with cloud computing. Cloud computing is a secure way to securely store and use the data. Algorithms such as ANN and decision trees are used here. They also used the data from the past to forecast future values. They got their train data from real data based on the meteorological data from time series given by the Dalian Meteorological Bureau. The test cases show adequate wind speed output and maximum or lowest temperature. They concluded that ANN can be a good choice for weather forecasting work.

Janani and Sebastian presented analysis on the weather forecasting and techniques [7]. In their work, they proposed that SVM should be able to perform better than traditional MLP trained with back-propagation for any order. They tried to predict the weather using a 3-layered neural network. An existing dataset was trained for the algorithm. They said this algorithm performs better, with few errors. They also said a fuzzy weather prediction system from KNN could boost this system's technique.

Yadav and Khatri "A Weather Forecasting Model using the Data Mining Technique" [8]. In their work, the authors have the two algorithms: the K-means clustering algorithm and the ID3 algorithm made an interesting comparison. They compared the algorithm performance, error rate, memory use, time of training, and time of prediction and offered a decision that clustering by K-means works better than algorithm ID3.

Kunjumon et al. presented survey on weather forecasting using data mining [9]. This research revolves around data mining. The authors used both classification techniques and clustering techniques to get the best accuracy and limitations. They then connected with each other. The algorithms used are artificial neural network (ANN), SVM, growth algorithm for FP, algorithm for K-medoids, algorithm for Naïve Bayes, and algorithm for decision tree. After estimating, the authors concluded that the SVM has the highest accuracy (about 90%). Yet, this study is more of a summary article, as the authors just compared the findings of the different available studies.

Jakaria et al. presented smart weather forecasting using machine learning: a case study in Tennessee [10]. By building models using a high-performance computing environment, they used smart forecasting systems. They applied machine learning techniques to them by gathering datasets from local workstations. Machine learning techniques such as the ridge regression, SVM, multilayer perceptron regressor, RFR, and ETR are used for the model construction. They found that MLPR yields high RMSE value, while RFR and ETR yield low RMSE. From their observation, they said that on larger datasets, the techniques would work better. They plan to work on the use of low-cost IoT devices in the future.

Scher and Messori presented weather and climate forecasting with neural networks, using general circulation models (GCMs) with different complexity as a study ground [11]. To train their model, they used a deep, neural network. They also used general circulation models such as GCM to do that. The network is trained in the model, with deep convolutionary architecture. They believed that the model yielded satisfactory performance. This approach will work well with larger datasets, the authors have said.

Jyosthna Devi et al. presented ANN approach for weather prediction using backpropagation. In their work, they used a back-propagated model building neural network technique. Backpropagated neural networks perform on broader functions better. Neural networks recover entity relationships by measuring established parameters. For this experiment, a neural network of three layers is used to find the relationship between known nonlinear parameters. They said the model could more accurately forecast future temperatures.

Riordan and Hansen presented a fuzzy case-based system for weather prediction [12]. In their work, they used a fuzzy c-means technique with KNN. This system is for airport weather forecasting. K-nearest neighbors are used to predict the weather. They select a value of $k = 16$ which gives the best accuracy in addition to others. If k is small, then the accuracy is decreased, and if k is larger, then the model seems to be overfitted. The authors said they will continue to work for other airports with other parameters.

Singaravel et al. presented component-based machine learning modeling approach for design stage building energy prediction: weather conditions and size [13]. The authors have introduced a method which is called as component-based machine learning modeling. They said that the previous machine learning techniques have some limitations that is why they used this technique. They collected data from local stations and build a model with a simple box building. They said that the machine learning modeling can predict the box building under different conditions. The estimated accuracy of the model is approximately 90%. From their observation, they concluded that the model works better than previous technologies.

Abhishek et al. presented weather forecasting model using artificial neural network [14]. They used ANN in their research to forecast the weather, based on different weather parameters. They have seen some overfitting of some inputs, hidden layer functions, and some sample variables on their layout. They concluded that model accuracy would increase by increasing parameters.

Salman et al. presented weather forecasting using deep learning techniques [15]. To forecast the weather, they used deep learning techniques such as RNN, conditionally limited Boltzmann system, and CNN. They noticed that RNN can provide the rainfall parameter with good precision. They hope to be working on CRBM and CNN in the next trial.

Torabi and Hashemi presented a data mining paradigm to forecast weather sensitive short-term energy consumption [16]. The authors have researched weather forecasting along with power consumption based on data mining technology. ANN and SVM were used to find out the pattern of energy consumption. They found that in the summer season the electricity consumption increases compared to other seasons. To do so, ANN gives the best accuracy, said by the authors.

Ashfaq presented machine learning approach to forecast average weather temperature of Bangladesh [17]. In their work, the authors have used several machine learning techniques like linear, polynomial, isotonic regression, and SVM to predict the average temperature of Bangladesh. They said that the isotonic regression gave good accuracy over train data but not in test data. So, polynomial regression or SVM was recommended by them.

3 Methodology

The goal of this work is to use ML techniques to predict the weather of the next day in Dhaka City. So, in this section, there will be some detailed descriptions of the research work. For more clarification, research subject and instrumentation will be explained shortly. Data processing is a very important part of machine learning, so it will be described after that.

3.1 *Research Subject and Instrumentation*

Because this is a research job, it needs to be very well understood. Not only that, but analysis will also vary from the study because it can alter the result at any time. So work is really effective in interpreting those variations correctly. And instrumentation refers to the instruments or devices used in this investigation.

3.2 *Data Processing*

In machine learning, no work can be done without data. So, the key part of this research was collecting data, and it is difficult part too. Since locating or obtaining data is not as straightforward as it would seem. No source for all of the data was available. The previous weather data will be gathered between August 1, 2018, and July 31, 2019, and divided into three seasons: summer (April, May, June, July), fall (August, September, October, November), and winter (December, January, February, March). Fall and winter data are from the year 2018. But this work is on the "Summer." The techniques can be extended to the remaining datasets as well. Prediction can be done on any other variables like humidity, pressure, and wind speed from here. But here the research has done only for predicting temperature.

3.3 *Statistical Analysis*

When working with data, there were some errors regarding some missing data in the dataset. Those errors are needed to fix because the successful implementation of machine learning algorithms is dependent on correct pre-processing data. So, fixing the dataset becomes the main responsibility then. In Fig. 1, there is a flowchart which is about the working process of the research.

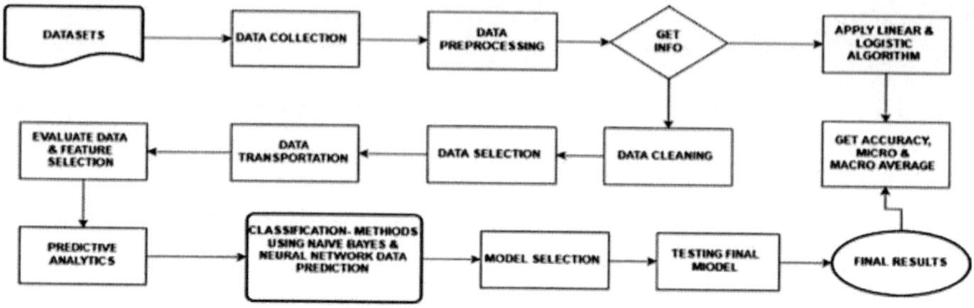

Fig. 1 Flowchart for analysis

3.4 Experimental Result and Discussion

After training the data with four algorithms, the experimental model is built. Dataset was missing some values and needed to get them filled by using panda's method. So that data can be more accurate. To build the regressor model, the dataset was separated into two parts:

- Training Dataset
- Testing dataset.

Ratio 5:1 is used for building the model. The four portions of the dataset are used as train data and rest 1 portion is used as test data. There were 122 data points and 15 attributes in the dataset. For training, 97 data points were taken and rest 25 data points were used for testing. To build the desired model, three different algorithms were used: linear regression, logistic regression, and Naïve Bayes.

4 Experimental Result Between Various Algorithms

After building models Accuracy, predicted the weather of different algorithms is added below. Algorithm details are given below:

4.1 Linear Regression

Linear regression is one of the basic and renowned machine learning algorithms. It is a way of shaping relationships among variables. Linear regression comprises two kinds of variables: continuous variable and independent variable. The line's slope is m and c is the intercept (the y value if $x = 0$). MaxTemp and MinTemp appear in Fig. 2.

In Fig. 3, the data are plotted in scatterplot so that the relationship can be visualized

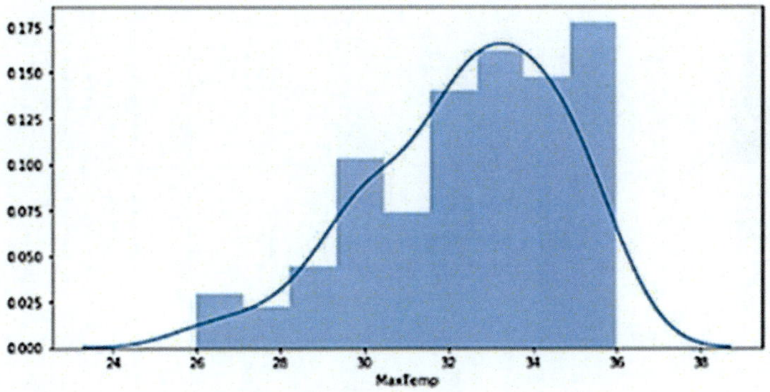

Fig. 2 MaxTemp

better.

Here is the visualization of how the model predicted tomorrow's temperature in Fig. 4.

Linear regression provides a comparison of the expected value to the actual value. This is shown in Table 1.

4.2 Logistic Regression

Logistic regression is also a popular and frequently used algorithm for solving classification problem which linear regression cannot handle. Like if someone wants to separate positive and negative values from some given random values, then they need to use logistic regression. In this research, logistic classifier is used for classifying a mid-value from MaxTemp and MinTemp that is shown in Fig. 5.

The logistic function takes any value between 0 and 1. The function is,

$$\sigma(t) = \frac{e^t}{e^t + 1} = \frac{1}{1 + e^{-t}} \quad (1)$$

In a univariate regression model, let us consider t as a linear function.

$$t = \beta o + \beta 1x \quad (2)$$

The logistic equation would then become,

$$p(x) = \frac{1}{1 + e^{-(\beta o + \beta 1x)}} \quad (3)$$

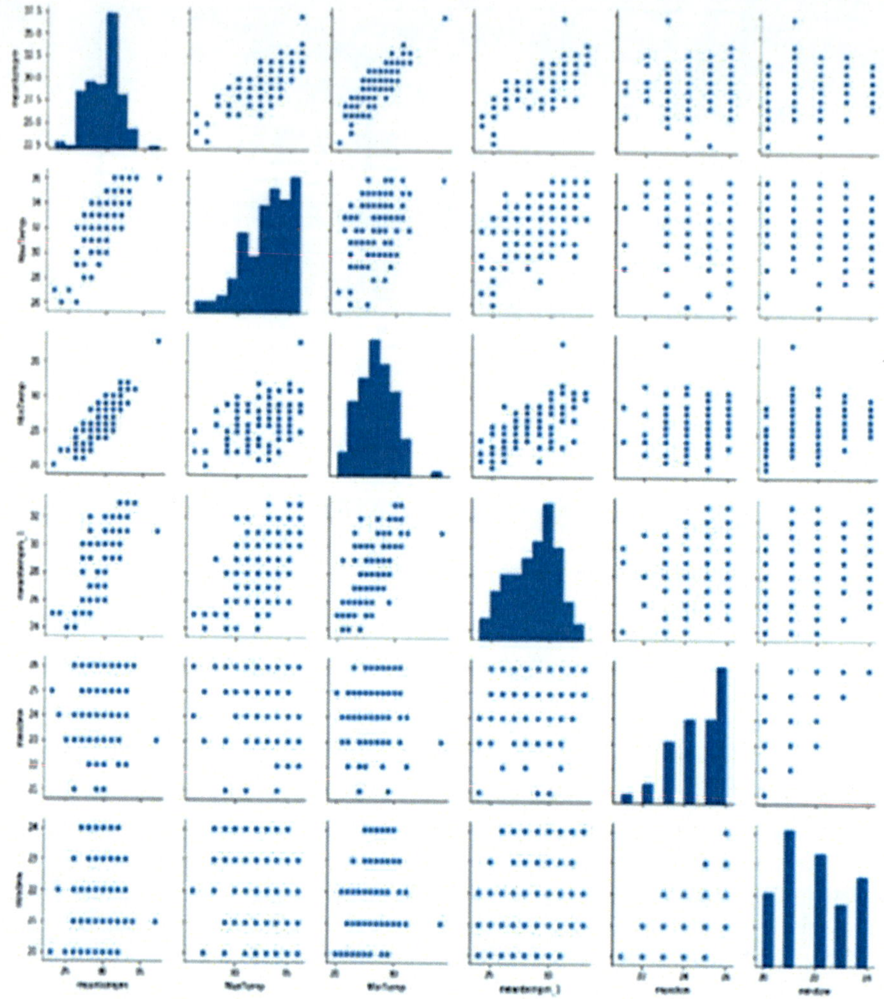

Fig. 3 Scatterplot result

4.3 Naïve Bayes

It is a very popular classification technique based on Bayes probability theorem. Bayes theorem assumes that every parameter is independent in the analysis. This theorem is very useful for large datasets.

Bayes theorem measures $P(a)$ from $P(a)$, $P(b)$, and $P(b)$ as the likelihood. Here is the Bayes theorem equation:

$$P(a|b) = \frac{P(b|a)P(a)}{P(b)} \tag{4}$$

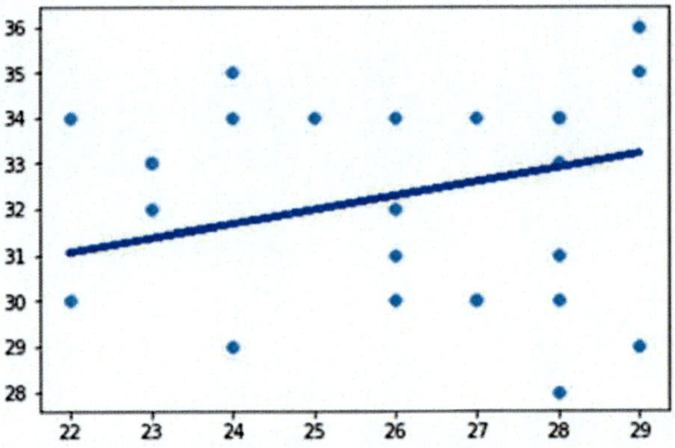

Fig. 4 Visualization about prediction

Table 1 Actual value versus predicted value

	Actual value	Predicted value
0	36	33.212321
1	30	32.597770
2	33	31.368670
3	29	33.212321
4	33	33.905045

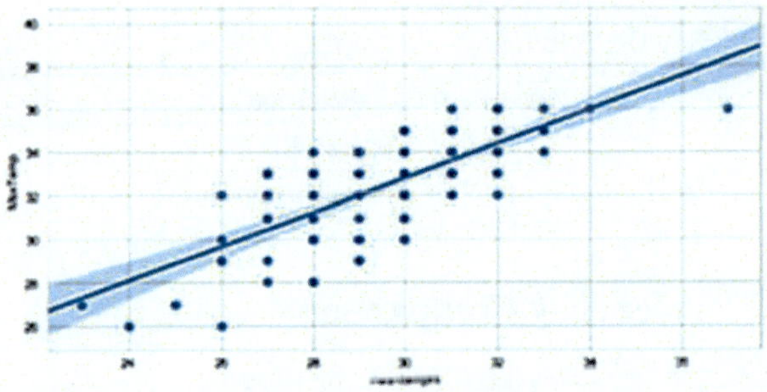

Fig. 5 Logistic regression classifier

Then,

$$P(a|b) = P(b_1|a) \times P(b_2|a) \times \cdots \times P(b|a) \times P(a) \tag{5}$$

To predict accuracy, a similar Naïve Bayes approach is used to predict the likelihood of different groups based on different data, as this algorithm is mainly used in the classification of text and with multiple class problems. The column was measured by the precipm and is visualized in Fig. 6. The Naïve Bayes algorithm is visualized.

In Table 2, there is the performance measurement of the Naïve Bayes classifier. The precision of a classifier defines the correctness and recall defines the entirety of the model. So, here the precision is 59% and recall is 25%.

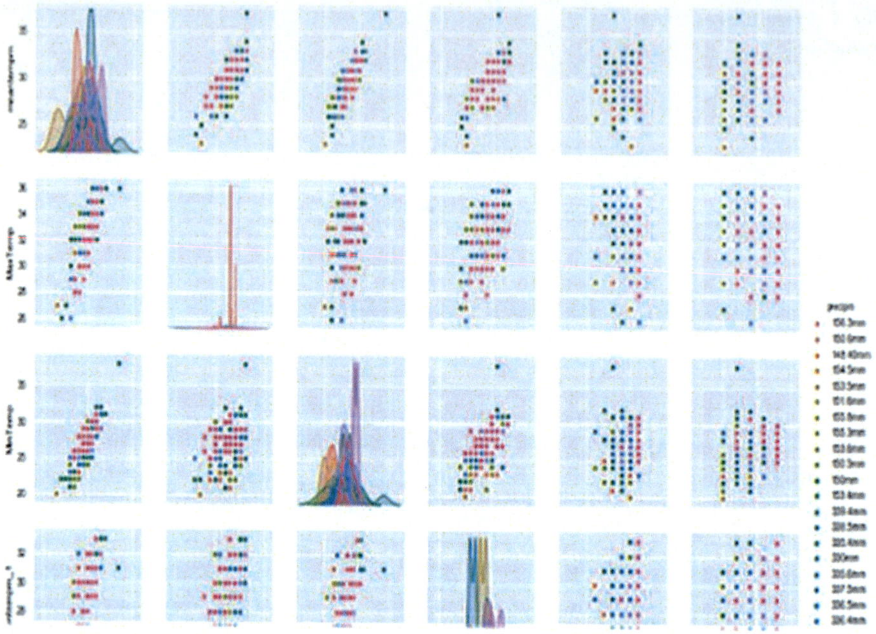

Fig. 6 Implementation of Naïve Bayes algorithm

Table 2 Naive Bayes accuracy score

Matric	Score (%)
Precision	58.70
Recall	25
Accuracy	29

5 Accuracy Comparison

In Fig. 7, there is an accuracy comparison between three algorithms, which are used to build the model. It is shown that linear regression gives the best accuracy compared to the rest.

6 Conclusion and Future Work

The objective of this research is to create a model of weather prediction which will offer a forecast of the weather in Dhaka City tomorrow. The summer dataset is currently being introduced into the model as it is now summer in Dhaka. Three algorithms were applied in this study. Linear regression, among them, provided

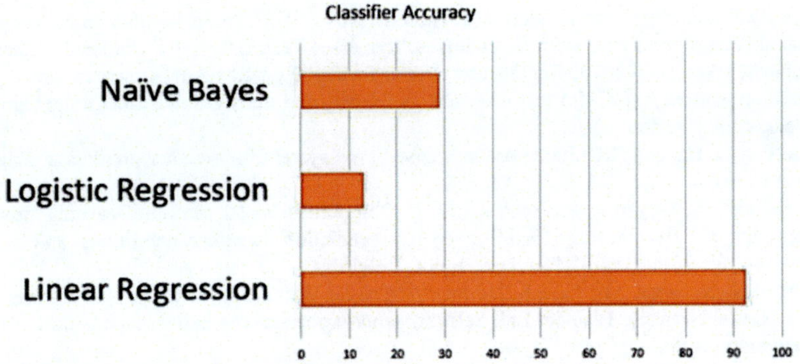

Fig. 7 Accuracy comparison between algorithms

adequate accuracy compared to others. For the future, the key concern will be the deployment of the remaining two datasets (fall and winter). Maybe along with these three algorithms, there will be a few other algorithms too, so that we can consider better weather forecasting techniques. We believe it would be important for potential weather data researchers to use more machine learning techniques.

References

1. Timeanddate.com. (2019) timeanddate.com. [online] Available at: https://www.timeanddate.com/. Accessed 6 Dec 2019
2. Medium (2019) What is the weather prediction algorithm? How it works? What is the future? [online] Available at: https://medium.com/@shivamtrivedi25/what-is-the-weather-prediction-algorithm-how-it-works-what-is-the-future-a159040dd269. Accessed 6 Dec 2019
3. Holmstrom M, Liu D, Vo C (2016) Machine learning applied to weather forecasting. Stanford University
4. Chauhan D, Thakur J (2013) Data mining techniques for weather prediction: a review, Shimla 5, India: ISSN
5. Biswas M, Dhoom T, Barua S (2018) Weather forecast prediction: an integrated approach for analyzing and measuring weather data. (International Journal of Computer Applications). BGC Trust University, Chittagong, Bangladesh
6. Wang ZJ, Mazharul Mujib ABM (2017) The weather forecast using data mining research based on cloud computing. (IOP Conf. Series: Journal of Physics: Conf. Series 910). Dalian University of Technology, Liaoning, China
7. Janani B, Sebastian P (2014) Analysis on the weather forecasting and techniques. In: International journal of advanced research in computer engineering & technology. Department of CSE, SNS College of Engineering, India
8. Yadav RK, Khatri R (2016) A weather forecasting model using the data mining technique. In: International journal of computer applications. Vikrant Institute of Technology & Management, Indore, India
9. Kunjumon C, Sreelekshmi SN, Deepa Rajan S, Padma Suresh L, Preetha SL (2018) Survey on Weather Forecasting Using Data Mining. In: Proceeding IEEE conference on emerging devices and smart systems. University of Kerala, Kerala, India

10. Jakaria AHM, Mosharaf Hossain Md, Rahman MA (2018) Smart weather forecasting using machine learning: a case study in tennessee. Tennessee Tech University, Cookeville, Tennessee
11. Scher S, Messori G (2019) Weather and climate forecasting with neural networks: using general circulation models (GCMs) with different complexity as a study ground. Stockholm University, Stockholm, Sweden
12. Riordan D, Hansen BK (2002) A fuzzy case-based system for weather prediction. Eng Intell Syst Canada
13. Singaravel S, Geyer P, Suykens J (2017) Component-based machine learning modelling approach for design stage building energy prediction: weather conditions and size. In: Proceedings of the 15th IBPSA conference. Belgium
14. Abhishek K, Singh MP, Ghosh S, Anand A (2012) Weather forecasting model using Artificial Neural Network. Elsevier Ltd. Selection and/or peer-review under responsibility of C3IT, Bangalore, India
15. Salman MG, Kanigoro B, Heryadi Y (2015) Weather forecasting using deep learning techniques. IEEE Jakarta, Indonesia
16. Torabi M, Hashemi S (2012) A data mining paradigm to forecast weather sensitive short-term energy consumption. In: The 16th CSI international symposium on artificial intelligence and signal processing. Shiraz, Iran
17. Ashfaq AS (2019) Machine learning approach to forecast average weather temperature of Bangladesh. Global J Comput Sci Technol: Neural Artif Intell, Dhaka, Bangladesh
18. Williams JK, Ahijevych DA, Kessinger CJ, Saxen TR, Steiner M, Dettling S (2008) National Center for Atmospheric Research. In: Boulder C (ed) A machine learning approach to finding weather regimes and skillful predictor combinations for short-term storm forecasting, Colorado

Image Processing: What, How and Future

Mansi Lather and Parvinder Singh

Abstract There is a well-known saying: an image is worth more than a thousand words. The wonders of this proverb are very well visible in our day-to-day life. In this paper, we have presented the current and trending applications of imaging in our day-to-day life having a wide scope of research. The digital image processing has revolutionized the field of technical endeavor and there is a lot more yet to be researched in this field. A huge amount of work can be carried out in the area of image processing. This paper summarizes the fundamental steps involved in digital image processing and focuses on the applicative areas of image processing in which research can be carried out for the betterment and quality improvement of human life.

Keyword Biomedical imaging · Digital image processing · Image · Image processing · Imaging applications

1 Introduction

An image is generally a 2D function $f(x, y)$ where x and y are spatial coordinates and the magnitude of f at (x, y) is known as gray/intensity level of an image. An image is known as a digital image when the values of x, y and f are all finite. A digital image is made up of a finite number of entities called pixels, each having a particular location and value [1]. An image can be processed to get a better understanding of useful information contained in the image or to get an enhanced image. This process is called image processing. It is a kind of signal disbursement having an image as input and producing some image characteristics as output [2]. Image processing is of two types: analog and digital image processing. Analog image processing is used for

M. Lather (✉) · P. Singh
Department of Computer Science and Engineering, DeenBandhu Chhotu Ram University of Science and Technology, Murthal, Sonipat 131039, India
e-mail: mansi.schcse@dcrustm.org

V. Singh et al. (eds.), *Computational Methods and Data Engineering*, Advances in Intelligent Systems and Computing 1227,
https://doi.org/10.1007/978-981-15-6876-3_23

taking photographs and printouts, that is, for hard copies. On the other hand, when images are manipulated by digital computers, it is known as digital image processing [2].

The major focus of digital image processing is on two things:

- Enhancement of image data for human evaluation;
- Image data processing for communication, caching and representation for uncontrolled machine perception [1].

The fundamental steps involved in digital image processing are shown in Fig. 1 [1]. It is not necessary to apply all the steps to each and every type of image. The figure shows all the steps that can be applied to images, but the steps are chosen depending on the purpose and objective of the image. The description of all the steps is as follows [1]:

- **Image Acquisition**: This step involves getting the image that needs to be processed. The image can be acquired using sensor strips, sensor arrays, etc.
- **Image Enhancement**: Image is enhanced to focus on certain characteristics of interest in an image or to get out the hidden details from an image. Image enhancement can be done using frequency and spatial domain techniques. The spatial domain technique focuses on direct pixel manipulation. Frequency domain methods, on the other hand, focus on the modification of the Fourier transform of an image.
- **Image Restoration**: It is an objective process that improves the image appearance by making use of probabilistic and mathematical models of image degeneration. This step restores the degraded image by making use of earlier knowledge of

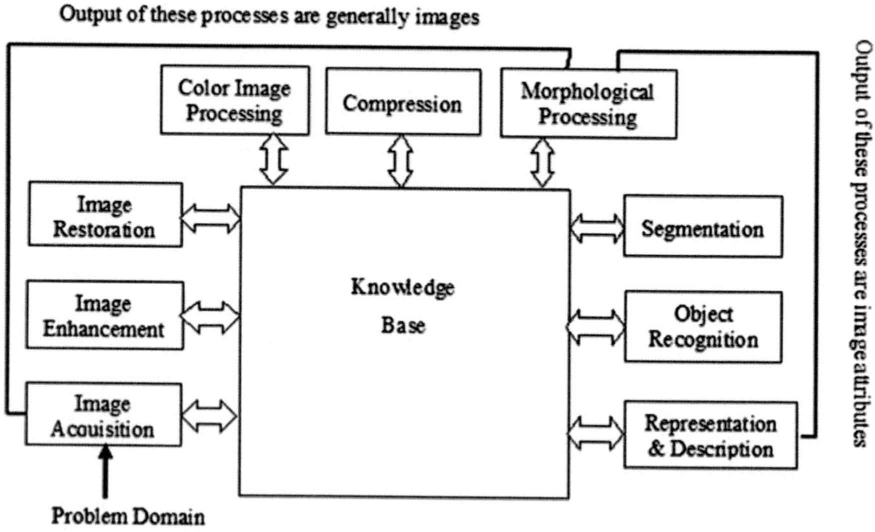

Fig. 1 Fundamental steps in digital image processing

the degradation phenomenon. Noise removal from images by using denoising techniques and blur removal from images by using deblurring techniques come under image restoration.

- **Color Image Processing**: This is basically of two types—full-color and pseudo-color processing. In the former case, images are captured through full-color sensors like a color scanner. Full-color processing is further divided into two categories: In the first category, each component is processed individually and then a composite processed color image is formed, and in the second category, we directly manipulate color pixels. Pseudo-color or false color processing involves color assignment to a particular gray value or range of values on the basis of a stated criterion. Intensity slicing and color coding are the techniques of pseudo-color processing. Color is used in image processing because of the human ability to differentiate between different shades of color and intensities in comparison with different shades of gray. Moreover, color in an image makes it easy to extract and identify objects from a scene.
- **Image Compression**: It means decreasing the quantity of information required to express a digital image by eliminating duplicate data. Compression is done in order to reduce the storage requirement of an image or to reduce the bandwidth requirement during transmission. It is done prior to storing or transmitting an image. It is of two types—lossy and lossless. In lossless compression, the image is compressed in such a way that no information is lost. But, in lossy compression, to achieve a high level of compression, loss of a certain amount of information is acceptable. The former is useful in image archiving such as storing medical or legal records, while the latter is useful in video conferencing, facsimile transmission and broadcast television. Lossless compression techniques involve variable length coding, arithmetic coding, Huffman coding, bit-plane coding, LZW coding, run-length coding and lossless predictive coding. Lossy compression techniques involve lossy predictive coding, wavelet coding and transform coding.
- **Morphological Image Processing**: It is the technique for drawing out those parts of an image that can be used to represent and describe the morphology, size and shape of an image. The common morphological operators are dilation, erosion, closing and opening. The principal applications of morphological image processing include boundary extraction, region filling, convex hull, skeletons, thinning, extraction of connected components, thickening and pruning.
- **Image Segmentation**: It is the process of using automated and semi-automated means to extract the required region from an image. The segmentation methods are broadly categorized as edge detection methods, region-based methods (includes thresholding and region growing methods), classification methods (includes K-nearest neighbor, maximum likelihood methods), clustering methods (K-means, fuzzy C-means, expectation-maximization methods) and watershed segmentation [3].
- **Representation and Description**: The result of the segmentation process is raw data in the form of pixels that needs to be further compacted for representation and description appropriate for additional computer processing. A region can be represented either in terms of its external features such as boundary

or in terms of its internal features such as pixels covering the region. Representation techniques include chain codes and polygonal approximations. In the next task, on the basis of the chosen representation, the descriptor describes the region. Boundary descriptors are used to describe the region boundary and are of the following types—length, diameter, curvature, shape numbers, statistical moments and Fourier descriptors. Regional descriptors, on the other hand, are used to describe the image region and are of the following types—area, compactness, mean and median of gray levels, the minimum and maximum values of gray levels and topological descriptors.

- **Object Recognition**: It involves recognizing the individual image regions known as patterns or objects. There are two approaches to object recognition—decision-theoretic and structural. In the former case, quantitative descriptors are used to describe patterns like texture, area and length. But in the latter case, qualitative descriptors are used to describe the patterns like relational descriptors.

2 Applications of Digital Image Processing

Digital image processing has influenced almost every field of technical inclination in one way or the other. The application of digital image processing is so vast and diverse that in order to understand the broadness of this field we need to develop some form of organization. One of the easiest ways to organize the applications of image processing is to classify them on the basis of their sources such as X-ray and visual [1].

2.1 Gamma-Ray Imaging

Nuclear medicine and astronomical observations are the dominant uses of imaging based on these rays. The entire bone scan image obtained using gamma-ray imaging is shown in Fig. 2. These kinds of images are used for locating the points of bone pathology infections [1].

2.2 X-Ray Imaging

X-rays are dominantly used in medical diagnostics, industry and astronomy [1]. Figure 3 shows the chest X-ray.

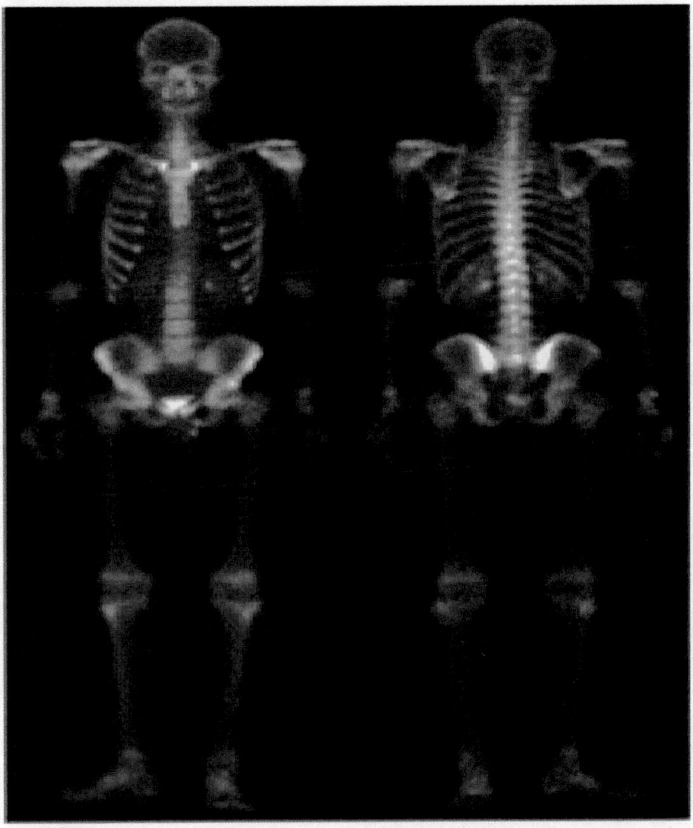

Fig. 2 Example of gamma-ray imaging [1]

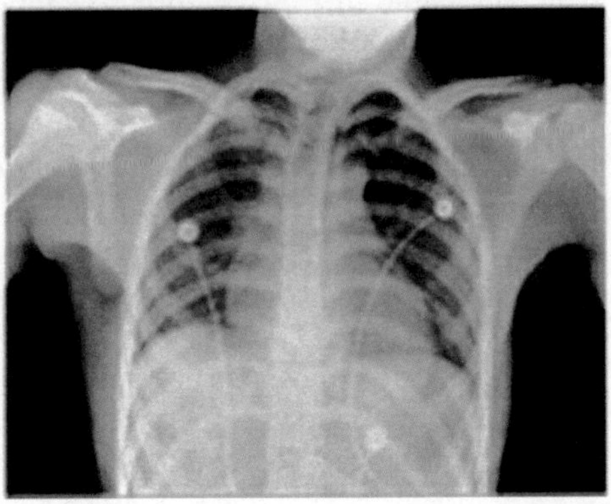

Fig. 3 Example of X-ray imaging: chest X-ray [1]

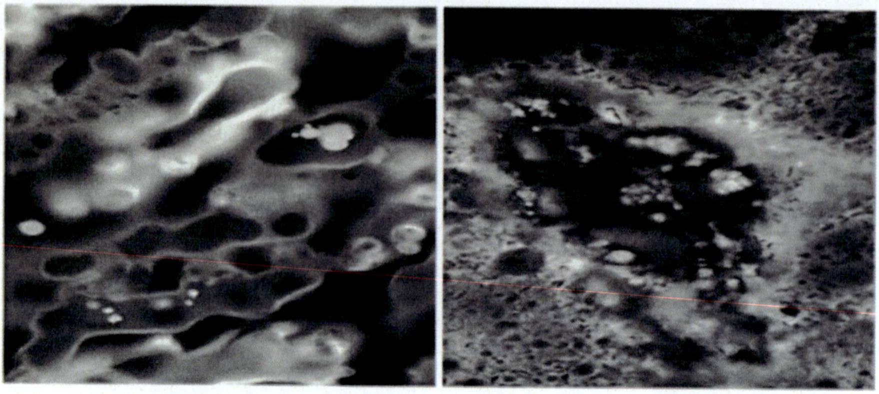

(a) (b)

Fig. 4 Examples of ultraviolet imaging: **a** normal corn; **b** smut corn [1]

2.3 *Ultraviolet Band Imaging*

Lithography, lasers, astronomical observations, industrial inspection, biological imaging and microscopy are the main applications of ultraviolet light [1]. The capability results of fluorescence microscopy are shown in Fig. 4a and b.

2.4 *Visible and Infrared Bands Imaging*

The main applications include light microscopy, industry, astronomy, law enforcement and remote sensing [1]. Some examples of imaging in this band are shown in Fig. 5. CD-ROM device controller board is shown in Fig. 5a. The objective here is to inspect the board for missing parts. Figure 5b shows an image of a pill container. The task is having a machine to identify the missing pills. The objective of Fig. 5c is to identify the bottles not filled up to a satisfactory level.

Some other examples of imaging in the visual spectrum are shown in Fig. 6. A thumbprint is shown in Fig. 6a. The objective here is to process the fingerprints using a computer either for enhancing the fingerprints or using them as security aid in bank transactions. Figure 6b shows the paper currency. The objective here is to automate the currency counting and is used in law enforcement by reading the serial numbers so as to keep track and identify the bills. Figure 6c shows the use of image processing in automatic number plate reading of vehicles for traffic monitoring and surveillance.

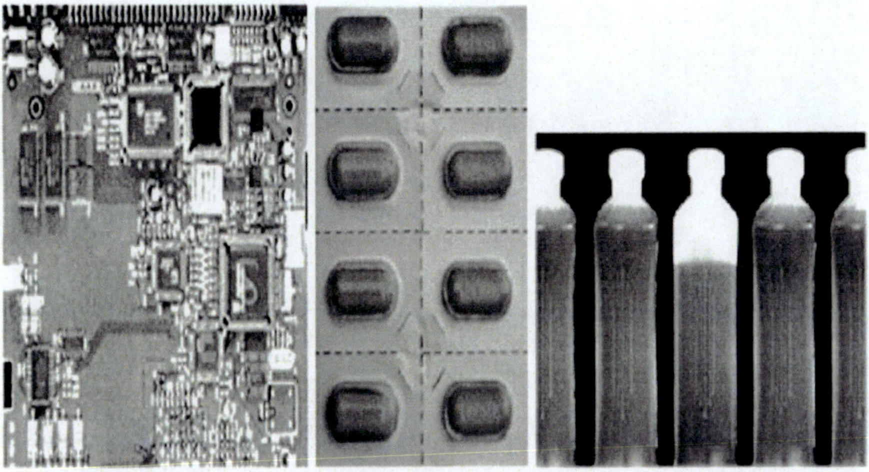

Fig. 5 Examples of manufactured goods often checked using digital image processing: **a** circuit board controller; **b** packaged pills; **c** bottles [1]

2.5 *Microwave Band Imaging*

Radar is the major use of imaging in a microwave band. The exclusive characteristics of radar imaging are its data-gathering capability relatively at any time and at any place, irrespective of lighting and weather conditions [1]. The spaceborne radar image of the rugged mountainous area of southeast Tibet is shown in Fig. 7.

2.6 *Imaging in Radio Band*

The main application of imaging in radio band is in medicine and astronomy. In medicine, magnetic resonance imaging (MRI) uses radio waves [1]. MRI images of the human knee and spine are shown in Fig. 8.

3 Imaging Applications

3.1 *Intelligent Transportation System*

Intelligent transportation system (ITS) combines the conventional transportation infrastructure with the advances in information systems, sensors, high technology,

Fig. 6 Some additional examples of imaging in visual spectrum: **a** thumbprint; **b** paper currency; **c** automated license plate reading [1]

controllers, communication, etc., and their integration alleviates the congestion, boosts productivity and increases safety [4].

In [5], a bi-objective urban traffic light scheduling (UTLS) problem is addressed to minimize the total delay time of all the pedestrians and vehicles.

Another important application of ITS is in the shared bike system. In order to save the time spent waiting for bikes at the bike stations, the bike-sharing system's operator needs to dispatch the bikes dynamically. For this, a bike repository can be optimized by forecasting the number of bikes at every station. The solution to this issue of predicting the number of bikes is given in [6].

Fig. 7 Spaceborne radar image of mountains in Tibet [1]

(a) (b)

Fig. 8 MRI images of a human **a** knee; **b** spine [1]

3.2 Remote Sensing

In this application, pictures of the earth's surface are captured using remote sensing satellites mounted on aircraft and these pictures are then sent to the earth station for processing. This is useful in monitoring agricultural production, controlling flood, mobilizing the resources, city planning, etc. [2].

In [7], remote sensing imagery is used to identify soil texture classes. Soil texture is very significant in figuring out the water-retaining capacity of the soil and other hydraulic features and thereby affecting the fertility of the soil, growth of the plants and the nutrient system of soil.

Another important application of remote sensing is to detect the center of tropical cyclones so as to prevent the loss of life and economic loss in coastal areas [8].

3.3 Moving Object Tracking

The main task of this application is to access the locomotive parameters and visual accounts of moving objects [2]. Motion-based object tracking basically relies on recognizing the moving objects over time using image acquisition devices in video sequences. Object tracking has its uses in robot vision, surveillance, traffic monitoring, security and video communication [9]. An automated system to create 3D images and object tracking in the spatial domain is presented in [9].

3.4 Biomedical Imaging System

This application uses the images generated by different imaging tools like X-ray, CT scan, magnetic resonance imaging (MRI), positron emission tomography (PET) and ultrasound [1]. The main applications under this system include the identification of various diseases like brain tumors, breast cancer, epilepsy, lung diseases, heart diseases, etc.

The biomedical imaging system is widely being used in the detection of brain tumors. The brain is regarded as the command center of the human nervous system. It is responsible for controlling all the activities of the human body. Therefore, any abnormality in the brain will create a problem for one's personal health [10]. The brain tumor is an uncontrolled and abnormal propagation of cells. It not only affects the immediate cells of the brain but can also damage the surrounding cells and tissues through inflammation [11]. In [12], an automated technique is presented to detect and segment the brain tumor using a hybrid approach of MRI, discrete wavelet transform (DWT) and K-means, so that brain tumor can be precisely detected and treatment can be planned effectively.

Another application of medical imaging is gastrointestinal endoscopy used for examining the gastrointestinal tract and for detecting luminal pathology. A technique to automatically detect and localize gastrointestinal abnormalities in video frame sequences of endoscopy is presented in [13].

3.5 Automatic Visual Inspection System

The important applications of automatic visual inspection system include [14]:

- Online machine vision inspection of product dimensions,
- Identifying defects in products,
- Inspecting quantity of material filled in the product,
- Checking proper installation of airbags in cars,
- License plate reading of vehicles,
- To ensure proper manufacturing of syringes,
- Irregularity detection on flat glasses,
- Person recognition and identification,
- Dimensionality checking and address reading on parcels,
- Inspection of label printing on the box,
- Surface inspection of bathtubs for scratches and so on.

The benefits of an automatic visual inspection system include speedy inspection with less error rate and with no dependability on manpower [14].

3.6 Multimedia Forensics

Multimedia is data in different forms like audio, video, text and images. Multimedia has become an essential part of everyday life. A huge amount of multimedia content is being shared on the Internet every day by online users because of the high use of mobile devices, availability of bandwidth and cheaper storage [15]. Multimedia forensics deals with the detection of any kind of manipulation in multimedia content as well as the authenticity of the multimedia content. Multimedia forensics is about verifying the integrity and authenticity of multimedia content [16]. It follows the virtual traits to disclose the actions and intentions of hackers and to detect and prevent cybercrime. Watermarking and digital signature are used in multimedia forensics. The biggest challenge in multimedia forensics is that the amount of multimedia data is so massive that it has surpassed the forensic expert's ability of processing and analyzing it effectively. The other challenges are limited time, dynamic environment, diverse data formats and short innovation cycles [15].

Every day, a huge amount of image content is shared over the Internet. Thus, the integrity of image data is doubtful because of the easy availability of image manipulation software tools such as Photoshop. In order to tamper an image, a well-known technique of replicating a region somewhere else in the same image to imitate or hide some other region called copy–move image forgery is being used. The replicated regions are invisible to the human eye as they have same texture and color parameters. In [17], a block-based technique employing translation-invariant stationary wavelet transform (SWT) is presented to expose region replication in digital images so that the integrity of image content can be verified. In [18], a copy–move image forgery is detected by using a discrete cosine transform (DCT). DCT has the ability of accurately detecting the tampered region.

4 Conclusion

Image processing has a wide range of applications in today's world of computer and technology. It has impacted almost every field of technical endeavor. The impact of digital image processing can also be seen in human life to a great extent. Imaging applications have a wide scope of research. There is a lot yet to be developed in this field. The power of modern computer computation can be utilized to automate and improve the results of image processing and analysis. Human life has achieved great heights and can become better in the years to come through the intervention of computer technology in imaging applications.

References

1. Gonzalez RC, Woods RE (2001) Digital image processing. 2nd ed. Upper saddle river, Prentice Hall, New Jersey
2. What is Image Processing : Tutorial with Introduction, Basics, Types & amp; Applications. https://www.engineersgarage.com/articles/image-processing-tutorial-applications
3. Lather M, Singh P (2017) Brain tumour detection and segmentation techniques : a state-of-the-art review. Int J Res Appl Sci Eng Technol 5(vi):20–25
4. Lin Y, Wang P, Ma M (2017) Intelligent transportation system (ITS): concept, challenge and opportunity. In: 2017 IEEE 3rd international conference on big data security on cloud, pp 167–172
5. Gao K, Zhang Y, Zhang Y, Su R, Suganthan PN (2018) Meta-heuristics for Bi-objective Urban traffic light scheduling problems. In: IEEE transactions on intelligent transportation systems, pp 1—12. https://doi.org/10.1109/TITS.2018.2868728
6. Huang F, Qiao S, Peng J, Guo B (2018) A bimodal gaussian inhomogeneous poisson algorithm for bike number prediction in a bike-sharing system. IEEE Trans Intell Transp Syst 1—10
7. Wu W, Yang Q, Lv J, Li A, Liu H (2018) Investigation of remote sensing imageries for identifying soil texture classes using classification methods. IEEE Trans Geosc Remote Sens 1–11
8. Jin S, Li X, Yang X, Zhang JA, Shen D (2018) Identification of tropical cyclone centers in SAR imagery based on template matching and particle swarm optimization algorithms. IEEE Trans Geosc Remote Sens 1–11
9. Hou Y, Chiou S, Lin M (2017) Real-time detection and tracking for moving objects based on computer vision method. In: 2017 2nd international conference on control and robotics engineering (ICCRE) pp 213–217

10. Tanya L, Staff W (2016) Human brain: facts, functions and anatomy. http://www.livescience.com/29365-human-brain.html
11. Ananya M (2014) What is a brain tumor? http://www.news-medical.net/health/What-is-a-Brain-Tumor.aspx
12. Singh P, Lather M (2018) Brain tumor detection and segmentation using hybrid approach of MRI, DWT and K-means. In: ICQNM 2018: the twelfth international conference on quantum, Nano/Bio, and micro technologies, pp 7–12
13. Iakovidis DK, Georgakopoulos SV, Vasilakakis M, Koulaouzidis A, Plagianakos VP (2018) Detecting and locating gastrointestinal anomalies using deep learning and iterative cluster unification. IEEE Trans Med Imaging, pp 1–15. https://doi.org/10.1109/tmi.2018.2837002
14. Automatic Online Vision Inspection System. http://www.grupsautomation.com/automatic-online-vision-inspection-system.html
15. Computer Forensics: Multimedia and Content Forensics. https://resources.infosecinstitute.com/category/computerforensics/introduction/areas-of-study/digital-forensics/multimedia-and-content-forensics/#gref
16. Böhme R, Freiling FC, Gloe T., Kirchner M (2009) Multimedia forensics is not computer forensics. In: Geradts ZJMH, Frake KY, Veenman CJ (eds) Computational forensics. IWCF 2009. LNCS, vol 5718. Springer, Berlin, Heidelberg, pp 90–103
17. Mahmood T, Mehmood Z, Shah M, Khan Z (2018) An efficient forensic technique for exposing region duplication forgery in digital images. Appl Intell 48:1791–1801. https://doi.org/10.1007/s10489-017-1038-5
18. Alkawaz MH, Sulong G, Saba T, Rehman A (2018) Detection of copy-move image forgery based on discrete cosine transform. Neural Comput & Applic 30(1):183–192. https://doi.org/10.1007/s00521-016-2663-3

A Study of Efficient Methods for Selecting Quasi-identifier for Privacy-Preserving Data Mining

Rigzin Angmo, Veenu Mangat, and Naveen Aggarwal

Abstract A voluminous amount of data regarding users' location services is being generated and shared every second. The anonymization plays a major role in data sanitization before sharing it to the third party by removing directly linked personal identifiers of an individual. However, the rest of the non-unique attributes, i.e., quasi-identifiers (QIDs) can be used to identify unique identities in a dataset or linked with other dataset attributes to infer the identity of users. These attributes can lead to major information leakage and also generate threat to user data privacy and security. So, the selection of QID from users' data acts as a first step to provide individual data privacy. This paper provides an understanding of the quasi-identifier and discusses the importance to select QID efficiently. The paper also presents the different methods to select quasi-identifier efficiently in order to provide privacy that eliminates re-identification risk on user data.

Keywords Quasi-identifier · Anonymization · Privacy · Adversary

1 Introduction

The digitization of every data is providing an important role in today's scenario to analyzing, mining, discovering knowledge, business, etc., by the government, researchers, analysts, or other third parties. The released data is credible only if it is used in an authorized and for specified limited level so the users' data can be secured as well as useful for them. Even the released data can be vulnerable to privacy and

R. Angmo (✉) · N. Aggarwal
Department of Computer Science and Engineering, UIET, Panjab University, Chandigarh, India
e-mail: rigzin.btech9@gmail.com

N. Aggarwal
e-mail: navagg@gmail.com

V. Mangat
Department of Information Technology, UIET, Panjab University, Chandigarh, India
e-mail: veenumangat@yahoo.com

V. Singh et al. (eds.), *Computational Methods and Data Engineering*, Advances in Intelligent Systems and Computing 1227,
https://doi.org/10.1007/978-981-15-6876-3_24

security threats. Therefore, there is a need for technical guarantee by data owner before the publication of private data of individual users. The objective is that the users' data cannot be re-identified to provide privacy as well as utility.

In 1986, Dalenius [1] introduces the term quasi-identifier (in short QID) [2, 3]. Since then, QID has been used for re-identification attacks to the released dataset. QID is an entity that is part of the information. QIDs cannot be identified uniquely but can be used to generate a distinctive identifier by effectively associating it with other entities or combining with other QIDs. This combination can be used by an adversary or other third parties to find out details about an individual. For example, an adversary can use certain users' details such as date of birth, zip code, and location that can lead to an identification of an individual by linking it with the released public datasets. This information of an individual can be also used for negative means as well to harm user's privacy which leads to an embarrassment.

Another example is given by Sweeney [4], in which they describe that gender, age, DOB, postal codes are not a unique identifier but the combination of these attributes can be used to identify an individual uniquely. According to a statistical example given by Sweeny [4], the arrangement of zip code, DOB, gender has taken from US census database can adequately identify 87% of individuals. Sweeney tries to locate the then governor of Massachusetts' health record by linking health record with publicly available information, and Sweeney et al. [5] used publicly available voter data to distinguish contributors in the personal genome project. Similarly, the same Massachusetts Governor's health record has been linked earlier with the insurance data to represent data identification risk and privacy protection on data [6]. Furthermore, Narayanan and Shmatikov also made use of QIDs to de-anonymize released Netflix data [7]. The probable privacy breaches had been indicated by Motwani et al. [8] that are being permitted by the publication of large volumes of user data containing QIDs owned by government and business. So, QID plays a major role in identifying an individual.

Examples of common quasi-identifiers (QIDs) in the context of user information related to health care, location-based services, social networking, advertisement agencies, mobile applications, sensor networking, cell phone networking, etc., are dates, namely admission, birth, death, discharge, visit; locations information such as zip codes, regions, location coordinates; ethnicity, languages spoken, race, gender, profession, and so on. An adversary can infer a lot about an individual only by analyzing the combination of these quasi-identifiers.

So, from the above example and literature, we can understand that it is very important to select quasi-identifier cautiously, so it cannot be misused for re-identification, avoid attribute disclosure risk and efficiently provide a balance between information loss and privacy of individuals.

2 Importance of Efficient Selection of Quasi-identifier (QID)

Nearly every person in the world has access to the digitization and at least one fact about them is stored in a database server in a digital form that an adversary can use for privacy and security threat that leads to embarrassment or inconvenience or security threats such as steal the identity, take the character of the person in question, harassment, blackmail, physical or mental threat, etc. The most well-known approach to protect users' data from such threats and exploitation is anonymization. The anonymization is the process where one removes sensitive attributes of user data from the dataset before releasing it publically or given to the third party. However, the remaining attribute which is called quasi-identifier may also contain information like age, sex, zip code, location information that can be used to link with other data or combination of two or more QIDs that lead to infer user identity by an adversary. Further, the paper illustrates why traditional anonymization approaches are not efficient enough to protect user privacy with the help of two case studies, namely AOL case [9, 10] and Netflix case [11, 12], based on leaking anonymized dataset.

In the case of AOL (America Online) [9, 10], an American web portal and online service provider companies leaked their searched data. Although for the user data privacy, they anonymized user IDs by replacing it with a number, and at that time, it seemed to be a good idea as it allowed researchers to use data where they can see the complete person's search queries list; however, it also created problems. As, those complete lists of search queries can be used to track down simply by what they had searched for. On the other hand, in the case studies of Netflix [11, 12] illustrates another principle that indicates that the data might seem to be anonymous, although the re-identification could be possible by combining anonymized data with other existing data. As Narayanan and Shmatikov [12] famously proved this point by combining Internet Movie Database (IMDb) with the Netflix database and were able to do user identification from the Netflix data despite the anonymization. The re-identification can lead to privacy and security threat. These re-identification and leaking of data are not only limited to the AOL and Netflix data, but these can happen to any other data too, such as health data re-identification studied in [5, 6].

In America Online (AOL) [10, 13], researchers released a massive anonymized searched query dataset by anonymizing user identities and IP addresses. Netflix [11] also did the same to make a huge database of movie suggestions available for analysis. Despite scrubbing identifiable information from the data, computer scientists were able to identify the individual user in both datasets by linking it with the other entities or database.

3 Quasi-identifier Selection Methods

In this section, we will discuss some of the quasi-identifier selection methods to provide privacy protection on user data so that the linking of such released data cannot be used for privacy and security breach. The section will discuss three types of QID selection methods that can be used to minimize the risk of privacy breach by appropriate selection of a QID.

3.1 *Greedy Minimum Key Algorithm*

There are various approaches to avoid linkage attack through released anonymized data where one can aggregate the results and release interactive report only. However, such techniques restrict the usefulness of the data.

So we need to select QID in such a way that it should not be used by an adversary to invade the individual by linking release QID as background knowledge and other publicly available data.

For this purpose, there are various methods are proposed based on the greedy minimum key algorithm: To avoid linking attack by via quasi-identifiers the concept of k-anonymity is introduced [14]. Generally, the greedy algorithm [8] deals with finding the least number of tuples from some generalized hierarchy with former value. It is the best approximation algorithm to solve the minimum key problem which is an NP-hard problem. The algorithm works as the greedy set cover algorithm, starting with an empty set of attributes. Further, adding the attributes gradually until the separation of all tuple pairs is done. Although the solution gives $O(\ln n)$—approximation solution, the algorithm requires multiple scans that make it expensive for a larger dataset.

3.2 *(ε, δ) Separation Minimum Key Algorithm and (ε, δ) Distinct Minimum Key Algorithm [8]*

The greedy algorithm is optimal for approximation ratio but it requires multiple scans, i.e., $O(m^2)$ [8] of the table which is an expensive task. So, another algorithm for the minimal key problem is designed by allowing quasi-identifiers with approximate guarantees. It is based on random sampling. For this algorithm, firstly it takes a random sample of k-elements or k-tuples, then calculates the input set cover instances, and reduces it to the smaller set cover instances (key) containing only sampled elements to give approximate solutions. The number k is carefully chosen so that the error of probability is bounded.

Table 1 [8] represents a comparative analysis of time and utility between the greedy and the greedy approximate algorithm by running on a random sample of

Table 1 Results of algorithm analysis for masking 0.8-separation QIDs [8]

Dataset	Table census size (k)	Greedy algorithm		(ε, δ) Separation algorithm greedy approximation algorithm by using random sampling	
		Time (s)	Utility	Time (s)	Utility
Adult	10 million	36	12	–	–
Idaho	8867	172	33	–	–
Texas	141,130	880	35	630	33
California	233,687	4628	34	606	32
Washington	41,784	880	34	620	33

30,000 tuples on different datasets. As we can see in Table 1, by decreasing the *k*-sample size, the running time and utility of data decrease as well. Whereas the running time decreases linearly while the utility is dropping slowly.

So, the above example shows how using random sampling for selecting minimal QID and masking QID is an effective way to solve the problem of selecting and providing masking to the attribute and output by minimizing time without degrading the output result. However, information can be lost as it contains a small set of randomly sampled data.

3.3 *Selective Algorithm and Decomposed Attribute Algorithm*

In privacy preservation, the most common approach is anonymization, i.e., removing all information that can directly associate data items with individuals. But remaining attribute which is quasi-identifier may also contain information that can be used to link with other data to infer identity of user. For example-age, gender, zip code can lead to loss of information and privacy as well. The algorithm proposed [15] is used effectively to select the quasi-identifier so that the balance between information loss and privacy is achieved. The proposed algorithm introduced enhancement of formal selection of quasi-identifier attribute by Motwani et al. [8] followed by a decomposition algorithm deployed to achieve a balance between information loss and privacy.

For the selection of the QID attribute, four steps have been introduced in selective algorithm to minimize the loss of information.

Step 1: Nominate the attributes.
Step 2: Generate power set, *P*(*S*) from nominated attributes.
Step 3: Generate the table with the help of *P*(*S*) and the number of tuples corresponding to the power set attributes.

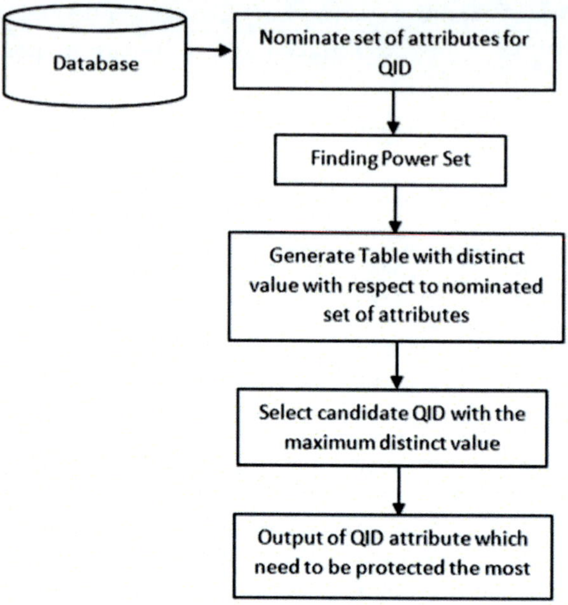

Fig. 1 Selection algorithm for quasi-identifier

Step 4: Select the candidate element from the power set with the maximum tuple value from table which has high chances to identify records distinctly (Fig. 1).

From the above selective algorithm, one can find the QID which can be used to link by an adversary, now we need to represent this frequent attribute in QID so as to avoid information loss and provide privacy. For this, the selective algorithm is followed by decomposition algorithm.

Example of the selection algorithm in Census-income dataset presented by [15] with a total number of 32,561 tuples.

Step 1: Nominated set (Zip, Gender, Race)
Step 2: *P*(*S*): {Zip, Gender, Race, {Zip, Gender}, {Zip, Race}, {Gender, Race}, {Zip, Gender, Race}}
Step 3: Calculate the number of tuples with respect to the selected set attribute as presented in Table 2.

From Table 2, it is found that only one attribute named zip has the highest probability to infer the identity of the user in the database by joining it with other attribute or attributes, in Census dataset, and it is one of the continuous attributes. To overcome the problem of the continuous attribute, a decomposed attribute algorithm has been formulated [15] and the following is an example of a decomposition algorithm [15].

Decomposition algorithm: For an efficient representation of selective QID, decomposed attribute algorithm is used. The two scenarios are applied, one is generalization class and other is code system for numbering. By applying this algorithm,

Table 2 Selective algorithm [15]

Element 1	Number of tuples
Gender	2
Race	5
Gender, race	10
Zip	21,648
Zip, gender	22,019
Zip, race	21,942
Zip, gender, race	22,188

one can efficiently reduce information loss as well as provide privacy as shown in Fig. 2 and Table 3.

As a result of the decomposition algorithm, state code attribute can be substituted by an identification number in the separated table, and new zip code with less number of digits can be generalized or used in data anonymity. Finally, after decomposition, count of the distinct values (Table 4) in each column is obtained from Table 3.

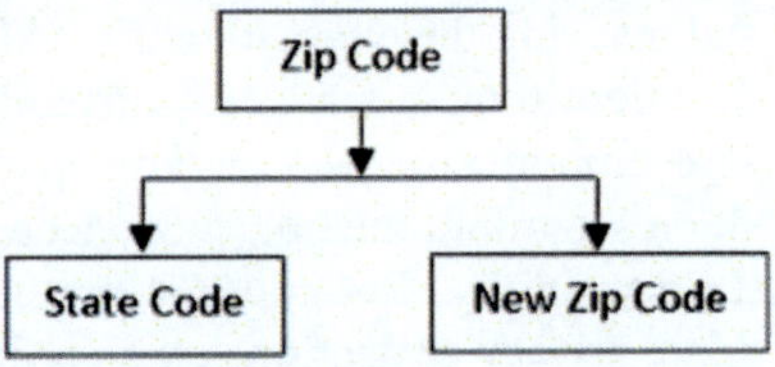

Fig. 2 Decomposition of zip code attribute

Table 3 Decomposition algorithm of zip code structure [15]

S. No.	Old zip code (actual zip code)	State code	New zip code
1	28496	284	96
2	32214	322	14
3	32275	322	75
4	28781	287	81
5	51618	516	18
6	51835	518	35
7	54835	548	35
8	54835	548	35
9	54352	543	52
10	59496	594	96

Table 4 Distinct value after decomposition algorithm [15]

Zip code	State code	New zip code
10	8	7

Table 4 shows that the ability to identify each tuple with zip is 100%, but when we split the zip-to-state code and zip code, the ability is decreased to 80% in state code and 70% in new zip code [15]. The percentile of distinct values can vary according to the decomposition or splitting of zip code into state or zip code and the size of the database.

4 Conclusion

In this paper, we discussed how remaining set of the attribute, i.e., quasi-identifier (QID) can be used to link with the other attributes or itself and lead to re-identification of individual users' identity and a threat to individual privacy. We have also discussed with the help of an example of how anonymization of certain attributes or tuples is not a satisfactory solution for this problem as an adversary can infer private information from the remaining attributes as well. So, we need to select the attribute carefully, which leads to less information loss and protects privacy as well. We discuss efficient algorithms by finding a small set of quasi-identifiers with provable size. We have also shown the greedy and random sampling approach can be used for selecting and masking the quasi-identifier. Also, we have discussed selective and decomposition algorithm, in which selective algorithm is minor enhancement of formal simple algorithm by random sampling approach. The results of the selection and decomposition algorithm method show decreasing loss of information which directly affects the data utility. But still, the minimal set of QID does not imply the most appropriate privacy protection setting because it does not consider background knowledge that an adversary has. And through this background knowledge, an adversary can launch linkage attack that might target victim beyond the minimal set. So, the issue of selecting QIDs efficiently is an open research challenge.

Acknowledgements The authors are grateful to the Ministry of Human Resource Development (MHRD) of the Government of India for supporting this research under the Design Innovation Center (MHRD-DIC) under the subtheme "Traffic Sensing and IT."

References

1. Dalenius T (1986) Finding a needle in a haystack or identifying anonymous census records. J Off Stat 2(3):329
2. Wikipedia Contributors (2019) Quasi-identifier. In: Wikipedia, the free encyclopedia. Retrieved 09:18, 21 Oct 2019, from https://en.wikipedia.org/w/index.php?title=Quasi-identifier&oldid=922082472
3. Vimercati SDCD, Foresti S (2011) Quasi-identifier. In: Encyclopedia of cryptography and security, pp 1010–1011
4. Sweeney L (2000) Simple demographics often identify people uniquely. http://dataprivacylab.org/projects/identifiability/paper1.pdf

5. Sweeney L, Abu A, Winn J (2013) Identifying participants in the personal genome project by name (a re-identification experiment). arXiv preprint arXiv:1304.7605
6. Barth-Jones D (2012, July) The 're-identification' of Governor William Weld's medical information: a critical re-examination of health data identification risks and privacy protections, then and now. In: Then and now
7. Narayanan A, Shmatikov V (2008) Robust de-anonymization of large datasets (how to break anonymity of the Netflix prize dataset). University of Texas at Austin
8. Motwani R, Xu Y (2007) Efficient algorithms for masking and finding quasi-identifiers. In: Proceedings of the conference on very large data bases (VLDB), pp 83–93
9. Ramasastry A (2006) Privacy and search engine data: a recent AOL research project has perilous consequences for subscribers. Law Technol 39(4):7
10. Barbaro M, Zeller T, Hansell S (2006, 2008) A face is exposed for AOL searcher no. 4417749. New York Times 8(2006), 9 (2008)
11. Bennett J, Lanning S (2007, August) The Netflix prize. In: Proceedings of KDD cup and workshop, vol 2007, p 35
12. Narayanan A, Shmatikov, V (2006) How to break anonymity of the Netflix prize dataset. arXiv preprint cs/0610105
13. Anderson N (2008) "Anonymized" data really isn't—and here's why not. https://arstechnica.com/tech-policy/2009/09/your-secrets-live-online-in-databases-of-ruin/. Accessed 21 Oct 2019
14. Sweeney L (2002) k-anonymity: a model for protecting privacy. Int J Uncertain Fuzziness Knowl Based Syst 10(05):557–570
15. Omer A, Mohama B, Murtadha M (2016) Simple and effective method for selecting quasi-identifier. J Theoret Appl Inf Technol 89(2)

Day-Ahead Wind Power Forecasting Using Machine Learning Algorithms

R. Akash, A. G. Rangaraj, R. Meenal, and M. Lydia

Abstract As of late, natural contemplations have incited the utilization of wind power as a maintainable energy resource. Still, the biggest test in coordinating wind power into the electric grid is its irregularity. One procedure to manage wind irregularity is anticipating future estimations of power generated by wind. The power generation relies on the fluctuating speed of the wind. The paper displays the correlation of different wind power forecasting (WPF) based on machine learning algorithms, i.e., multiple linear regression (MLR), decision tree (DT) and random forest (RF). Python (Google Colab) an open-source tool is used to find the result of these models. The exactness of the model has been estimated utilizing three execution measurements to be specific: mean absolute error (MAE), mean outright percentage error (MAPE) and root mean square error (RMSE). To implement these models, we have taken wind speed and corresponding power data of four different sites from National Renewable Energy Laboratory (NREL).

Keywords Wind power · Multiple linear regression · Decision tree · Random forest · MAE · MAPE · RMSE

R. Akash (✉) · R. Meenal
Department of Electrical and Electronics Engineering, Karunya Institute of Technology and Sciences, Coimbatore 641114, India
e-mail: shaneroysten98@gmail.com

R. Meenal
e-mail: meenal@karunya.edu.in

A. G. Rangaraj
National Institute of Wind Energy (NIWE), Chennai, India
e-mail: rangaraj.niwe@nic.in

M. Lydia
Department of Electrical and Electronics Engineering, SRM University, Delhi-NCR, Sonepat, Haryana 131029, India
e-mail: lydiaedwin.05@gmail.com

V. Singh et al. (eds.), *Computational Methods and Data Engineering*, Advances in Intelligent Systems and Computing 1227,
https://doi.org/10.1007/978-981-15-6876-3_25

1 Introduction

Wind power (WP) is a discontinuous vitality with enormous stochastic nature. The expanding advancement of WP makes incredible difficulties in the solidness and security of the power system. Wind power is growing swiftly all over the world, particularly in countries like America, China and many European countries. India also has passable potential for wind energy. As of March 31, 2019, the total installed wind power capacity in India was 36.625 GW, the fourth largest installed wind power capacity in the world [1]. A viable approach to beat the difficulties is wind power forecasting (WPF). The variation in the wind power can be distinguished ahead of schedule by forecasting. Conferring to the forecast results, an acceptable generation of power can be planned, which can decrease the pivot hold in power system. WPF is becoming more vital in financial dispatching of the power system with increase in integration of wind power in the system.

Accurate forecast is significant in specialized business, expert, exchange and different backgrounds. Currently, numerous strategies have been proposed for the WPF, such as analysis of time series, statistical study, physical models and machine learning (ML) methods. Analysis of time series is cooperative in portraying the data through graphical methods. Analysis of time series also supports in resembling the forthcoming values of the series. If $\boldsymbol{y}_1, \boldsymbol{y}_2, \ldots, \boldsymbol{y}_t$ be the witnessed time series and if a forecast is made for a future estimation $\boldsymbol{y}_{t+h}$, then the integer h is known as the lead time or the forecasting horizon and the forecast of y_{t+h} made at time t for h steps ahead is denoted by $\bar{y}_t(h)$. Decent and precise forecast will help specialists and experts to locate the most proper method for observing a given procedure [2].

Forecasting wind at different time possibilities has included criticalness in the ongoing days. Wind power forecast plays dynamic job for the activity and upkeep of wind farms and for combination into power systems just as in delivery. Accessibility of precise wind power forecast will certainly aid in enlightening the power grid security, expanding firmness in the operation of power system of and market financial matters and help fundamentally upgrading the entrance of power. This will unquestionably bring about huge-scale decline of ozone harming substance creation and different toxins discharged during the utilization of reducing regular vitality assets.

This paper proposes a random forest (RF) approach arriving at the forecast of the WP for day ahead. The actualized technique is an immediate strategy. The RF is picked for every one of the prizes referenced of AI procedures and was favored over artificial neural network (ANN) and support vector regression (SVR) along these lines; it need not bother with any improvement [3]. The usage of a nonparametric procedure, for example, RF spins the serious issue regularly experienced with man-made consciousness techniques, which is hyperparameter tuning. The use of a nonparametric method, such as RF, is a serious problem frequently encountered with man-made consciousness methods , i.e. hyperparameter tuning.

2 Related Works

Nowadays, several machine learning-based algorithms have been developed for future estimation of wind power on day-ahead basis. Artificial neural network (ANN), support vector regression (SVR), *k*-nearest neighbor (KNN) and least absolute shrinkage selector operator (LASSO) are some of the latest algorithms used [4]. Statistical models like AR (autoregression), ARIMA (autoregressive integrated moving average), ARIMAX and SARIMA are also used. A ton of crossover models are additionally rising like ANFIS, CNN, hybrid models using wavelets, etc., which is enhancing the accuracy of the forecast [5, 6]. These methodologies give solid preferences contrasted with traditional neural systems or other measurable or physical procedures. The module is redesigned with online alteration capacities for perfect execution. All things considered, the need for progressively exact forecasting models is not placated.

3 Regression Models

In this paper, it has been evaluated a day-ahead forecasting of wind power using three distinctive regression models to be specific multiple linear regression (MLR), decision tree (DT) and random forest (RF). In light of the presentation measurements for four distinct destinations, the models are grouped.

3.1 Multiple Linear Regression

Multiple linear regression, additionally referred to multiple regression, is a factual system that uses a few informative factors to estimate the result of a response variable [7]. Multiple linear regression attempts to model thc association between two or more features and a response by fitting a linear equation to practical data.

$$y_i = \beta_0 + \beta_1 X_{i1} + \beta_2 X_{i2} + \cdots + \beta_p X_{ip} + \in \tag{1}$$

where $\boldsymbol{y_i}$ is the reliant variable and in our case, it is the wind power, $\boldsymbol{X_{i1}}$ is the free factor which is wind speed, $\boldsymbol{i}$ denotes the observation, $\boldsymbol{\beta_0}$ is the *y*-intercept (steady term), $\boldsymbol{\beta_p}$ is the slope coefficients for each logical variable, $\in$ denotes error term of the model. The phases to play out multiple regression are practically like that of basic linear regression [8]. In the evaluation lies the difference. It can be used to discover which factor has the most elevated effect on the anticipated output and how different variables relate to each other. The following are the steps to forecast using MLR

Step 1: Data pre-processing,
Step 2: Fitting multiple linear regression to the training set,
Step 3: Forecast the test set results.

3.2 *Decision Tree*

A decision tree, likewise called classification and regression tree (CART), is a statistical model exhibited in the year 1984 by Breiman [7]. It delineates the assorted classes or qualities that an output may take as far as a lot of input features. Usually, the nodes and branches prearranged in order with no loops are known as tree. A decision tree is a tree whose nodes stock a test function to be functional to received data. The tree leaves are nothing but fatal nodes, and the ultimate test result is recorded in individual leaves [7]. The decision tree is durable, insusceptible to inputs which are inappropriate and give great interpretability.

The residue of this segment is constrained to regression glitches, since the result is a sort of regression. Let *Wind Speed* (X) with m features be an input vector, Y an output scalar and S_n a training set comprising n annotations (X_i, Y_i) as shown in the formula (2, 3, 4) below:

$$X = \left(x_1, \ldots, x_j, \ldots, x_n\right)T, \quad X \in R \tag{2}$$

$$Y \in R \tag{3}$$

$$S = \{(X_1, Y_1), \ldots, (X_m, Y_m)\} \tag{4}$$

The procedure for training consists in predictor h construction by recursively dividing the features into nodes with various marks Y until a specific end condition is met [9]. Whenever it is not likely to have children nodes with different labels, this criterion is used (Fig. 1).

3.3 *Random Forest*

The RF regression the is improvement of decision trees projected by the same person in the year 2001, Breiman et al. [10]. This method joints forecast results of feeble predictor h_i. The most noteworthy parameters are number of trees *ntree* and the quantity of factors to segment at every hub *mtry*. A random forest is an ensemble technique capable of executing both classification and regression errands with the utilization of various choice trees and a system called bootstrap aggregation, frequently known as bagging [11]. The fundamental thought behind this is to join various choice trees in forming the last yield instead of trusting on discrete choice trees.

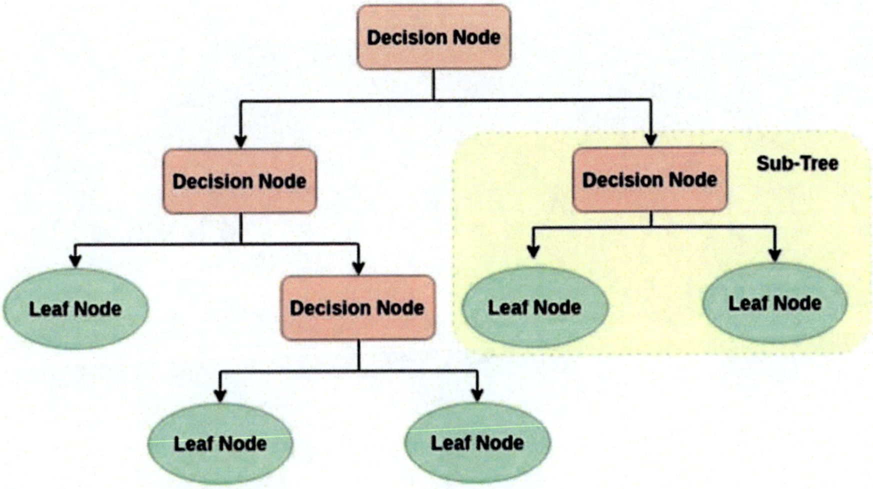

Fig. 1 Decision tree—block diagram

$$Y = h(X) = \frac{1}{ntree} \sum_{i=1}^{ntree} h_i(X) \tag{5}$$

The main advantage of using random forest regression is predictive performance which can compete with the best supervised learning, and they provide reliable feature importance. Here is the step-by-step implementation of random forest regression for forecasting [12].

Step 1. Import the required libraries,
Step 2. Import and visualize the dataset,
Step 3. Select the rows and columns for X and Y,
Step 4. Fit RF regressor to the dataset,
Step 5. Forecast a new result.

The Random Forest results show improvement with MLR and DT (Fig. 2).

4 Data

The performance of the regression models has been validated using four different datasets that consist of hourly series ranging from January 1, 2006, to January 1, 2007. It also includes wind speeds at 80 and 100 m hub heights. The datasets are separated into training and test data.

The selected data for training are used for performance validation by fine-tuning numerous hyperparameters for various algorithms. The optimal hyperparameter settings for each algorithm are fixed based on the performance of the training data.

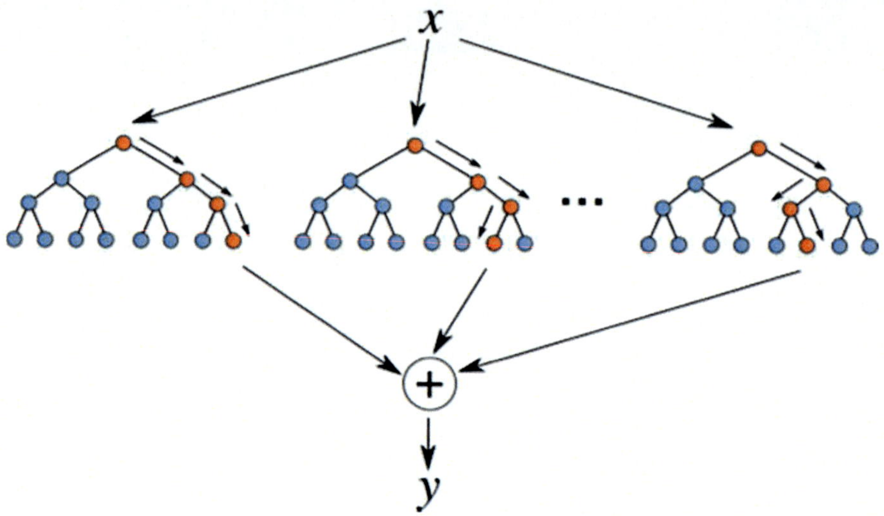

Fig. 2 Random forest—tree diagram

After selecting the hyperparameter for each algorithm, all models are retrained on the training set and test performance is determined by forecasting the time series on the test set.

Since finding an optimal window length for the training set is also a hyperparameter, there is no predefined training period. It varies from model to model and site to site. All the above-mentioned data are downloaded from NREL web portal. Table 1 gives a clear picture of the datasets, and Figs. 3 and 4 represent wind speed data of the SITE_4281 at different hub heights. Figure 5 shows the generated power data for

Table 1 Descriptive statistics of datasets

SITE No.	Variables	Units	Mean	Median	SD	Min	Max
SITE_3975	Speed (80 m)	m/s	7.7	7.75	3.23	0.35	18.05
	Speed (100 m)	m/s	8.13	8.10	3.50	0.35	19.23
	Power	kW	45.03	37.7	36.83	0	122.2
SITE_4281	Speed (80 m)	m/s	7.62	7.53	3.31	0.30	17.88
	Speed (100 m)	m/s	8.06	7.90	3.61	0.27	18.96
	Power	kW	65.65	51.2	56.30	0	183.2
SITE_4810	Speed (80 m)	m/s	7.47	7.40	3.21	0.37	18.34
	Speed (100 m)	m/s	7.89	7.74	3.48	0.30	19.25
	Power	kW	74.08	61.5	59.73	0	182.4
SITE_5012	Speed (80 m)	m/s	7.38	7.26	3.29	0.29	20.17
	Speed (100 m)	m/s	7.80	7.63	3.57	0.37	21.59
	Power	kW	50.88	40.98	41.98	0	128.5

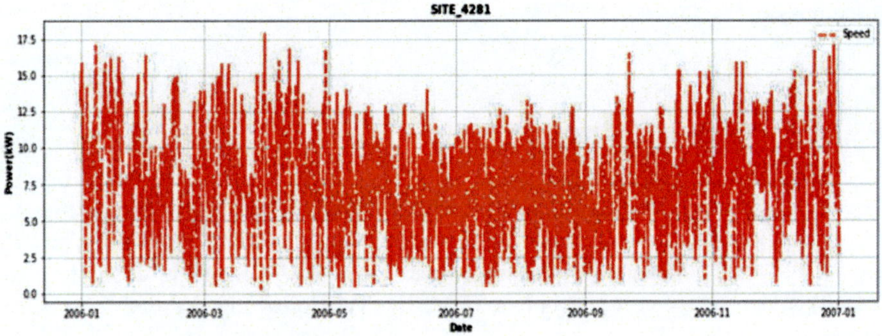

Fig. 3 Dataset (speed)—SITE_4281

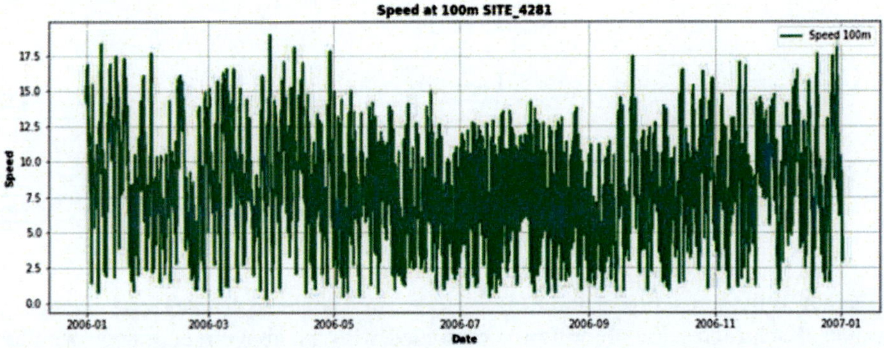

Fig. 4 Dataset (speed @ 100 m)—SITE_4281

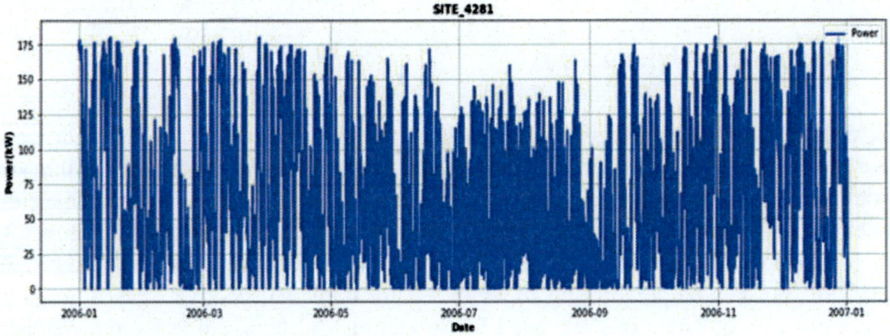

Fig. 5 Dataset (power)—SITE_4281

the same site.

5 Performance Metrics

In general, accurate and reliable forecasting models of wind power forecasting are recognized as a major involvement for increasing wind dispersion. Habitually, models are judged using mean absolute percentage error (MAPE), mean absolute error (MAE) and root mean square error (RMSE). The following are their respective formulas (6, 7, 8) [13]:

$$\text{MAE} = \frac{1}{n}\sum_{i=1}^{n}|A_i - F_i| \tag{6}$$

$$\text{MAPE} = \frac{1}{n}\sum_{i=1}^{n}\left|\frac{A_i - F_i}{A_i}\right| * 100 \tag{7}$$

$$\text{RMSE} = \sqrt{\frac{1}{n}\sum_{i=1}^{n}\left|\frac{A_i - F_i}{A_i}\right|} \tag{8}$$

where A_i is the actual power and F_i is the forecasted power, n denotes total number of values. Each regression algorithm is evaluated with the above-mentioned formulas for all the four sites.

6 Results and Discussion

To evaluate the selected regression models on the chosen data, a day-ahead forecast is done on four different sites. The performance of the three regression algorithms is compared and tabulated. Table 2 gives us the comparison of the performance metrics at 80 m hub height, and Table 3 provides the same with 100 m hub height. Figures 3, 4, 5, 6, 7 and 8 show the best day-ahead forecasted results. It is very clear that random forest is outperforming other two algorithms irrespective of hub heights. Different sets of equations are used for modeling and forecasting by the random forest regressor. The default number of trees is set (Figs. 9, 10, 11, 12 and 13).

Table 2 Performance metrics at 80 m hub height

SITE No.	Metrics	Multilinear regression	Decision tree	Random forest
SITE_3975	MAE	6.87	8.46	**2.35**
	MAPE	20.32	18.5	**5.73**
	RMSE	7.72	11.08	**2.80**
SITE_4281	MAE	10.01	11	**1.73**
	MAPE	21.73	17.32	**3.49**
	RMSE	11.1	14.53	**1.95**
SITE_4810	MAE	9.46	10.33	**3.07**
	MAPE	13.47	15.9	**4.31**
	RMSE	11.96	13.5	**4.14**
SITE_5012	MAE	6.27	7.76	**2.16**
	MAPE	9.43	11.47	**3.03**
	RMSE	7.54	10.3	**2.95**

Table 3 Performance metrics at 100 m hub height

SITE No.	Metrics	Multilinear regression	Decision tree	Random forest
SITE_3975	MAE	7.7	6.48	**2.09**
	MAPE	24.22	13.61	**5.49**
	RMSE	9.00	9.00	**2.45**
SITE_4281	MAE	11.45	9.51	**2.32**
	MAPE	26.69	15.7	**4.16**
	RMSE	13.01	12.05	**3.06**
SITE_4810	MAE	9.92	10.16	**3.15**
	MAPE	14.60	13.30	**4.50**
	RMSE	11.92	13.82	**3.81**
SITE_5012	MAE	6.48	7.05	**2.65**
	MAPE	9.20	10.07	**4.08**
	RMSE	7.78	9.83	**3.38**

7 Conclusion and Future Work

A day-ahead forecast of wind power was demonstrated in this work, for the NREL sites with three different regression algorithms to be specific MLR, DT regression and RF regression. The results are compared and showed that the random forest regression outperforms the other two regressions. It was likewise demonstrated that the performance score of RF was improved by joining wind speed information at different center height.

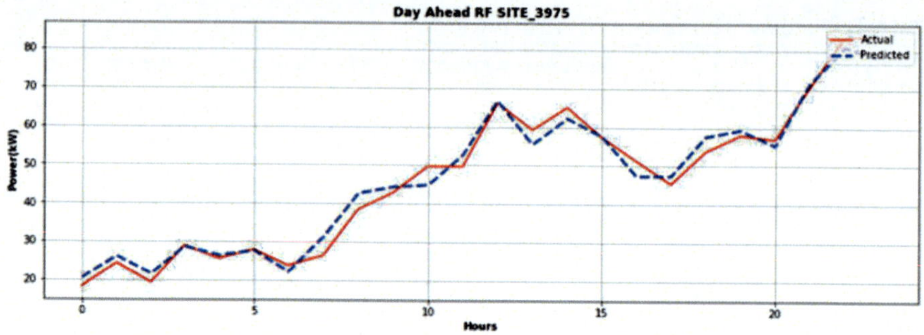

Fig. 6 Day-ahead forecast (RF)—SITE_3975 at 80 m height

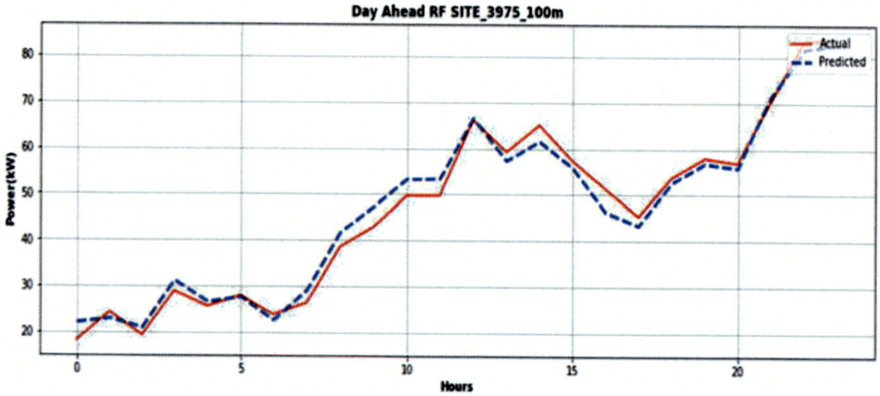

Fig. 7 Day-ahead forecast (RF)—SITE_3975 at 100 m height

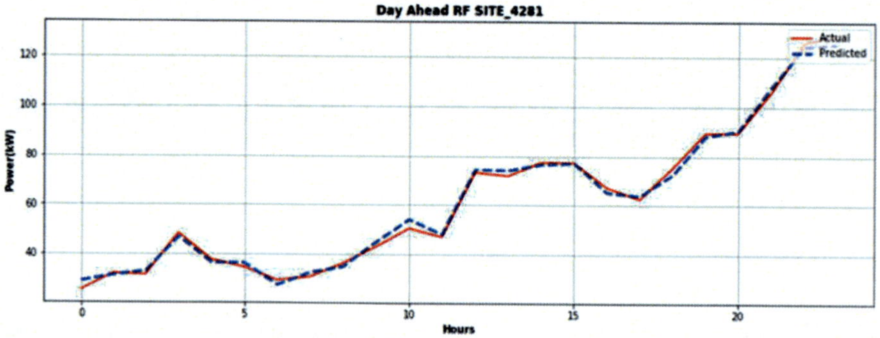

Fig. 8 Day-ahead forecast (RF)—SITE_4281 at 80 m height

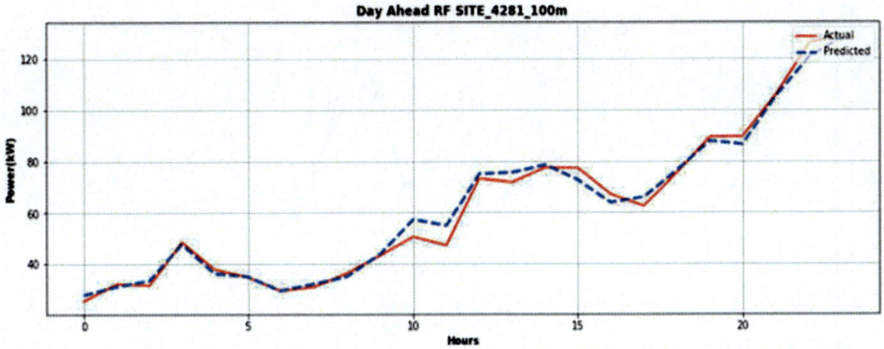

Fig. 9 Day-ahead forecast (RF)—SITE_4281 at 100 m height

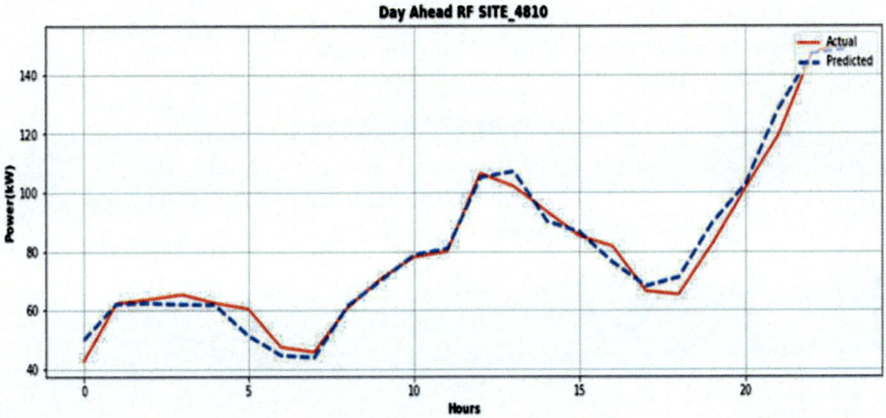

Fig. 10 Day-ahead forecast (RF)—SITE_4810 at 80 m height

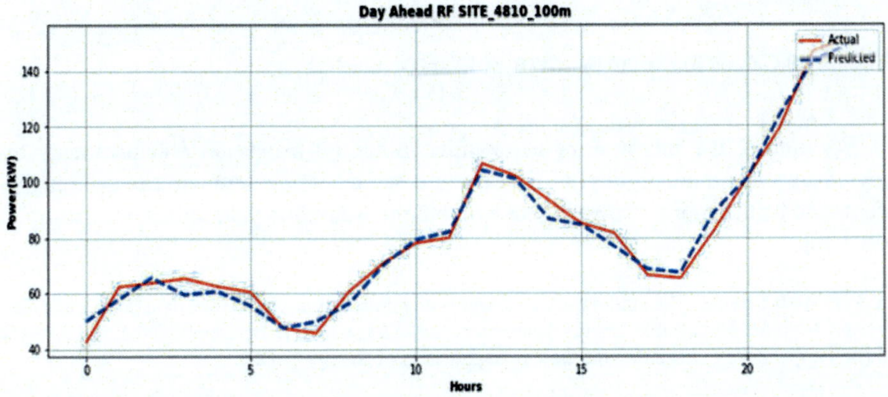

Fig. 11 Day-ahead forecast (RF)—SITE_4810 at 100 m height

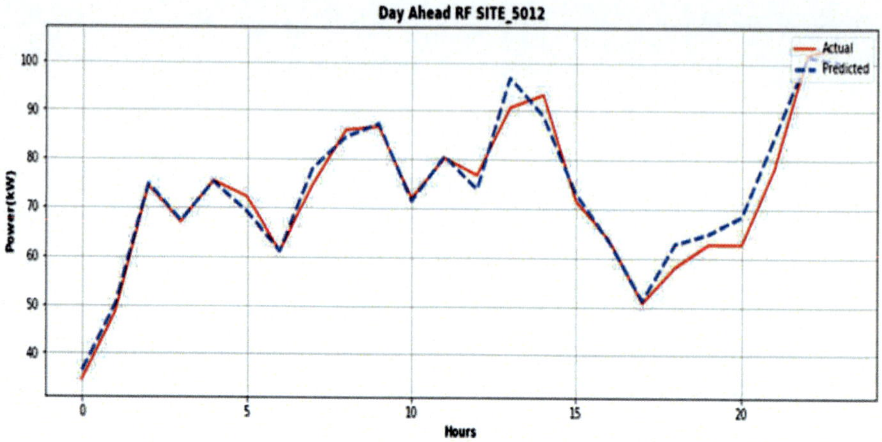

Fig. 12 Day-ahead forecast (RF)—SITE_5012 at 80 m height

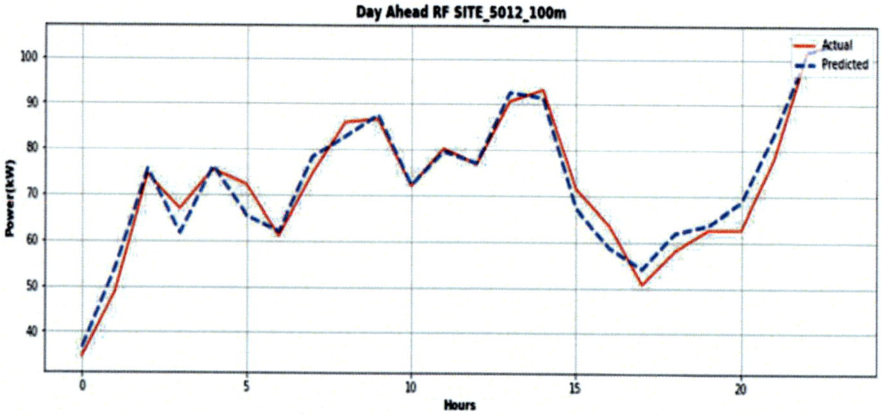

Fig. 13 Day-ahead forecast (RF)—SITE_5012 at 100 m height

For future, we might want to examine potential execution enhancements by including some more highlights, for example, wind bearing, temperature and mugginess, and furthermore, to investigate with other ensembling procedures, for example, boosting.

Acknowledgements The authors acknowledge with gratitude the wind power data (Data) provided by the National Renewable Energy Laboratory (NREL), which is operated by the Alliance for Sustainable Energy (Alliance) for the US Department of Energy (DOE).

References

1. https://en.wikipedia.org/wiki/Wind_power_in_India
2. Lydia M, Suresh Kumar S, Immanuel Selvakumar A, Edwin Prem Kumar G (2016) Linear and non-linear autoregressive models for short-term wind speed forecasting. In: Energy conversion and management, vol 112, pp 115–124. https://doi.org/10.1016/j.enconman.2016.01.007
3. Lahouar A, Ben Hadj Slama J (2015) Random forests model for one day ahead load forecasting. In: 2015 6th International renewable energy congress (IREC 2015), Institute of Electrical and Electronics Engineers, Sousse, Tunisia, 24–26 Mar 2015
4. Demolli H, Dokuz AS, Ecemis A, Gokcek M (2019) Wind power forecasting based on daily wind speed data using machine learning algorithms. Energy Convers Manag 198:111823
5. Hong Y-Y, Rioflorido CLPP (2019) A hybrid deep learning-based neural network for 24-h ahead wind power forecasting. Appl Energy 250:530–539
6. Zhao X, Liu J, Yu D, Chang J (2018) One-day-ahead probabilistic wind speed forecast based on optimized numerical weather prediction data. Energy Convers Manag 164:560–569
7. Lahouar A, Ben Hadj Slama J (2017) Hour-ahead wind power forecast based on random forests. In: Renewable energy, vol 109, pp 529–541. https://doi.org/10.1016/j.renene.2017.03.064
8. https://www.investopedia.com/terms/m/mlr.asp
9. https://www.analyticsvidhya.com/blog/2015/01/decision-tree-simplified/2/
10. Breiman L et al (1984) Classification and regression trees. Chapman & Hall, New York
11. Breiman L (2001) Random forest. Mach Learn **45**(1):5–32. https://doi.org/10.1023/A:1010933404324
12. https://www.geeksforgeeks.org/random-forest-regression-in-python/
13. https://ibf.org/knowledge/posts/forecast-error-metrics-to-assess-performance-39

Query Relational Databases in Punjabi Language

Harjit Singh and Ashish Oberoi

Abstract Public relational databases are accessed by end users to get the information they require. Direct interaction with relational databases requires the knowledge of structured query language (SQL). It is not feasible for every user to learn SQL. An access through an application limits the query options. An end user can ask a query very easily in a natural language. To provide the full advantages of public access, the users should be allowed to query the required data through natural language questions. It is possible by providing natural language support to query relational databases. This paper presents the system model, design and implementation to query relational databases in Punjabi language. It allows human–machine interaction in Punjabi language for information retrieval. It accepts a Punjabi language query in flexible format, uses pattern matching techniques to prepare an SQL query from it, maps data element tokens of the query to actual database objects and joins multiple tables to fetch the required data.

Keywords Intelligent information retrieval · Human–machine interaction in natural language · Punjabi language database query · Database access

1 Introduction

Every organization has some data and that data is maintained in a relational database. Relational databases are capable of storing huge amount of data in tables with relationships. Storing whole data in a single table is impractical because it results in redundancy of data [1]. So the table is normalized by splitting it into two or more tables as per the rules of different normal forms such as first normal form (1NF) and second normal form (2NF). It reduces redundancy of data but at the same time data

H. Singh (✉)
Punjabi University Patiala, Patiala, India
e-mail: hjit@live.com

A. Oberoi
RIMT University, Mandi-Gobindgarh, India

V. Singh et al. (eds.), *Computational Methods and Data Engineering*, Advances in Intelligent Systems and Computing 1227,
https://doi.org/10.1007/978-981-15-6876-3_26

gets divided into multiple tables [2]. To fetch required data, it may require joining of two or more tables temporarily. All this is done by using a special language called structured query language (SQL). SQL is a language for relational databases to store and retrieve data [3].

Many government agencies and nonprofit and public organizations provide open access to their databases for public use [4]. For an end user, it is not feasible to learn SQL to interact directly with relational databases. Access to relational database through an application limits the query options [5]. The end user can take full benefits if he/she is allowed to ask any query or question that comes in his/her mind. The most appropriate answer can be given in the response [6]. It is possible by providing natural language support to query relational databases. The end user can query the required data through a natural language question [7].

This paper presents the system model, design and implementation to query relational databases in Punjabi language. The system is developed to accept a Punjabi language query in flexible format for the database related to any domain. So, it is a domain-independent system for which the input query does not need a fixed format. In a natural language, same query or question can be asked in a number of ways [8]. It accepts a Punjabi language query in flexible format, uses pattern-matching techniques to prepare an SQL query from it, maps data element tokens of the query to actual database objects and joins multiple tables to fetch the required data. The system can be linked to any database domain without modifications. This paper presents the complete architecture with implementation and testing in the following sections. Section 2 highlights related work, Sect. 3 presents the system model, Sect. 4 presents implementation details with testing and Sect. 5 concludes the research.

2 Related Work

Since everyone is not able to use SQL to query databases, much research has been done by researchers to make an easy alternative for common users. Most of the efforts have been done for English language.

A domain-independent system was developed by Wibisono which was later improved by Reinaldha and Widagdo [9]. Stanford dependency parser and ontology were used for processing the query. Various tasks performed during processing included question analysis, parsing and query generation. The query was generated in parts and those parts were combined. Ontology building used meta-data from the target database [10].

Database Intelligent Querying System (DBIQS) was proposed by Agrawal et al. [11]. The information about column names and relationships was taken from database to build semantic map. An intermediate representation was used to translate user query to SQL query. Multiple SQL queries were produced, out of which the best one was chosen for execution.

A system for Hindi language queries was proposed by Kataria and Nath based on Computational Paninian Framework [12]. The query was parsed to get base words, remove undesirable words and find data elements based on Hindi language case symbols. The data element tokens were translated to English which were used for SQL query generation.

Aneesah was proposed by Shabaz et al. based on pattern matching approach [13]. The controller component was designed to communicate with the user and to validate the query. The valid input query was pattern matched by pattern matching engine using a knowledge base for mapping database elements. These database elements were used to formulate SQL query. The knowledge base was implemented in four layers. Each layer was used to handle different types of queries.

Sangeeth and Rejimoan developed an information extraction system for relational databases [14] based on Hidden Markov Model (HMM) [15]. The linguistic module was developed to identify predicates and constraints from input query. These predicates and constraints were used by database module to map and generate SQL query. The system was implemented using the MySQL database and the C# dot NET programming language. It was tested with the GEO-query database [16].

A system for Hindi language was implemented by Virk and Dua using machine learning approach [17]. The linguistic module was developed to parse the query, generate tokens and discard non-useful tokens. The useful tokens were used to identify data elements such as table names, column names and condition. The query translator module was capable of correcting incomplete and misspelled words using Smith Waterman similarity function [18]. *K*-nearest neighbor algorithm was used for classification and the classified output was used to generate SQL query [19]. The system was domain-dependent.

3 The System Model

Punjabi is a low resource availability language. Due to the lack of quality tools and resources, it is more difficult to process Punjabi text than English. The research and development of this system was started almost from scratch. The system model is shown in Fig. 1. The system takes Punjabi language query in the Gurmukhi script as input and passes it through various phases to generate the equivalent SQL query. For explanation of each module, the following Punjabi query is taken as example. Along with the Punjabi language query, pronunciation and English translation are given for reference:

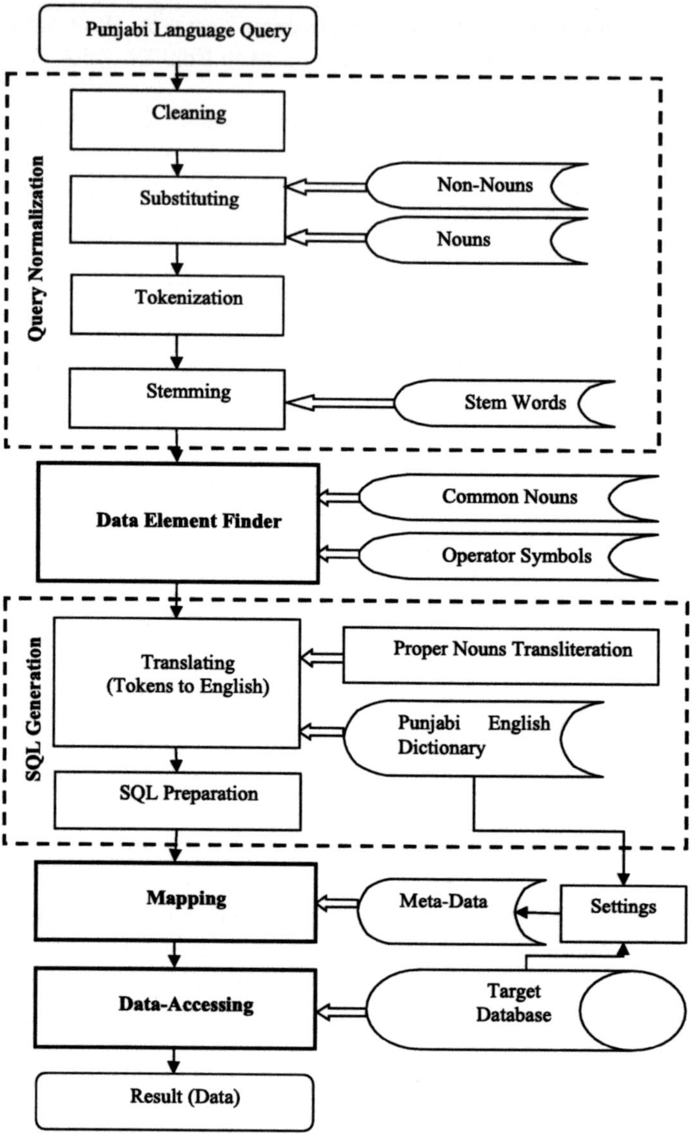

Fig. 1 System model

Punjabi: ਉਹਨਾਂ ਵਿਦਿਆਰਥੀਆਂ ਦਾ ਨਾਮ, ਪਤਾ, ਵਿਸ਼ਾ ਤੇ ਅੰਕ ਦੱਸੋ ਜਿਹਨਾਂ ਦੇ ਅੰਕ 50 ਨਾਲੋਂ ਜਿਆਦਾ ਅਤੇ ਸ਼ਹਿਰ ਮਲੇਰਕੋਟਲਾ ਹੈ ।

Pronunciation: Uhanāṁ vidi'ārathī'āṁ dā nām, patā, viśā tē aṅk dasō jihanāṁ dē aṅk 50 nālōṁ ji'ādā atē śahir malērkōṭlā hai.

English (word-by-word): those, students, of, name, address, subject, and, marks, tell, whose, of, marks, 50, than, more, and, city, Malerkotla, is

English (words-rearranged): tell name, address, subject and marks of those students whose marks more than 50 and city is Malerkotla.

The word-by-word English translation is given to understand the methodology without the knowledge of Punjabi language. The English translation (made with words rearranged) is given to understand the meaning of input query. It may not be grammatically correct.

The system uses following modules:

3.1 *Query Normalization*

The first module is 'Query Normalization' module, which takes the Punjabi language query as input and performs various operations to normalize the query sentence so that it can be processed by the second phase. 'Query Normalization' module normalizes the input query through four separate sub-modules named 'Cleaning', 'Substituting', 'Tokenization' and 'Stemming'. These sub-modules are explained below:

Cleaning 'Cleaning' sub-module of 'Query Normalization' is the first step of processing that is applied to input. It takes the Punjabi language query as input and removes unwanted characters and noise from the input query text.

Substituting The cleaned query sentence is processed to replace some words or multiword expressions with substitute words to make the query sentence simpler for further processing. Substitution used two database tables.

To create noun-substitution database table, dataset was taken from IndoWordNet.[1] IndoWordNet is a multilingual WordNet for Indian languages [20]. In input query, the user may use some complex words in place of a popular and simple word. This database table is used to replace such words with their synonym simple words. For example, ਸਿਰਨਾਵਾਂ (Siranāvāṁ means address) is a synonym of commonly used wordਪਤਾ (patā means address). So, ifਸਿਰਨਾਵਾਂ (Siranāvāṁ) is found in the Punjabi language query, it is replaced with ਪਤਾ (patā). The non-noun-substitution database table was specifically created by manually identifying a total of 408 non-noun substitutions.

[1] http://www.cfilt.iitb.ac.in/indowordnet/.

In the example query, the following replacements are made:

ਨਾਮ (nām means name) → ਨਾਂ (nāṁ means name)
ਦੱਸੋ (dasō means tell) → ਦਿਖਾ (dikhā means show)
ਜਿਹਨਾਂ ਦੇ (jihanāṁ dē means whose) → ਜਿਸਦਾ (jisadā means whose)
ਨਾਲੋਂ ਜਿਆਦਾ (nālōṁ ji'ādā means more than) → ਵੱਧ (vadh means greater, than is not required)
ਅਤੇ (atē means and) → ਤੇ (tē means and)

Tokenization The third step of 'Query Normalization' module is tokenization, which splits the query sentence into individual words using white space as a word separator. These words are called tokens and are stored in an array for fast processing and easy traversal of tokens back and forth.

Stemming The tokens are available in single-dimensional array and are processed one by one. Some of the tokens, mostly those that were not processed by earlier steps, may have suffixes attached to them. The suffixes generate many variants of a word, and processing each variation separately is a very difficult task and reduces the performance of text processing. So, the better way is to strip off any suffixes from the words and then process their stem forms. For example, the words ਘੋੜੇ (ghōṛē means horses) and ਘੋੜਿਆਂ (ghōṛi'āṁ means horses) can be stemmed to the same word ਘੋੜਾ (ghōṛā means horse).

In the example query, following words are stemmed:

ਵਿਦਿਆਰਥੀਆਂ (vidi'ārathī'āṁ means students) → ਵਿਦਿਆਰਥੀ (vidi'ārathī means student)

The stemming module uses two step-stemming processes, which are table lookup-based stemming and rule-based stemming. In table lookup stemming approach, a database table contains a collection of Punjabi stem words along with their inflated forms. To create this database table, dataset was taken from IndoWordNet [20]. If a match found, then the word from the database table is fetched and is taken as a stem word. If no match occurs, the control is transferred to rule-based stemming. The system uses the rule-based stemming approach presented by Gupta and Lehal [21]. After query normalization, the example Punjabi language query is:

Punjabi: ਉਹਨਾਂ ਵਿਦਿਆਰਥੀ ਦਾ ਨਾਂ ਪਤਾ ਵਿਸ਼ਾ ਤੇ ਅੰਕ ਦਿਖਾ ਜਿਸਦਾ ਅੰਕ 50 ਵੱਧ ਤੇ ਸ਼ਹਿਰ ਮਲੇਰਕੋਟਲਾ ਹੈ

Pronunciation: Uhanāṁ vidi'ārathī dā nāṁ patā viśā tē aṅk dikhā jisadā aṅk 50 vadh tē śahir malērkōṭlā hai

English (word-by-word): those student of name address subject and marks show whose marks 50 greater and city Malerkotla is

3.2 Data Element Finder

Data element finder is a module that extracts the data-related tokens from the list of normalized tokens. It uses a rule-based approach to find data elements. The rules are applied by traversing the tokens back and forth to extract data elements such as entity, attributes and conditions. Since the tokens are stored in memory in a single-dimensional array, traversal of tokens back and forth is fast and improves the response time.

Various rules were generated based on pattern matching to identify the appropriate data elements from the token list. The rules are based on the words that appear before and after the word under scan. As an example, it was analyzed that if a word is not a stop word [22], not a comparison word, not listed in the non-noun-substitution database table and appears after the Punjabi word 'ਦਾ' (dā means of), then it is the name of some entity about which the data is demanded in the Punjabi query. It is tagged as {EN}. Continuing this rule further, if the word that appears after the Punjabi word 'ਦਾ' is not a stop word [22], not a comparison word and not listed in the non-noun-substitution database table, then it is the name of some attribute of found entity. So, it is tagged as {AT1} for first attribute. There may be multiple attribute words in a query, so the rules were generated to extract and tag those attributes one by one until the Punjabi token 'ਤੇ' (tē means and) appears, after which the next token is the last attribute which is also extracted and tagged. All extracted attribute tokens are tagged in sequence as {AT1}…{ATn}. It is an example of a simple rule; many such rules were generated according to the possible sentence formats that a user could enter as input Punjabi query. For example, if the query is:

50 ਤੋਂ ਵੱਧ ਅੰਕ ਲੈਣ ਵਾਲੇ ਵਿਦਿਆਰਥੀ ਕਿਹੜੇ ਹਨ

The above-discussed rule cannot be applied to this query because in this query the word ਦਾ (dā means of) does not exist. So, the rules set differ depending upon the format of query sentence. As an example of a condition extraction rule, a token 'ਜਿਸਦਾ' (Jisadā means whose) is searched. The condition(s), if specified in the Punjabi query, appears after the token 'ਜਿਸਦਾ'. In most Punjabi query formats, the word after the token 'ਜਿਸਦਾ' is condition attribute and then condition value with comparison word. For example, considering the following tokens in the array after 'Query Normalization':

ਜਿਸਦਾ ਅੰਕ 50 ਵੱਧ

It specifies a condition attribute, a condition value and at the end a comparison word. These are tagged as {CA1} for first condition attribute, {CV1} for first condition value and {CO1} for first condition operator. The comparison words such as 'ਵੱਧ', 'ਘੱਟ' and 'ਬਰਾਬਰ' ('vadh', 'ghaṭ', 'barābar' means 'greater', 'less', 'equal', respectively) are replaced with their symbol equivalents (>, <, = etc.). Similarly, the rules are extended to find other conditions in the query tokens. All condition attributes are tagged in sequence as {CA1}…{CAn}, condition values are tagged as {CV1}…{CVn} and condition operators are tagged as {CO1}…{COn}. When multiple conditions are given in a Punjabi query, they are combined with the words 'ਜਾਂ' or 'ਤੇ' ('jāṁ' or 'tē' means 'or' or 'and') which are used to form logical operators 'OR' or 'AND'. Also, the word 'ਨਹੀਂ' (Nahīṁ means not), if used in conditions, shows a negation which is used to form logical 'NOT' operator. These tokens are replaced with their equivalent logical operator symbols (OR, AND, NOT). This module is responsible to handle all these tokens by using rule-based approach. Every extraction uses a set of rules to correctly tag the token. The tagged elements in the example Punjabi language query are:

Punjabi: ਵਿਦਿਆਰਥੀ{EN} ਨਾਂ{AT1} ਪਤਾ{AT2} ਵਿਸ਼ਾ{AT3} ਅੰਕ{AT4} ਅੰਕ{CA1} >{CO1} 50{CV1} AND{LO1} ਸ਼ਹਿਰ{CA2} ={CO2} ਮਲੇਰਕੋਟਲਾ{CV2}

Pronunciation: Vidi'ārathī{EN} nāṁ{AT1} patā{AT2} viśā{AT3} aṅk{AT4} aṅk{CA1} >{CO1} 50{CV1} AND{LO1} śahir{CA2} ={CO2} malērkōṭlā{CV2}

English: student{EN} name{AT1} address{AT2} subject{AT3} marks{AT4} marks{CA1} >{CO1} 50{CV1} AND{LO1} city{CA2} ={CO2} Malerkotla{CV2}

3.3 SQL Generation

Overall function of this module is to translate Punjabi language query to SQL query. This is not a final SQL query; instead, it is a prepared SQL using available tokens. The module is composed of two sub-modules that are 'Translating' sub-module and 'SQL Preparation' sub-module.

Translating The tagged data elements are translated to equivalent English language tokens. 'Translating' sub-module uses a Punjabi to English dictionary to directly translate entity, attribute and condition-attribute tokens to the equivalent English words. The Punjabi to English dictionary was created using the dataset taken from IndoWordNet [20]. The condition-value token may contain proper nouns of cities, persons, places, etc., so these require transliteration rather than translation. It is called the Romanization of Punjabi proper nouns. A rule-based approach is used for Romanization. A character map from Gurmukhi character set to equivalent Roman character set was stored in a database table. To fasten the processing, the database table is first loaded into a two-dimensional array in memory. It is also advantageous for traversing the characters back and forth during transliterating process. All Gurmukhi characters are first mapped to their corresponding Roman character combinations directly. It does not produce proper transliteration, so the Roman character 'a' needs to be adjusted in between the characters of produced word to correct it. Three sets of rules were generated to cover each possibility of transliteration that can be used in pattern matching to find a match. For example, a proper noun in Punjabi such as 'ਗੁਰਮੀਤ' (Gu-ramīt) may be spelled in English as 'Gurmeet' or 'Gurmit'. Both are the correct forms of transliteration, so both are needed to try for a match using pattern matching. Therefore, three sets of transliteration rules were generated to produce three transliterations of each Punjabi proper noun and are grouped together with comma as separator. If duplicate transliterations are generated, those are removed from the set.

After translating, the example Punjabi language query is:

Student{EN} Name{AT1} Address{AT2} Subject{AT3} Marks{AT4} Marks{CA1} >{CO1} 50{CV1} AND{LO1} City{CA2} ={CO2} malerakotala, malerkotla{CV2}.

SQL Preparation The 'SQL Preparation' sub-module prepares the SQL query using English language tokens translated by 'Translating' sub-module. The following SQL template is used by the module:

SELECT <attributes> FROM <table> WHERE <conditions>

The <attributes> is replaced with the attribute list, <table> is replaced with the table name and <conditions> is replaced with conditions. The resultant SQL query is a prepared SQL which requires mapping of attributes, table and condition-attributes to the actual database objects. After SQL preparation, the example Punjabi language query is:

SELECT Name, Address, Subject, Marks FROM Student WHERE Marks > 50 AND City IN ('malerakotala', 'malerkotla').

3.4 Mapping

The previous module provides a set of English language tokens to 'Mapping' module. The tokens were extracted from Punjabi language query as they were specified by the user. These tokens are not exact database objects that are used in the target database for data storage. For example, a token 'Student' may not be the exact name of the table in target database that contains the records. The actual name of table may differ; it may be 'Students', 'StudentRecords', 'Student_Data' or any other name. 'Mapping' module is responsible to map the extracted data element tokens to their equivalent actual database objects in the target database.

Mapping requires the list of all tables, fields in each table and relationships between tables from the target database. This information is called meta-data and is maintained by the relational database management system (RDBMS). Mapping module uses this information for mapping data element tokens to actual database objects.

Pattern matching is used to map data element tokens to database objects. For example, if the data element token is 'Student', it is pattern matched with available table names of the target database fetched from meta-data. Through pattern matching, the match may be 'Students', 'StudentRecords', 'Student_Data' or any other name which has the pattern 'Student'. It is done automatically, if the exact pattern is present as a part of the table name. But if it is something unmatchable, such as the table name 'SRecords', then it requires some manual settings. Manual settings for every Punjabi query are not suitable solution. So, it is done using 'Settings' module automatically.

Settings This module is used to make the settings once at the time of linking the system to the target database. It makes the system domain-independent; that is, it can be linked to any database to fetch the records using Punjabi language query. It fetches the meta-data from the target database. The fetched information is preserved in a database table for use by 'Mapping' module. The table names and field names stored in meta-data are pattern matched with the English words of dictionary. The module maps the database objects with dictionary words automatically. In case of some incorrect mapping, the interface allows to select another dictionary word. Figure 3 shows the interface. This is required only once at the time of linking the system to the target database.

Now continuing with the 'Mapping' module, the data element tokens tagged as {AT1}...{ATn} and {CA1}...{CAn} are mapped with appropriate field names using the information generated by the 'Settings' module. 'Mapping' module may map the attributes to the field names, those belong to separate tables. So, joining of two more tables is done by using relationship information of meta-data. The related tables of those attributes are the actual table names from which the data is demanded in Punjabi language query. Mapped database objects and relationship information are used to generate the final mapped SQL query from prepared SQL. After mapping, the example Punjabi language query is:

SELECT DISTINCT Student_Data.name, Student_Data.address, Subjects.Subject, Results.marks FROM Results, Student_Data, Subjects WHERE Results.marks > 50 AND Student_Data.city IN ('malerakotala', 'malerkotla') AND (Student_Data.roll_no = Results.roll_no AND Subjects.Sub_Code = Results.sub_code).

3.5 Data Accessing

Mapped SQL query generated by 'Mapping' module is used by 'Data-Accessing' module to fetch data from the database and show it to the user. The resultant SQL query is submitted to the RDBMS for execution, which in turn returns the data records. The 'Data Accessing' module displays the returned data to the user.

4 Implementation and Results

Figure 2 shows the end user interface of the system. For an end user, a user-friendly interface without the low-level details was developed, showing only input query, SQL query and output data. Figure 3 shows the interface of 'Settings' module.

The system was tested with three different database domains. All the databases used in this test were related to the university where the research was carried out. Some students were selected and divided into three different groups. Each group is asked to write Punjabi language queries related to a particular domain. The queries were taken from the students in three Unicode text files. The Punjabi queries were manually translated to the equivalent SQL queries in same sequence and were stored

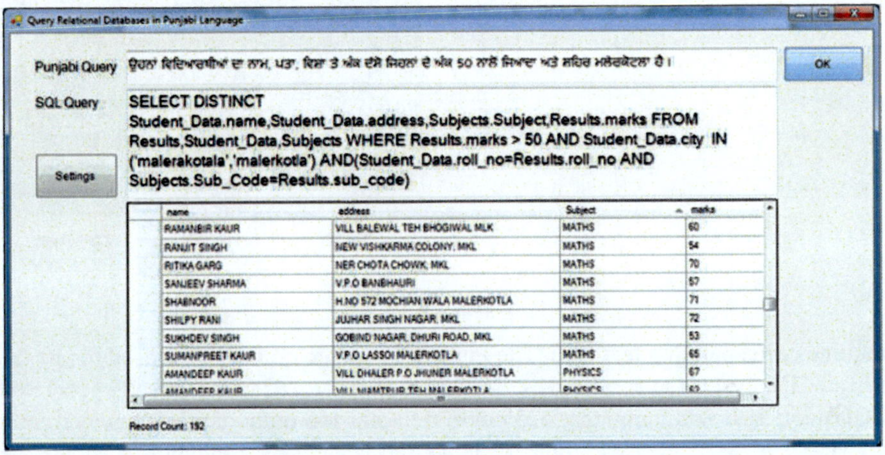

Fig. 2 End user interface of the system

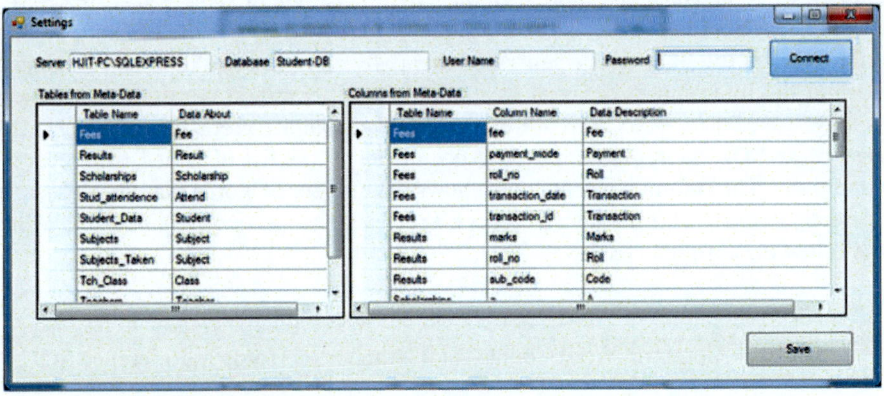

Fig. 3 Settings module showing automatically mapped database objects

Table 1 Test results of the system

Database domain	Number of test queries	Number of correctly generated SQL queries	Accuracy
Student database	160	156	97.5
Library database	136	123	90.44117647
Employee database	152	146	96.05263158
Average accuracy	**–**	**–**	**94.6646**
Overall accuracy	**448**	**425**	**94.86607143**

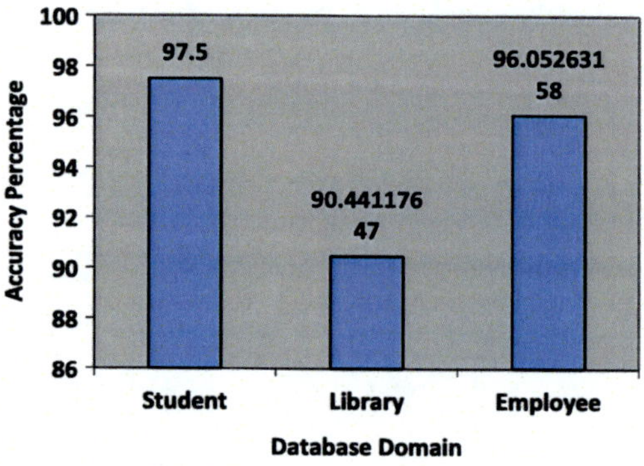

Fig. 4 Graphical representation of accuracy for different database domains

in three corresponding text files. The Punjabi language queries were used to test the system. The system-generated SQL queries were compared with manually generated SQL queries. It was found that 425 out of the total 448 input queries were correctly converted to SQL queries. Table 1 shows the test results, and Fig. 4 shows the graphical representation of accuracy for different database domains.

The system was further evaluated using precision, recall and F1-score. To evaluate the system, a database table with the structure shown in Table 2 was created. In this table, 'Candidate SQL' is the SQL query generated by the system and 'Reference SQL' is the SQL query generated manually for reference to check accuracy. To find the intersection between two pairs of text, common words were counted among 'Candidate SQL' and 'Reference SQL' through pattern matching [23]. In case of an SQL query, the order does not matter in the list of tables, attributes or conditions. The common words count shows the number of words (database objects) correctly mapped in resultant SQL query. Precision and recall were calculated using the following formulas [23]:

$$\text{Precision} = \text{Common Word Count}/\text{Total Number of Words in Candidate SQL}$$
$$\text{Recall} = \text{Common Word Count}/\text{Total Number of Words in Reference SQL}$$

When a Punjabi query was evaluated, a record was saved in the database table as specified in Table 2. The test process automatically calculated the precision, recall

Table 2 Database table used to evaluate the system

Field name	Description
Database_Domain	Database domain used for evaluation
Query_Number	Unique ID for queries
Candidate_SQL	SQL generated by the system
WC_Candidate_SQL	Total number of words in Candidate SQL
Reference_SQL	SQL manually generated for reference
WC_Reference_SQL	Total number of words in Reference SQL
Intersection_Count	Common words in Candidate_SQL and Reference_SQL
Precision	Intersection_Count/WC_Candidate_SQL
Recall	Intersection_Count/WC_Reference_SQL
F1-score	2 * Precision * Recall/(Precision + Recall)

and F1-score by comparing the 'Candidate SQL' with 'Reference SQL'. The saved records were checked and found that out of the 448 input Punjabi queries, for 425 queries precision is 1, recall is 1 and so F1-score is 1. It means that all the words of 'Candidate SQL' and 'Reference SQL' were matched. In the remaining 23 queries, some tables, attributes or conditions were not properly mapped. Table 3 shows the precision, recall and F1-score as recorded for those 23 queries. Table 3 also shows the calculations of overall average precision, recall and F1-score.

The test results show an average accuracy of 94.6% and F1-score as 0.98 for the system.

5 Conclusion

This paper presented the system model, design and implementation to query relational databases in Punjabi language. The system was able to convert a Punjabi language query into equivalent SQL query with automatic joining of multiple tables to fetch the required data. It is a domain-independent system that can query a database related to any domain without modifying the system. It was tested with varied number of queries related to three different domains. The test results show 94.6% average accuracy in generating correct SQL queries from input Punjabi language queries. The F1-score was calculated as 0.98 for the system.

Table 3 Precision, recall and F1-score calculations

Database	Query number	Precision	Recall	F1-score
Student database	63	0.714286	0.769231	0.740741
	80	0.625	0.588235	0.606061
	82	0.5625	0.45	0.5
	113	0.769231	0.588235	0.666667
Employee database	17	0.625	0.5	0.555556
	56	0.6	0.545455	0.571429
	57	0.916667	0.846154	0.88
	58	0.8	0.666667	0.727273
	91	0.6	0.6	0.6
	114	1	0.866667	0.928571
	115	0.769231	0.714286	0.740741
	135	0.666667	0.555556	0.606061
Library database	10	0.714286	0.5	0.588235
	11	0.833333	0.588235	0.689655
	12	0.714286	0.5	0.588235
	44	0.555556	0.5	0.526316
	47	0.769231	0.769231	0.769231
	73	0.625	0.555556	0.588235
	81	0.714286	0.666667	0.689655
	83	0.769231	0.625	0.689655
	84	0.833333	0.714286	0.769231
	100	0.666667	0.555556	0.606061
	101	0.714286	0.5	0.588235
	135	0.714286	0.769231	0.740741
	136	0.625	0.588235	0.606061
Above three	For 425 queries	425 times 1	425 times 1	425 times 1
Average of each column		**0.985620701**	**0.980279049**	**0.982624648**

References

1. O'Neil P (1994) Database design. In: Database, pp 293–384. https://doi.org/10.1016/b978-1-4831-8404-3.50008-3
2. Arenas M (2018) Normal forms and normalization. In: Encyclopedia of database systems, pp 2509–2513. https://doi.org/10.1007/978-1-4614-8265-9_1237
3. Zhang P (2017) Joins. In: Practical guide to oracle SQL, T-SQL and MySQL, pp 137–145. https://doi.org/10.1201/9781315101873-10
4. Lips M (2019) Open and transparent government. In: Digital government, pp 106–131. https://doi.org/10.4324/9781315622408-5
5. Mabuni D (2017) User interfaces to databases—a survey. Int J Adv Res Comput Sci 8(9):710–713. https://doi.org/10.26483/ijarcs.v8i9.5201

6. Foster EC, Godbole S (2016) Database user interface design. In: Database systems, pp 139–153. https://doi.org/10.1007/978-1-4842-1191-5_6
7. Krägeloh K-D, Lockemann PC (n.d.) Access to data base systems via natural language. In: Natural language communication with computers, pp 49–86. https://doi.org/10.1007/bfb0031369
8. Juan B, González J, Rangel P, Cruz C IC, Héctor H, Fraire J, de Santos Aguilar L, Joaquín Pérez O (2006) Issues in translating from natural language to SQL in a domain-independent natural language interface to databases. In: MICAI 2006: advances in artificial intelligence, pp 922–931. https://doi.org/10.1007/11925231_88
9. Reinaldha F, Widagdo TE (2014) Natural language interfaces to database (NLIDB): question handling and unit conversion. In: 2014 International conference on data and software engineering (ICODSE). https://doi.org/10.1109/icodse.2014.7062663
10. Li M, Du X-Y, Wang S (2005) Learning ontology from relational database. In: 2005 International conference on machine learning and cybernetics. https://doi.org/10.1109/icmlc.2005.1527531
11. Agrawal R, Chakkarwar A, Choudhary P, Jogalekar UA, Kulkarni DH (2014) DBIQS—An intelligent system for querying and mining databases using NLP. In: 2014 International conference on information systems and computer networks (ISCON). https://doi.org/10.1109/iciscon.2014.6965215
12. Kataria A, Nath R (2015) Natural language interface for databases in Hindi based on Karaka theory. Int J Comput Appl 122(7):39–43. https://doi.org/10.5120/21716-4841
13. Shabaz K, O'Shea JD, Crockett K, Latham A (2015) Aneesah: a conversational natural language interface to databases. In: World congress on engineering. URL: http://www.iaeng.org/publication/WCE2015/
14. Sangeeth N, Rejimoan R (2015) An intelligent system for information extraction from relational database using HMM. In: 2015 International conference on soft computing techniques and implementations (ICSCTI). https://doi.org/10.1109/icscti.2015.7489594
15. Ghahramani Z (2001) An introduction to hidden Markov models and Bayesian networks. In: Hidden Markov models, pp 9–41. https://doi.org/10.1142/9789812797605_0002
16. Goodman S, BenYishay A, Lv Z, Runfola D (2019) GeoQuery: integrating HPC systems and public web-based geospatial data tools. Comput Geosci 122:103–112. https://doi.org/10.1016/j.cageo.2018.10.009
17. Virk ZS, Dua M (2016) An advanced web-based Hindi language interface to database using machine learning approach. Lecture notes in computer science, pp 381–390. https://doi.org/10.1007/978-3-319-40349-6_36
18. Munekawa Y, Ino F, Hagihara K (2008) Design and implementation of the Smith-Waterman algorithm on the CUDA-compatible GPU. In: 2008 8th IEEE international conference on bioinformatics and bioengineering. https://doi.org/10.1109/bibe.2008.4696721
19. Laaksonen J, Oja E (1996) Classification with learning k-nearest neighbors. In: Proceedings of international conference on neural networks (ICNN'96). https://doi.org/10.1109/icnn.1996.549118
20. Bhattacharyya P (2016) IndoWordNet. In: The WordNet in Indian languages, pp 1–18. https://doi.org/10.1007/978-981-10-1909-8_1
21. Gupta V, Lehal GS (2011) Punjabi Language Stemmer for nouns and proper names. In: Proceedings of the 2nd workshop on South and Southeast Asian natural language processing (WSSANLP), IJCNLP 2011, Chiang Mai, Thailand, pp 35–39, 8 Nov 2011. URL: https://www.aclweb.org/anthology/W11-3006
22. Kaur J, Saini JR (2016) Punjabi stop words. In: Proceedings of the ACM symposium on women in research 2016, WIR'16. https://doi.org/10.1145/2909067.2909073
23. Melamed ID, Green R, Turian JP (2003) Precision and recall of machine translation. In: Proceedings of the 2003 conference of the North American chapter of the association for computational linguistics on human language technology companion volume of the proceedings of HLT-NAACL 2003. Short papers, NAACL'03. https://doi.org/10.3115/1073483.1073504

Machine Learning Algorithms for Big Data Analytics

Kumar Rahul, Rohitash Kumar Banyal, Puneet Goswami, and Vijay Kumar

Abstract A machine learning algorithm (MLA) is an approach or tool to help in big data analytics (BDA) of applications. This tool is suitable to analyze a large amount of amount generated by an application for effective and efficient utilization of the data. Machine learning algorithms considered to find out meaningful data and information for industrial applications. It is one of the services under big data analytics (BDA). Big data analytics (BDA) is suitable for identifying risk management, cause of failure, identifying the customer based on their procurement detail records, detection of fraud, etc. So, this paper deals with the work done in this field to analyze the importance of machine learning tools and techniques, identify the field where it is suitable to use including industries such as marketing, human resource, healthcare, insurance, banking, automobile, etc. This paper identifies different challenges of machine learning tools and technologies including the current status of adoption in industries.

Keywords Machine learning · Big data · Healthcare · Filtering · *K*-Means

K. Rahul (✉) · R. K. Banyal
Department of CSE, Rajasthan Technical University, Kota, India
e-mail: kumarrahul.niftem@gmail.com

R. K. Banyal
e-mail: rkbayal@gmail.com

V. Kumar
Department of Basic and Applied Science, NIFTEM, Sonepat, India
e-mail: drvijay.niftem@gmail.com

P. Goswami
Department of CSE, SRM University, Sonepat, India
e-mail: goswamipuneet@gmail.com

V. Singh et al. (eds.), *Computational Methods and Data Engineering*,
Advances in Intelligent Systems and Computing 1227,
https://doi.org/10.1007/978-981-15-6876-3_27

1 Introduction

Machine learning is an important component of artificial intelligence. Machine learning is collecting data, analyzing data, and prediction data for industrial applications. Machine learning learns automatically from the data and behaves accordingly without explicitly programmed. Data analytical is required to evaluate and estimate the benefits of organizational goals. It is used to formulate equations, models and functions within systems. Data analytics required statistics, artificial intelligence, data mining, deep learning, prediction mechanism, and so on to evaluate the evaluation of data within an organization. The statistical analysis leads to work on the behavioral function of industries along with analysis of collecting data, presentation, processing, and visualization for different purposes including uses, re-uses, filtering, binning, etc. Machine learning (ML), deep learning (DL) provides jointly tools for evaluation of data reusability and it is also suitable for prediction and estimation of advanced analytics. The program uses data generated through different sources for training. It is a process of the scientific study of algorithm which is suitable for the development and designing of a computer program which is sufficient for accessing data. Machine Learning Algorithm (MLA) are applicable to different areas of research and industries.

Advanced machine learning uses supervised learning, semi-supervised learning, and unsupervised learning. Machine learning uses in finding predictions; identify results analytics, decision making, and modeling design form for any big data analytics based application. In the traditional system, data and program provide, output receives, whereas in machine learning concept, data and output provide, the program receives. It is categorized under supervised, unsupervised, and reinforcement learning. Supervised learning is a task of learning functions that map from input to an output. Under supervised learning, prepare data, choose an algorithm, and fit a model that can learn and predict. In supervised learning, predicting values with the known target. Machine learning is suitable for big data-based applications such as the healthcare sector where predicting health, risk analysis of patient health, diagnosis, alarm, and alerts for individual and patients and fraud detection can be carried out with the support of machine learning technologies. Apart from these applications, it is helpful in marketing and sales, cybersecurity, stock exchange, asset management, and so on. It gives output in forms of regression and classification methods.

Supervised learning includes a few algorithms including linear algorithm, Naïve Bayes algorithm, *K*-nearest neighbors polynomial regression, decision tree, supply vector machine (SVM), logistic regression, and so on. These are some common algorithm of supervised learning which applies to business as well. Supervised learning is suitable to find out to reduce cut health care cost, diagnosis, and medical image recognitions, radiologist and pathological, etc. for patients.

Classification is helpful to predict category and it is achieved in bucket form, i.e., Boolean or discrete form. In big data-based applications, e-mail generations are in voluminous form and need to handle properly, and therefore through machine learning or supervised learning algorithm e-mail is categorized properly as "spam".

Whereas unsupervised learning uses historical data and it does not have a target field. In unsupervised learning, the output is not known, data is imbalanced and unknown.

Under unsupervised algorithms, *k*-means, clustering, genetic algorithm, principal component analysis (PCA), apriori algorithm, independent component analysis, and so on are suitable to execute for any big data-based applications. Unsupervised learning is the algorithm where the user provides input data and no corresponding output. It's a major goal to help the model to the underlying structure.

K-means are the most suitable and familiar unsupervised algorithm to solve the clustering problem. Association rule is the unsupervised learning algorithm, which is famous in the big data-based applications for a market analysis for industries. In unsupervised learning, the system discovers the pattern itself. Different types of machine learning can be understood in Fig. 1.

Reinforcement learning is the process in which learning agents decide to act to perform the task. Reward granted to agents based on action or environment (Fig. 2).

The objective of reinforcement learning is to design a mathematical framework to solve the problem. Applications of reinforcement learning would be resource management in computer clusters, traffic light control, robotics, web system configuration, bidding and advertising, games, and so on. There are many facts to know before applying reinforcement learning including understanding the problem, simulated environment, Markov decision process (MDP), and algorithm.

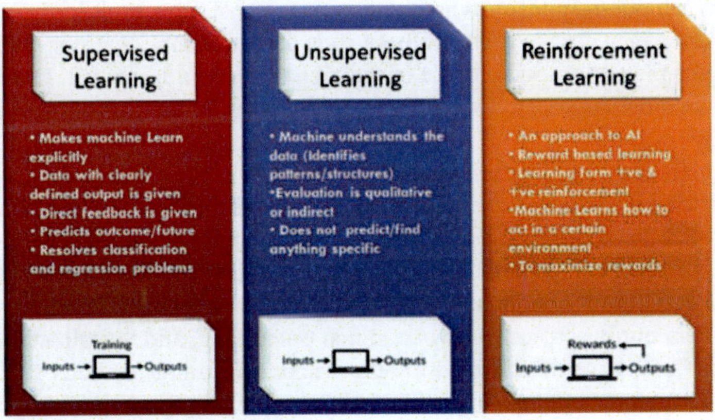

Fig. 1 Machine learning and it's type. *Source* [1]

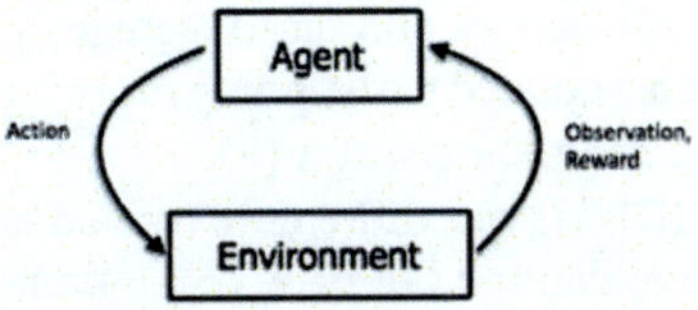

Fig. 2 Reinforcement learning. *Source* [2]

2 Related Work

The author described artificial intelligence and strength, weakness, and opportunities for machine learning applications [3]. In [3], it was defined in different terms such as machine learning, supervised and unsupervised learning, natural language processing, speech recognition, etc. In this paper [3], the authors defined the strength in radiology, automatic lesion detection, image-based diagnosis, etc. The authors in [3] defined weakness in the form of machine learning (ML) and deep learning (DL) requires lots of training sets of data. In the absence of correct validation, it may lead to noise than real data.

The authors identified unstructured data is hot nowadays to work upon and make it meaningful in all respect [4]. In paper [4], the authors discussed opportunities and challenges in healthcare applications of big data. The authors discussed various V's, tools for big data, methodology, and conclusion. In this paper [4], authors said, diverse amount and range of data varies from written notes to sensor data and which are shifting from prevention to cure. Authors [4] also design the architecture of big data processing. The author has discussed different mechanisms including deep learning, representation, transfer, parallel, kernel, and active learning in [5]. It has been observed that machine learning (ML) adoption rates are quite high from the last several years [5]. The inclusion of machine learning expands to the field of astronomy and biological-based applications [5]. The authors explained the importance of maintenance in a modern industry where maintenance cost goes high in the absence of lower maintenance cost [6]. The authors show how machine learning (ML) and its technologies play an important role in maintenance for industry [6]. Data analytics is a mathematical approach to optimize the process in industrial applications-based software [6].

Data analytics classification such as descriptive analysis, diagnostics analysis, predictive analytics, the perspective analysis given by the authors in [6]. Different methods such as search methods, classifications, filtering discussed in [6]. Data analytics are helpful to examine historical data and finding impact and efficiency in applications. The authors defined various data analytics techniques such as machine learning, data mining, neural network, signal processing, and visualization methods [7]. The authors have discussed different machine learning methods such as decision tree, deep learning, and ensemble methods which is suitable for improving the finding of predictions by utilization and merging of many other machine learning methods [8]. It explores methods such as regularization, unsupervised learning (neural network, factor analysis, *K*-means, principle component analysis (PCA), etc.), random forest with advantages and disadvantages [8]. The authors show how machine learning tools are associated with finding meaning endeavors (data empathy, data analyzing, etc.) in the healthcare sector [8].

Deep Neural Network (DNN) was difficult to trained and implement before 2006 as per [9]. The authors have defined big data, data mining and machine learning in different parts [10]. These parts include the computing environment, turning data into business value, and success stories at the end [10]. Turning data into business values shows predictive modeling, modeling techniques, segmentation, incremental response modeling, time-series data mining, recommendation systems, text analytics, etc. in detail [10]. The authors have surveyed the machine learning (ML) algorithm suitable for big data analytics (BDA) in [11]. This survey identifies the importance of machine learning in different types of big data-based applications [11]. There are various problems associated with machine learning which is used in big data [12]. These problems include learning various data types, learning incomplete and insecure data, high speed streaming data management, and large scale data generations [12].

The authors discussed the new approach of machine learning such as representation-learning, transfer-learning, deep learning, active learning, etc. [12]. Learning analytics methods, analysis of massive data with standard methods and technologies, MOOC (Massive Open Online Courses), methodological process, etc., explained by the authors in [13]. Potential strategies to optimize technologies for efficient and effective utilization under the healthcare system discussed in [14]. The optimizations for machine learning techniques are applicable to the intensive care unit (ICU) of healthcare systems [14]. Similarly, in the case of material design, applications of machine learning identified and discussed by the authors [15]. The authors identify workflow for material discovery and design based on regression, clustering, and classification [15].

3 Applications of Machine Learning Algorithm

There are different applications of machine learning technologies including of diagnostics, forecasting, customer segmentation, big data visualization, robot navigation, process optimization, real-time decisions, meaningful compression, fraud detections, games, finance for option pricing, remote sensing, business management, engineering, energy, consumption, healthcare, agriculture, and fault diagnosis, etc. In this paper, we have focused on related work based on machine learning applications in the healthcare sector. Machine learning to find out a hidden target in 2020 war game as per record and statement are given in [16]. The most favored industries where machine learning used include retail, travel, healthcare, finance, and media. In the healthcare sector, it is used to find drug discovery, personalized treatment, and medication, etc. in the finance sector, it is suitable to find fraud detections, and focused on the account holder, whereas in the retail sector, it is suitable to find product recommendations and improvement of customer services. Applications of machine learning in healthcare and medicine sectors include such as diseases identification, personalized treatment, drug discovery, clinical research, radiology and radiotherapy, smart electronic health record, epidemic outbreak prediction, etc. [17]. Artificial intelligence (AI) used in healthcare sector for different purposes including healthcare services, practices, data applications, data security, patient advocacy group, and social interests [18]. In [19], the authors focused on machine learning methods to extract information, image analysis, data analysis, predictions, etc.

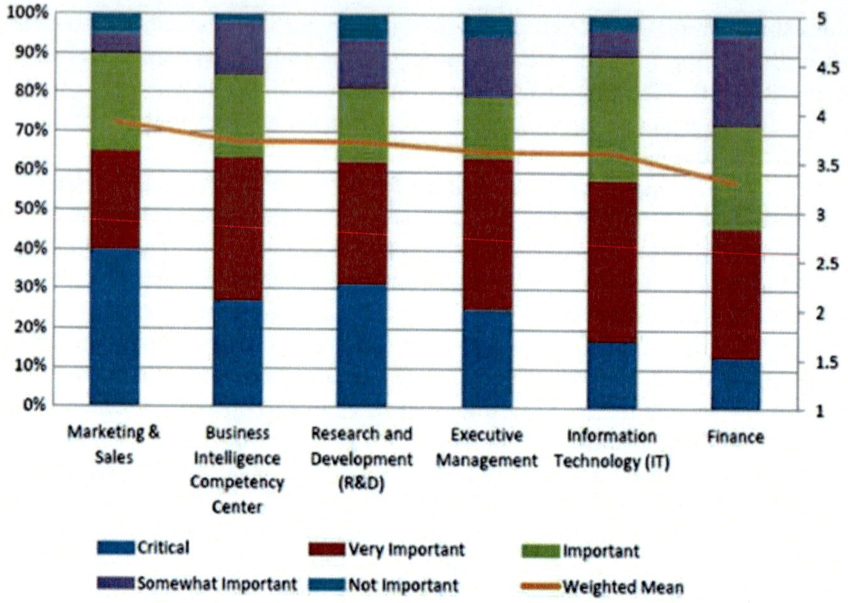

Fig. 3 Importance of data science and machine learning (ML) by functions copyright 2019-Dresner Advisory Services, *Source* [21]

The authors have shown physical health system infrastructure which moves towards digitized processes equipped with software [20]. The goal of cyber healthcare system framework is to make doctors, infrastructure, patients, emergency services, researchers access healthcare system effectively [20]. In cyber healthcare system framework, it includes patient priorities, patient monitoring, time-saving, cost-saving [20]. The authors suggested HADOOP and MapReduce are suitable for big health data sets [4]. It includes In the case of machine learning applications in travel, it is suitable to help to estimate dynamic pricing and sentiment analysis, etc. The adoption of machine learning can be understood with Fig. 3.

3.1 Machine Learning Algorithm in Healthcare Sector

Machine learning algorithm becomes crucial when a large number of datasets exist in healthcare sector. Because of 3V's of big data characteristics and growing computational efficiency, machine learning (ML) is keeping busy healthcare-related researchers to provide meaningful results with clinically reliable predictions [22]. In [23], the authors discussed applications and usability of machine learning in Alzheimer's diseases and mild cognitive impairment diagnosis. Electronic health records (EHR) are used to predict disease risk prediction and also used for the utilization of clinical data for different purposes [24]. Different examples of applications

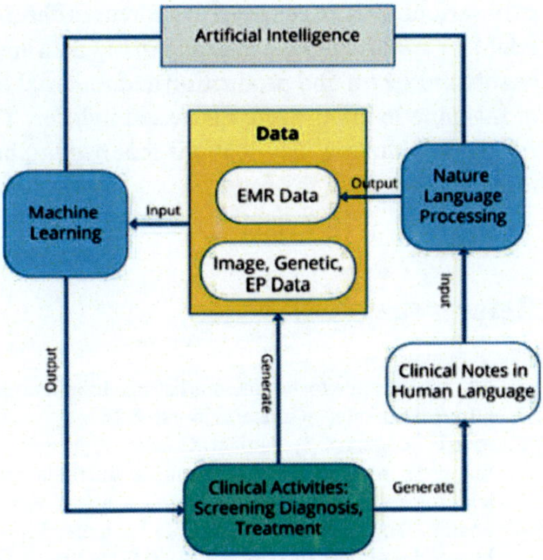

Fig. 4 Machine learning in the healthcare sector. *Source* [26]

of data mining and machine learning have been given in [24]. Authors have developed a machine learning model to improve the efficiency of the operating room with maximum optimization [25] (Fig. 4).

4 Challenges of Machine Learning Technologies

There are various challenges that exists of machine learning technologies comprises of the talent gap, expensive computational needs, black-box answers, data-hungry. However, these problems and challenges can be overcome with a commitment to technological aspects. Apart from these challenges, it also includes memory network, natural language processing (NLP), deep net settings, to understand the deep reinforcement learning, semantic segmentation, object detections; data is not free at all, technological upgradations, etc. Machine learning is a combination of statistical models and scientific study of algorithms applicable to any field for operations to analyze prediction or results [27]. Data meaning perspective, pattern training perspective, realization and application perspective, privacy and security perspective, and technique integration perspective, etc., are the open issue under machine learning for big data applications [5].

5 Conclusion

This paper briefly introduces the machine learning concept applicable to various applications. So far various works have been done in big data-based applications but still, various challenges need to address differently. This paper explained the different related work of machine learning in big data applications especially in the case of the healthcare sector and production and material segment. Applications and challenges of machine learning were discussed in brief. This paper focused on the open issues related to industrial applications that need to be focused and can be considered as a research agenda in the future.

References

1. https://www.newtechdojo.com/list-machine-learning-algorithms
2. http://rll.berkeley.edu/deeprlcourse-fa15
3. Martín Noguerol T, Paulano-Godino F, Martín-Valdivia MT, Menias CO, Luna A (2019) Strengths, weaknesses, opportunities, and threats analysis of artificial intelligence and machine learning applications in radiology. J Am Coll Radiol 16:1239–1247
4. Shaikh TA, Ali R (2019) Big data for better Indian healthcare. Int J Inf Tecnol 11:735–741. https://doi.org/10.1007/s41870-019-00342-6
5. Qiu J, Wu Q, Ding G, Xu Y, Feng S (2016) A survey of machine learning for big data processing. EURASIP J Adv Signal Process 1:2016
6. Baum J, Laroque C, Oeser B, Skoogh A, Subramaniyan M (2018) Applications of big data analytics and related technologies in maintenance—literature-based research. Machines 6:54
7. Dorepalli S (2018) Machine learning and statistical approaches for big data: issues, challenges and research directions. Int J App Eng Res 12
8. Doupe P, Faghmous J, Basu S (2019) Machine learning for health services researchers. Value Health 22(7):808–815
9. https://ethz.ch/content/dam/ethz/special-interest/gess/computational-social-science-dam/documents/education/Spring2017/ML/big_data.pdf
10. Dean J (2014) Big data, data mining, and machine learning: value creation for business leaders and practitioners. Wiley and SAS Business Series
11. Athmaja S, Hanumanthappa M, Kavitha V (2017) A survey of machine learning algorithms for big data analytics. International Conference on Innovations in Information, Embedded and Communication Systems (ICIIECS), Coimbatore, 2017, pp 1–4. https://doi.org/10.1109/ICIIECS.2017.8276028
12. Singh SP, Jaiswal UC (2018) Machine learning for big data: a new perspective. Int J Appl Eng Res 13(5):2753–2762
13. Hadioui A, El Faddouli N, Touimi YB, Bennani S (2017) Machine learning based on big data extraction of massive educational knowledge. Hadioui 12(11):151–167. https://doi.org/10.3991/ijet.v12.i11.7460
14. Núñez Reiz A, Armengol de la Hoz MA, Sánchez García M (2019) Big data analysis and machine learning in intensive care units. Big Data Analysis y Machine Learning en medicina intensiva. Med Intensiva 43(7):416-426. https://doi.org/10.1016/j.medin.2018.10.007
15. Zhou T, Song Z, Sundmacher K (2019) Big data creates new opportunities for materials research: a review on methods and applications of machine learning for materials design. Engineering. https://doi.org/10.1016/j.eng.2019.02.011
16. https://www.militaryaerospace.com/computers/article/14069203/artificial-intelligence-ai-machine-learning-military-applications
17. https://emerj.com/ai-sector-overviews/machine-learning-in-pharma-medicine
18. https://www.bsigroup.com/globalassets/localfiles/en-gb/about-bsi/nsb/innovation/mhra-ai-paper-2019.pdf
19. Dash S, Shakyawar SK, Sharma M, Kaushik S (2019) Big data in healthcare: management, analysis and future prospects. J Big Data

20. Bagula A, Mandava M, Bagula H (2018) A framework for healthcare support in the rural and low income areas of the developing world. J Netw Comput Appl 120(May):17–29
21. https://www.forbes.com/sites/louiscolumbus/2019/09/08/state-of-ai-and-machine-learning-in-2019/#2c0446581a8d
22. Nevin L, On behalf of the PLOS Medicine Editors (2018) Advancing the beneficial use of machine learning in health care and medicine: toward a community understanding. PLoS Med 15(11): e1002708. https://doi.org/10.1371/journal.pmed.1002708
23. Shen D, Wee C, Zhang D, Zhou L, Yap P (2014) Machine learning techniques for AD/MCI diagnosis and prognosis. In: Dua S, Rajendra U, Acharya, Dua P (eds) Machine Learning in Healthcare Informatics, pp. 147–179. Springer-Verlag Berlin Heidelberg.Intelligent Systems Reference Library 56, New York, United States. https://doi.org/10.1007/978-3-642-40017-9_8
24. Stewart WF, Roy J, Sun J, Ebadollahi S (2014) Clinical utility of machine learning and longitudinal EHR data. In: Machine learning in healthcare informatics, Intelligent Systems Reference Library 56, pp 209–227. https://doi.org/10.1007/978-3-642-40017-9_10
25. Fairley M, Scheinker D, Brandeau ML (2019) Improving the efficiency of the operating room environment with an optimization and machine learning model. Health Care Manag Sci 22(4):756–767. https://doi.org/10.1007/s10729-018-9457-3
26. https://medium.com/sciforce/top-ai-algorithms-for-healthcare-aa5007ffa330
27. https://mytechdecisions.com/facility/industrial-machine-learning-systems

Fault Classification Using Support Vectors for Unmanned Helicopters

Rupam Singh and Bharat Bhushan

Abstract This paper develops a fault classification algorithm using support vectors for monitoring operation anomalies and early fault detection in unmanned helicopters. In general, a fault classifier identifies the operating state of the system which can be further monitored by developing a necessary control action. Hence, to achieve the fault detection, the data between normal operating condition and faulty conditions is monitored. In this paper, a two-class classification method is developed using support vector data description for identifying the motor faults in unmanned helicopter system. The algorithm is developed by observing the data of pitch and yaw motors of a 2DoF helicopter. Further, the action of SVDDs during unknown faults is observed by developing a residual generator which improves the functioning of classifier. The results depicted 98.2% training accuracy and efficient results when tested with a faulty condition.

Keywords Fault detection and diagnosis · 2DoF helicopter · MLPNN · FPID controller · Discrete wavelet transform

1 Introduction

In present time, helicopters are most suitable way of transportation as they can easily take off and land even in small places. Due to which helicopter found so many applications in the field of construction, military, firefighting, tourism, search and rescue operation, law enforcement, news and media, medical transport, agriculture, and aerial observation. The helicopters are nonlinear in nature due to coupling of

R. Singh (✉) · B. Bhushan
Department of Electrical Engineering, Delhi Technological University, New Delhi, India
e-mail: singhrupam99@gmail.com

B. Bhushan
e-mail: bharat@dce.ac.in

V. Singh et al. (eds.), *Computational Methods and Data Engineering*,
Advances in Intelligent Systems and Computing 1227,
https://doi.org/10.1007/978-981-15-6876-3_28

yaw and pitch motion and subjected to various external disturbances which make it difficult to control. The control objective is to follow predefined trajectory by helicopter system without oscillations. Several controllers have been designed to control path following problem of helicopter system.

The controlling of pitch and yaw rotor has been done by using fractional order reference model-based autotuned PID controller [1] and in another approach cross-coupled PID designed for motion control of helicopter [2]. The control parameters of PID are modified as per the genetic algorithm to avoid the tuning problem and give precise value of control gain [3] and improve helicopter controlling without wasting time. The tracking error from pitch and yaw output is optimized by linear quadratic regulator (LQR) as it minimizes the certain quadratic function [4] using state feedback law and gives more precise output with less control effort as compared to PID control. LQR is complex to implement, and all states should be measurable while implementing it. This problem is overcome by intelligent controllers. Further, fuzzy LQR controller (PFLQR) distributed parallelly for both axis control of model [5], and gradient descent algorithm has been used in designing of the robust adaptive fuzzy controller (RAFC) [6]. Moreover, there have been various other controlling techniques that designed for controlling of pitch and yaw motion such as adaptive second-order sliding mode controller (ASMC) [7, 8], MPC [9], adaptive disturbance rejection controller (ADRC) [10], and dual boundary conditional integral back stepping controller [7].

In addition, the performance of helicopter is also affected when a fault occurs during real-time operation and makes system unstable. So, in order to provide consistency and protection to the system, the control system improvement is necessary. Here comes the importance of fault localization and classification. The ability of a system to accept faults through control design, using either passive or active approaches, is essentially a structural property of the system itself. A system with insufficient redundancy cannot be made successfully tolerant to fault irrespective of the control approach used. Therefore, a tool is needed to check the fault-tolerance capability before designing the controller. The concept of control reconfigurability was first developed by [11] for linear time-invariant systems. Later, the concept of coverage of fault tolerant control was developed in [12] for analyzing the reliability of the controller. These two concepts have been further developed and used by many researchers as analysis and design tools for fault tolerant control systems. Since the operational safety of helicopter is primary issue, the fault-tolerant control plays essential part in aero vehicle control [13]. To control breakdowns in higher-order flight control actuators, a decentralized fault-tolerant control system has been developed [14]. Further, the partial and complete failure that arises from actuator and sensor fault is controlled by H-infinity observer [15]. The external disturbances create huge error from the measurements of sensor and actuator output which approximated by intelligent adaptive fuzzy controller [16]. A comprehensive review of fault detection methods can be found in [17, 18].

Apart from the widely available classifiers, the true structure of the fault data is observed to be hard for developing a complete model. This resulted in ignoring the delicate details of the data. In order to overcome this and improve the classification performance, the combination of classifiers is suggested by many researchers [19]. But these methods proved to be costly and lacked while dealing with the delicate data and unknown faults.

In this research, the authors propose the development of support vector classifiers for fault classification in unmanned helicopters. The problem of handling delicate data is overcome by mapping the feature space into high dimension using hyperspheres and kernels. Further, unknown faults were identified by developing a residual generator which validates the classified fault with the data from actuator and the sensors. The major advantage with this classification process is that it can efficiently classify the problem where one class is sampled well, while the other class is under sampled. To achieve the development of the classifier, a model of 2DoF helicopter is developed in the MATLAB/Simulink as discussed in Sect. 2. The data required for developing the classifier is extracted using wavelet transforms, and the steps for developing the classifier are discussed in Sect. 3. Further, the methodology discussed in Sect. 3 is applied with the modeling discussed in Sect. 2, and the corresponding results are discussed in Sect. 4. In order to validate the classified fault and improve the performance of the classifier for unknown faults, a fault identification process is detailed in Sect. 5. The research is concluded in Sect. 6.

2 2DoF Helicopter System

For this research, two degree of freedom (2DoF) helicopter system [20] has been considered for experimental analysis. This setup is mainly designed for laboratory research and consists of two propellers which mounted on both end of helicopter body with fixed base. The DC motor is used to drive these propellers and stated as front-end propeller and beck end propeller. The front-end propeller is utilized for altitude (elevation) control about its pitch and back-end propeller is used for controlling the yaw axis for side motion. The freedom of movement along its center is provided by fixed base. 2DoF helicopter model is shown in Fig. 1. with the help of free-body diagram. Z-axis signifies yaw motion of helicopter and expressed as φ, Y-axis shows pitch control denoted as α.

The center of mass of helicopter system is obtained by taking distance between helicopter body center and two propeller models named as r_{pi} and r_{ya}. The linear model system dynamics by replacement of following values of $\alpha = 0, \varphi = 0, \dot{\alpha} = 0, \dot{\varphi} = 0$ and denoted as:

$$\ddot{\alpha} = \frac{k_{\text{ppi}} v_{\text{mpi}}}{J_{\text{eqpi}} + m_{\text{heli}} l_{\text{CM}}^2} + \frac{k_{\text{pya}} v_{\text{mya}}}{J_{\text{eqpi}} + m_{\text{heli}} l_{\text{CM}}^2} - \frac{B_{\text{pi}} \dot{\alpha} + m_{\text{heli}} g l_{\text{CM}}}{J_{\text{eqpi}} + m_{\text{heli}} l_{\text{CM}}^2} \tag{1}$$

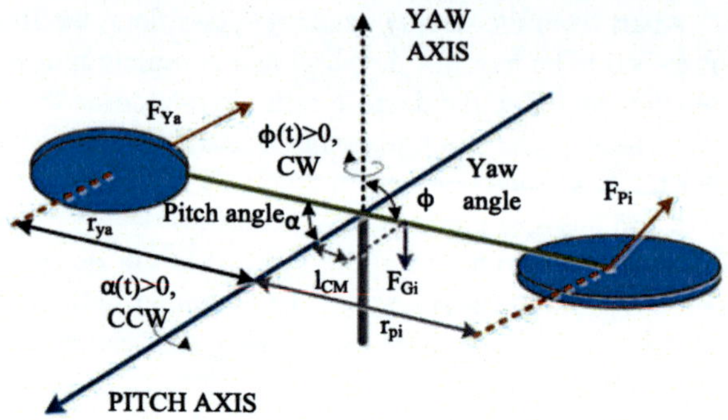

Fig. 1 Graphic representation of the helicopter model

$$\ddot{\varphi} = \frac{k_{ypi} v_{mpi}}{J_{eqya} + m_{heli} l_{CM}^2} + \frac{k_{yya} v_{mya}}{J_{eqya} + m_{heli} l_{CM}^2} - \frac{B_{ya} \dot{\varphi}}{J_{eqya} + m_{heli} l_{CM}^2} \tag{2}$$

The four state variables for the helicopter system considered in this research are pitch angle (α), yaw angle (φ), pitch velocity ($\dot{\alpha}$), and yaw velocity ($\dot{\varphi}$). The position sensors and encoders are implanted in the DC motor to assess pitch and yaw angles using the yaw and pitch velocities. The rotor speed is controlled by aerodynamical forces resulting from the DC motors, and so, the control input for the system is DC motor supply voltage. By using the input $u = \left[v_{mpi}, v_{mya}\right]^T$ and states as $x = [\alpha, \varphi, \dot{\alpha}, \dot{\varphi}]^T$, the state space representation for the 2DoF helicopter is given by:

$$\begin{bmatrix} \dot{\alpha} \\ \dot{\varphi} \\ \ddot{\alpha} \\ \ddot{\varphi} \end{bmatrix} = \begin{bmatrix} 0\,0 & 1 & 0 \\ 0\,0 & 0 & 1 \\ 0\,0 & \frac{-B_{pi}}{J_{eq,pi} + m_{heli} l_{CM}^2} & 0 \\ 0\,0 & 0 & \frac{-B_{ya}}{J_{eq,ya} + m_{heli} l_{CM}^2} \end{bmatrix} \begin{bmatrix} \alpha \\ \varphi \\ \dot{\alpha} \\ \dot{\varphi} \end{bmatrix} + \begin{bmatrix} 0 & 0 \\ 0 & 0 \\ \frac{k_{ppi}}{J_{eq,pi} + m_{heli} l_{CM}^2} & \frac{k_{pya}}{J_{eq,pi} + m_{heli} l_{CM}^2} \\ \frac{k_{ypi}}{J_{eq,ya} + m_{heli} l_{CM}^2} & \frac{k_{yya}}{J_{eq,ya} + m_{heli} l_{CM}^2} \end{bmatrix} \begin{bmatrix} v_{mpi} \\ v_{mya} \end{bmatrix} \tag{3}$$

The output vector for the system expressed as:

$$Y = \begin{bmatrix} 1\,0\,0\,0 \\ 0\,1\,0\,0 \end{bmatrix} \begin{bmatrix} \alpha \\ \varphi \\ \dot{\alpha} \\ \dot{\varphi} \end{bmatrix} \tag{4}$$

Nonlinear mathematical modeling of helicopter system makes it simple and easier to gather data for developing a fault detection system, as the datasets for analysis are effortlessly generated using Simulink. These datasets are obtained by simulating various fault scenarios with the Simulink model, which can be further trained to detect and categorize the type of faults that occur in the system and to develop a controller that can adapt to the faults accordingly. The voltages of pitch and yaw are directly proportional to the torque, which is responsible for changes in pitch motion and yaw motion due to direct and cross-coupling. This scenario further controls the yaw angle, pitch angle, yaw velocity, and pitch velocity. The various faults that have been considered in the paper are pitch and yaw motor collective faults such as pitch actuator torque, pitch actuator cross torque, yaw actuator torque, and yaw actuator cross torque, during flying conditions. These variations resulted in various disturbances throughout the state variables and input voltage parameters.

3 Adopted Methodology

The pitch and yaw operations in 2DoF helicopter are directly affected by the various fault conditions and responsible for variation in helicopter motion. This variation in fault condition is accountable for variation in output waveform with respect to its references signal. The frequency and time domain observed these changes in signal or waveform while extracting features. The wavelet analysis has the advantage of synchronized localization in time and frequency domain, and due to which, it is widely accepted in the area of pattern recognition and feature extraction.

3.1 *Feature Extraction*

A signal is decomposed into group of oscillatory function using wavelet transform (WT) [21] and named as wavelets. Due to wavelet localization in time, WT delivers time-frequency interpretation. The time-domain signal $f(t)$ of $L^2(R)$ space is formulated with the help of family of orthonormal wavelet functions. This gives the wavelet transformation and represented as:

$$[W_\psi f](s,\tau) = \frac{1}{\sqrt{|s|}} \int_R f(t)\psi^*\left(\frac{t-\tau}{s}\right)\mathrm{d}t \tag{5}$$

where $\psi(t)$—mother wavelet, $s = 1, \ldots, m$—scale parameters, $\tau = 1, \ldots, n$—translation parameters, and $[W_\psi f]f(t)$ represented as wavelet transformation [22].

The wavelets are used in two different manners which are scaling of the wavelet and translation of the wavelet. In scaling of the wavelet, the signal spread out of the wavelet, and in translation process, the signal moves along the time axis. The discrete wavelet transforms (DWT) [23] use these basic manipulations of the signal and able to implement DWT at numerous locations of signal for numerous scales of wavelet. Further, this process captures features that are local in frequency and local in time.

The discrete wavelet transforms of a discrete signal $x(t)$ is represented as:

$$T_{m,n} = \sum_{t=0}^{N-1} x(t)\frac{1}{\sqrt{2^m}}\Psi\left(\frac{t-n2^m}{2^m}\right) := \langle x, \Psi_{m,n}\rangle \tag{6}$$

where, —discrete signal length $x(t)$, $T_{m,n}$—detailed wavelet coefficients of scale-location grid with indexing (m, n) which are discrete dilation and translation parameters, respectively.

The convolution of signal $x(t)$ using dilated version and translated version of mother wavelet is responsible for discrete wavelet transforms. With the help of this, the wavelet coefficients $T_{m,n}$, for all scale-location indices (m, n) able to discover. In same manner, the inverse discrete wavelet transform reconstructs the original signal $x(t)$ using wavelet coefficients $T_{m,n}$, and given as:

$$x(t) = \sum_{m=1}^{M}\sum_{n=0}^{2^M-1} T_{m,n}\Psi_{m,n}(t) \tag{7}$$

where $M = \frac{\ln(N)}{\ln(2)}$—computed iterations number and $N = 2^M$—discrete input signal length $x(t)$.

Equations (6) and (7) may be defined in terms of decomposition–reconstruction method and explained as given below:

Decomposition process:

$$x(t) \rightarrow \langle x, \Psi_{m,n}\rangle \rightarrow T_{m,n}$$

Reconstruction process:

$$x(t) \leftarrow \sum_{m=1}^{M}\sum_{n=0}^{2^M-1} \langle x, \Psi_{m,n}\rangle\Psi_{m,n}(t) \leftarrow T_{m,n}$$

The smoothing process and high-frequency extraction process are made possible, due to the association of orthonormal dyadic discrete wavelets with wavelet functions and scaling function. Now to achieve approximation coefficients $S_{m,n}$ for all scale-location indices (m, n), the signal $x(t)$ is convoluted with the wavelet coefficients $T_{m,n}$ (6) using translated and dilated versions of the father wavelet, and obtained as:

$$S_{m,n} = \sum_{t=0}^{2^M-1} x(t)\frac{1}{\sqrt{2^m}}\emptyset\left(\frac{t-n2^m}{2^m}\right) := \langle x, \emptyset_{m,n}\rangle \tag{8}$$

where, $S_{m,n}$—approximation coefficient of scale-location grid with index *m*, *n*.

The chosen mother wavelet is responsible of decomposition process after signal discretization completed. Further, a feature matrix is developed by completing the process of reconstruction, decomposition, and feature extraction. This feature matrix is responsible for classification process and elaborated in classifier section given below.

3.2 Classifier

In this section, a boundary is obtained around the feature matrix by computing a hypersphere. The computed hypersphere deals with all the targeted objects, and a minimum outlier acceptance is achieved by minimizing the volume of the sphere. The process is similar to the conventional support vector classifier (SVC) developed by Vapnik in [24] and can be referred as the support vector data description (SVDD). This classification process maps the data into a new, high-dimensional feature space without much extra computational costs resulting in more flexible descriptions. The process of controlling the outlier sensitivity in a flexible way and finding an efficient description is explained as follows:

For any given signal i, consider that it has a set of d real valued measurements. This develops a feature vector X for the signal i in such a way that $X_i = \left(x_{i,1}, x_{i,2}, \ldots, x_{i,d}\right), x_{i,j} \in \mathbb{R}$. Similarly, each signal can be represented in the feature space χ as a single point. Further, let $f(X; w)$ be a model defined for a hypersphere, where w is the weight vector. In general terms, a sphere is defined by its radius R and center a, and it is necessary that this sphere contains all the training samples χ^{tr}. This defines that the empirical error for the training or sample data is 0. Hence, with reference to the classical support vector classifiers, the structural error is defined as:

$$\varepsilon_{\text{struct}}(R, a) = R^2 \tag{9}$$

Further, this needs to be minimized for different constraints:

$$||X_i - a||^2 \leq R^2, \ \forall i \tag{10}$$

To make the classifier robust, and to allow the possibility of outliers in the training set, the distance from objects X_i to the center a should not be strictly smaller than R^2. This means that the empirical error does not have to be 0 by definition. For any case, if an object X_i lies outside the sphere area at a distance $\xi_i > 0$. It defines that one object which is rejected by the description. The error now contains both a structural and an empirical error contributions. We introduce slack variables $\xi, \xi_i \geq 0, \forall i$ and the minimization problem changes into:

$$\varepsilon(R, a, \xi) = R^2 + C\sum_i \xi_i \tag{11}$$

With a constraint that almost all the objects are within the sphere:

$$||X_i - a||^2 \leq R^2 + \xi_i, \quad \xi_i \geq 0, \ \forall i \tag{12}$$

where C provides a tradeoff between the errors and the volume of the description.

Further, the constraints in (11) can be mapped with (12) with the help of a Lagrangian multiplier which is given by [25]:

$$L(R, a, \xi, \alpha, \gamma) = R^2 + C\sum_i \xi_i - \sum_i \alpha_i\{R^2 + \xi_i - (X_\text{i}.X_\text{i} - 2a.X_\text{i} + a.a)\} - \sum_i \gamma_i\xi_i \tag{13}$$

where α_i and γ_i are the Lagrangian multipliers defined for each signal X_i. Further, L needs to be minimized in reference with R, a, and ξ and maximized in reference with α and γ. This is further achieved by setting the partial derivatives to 0 and adding an extra constraint for α for the purpose limiting the influence of the signal on the final output. This corresponds to the final error L, given by:

$$L = \sum_i \alpha_i(X_i.X_i) - \sum_{ij} \alpha_i\alpha_j(X_i.X_j) \quad \text{with } 0 \leq \alpha_i \leq C, \ \forall i \tag{14}$$

As the expression for the center a of the hypersphere has been calculated, further, it can be elaborated for a new test with an object z which is within or exactly on the radius accepted by the description as:

$$||z - a||^2 = (z.z) - 2\sum_i \alpha_i(z.X_i) + \sum_{ij} \alpha_i\alpha_j(X_i.X_j) \leq R^2 \tag{15}$$

$$R^2 = (X_k.X_k) - 2\sum_i \alpha_i(X_i.X_k) + \sum_{ij} \alpha_i\alpha_j(X.X_j) \tag{16}$$

where $X_k \in 0 \leq \alpha_i \leq C$ (set of support vectors with Lagrange multipliers).

Further, this classification process can be given by:

$$f_{\text{SVDD}}(z; \alpha, R) \begin{aligned} &= I(||z - a||^2 \leq R^2) \\ &= I\left((z.z) - 2\sum_i \alpha_i(z.X_i) + \sum_{ij} \alpha_i\alpha_j(X_i.X_j) \leq R^2\right) \end{aligned} \tag{17}$$

where I is the identification factor given by:

$$I(A) = \begin{cases} 1 \text{ if } A \text{ is true} \\ 0 \text{ otherwise} \end{cases} \tag{18}$$

Since the hypersphere depicts a rigid model for defining the boundary of the data, it cannot fit the data as expected. But, by mapping the data to a new dimension, a better fit can be obtained between the hypersphere model and the actual boundary of the data. Considering a mapping Φ which can improve the fit for the data:

$$X^* = \Phi(X) \tag{19}$$

By applying this mapping to Eqs. (14) and (17), the corresponding error is given by:

$$L = \sum_i \alpha_i \Phi(X_i).\Phi(X_i) - \sum_{ij} \alpha_i \alpha_j \Phi(X_i).\Phi(X_j) \tag{20}$$

and

$$f_{\text{SVDD}}(z; \alpha, R) = I\left(\Phi(z).\Phi(z) - 2\sum_i \alpha_i \Phi(z).\Phi(X_i) + \sum_{ij} \alpha_i \alpha_j \Phi(X_i).\Phi(X_j) \le R^2\right) \tag{21}$$

From the equations, it can be observed that all the mapping $\Phi(X)$ can only occur with inner products in other mappings. Considering this, a new function with two new input variables can be defined as follows:

$$K(X_i, X_j) = \Phi(X_i).\Phi(X_j) \tag{22}$$

where $K(X_i, X_j)$ is a mercer kernel function which can be written as an inner product of two functions. This modifies L as follows:

$$L = \sum_i \alpha_i K(X_i, X_j) - \sum_{ij} \alpha_i \alpha_j K(X_i, X_j) \tag{23}$$

and

$$f_{\text{SVDD}}(z; \alpha, R) = I\left(K(z, z) - 2\sum_i \alpha_i K(z, X_i) + \sum_{ij} \alpha_i \alpha_j K(X_i, X_j) \le R^2\right) \tag{24}$$

Here, it can be observed that the mapping $\Phi(X)$ is only used for defining the kernel, but has never been explicitly used. This implies that a good kernel which is capable of defining the data into the feature space and outlier objects outside the feature space needs to be defined. This will form a hypersphere model which perfectly fits the data and achieves good classification performance. To achieve this, the authors in [26] proposed a kernel trick for replacing the mapping. This trick is mainly used with support vector machines, especially in the case for not linearly separable data. Apart from the kernel trick, several other kernels are also used with the support vector classifiers [27]. In this research, we adapt the Gaussian kernel for due its advantages with independency with position of the data and usage of distance between the objects. Mathematically, Gaussian kernel is given by:

$$K(X_i, X_j) = \exp\left(-\frac{||X_i - X_j||^2}{s^2}\right) \tag{25}$$

where s is the width of the Gaussian kernel.

Further, the development of SVDD classifier for fault identification in unmanned helicopters is explained in next section.

4 Algorithm Development

The 2DoF helicopter model developed in Sect. 2 is used to develop the fault classification algorithm for motor faults in unmanned helicopters. The methodology of wavelet transforms for feature extraction and support vector data description for classification as discussed in Sect. 3 is adapted here. Initially, the system is operated to track a sine trajectory. The sine input signal of 0.05 Hz with an amplitude of 5 has been taken for analysis purpose. The system is simulated with a basic linear quadratic controller with a control objective to maintain the system to track the desired trajectories. Further, the pitch motor faults are injected into the system by causing a sudden decrease in the pitch voltage which results in a thrust torque due to sudden variation helicopter position. In addition, the yaw motor fault is injected in a similar way to the pitch motor fault. At this condition, both the normal and fault operating outputs of pitch and yaw motors are extracted as depicted in figure.

From Fig. 2, the normal operating curve depicts the operation of LQR controller on the helicopter for tracking the desired reference trajectories. Simultaneously, the effect of fault can be observed in the fault waveforms. When a fault occurs, the pitch angle suddenly increases to 9.7°. This generates a huge movement in the vertical position of the helicopter. Similarly, for horizontal motion, the angle variation by yaw is not sufficiently high during the faulty condition, which causes a huge coupling misalignment between yaw and pitch motions. These differences in the output signals can help in developing the classifier by extracting the relevant features.

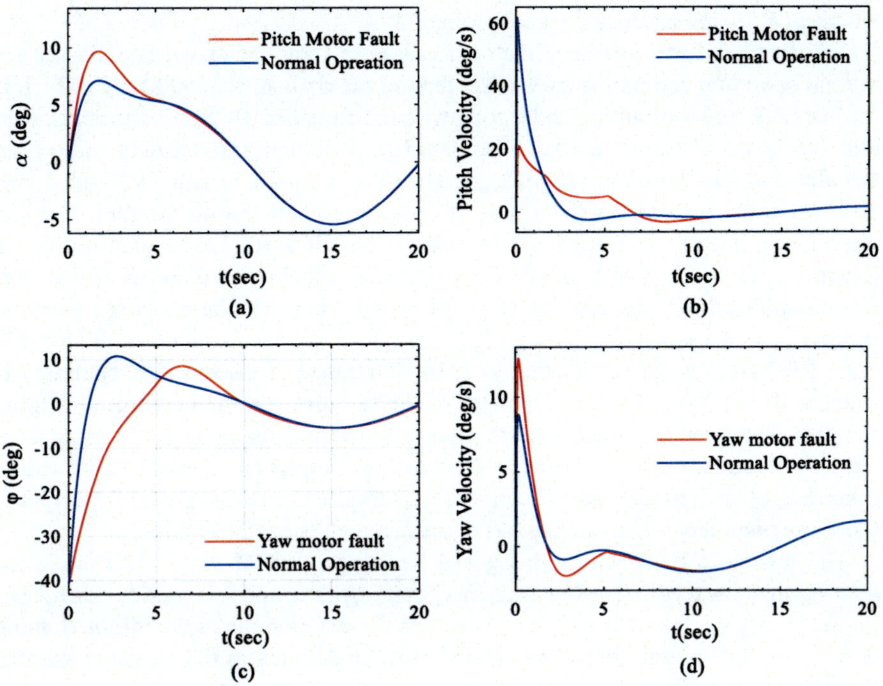

Fig. 2 Output characteristics of pitch and yaw motors for normal and fault operating conditions. **a** Pitch angle, **b** pitch velocity, **c** yaw angle, **d** Yaw velocity

By observing the pattern of the results during normal and motor faults, here, the 'db5' mother wavelet has been selected. Due to less impact of noise on signal and proper assessment of robustness of signal, a fifth level of wavelet transform is chosen. The wavelet coefficients values and energy values are calculated by (26) and showing huge difference for each fault.

$$E = \sqrt{\left[\sum_{i=1}^{i=N} \frac{abs^2(d_m(i))}{N}\right]} \tag{26}$$

where E—energy, —wavelet coefficients number for each power cycle, and $d_m(i)$—detailed coefficient of wavelet, at level i.

This energy feature of signal is main concern but the entropy ($H(R_s)$) also plays a key tool in the field of information theory. This assesses which type of wavelet is appropriate for reconstruction and decomposition of a signal. The entropy is expressed as:

$$H(R_s) = -\sum_{i=1}^{N} p(R_s) \log_{10} p(R_s) \tag{27}$$

where $p(R_s)$—reconstructed signal probability.

The features form a feature matrix along with their corresponding classes for normal operation and motor faults. The feature matrix is of size 5000×2, which is further subjected for training and testing with the classifier. To develop the classifier, initially the data (feature matrix) is prepared by dividing it into training and testing samples and labeling then accordingly. The feature matrix is equally divided into different matrices with 2500 training samples and 2500 testing samples. Further, the training process is carried out by setting the width for Lagrangian multiplier C and kernel width. The value of C is chosen depending on the size of the data. In this application, the value of C is set at 0.8 such that the α_i varies between $0 \leq \alpha_i \leq 0.8$. Further, the value of the kernel matrix is calculated as depicted in (25). The kernel width is adjusted to $1/16$. The range of the decision boundary is calculated with the help of total length of the samples and the number of training samples. This produces the bias term ρ and sets the grid range for the training data. Once the boundary conditions are set, the training samples are trained and the testing samples are tested based on the distance of the samples from the center of hypersphere (decision boundary). The corresponding results are depicted in figure.

The results visualize the training and testing process of the SVDD model by plotting the bias term, decision boundary, training data, and the prediction process. Figure 3a depicts the contour of the bias term for defining the region of the training data. From the contour, it can be defined that the data fits in the classifier without excluding any outliers or the delicate details of the data. In order to get a clear picture of the contour, the grid lines are used as depicted in Fig. 3b to get an exact

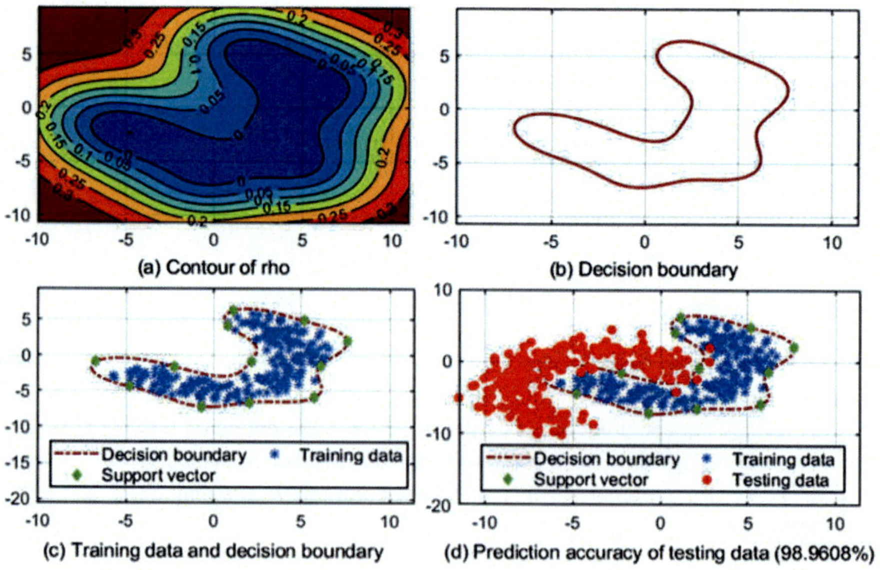

Fig. 3 Results of support vector data description classifier. **a** bias term, **b** decision boundary, **c** training data with decision boundary and support vectors, **d** classification process of SVDD

Table 1 Comparison of classification parameters

Parameters	Model type		
Preset	Support vector data description	Support vector machine classifier	K-nearest neighbor classifier
Network structure	2 predictors- 2 responses	2 predictors- 2 responses	2 predictors- 2 responses
No. of training Samples	2500 samples/feature of both classes	2500 samples/feature of both classes	2500 samples/feature of both classes
Training algorithm	Gaussian kernel	Fine Gaussian kernel	Coarse K-NN
Accuracy (%)	98.96	88.6	81.4

representation of the hypersphere. In Fig. 3c, the training data is scattered in the hyperplane and it is observed that the data is trained to lie within the support vectors. Further, in Fig. 3d the behavior of the classifier with the testing data is depicted. The classifier accurately predicts the data with the help of the hypersphere with an accuracy of 98.96%.

To analyze the classification efficiency of the developed classifier, the same input feature matrix is subjected to training with various classifiers using classification learner application of MATLAB. The analysis depicted that the developed classification approach has better efficiency than the conventional classifiers. The test results were depicted in Table 1.

From the results, it is observed that the classifier efficiently trains the feature data to develop the fault classifier for the helicopter system.

5 Fault Validation

For a helicopter system operating under various factors and conditions, there are many potential faults that can impact its operation and performance. Let $\mathcal{C} = \{c_1, c_2\}$ be the set of normal operating condition and motor fault condition, which can be monitored by the previously developed fault classifier. However, the set $\mathcal{C} \subseteq \mathcal{C}^*$ only deals with known subset of all possible motor fault classes $\mathcal{C}^*$ in helicopter. This implies that the set $\mathcal{C}$ can vary over time whenever a new type of fault is observed in the system. In most of the cases, the possibility for multiple faults to affect the operation of the system needs to be considered. Hence, in order to define the operating state of the system, a fault mode C' is considered to represent all the set of motor faults that can be present in a system $C' \subseteq \mathcal{C}$.

For identifying or validating the presence of a classified fault in system, which is set of residual generators $\mathcal{R} = \{r_1, r_2, r_3, \ldots, r_{n_r}\}$ are computed. These residual generators act as a function of actuator and sensor data is zero for an ideal or fault-free condition. This implies that, for any fault c_i the value of the residual generator is nonzero, and for any condition if the residual generator is not zero for fault c_i, then the fault is considered to be decoupled. All these scenarios of fault sensitivity and residual generators are described on the basis of the ideal case of the system. However, the performance of fault detection or classification is complicated by different measurement noises and model uncertainties. Hence, any change in the output of residual generator needs to be evaluated initially either by thresholding or with a test quantity. In order to monitor different parts of the system and generate various fault subsets, multiple residual generators are designed. Therefore, for every fault mode $C_i' \subseteq C$ which can be structurally detectable, a residual generator $r_k \in \mathcal{R}$ is always available which is sensitive to at least one fault $c \in C_i'$.

Initially, consider that all the possible subsets of C including the fault-free operation have a set of possible classification or diagnosis $\mathcal{D}$. This helps in explaining the current system state, whenever a residual is triggered for a class. For any case, if the diagnosis candidates cannot describe the residual trigger, they are refused to reduce the samples in the set. All the classified samples or events, where the subset of faults is a not a part of the classifier or diagnosis, are considered as minimal diagnosis candidates D_{min}. Generally, these minimal diagnosis candidates are considered as a fault-free case if they do not trigger any residual. Since different fault magnitudes have different effects on the residual outputs, a bounded residual uncertainty is assumed which can be described by a specific fault type $C_l' \subseteq C$ is characterized by a set $\phi^*(C_l') \subseteq \mathbb{R}^{n_r}$. Further, various fault modes or classes C_l' can describe various sets of residual outputs as $\phi^*(C_l')$. For explaining the multiple fault modes, the residual outputs are given as $\phi^*(C_{l1}') \cap \phi^*(C_{l2}') \neq \emptyset$.

Hence, each fault class C_l' is modeled by the classifier as:

$$\mathbb{C}^R_{C_l'} : \mathbb{R}^{|R|} \rightarrow \{0, 1\} \tag{28}$$

where $R \subseteq \mathcal{R}$ is the residual set utilized by the classifier. Since the adapted classifier does not consider the distribution and solely depends on modeling the data, the minimal diagnosis candidates can be ranked depending on the number of residual samples classified by $\mathbb{C}^R_{C_l'}$. Let $r_1, r_2, r_3 \ldots, r_n$, be N samples of the residuals for any classified fault. If $C_l' \in D_{\text{min}}$, its rank is computed as

$$\text{rank}(C_l') = \frac{1}{N} \sum_{k=1}^{N} \mathbb{C}^R_{C_l'}(r_k) \tag{29}$$

In case of untrained or undetermined faults, i.e., fault assumptions not protected by the computed minimum analysis candidates, it is essential to recognize the probability that an unidentified fault has happened. Let D_{min} denote the group of minimum analysis candidates, eradicating the unidentified fault issue. Further, the ranking of an unidentified fault $C_x' = \{c_x\}$ is presented in the following manner

$$\text{rank}(C_x') = \frac{1}{N} \sum_{k=1}^{N} \left(1 - \bigwedge_{\forall C_l \in D_{\text{min}}} \mathbb{C}^R_{C_l'}(r_k)\right) \tag{30}$$

This represents the ratio of the samples which have no belonging for any kind of known fault mode. For an unknown fault, a high rank represents identified possible locations of the fault using model support of the prompted residuals.

6 Conclusion

Fault detection and identifying the severity of a fault to enhance the operation and performance are crucial factors for unmanned helicopters. While having enough information about a fault, it is necessary to develop a fault classifier which can help in enhancing the control of the helicopter. In this paper, a support vector data description approach is developed for motor fault classification of unmanned helicopters. The faults were classified depending on their response with respect to normal operating conditions. The system is exploited by creating yaw and pitch motor failures and by extracting the fault and normal operating data. Further, the wavelet-based feature extraction is adapted and the energy and entropy features of the output data are extracted. This feature vector is mapped into a high-dimensional feature space using support vector data descriptor and the Gaussian kernel. The developed classifier depicted 98.96% accuracy. In order to validate the classifier during unknown fault conditions, the actuator and sensor data are used to develop a residual generator which further ranks the output of the system to validate the fault. This classification and validation approach depicted better performance when compared with the conventional classification processes.

References

1. Alagoz BB, Ates A, Yeroglu C (2013) Auto-tuning of PID controller according to fractional-order reference model approximation for DC rotor control. Mechatronics 23:789–797. https://doi.org/10.1016/j.mechatronics.2013.05.001
2. Ramalakshmi APS, Manoharan PS (2012) Non-linear modeling and PID control of twin rotor MIMO system. In: 2012 IEEE international conference on advanced communication control and computing technologies (ICACCCT). IEEE, pp 366–369
3. Juang JG, Te Huang M, Liu WK (2008) PID control using presearched genetic algorithms for a MIMO system. IEEE Trans Syst Man Cybern Part C Appl Rev 38:716–727. https://doi.org/10.1109/TSMCC.2008.923890
4. Phillips A, Sahin F (2014) Optimal control of a twin rotor MIMO system using LQR with integral action. In: 2014 world automation congress (WAC). IEEE, pp 114–119
5. Tao CW, Taur JS, Chen YC (2010) Design of a parallel distributed fuzzy LQR controller for the twin rotor multi-input multi-output system. Fuzzy Sets Syst 161:2081–2103. https://doi.org/10.1016/j.fss.2009.12.007
6. Jahed M, Farrokhi M (2013) Robust adaptive fuzzy control of twin rotor MIMO system. Soft Comput 17:1847–1860. https://doi.org/10.1007/s00500-013-1026-6
7. Mondal S, Mahanta C (2012) Adaptive second-order sliding mode controller for a twin rotor multi-input–multi-output system. IET Control Theory Appl 6:2157–2167. https://doi.org/10.1049/iet-cta.2011.0478

8. Brahim AB, Dhahri S, Hmida FB, Sellami A (2018) Multiplicative fault estimation-based adaptive sliding mode fault-tolerant control design for nonlinear systems. Complexity 2018. https://doi.org/10.1155/2018/1462594
9. Gao J, Wu P, Li T, Proctor A (2017) Optimization-based model reference adaptive control for dynamic positioning of a fully actuated underwater vehicle. Nonlinear Dyn 87:2611–2623. https://doi.org/10.1007/s11071-016-3214-2
10. Ghabraei S, Moradi H, Vossoughi G (2015) Multivariable robust adaptive sliding mode control of an industrial boiler-turbine in the presence of modeling imprecisions and external disturbances: a comparison with type-I servo controller. ISA Trans 58:398–408. https://doi.org/10.1016/j.isatra.2015.04.010
11. Wu NE, Zhou K, Salomon G (2000) Control reconfigurability of linear time-invariant systems. Automatica 36:1767–1771. https://doi.org/10.1016/S0005-1098(00)00080-7
12. Wu NE (2004) Coverage in fault-tolerant control. Automatica 40:537–548. https://doi.org/10.1016/j.automatica.2003.11.015
13. Xu Y, Jiang B, Tao G, Gao Z (2011) Fault tolerant control for a class of nonlinear systems with application to near space vehicle. Circuits, Syst Signal Process 30:655–672. https://doi.org/10.1007/s00034-010-9239-8
14. Boskovic JD, Mehra RK (2010) A decentralized fault-tolerant control system for accommodation of failures in higher-order flight control actuators. IEEE Trans Control Syst Technol 18:1103–1115. https://doi.org/10.1109/TCST.2009.2033805
15. Rao VS, George VI, Kamath S, Shreesha C (2015) Implementation of reliable H infinity observer-controller for TRMS with sensor and actuator failure. In: 2015 10th Asian control conference (ASCC). IEEE, pp 1–6
16. Bounemeur A, Chemachema M, Essounbouli N (2018) Indirect adaptive fuzzy fault-tolerant tracking control for MIMO nonlinear systems with actuator and sensor failures. ISA Trans 79:45–61. https://doi.org/10.1016/j.isatra.2018.04.014
17. Qi X, Qi J, Theilliol D et al (2014) A review on fault diagnosis and fault tolerant control methods for single-rotor aerial vehicles. J Intell Robot Syst Theory Appl 73:535–555. https://doi.org/10.1007/s10846-013-9954-z
18. Gao Z, Cecati C, Ding SX (2015) A survey of fault diagnosis and fault-tolerant techniques-part 1&2: fault diagnosis with knowledge-based and hybrid/active approaches. IEEE Trans Ind Electron 62:3768–3774. https://doi.org/10.1109/TIE.2015.2419013
19. Huang J, Wang M (2010) Multiple classifiers combination model for fault diagnosis using within-class decision support. In: 2010 international conference of information science and management engineering. IEEE, pp 226–229
20. Quanser Innovate Educate Position Control 2-DOF Helicopter Reference Manual
21. Dremin IM, Ivanov OV, Nechitailo VA (2001) Wavelets and their use. https://doi.org/10.1070/PU2001v044n05ABEH000918
22. Mallat S (2008) A wavelet tour of signal processing, 3rd edn. The Sparse Way
23. Discrete THE, Transform W The discrete wavelet transform. 6–15
24. Vapnik VN (2000) The nature of statistical learning theory. Springer, New York, NY
25. Strang G (2010) Linear algebra and its applications, 2nd edn. Harcourt Brace Jovanovich College Publishers
26. Vapnik VN (1999) An overview of statistical learning theory. IEEE Trans Neural Netw 10:988–999. https://doi.org/10.1109/72.788640
27. Smola AJ, Schölkopf B, Müller K-R (1998) The connection between regularization operators and support vector kernels. Neural Netw 11:637–649. https://doi.org/10.1016/S0893-6080(98)00032-X

EEG Signal Analysis and Emotion Classification Using Bispectrum

Nelson M. Wasekar, Chandrkant J. Gaikwad, and Manoj M. Dongre

Abstract In this work, three emotional states have been classified using EEG signals. Here, we analyzed different EEG signals and bispectrums of various emotions. Bispectral differential entropy, differential asymmetry and rational asymmetry features are computed. This work uses neural network-based approach for emotion classification using higher-order statistical features. Finally, we show that proposed approach using higher-order statistical features performed better than the method available in the literature.

Keywords EEG · Emotion recognition · Bispectrum · Neural network

1 Introduction

Emotion connects human-to-human communication and interaction and understands the real scenario behind the human ideology [1]. The goal of emotion recognition is to classify user's temporal emotional state based on some input data. Emotion can be defined in multiple ways [2].

There are various cultures and different languages available in the country and all over the world. From an application point of view, it is difficult to analyze the real-time situation, in that time emotion and emotion recognition play an important role. EEG signals are a more challenging problem than the speech and image signals. To calculate the features set of EEG signals for different models is a very big task. In this work, we will explore the use of higher-order spectra (HOS) in emotion classification using EEG signals. HOS have potential for application in the fields of communication, pattern recognition, economics, speech, seismic data processing, plasma physics and optics.

N. M. Wasekar (✉)
Ramrao Adik Institute of Technology, Navi Mumbai, India
e-mail: nelsonwasekar555@gmail.com

C. J. Gaikwad · M. M. Dongre
Department of Electronics and Telecom, Ramrao Adik Institute of Technology, Navi Mumbai, India

V. Singh et al. (eds.), *Computational Methods and Data Engineering*, Advances in Intelligent Systems and Computing 1227,
https://doi.org/10.1007/978-981-15-6876-3_29

In the literature, we find many works and emotion classifications based on EEG signals. Mohammad Soleymani et al. proposed a user-independent emotion recognition method using EEG recordings. In this work, eye gaze and pupillary response have been used. They achieved 68.5 percentage accuracy for three labels of valence and 76.4 percentage accuracy for three labels of arousal across 24 subjects [3]. Using EEG recording of 26 subjects, Yuan Pin Lin et al. classified four emotions (joy, anger, sadness and pleasure) and obtained 82.29% accuracy [4]. Here, the authors used music recordings for excitation of emotions. In this work, EEG bands were computed and features such as differential asymmetry, rational asymmetry and power spectral density based were used. Using Bispectral features of 32 participants, Nitin Kumar et al. classified four emotions (high and low valence, high and low arousal). They got accuracies of 64.84% for low/high arousal classification and 61.17% for low/high valence of the DEAP data set [5].

In this work, we consider three basic emotions positive, neutral and negative. We analyze the EEG signals for various emotional states using bispectrum. For feature extraction, we calculate differential entropy, differential asymmetry and rational asymmetry from the data samples. These features are computed based on the bispectrum of each channel of individual samples. By using neural network, we classified three emotional states.

2 Overview: Higher-Order Statistics

A major limitation of signal analysis using the autocorrelation is that the phase information is lost. Signal may be reconstructed from its magnitude and phase spectrum only. A set of parameters known as cumulants may be related to the moments; the Fourier transform of the cumulants provides the corresponding HOS. Higher-order spectra is also called as polyspectra, defined in terms of HOS; cumulants of a signal do contain such information.

The Fourier transform of the second-order, third-order and fourth-order cumulants is known as the power spectrum, bispectrum, trispectrum, etc. A biomedical engineer can use HOS to derive features from EEG or ECG recordings which will be very useful in many applications as HOS gives extra information about phase and nonlinearities present in the signals.

3 Fundamentals of EEG Signals

For measuring the activity of the brain, EEG has been found to be a very powerful tool in the field of neurology and clinical neurophysiology. EEG records the relative strengths and positions of electrical activity in different brain regions.

Different EEG rhythms are observed during various states from sleep to wakefulness. They are deeply asleep, drowsy, relaxed, asleep and excited. In general,

Table 1 Different bands

Types of rhythms	Frequency range (Hz)
Delta	Below than 4
Theta	4–7
Alpha	8–13
Beta	14–30
Gamma	Higher than 30

the amplitude of EEG signals ranges from a few microvolts to hundred millivolts. Depending on the age and mental state of the participants, the frequency content of EEG signals ranges from 0.5 to 40 Hz [6, 7]. However, to provide a clinically insightful characterization of these signals, the EEG rhythms are often classified into the following five different frequency bands:

In many works, we find the extraction of these bands has been used to compute the features for classification of emotional states [8].

4 Three Basic Emotions

In this work, we consider three types of emotions [8]

(1) Positive: Positive emotions are those when we feel there is a lack of negativity, such that no pain in that feeling is experienced by the participants.
(2) Neutral: It is a middle range between the pain and pleasure; they are neither distinctly painful nor clearly pleasant. Neutral emotions include surprise, astonishment, bored and drained.
(3) Negative: This is a feeling which causes you to be miserable and sad. These emotions will create a thought of dislike yourself, other and life in general. It also takes out your confidence and makes your mind unclear.

5 EEG Signal Preprocessing

In this work, we use the SEED dataset [8]. The sampling frequency of raw data is 200 Hz. The artifacts are removed using a bandpass filter. It is applied to EEG data from 0.3 to 50 Hz [8]. Electroencephalogram (EEG) is a highly complex signal, which contains information relating to the different physiological states of the brain. Hence, the EEG signal is an invaluable measurement for the purpose of accessing brain activities.

It is clear from Fig. 1 that the EEG signals are captured from the brain activity and the signal is preprocessed, and features are extracted and classifier model is trained using these features.

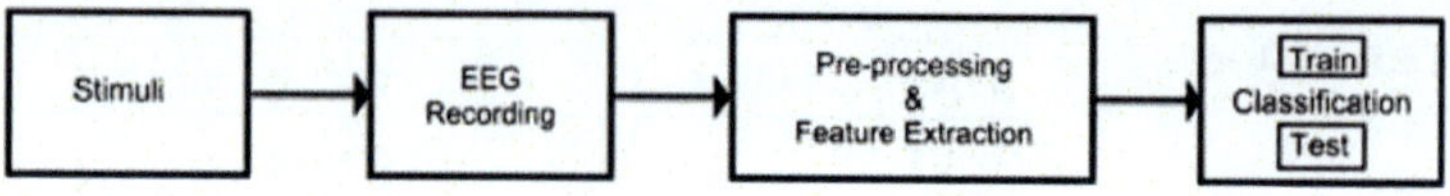

Fig. 1 Emotional brain–computer interface cycle [8]

A. Subjects
   In the experiment, recorded data consist of 15 subjects, out of them 7 are males and 8 are females [8]. All of them are students of Shanghai Jiao Tong University and they were right-handed. Subjects are selected according to Eysenck Personality Questionnaire (EPQ). It consists of only questions for the subjects and the participants are informed about the process.
B. Protocol
   For each session, the following method is used.
   Figure 2 shows the detail protocol; in the morning and afternoon, this experiment was performed in quiet environment. First, they recorded the EEG from 62 channel using electrode caps by following the international 10–20 system using ESI NeuroScan system of sampling rate 1000 Hz. EEG electrode positions are shown in Fig. 3.
   Every electrode has a tag with one alphabet letter and one number. It represents the area of brain. Even numbers represent the right side of the head and odd numbers represent the left side of the head. They mounted the camera on the face and recorded videos of the frontal face portion. In one experiment, there

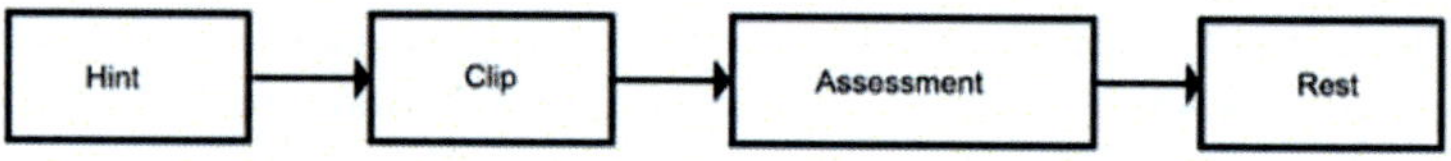

Fig. 2 Protocol of the EEG experiment [8]

Fig. 3 62 electrodes of EEG cap [8]

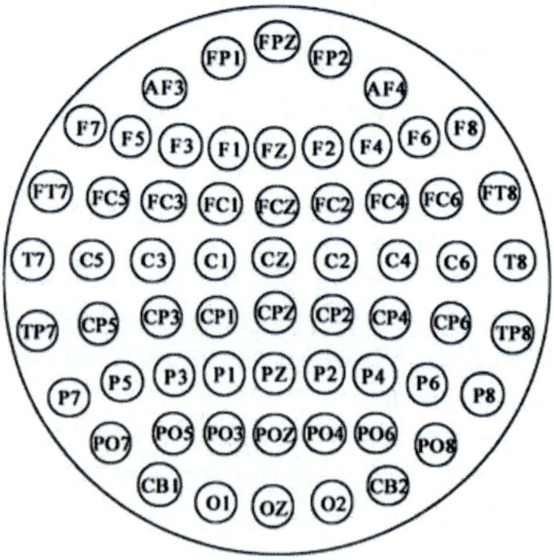

are 15 sessions and 5 s hint before each video clip, for questioning session were given 45 s and 15 s for rest after each in one session [8]. Following questions were asked to the subjects as follows:

(1) What is the output after seen such types of film clips?
(2) Have they watched these films before?
(3) Can these films understandable to all?
EEG signal was recorded by placing some electrodes on the scalp. These recordings are used further for analysis.

6 EEG Signal Analysis

We analyzed the considered emotions by plotting different EEG signals and bispectrums. These emotions are further analyzed using different channels.

A. EEG Signal
In this section, we analyzed different EEG signals. Figure 4 shows the EEG signal of a positive emotion. The spectrum of EEG signal is shown in Fig. 5, Figure 6 shows the EEG signal of a neutral emotion and its spectrum as shown in Fig. 7,
Figure 8 shows the EEG signal of a negative emotion and its spectrum is shown in Fig. 9, and this emotion is different from the other two. It denotes a single channel of EEG signal.
By taking the emotion samples of data, we analyzed the spectrums of these EEG signals. As can be observed from these spectrums, clear separation of emotions is not visible in these spectrums. In the next section, we analyze bispectrum of these EEG signals.

Fig. 4 EEG of positive emotion

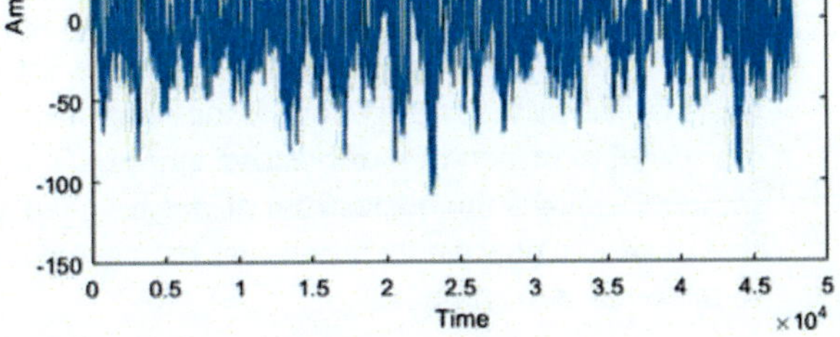

Fig. 5 FFT of positive emotion

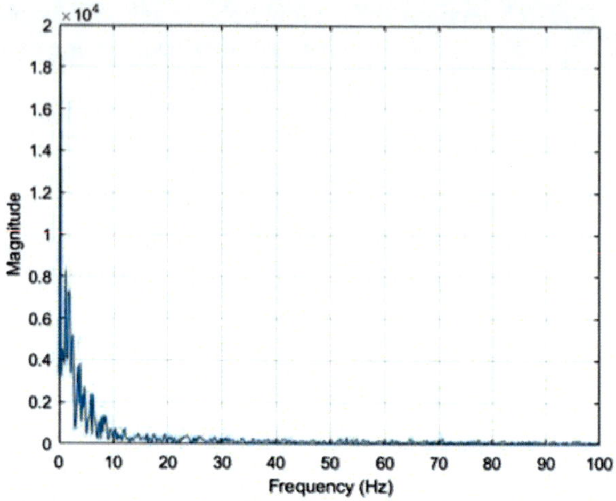

Fig. 6 EEG of neutral emotion

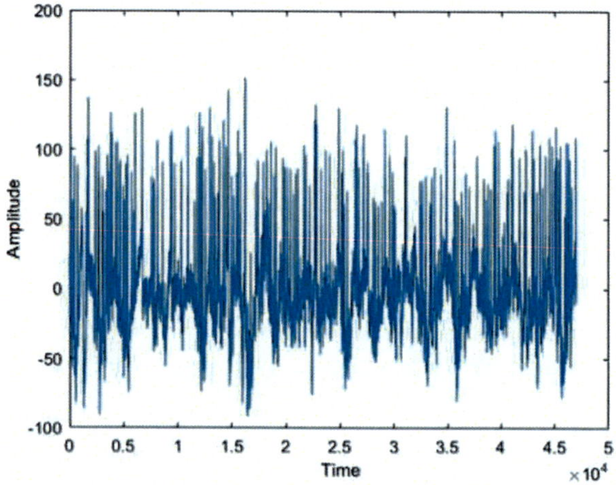

B. Bispectrum of Different Emotions

Various samples of EEG signals are used for calculating bispectrum. Figure 10 shows the bispectrum of positive emotion. Bispectrum computes the quadratic phase relation and this information is shown in the bispectrum, see Fig. 10. Here, quadratic phase relations are centered around low frequencies.

Figure 11 shows the bispectrum of neutral emotion. The bispectrum of neutral emotion shows the stronger quadratic relations for wide range of frequencies as compared to bispectrum of positive emotion.

Figure 12 shows the bispectrum of negative emotion. The figure clearly shows that quadratic phase interactions are stronger at higher frequencies as compared to the other two emotions.

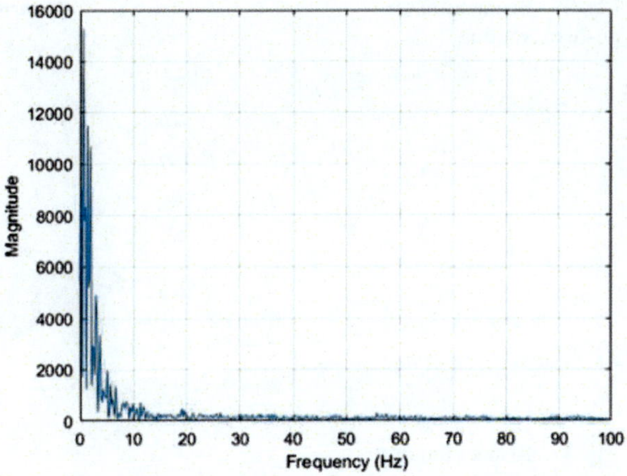

Fig. 7 FFT of neutral emotion

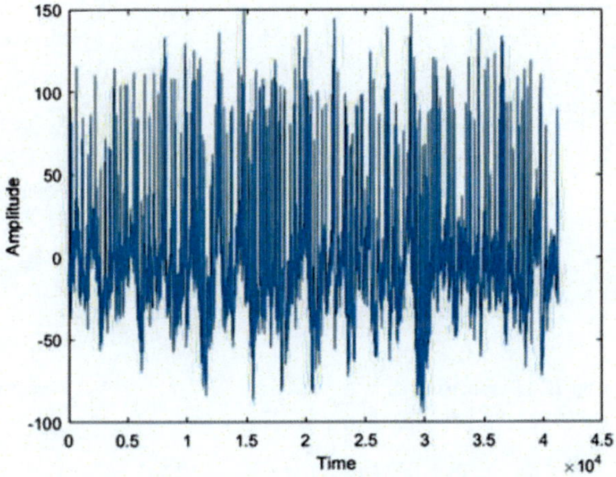

Fig. 8 EEG of negative emotion

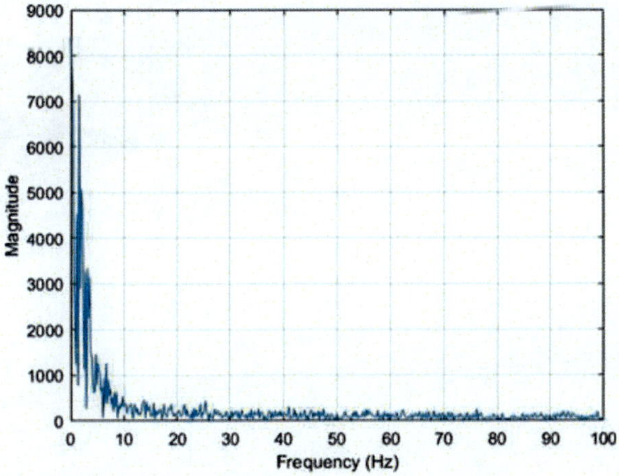

Fig. 9 FFT of negative emotion

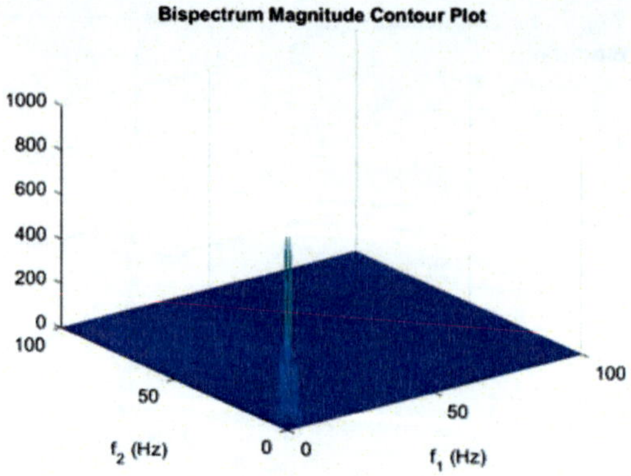

Fig. 10 Bispectrum of positive emotion

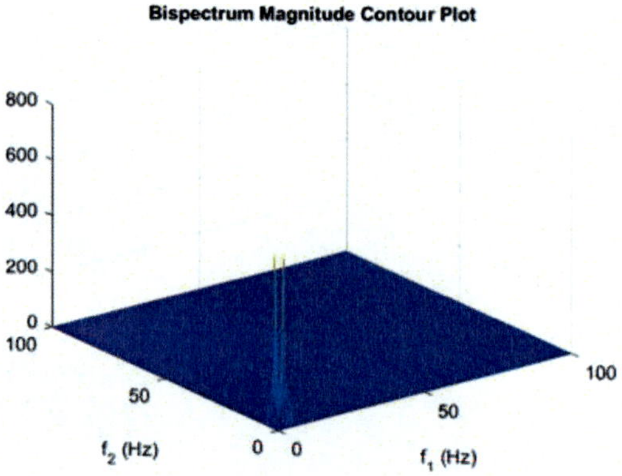

Fig. 11 Bispectrum of neutral emotion

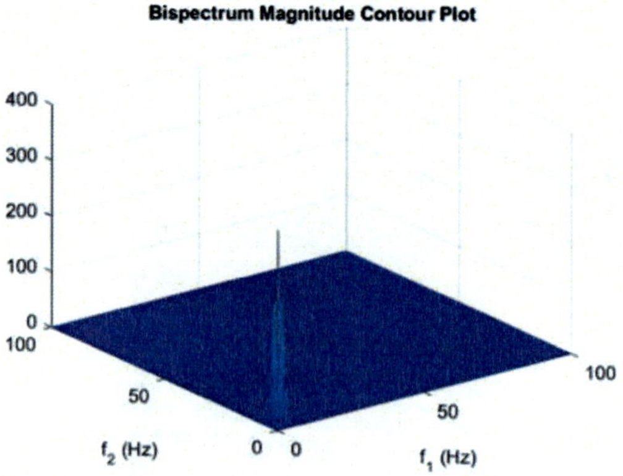

Fig. 12 Bispectrum of negative emotion

This analysis indicates that bispectrum captures various emotions. This information is present in magnitudes of bispectrum values and their frequency locations. In the following sections, we use the relative measurement of this as features for classification of various emotions.

7 Feature Extraction

The feature extraction is the first step in classification analysis for various types of applications. Here, we find bispectrum-based features such as differential entropy (DE), differential asymmetry (DASM) and rational asymmetry (RASM) [8].

A. Bispectrum
Third-order spectrum is called as bispectrum which is, by definition, the Fourier transform of the third-order statistics. Bispectrum (third-order cumulant) is used to identify and characterize the nonlinear effects (quadratic nonlinearity, phase coupling between two frequency components). Bispectrum can be calculated by using the Fourier transform of a signal evaluated at f_1, f_2 and f_3. The bispectrum of x is shown as follows:

$$B(f_1, f_2) = \frac{1}{k}\sum_{i=1}^{k} X_i(f_1)X_i(f_2)X_i * (f_1 + f_2) \tag{1}$$

where X denotes the Fourier transform of x.

B. DE
First bispectrum of each channel is calculated. After that, we take variance of individual bispectrum and then we find out differential entropy (DE). It is based on Shanon entropy, and it also calculates the complexity of a continuous random variable. If a random variable follows the Gaussian distribution, the DE is calculated by the following formula [8]

$$h(X) = -\int_{-\infty}^{\infty} \frac{1}{\sqrt{2\pi\sigma^2}} \exp\frac{(x-\mu^2)}{2\sigma^2} \log\frac{1}{\sqrt{2\pi\sigma^2}}$$
$$\exp\frac{(x-\mu^2)}{2\sigma^2}\mathrm{d}x = \frac{1}{2}\log\left(2e\pi\sigma^2\right) \tag{2}$$

From the above equation, for a fixed segment, DE is equal to the logarithmic energy spectrum. In many works, EEG PDF is considered to be Gaussian, though it is controversial to assume it to be Gaussian [7, 9, 10].

C. DASM
For the emotion processing, asymmetrical brain activity of left-right direction and frontal-posterior is quite effective. That is the reason the differential asymmetry is a good feature for classification. DASM is the difference of DE features

of 27 pairs by calculating individual bispectrum of each channel of the samples. We have used 27 pairs of the hemispheric asymmetry electrodes of the left hemisphere and right hemisphere for the calculation:

$$\mathrm{DASM} = \mathrm{DE}(X_{\mathrm{left}}) - \mathrm{DE}\left(X_{\mathrm{right}}\right) \tag{3}$$

where X_{left} and X_{right} are pairs of electrodes on left and right hemisphere.

D. RASM

RASM is the simple ratio of DE features of 27 pairs by calculating individual bispectrum of each channel of left and right electrodes.

$$\mathrm{RASM} = \frac{\mathrm{DE}(X_{\mathrm{left}})}{\mathrm{DE}\left(X_{\mathrm{right}}\right)} \tag{4}$$

where X_{left} and X_{right} are pairs of electrodes on left and right hemisphere.

E. Neural Network

Artificial neural network (ANN) is inspired by biological nervous system. It consists of interconnected elements (neurons) to solve difficult problems. Recent works have explored ANN for emotion recognition and clustering. In this work, we developed ANN model for emotion classification using bispectrum-based features. The model consists of two layers (3L-10 N-9 N-3L). The model uses eighty percent data for training and remaining for testing and validation.

8 Simulation Results

Training performance of ANN models was compared using confusion matrix. Using bispectral analysis, we calculate the asymmetrical spatial pattern features (DASM, RASM), differential entropy and obtained maximum accuracy of 75%. Table 2 shows three quantitative measures: precision, recall and F-score, for each emotion. The recognition performance for negative emotions is highest followed by neutral and positive emotions, respectively. Table 2 also shows that the model was able to predict best interclass separation for negative emotion, while the positive emotion is mixed with the other two.

Table 2 Results

Parameters	For Class 1 (positive)	For Class 2 (neutral)	For Class 3 (negative)
Precision	0.704	0.732	0.853
Recall	0.754	0.732	0.777
F-score	0.728	0.732	0.812

9 Conclusion

In this work, bispectral analysis of three emotions is carried out. Bispectrum-based features, namely differential entropy, differential asymmetry and rational asymmetry, are computed. We have applied pattern recognition neural network model to classify the three emotions (positive, neutral and negative). This new approach of classifying emotions using EEG signals is proposed. It computes DASM, RASM and DE based on bispectral values. This method is evaluated for 225 samples, and accuracy of 75% has been achieved. In our future work, larger dataset will be explored.

References

1. Cambria E (2016) Affective computing and sentiment analysis. IEEE Intell Syst 31(2):102–107
2. Liu Y, Sourina O, Nguyen MK (2011) Real-time EEG-based emotion recognition and its applications. In: Transactions on computational science XII. Springer, Berlin, Heidelberg, pp 256–277
3. Soleymani M, Pantic M, Pun T (2011) Multimodal emotion recognition in response to videos. IEEE Trans Affect Comput 3(2):211–223
4. Lin Y-P et al (2010) EEG-based emotion recognition in music listening. IEEE Trans Biomed Eng 57(7):1798–1806
5. Kumar N, Khaund K, Hazarika SM (2016) Bispectral analysis of EEG for emotion recognition. Procedia Comput Sci 84:31–35
6. Cohen A (1986) Biomedical signal processing: time and frequency domains analysis, vol I. CRC-Press
7. Srnmo L, Laguna P (2005) Bioelectrical signal processing in cardiac and neurological applications, vol 8. Academic Press
8. Zheng W-L, Lu B-L (2015) Investigating critical frequency bands and channels for EEG-based emotion recognition with deep neural networks. IEEE Trans Auton Mental Dev 7(3):162–175
9. Pardey J, Roberts S, Tarassenko L (1996) A review of parametric modelling techniques for EEG analysis. Med Eng Phys 18(1):2–11
10. Pritchard WS, Duke DW (1995) Measuring chaos in the brain-a tutorial review of EEG dimension estimation. Brain Cognit 27(3):353–397
11. Sanaullah M (2013) A review of higher order statistics and spectra in communication systems. Glob J Sci Front Res Phys Space Sci 13(4)

Slack Feedback Analyzer (SFbA)

Ramchandra Bobhate and Jyoti Malhotra

Abstract Using Slack for communication and teamwork is a common practice in the IT industry. It provides multiple Application Program Interfaces (APIs), which help developers significantly in writing a set of routines and protocols for software. IT companies conduct various sessions to educate, and to motivate their employees. The quality of those sessions can be improved by taking feedback and act accordingly. Slack APIs can be useful in taking such feedback. To simplify the feedback collection method, we came up with the 'Slack Feedback Analyzer (SFbA)', which collects feedback on session content as well as presenter's presentation skills and the knowledge expertise on that topic. With the help of this application, we are presenting the most common keywords, audience focused on, and the number of positive, negative, and neutral feedback.

Keywords Slack app · Sentiment analysis · IBM cloud · IBM Watson Natural Language Understanding

1 Introduction

Slack is team collaboration and communication applications; teams use irrespective of their location. This software provides many APIs (Application Programming Interfaces), which help developers significantly. IT companies also conduct many sessions to educate or to motivate developers. Those sessions can be improved by taking feedback from audiences and act accordingly. There are many Slack APIs available, which collect anonymous feedback. One of the issues with those APIs is we cannot take multiple feedback at the same time. Suppose, we need feedback on how the session went and how were presenter's presentation skills. But, with Slack

R. Bobhate · J. Malhotra (✉)
Department of CSE, MIT School of Engineering, MIT ADT University, Pune, India
e-mail: jyoti.j.malhotra@gmail.com

R. Bobhate
e-mail: gbobhate@gmail.com

V. Singh et al. (eds.), *Computational Methods and Data Engineering*, Advances in Intelligent Systems and Computing 1227,
https://doi.org/10.1007/978-981-15-6876-3_30

APIs, we cannot take feedback for session and presenter's skills at the same time. To get both the feedbacks, we need to ask for the feedback twice. Also, we need to collect feedback manually.

To overcome this issue, we came up with the 'Slack Feedback Analyzer (SFbA)', which collects session content feedback as well as presenter's presentation skills feedback. This app also performs sentiment analysis using IBM Watson Natural Language Understanding [1] and return those keywords, which audiences focused on majorly. It also shows the number of positive, negative, and neutral feedback given by the audience.

Further sections cover the related work, working methodology, results followed by a conclusion and future scope.

2 Related Work

Slack is used worldwide for team collaboration in IT companies. It not only supports team messaging but also supports a wide range of integration and bots. This integration can now obtain through Slack apps. In the manuscript [2], the authors studied how this Slack can be used by developers and what advantages they can get from integrations and bots. For this paper, the authors did two surveys, one for developers who were using Slack. In the second survey, authors filtered for those developers who were using Slack bots. In those surveys, authors found that Slack integrations played a remarkable role in the development process.

We are using messengers for a long period of time, but as an IT professional, we do not have that much knowledge of the capabilities of modern messengers. These messengers save developer's time needed for meetings and knowledge transfer. Knowledge transfer or connecting to other people can be done on these messengers easily and efficiently, which saves time. In paper [3], authors have discussed how group chats should be used for optimization. Authors also have developed a model to show efficient use of group chats.

Nowadays, people prefer one application with multiple functionalities. Similarly, people need messages with functionalities like file sharing, video calling, etc. But, in some cases, they face issues like installation. People may find it difficult to add functionalities to the messengers. To overcome this issue, the authors came up with a library as mentioned in [4], which adds or installs functionality to the messenger. This library supports multiple messengers like Slack, Facebook, Telegram. Authors have developed this library in Java and in the future they are planning to add other platforms as well.

As we know, Slack integration plays an important role in IT companies. But, nowadays it has entered in other domains as well. One of the major domains is healthcare. In paper [5], authors have focused on the Slack bot and the healthcare domain. Authors have provided a direction on how Slack bots can be used in the healthcare domain for a personalized subscription for patients. Currently, they have implemented Slack bot for one of the fields of healthcare, but they are aiming to expand this bot for the use of other fields of healthcare as well.

3 Methodology

Figure 1 illustrates the architecture of Slack Feedback Analyzer (SFbA) consisting of five components: Slack [6], IBM Cloud [7], IBM Watson Natural Language Understanding (NLU), Gmail, and web browser. As shown in Fig. 1, the user/presenter enters a Slack slash command [8] on Slack. This request is sent to Python-Flask [9] app hosted on IBM Cloud. On receiving this request, IBM Cloud sends a dialog box as a response [10]. This dialog box is available for all users/audiences to submit feedback. All these feedback are stored on IBM Cloud.

The user/presenter can see analysis by clicking on the 'Show analysis' button. When IBM Cloud receives this request, it forwards all feedback data to IBM Watson Natural Language Understanding (NLU). NLU performs sentiment analysis on each feedback and returns sentiment value and keywords of that feedback. SFbA then performs analysis on this data and sends this analysis to the user/presenter through Gmail with the help of SMTP (Simple Message Transfer Protocol).

SFbA's complete process is shown in Fig. 2. This process starts with the creation of the Slack slash command. For a slash command, you must have a Slack app. **Algorithm 1** shows how to create a Slack app.

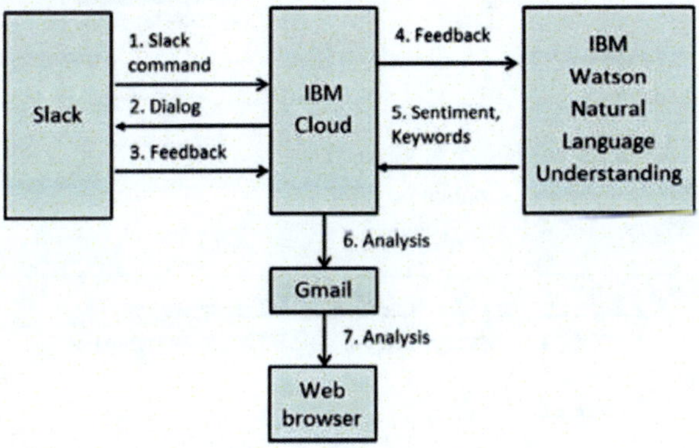

Fig. 1 Slack Feedback Analyzer (SFbA) Architecture

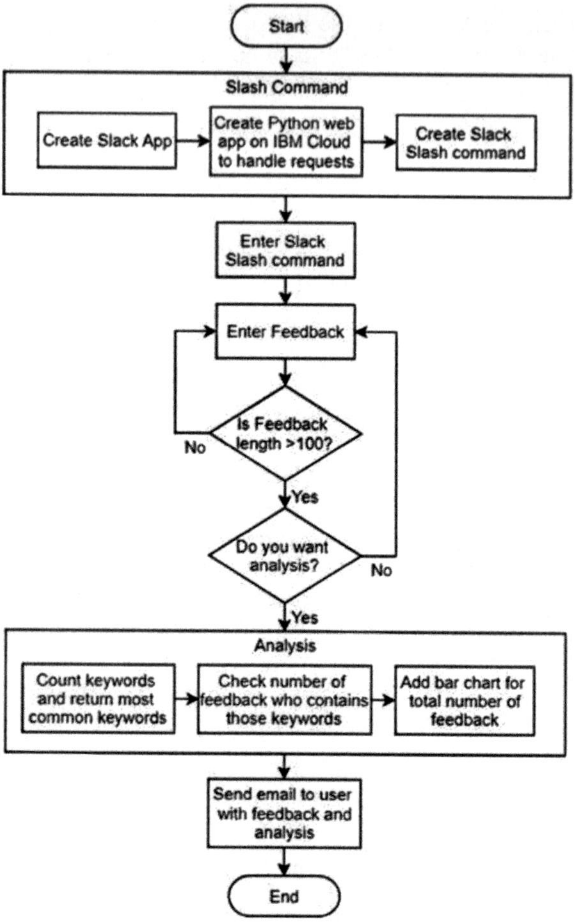

Fig. 2 SFbA's complete process

```
Algorithm 1: Create Slack App
  If already have a Slack account
    Login
  Elif
    Create a Slack account
  End If
  Goto manage apps
  Click on 'Create new app'
  enter Slack app name and select workspace
  select required components
  create app
End
```

After creating a Slack app, a Python-Flask app is created using the IBM Cloud. That app's web address is given to the Slack slash command. When slash command is entered in Slack, it sends a request to the web address given to the slash command. The presenter will enter slash command '/take feedback' along with the session name, on which presenter is seeking feedback. When the presenter enters the slash command, the request is sent to the Python-Flask app (SFbA) hosted on IBM Cloud. This app sends a dialog box to the Slack as a response. This dialog box is visible to all users on the channel. All users can click on 'Submit feedback' to submit their feedback.

When other users/audience submit their feedback, SFbA returns two options as a response: one for feedback submission and another to get analysis. When presenter clicks on 'Show analysis', SFbA perform algorithm shown as an Algorithm 2.

Algorithm 2: Algorithm For Analysis

```
doAnalysis():
  Group all keywords from positive, negative, and
  neutral feedback
  Count how many times each keyword is repeated
  Take 20 most repeated keywords
  foreach keywords:
    countKW["pos"]= # positive feedback for that keyword
    countKW["neg"]= # negative feedback for that keyword
    countKW["neu"]= # neutral feedback for that keyword
  end for
  countFB["pos"] = total number of positive feedback
  countFB["neg"] = total number of negative feedback
  countFB["neu"] = total number of neutral feedback
  return countFB, countKW
End
```

SFbA sends this data to the presenter through Gmail using Simple Mail Transfer Protocol (SMTP). This data is presented in tabular form. SFbA shows session content and presenter's skills feedback and its sentiment values in a tabular form. It shows the most repeated keywords and number of positive, negative, and neutral feedback they appeared in. It draws a bar chart for a total number of positive, negative, and neutral feedback for session content as well as presenter's presentation skills and his/her knowledge on that topic.

All analyses and feedback data are sent to the presenter through Gmail. Figure 3 shows the sample script used to add analysis and feedback data as an email body and sending email to the presenter.

```
body = """
    <html>
    <head>
    </head>
    <body> """ + e_body + """
    <br><br> Please, click on below link to see all feedbacks we received and their analysis, on '
    """ + session_name + """' till {}""".format(date_time) + """.
    <br><br> <a href="https://slackfeedbackanalyzer.eu-gb.mybluemix.net/display_data">
    Click here for feedbacks and its analysis</a>
    </body>
    </html>

    """
part = MIMEText(body, "html")
message.attach(part)
msg = message.as_string()

mail = smtplib.SMTP('smtp.gmail.com',587)
mail.starttls()
mail.login(user_from,password)
mail.sendmail(user_from,user_to,msg)
mail.quit()
```

Fig. 3 Creating email body and sending an email

4 Results

For SFbA, we defined '/take feedback' as a Slack slash command. While entering a slash command, we need to enter a session name as well. For testing, we took feedback on the 'IBM Cloud Pak System for Data'. When the slash command is entered, it sends a request to SFbA hosted on IBM Cloud. SFbA sends an interactive button as a response (as shown in Fig. 4). User needs to click on this button to submit their feedback.

When a user clicks on the interactive button 'Submit your feedback', again this request is sent to SFbA and it sends a dialog box as a response, where the audience/users can submit their feedback. (as shown in Fig. 5). Through this dialog box, the audience/users can submit feedback for session content as well as presenter's presentation skills and his/her knowledge in that area. Session feedback is mandatory and presenter's skills feedback is optional. Also, we kept the upper limit of 500 characters to prevent misuse of resources.

When the user/audience submits feedback, this request is sent to SFbA along with feedback. SFbA stores this feedback on IBM Cloud and sends two interactive buttons in response (as shown in Fig. 6). The first button is 'Submit your feedback', where other users/audiences can submit their feedback and the second button is 'Show analysis'. Using the 'Show analysis' button, the presenter can get the feedback

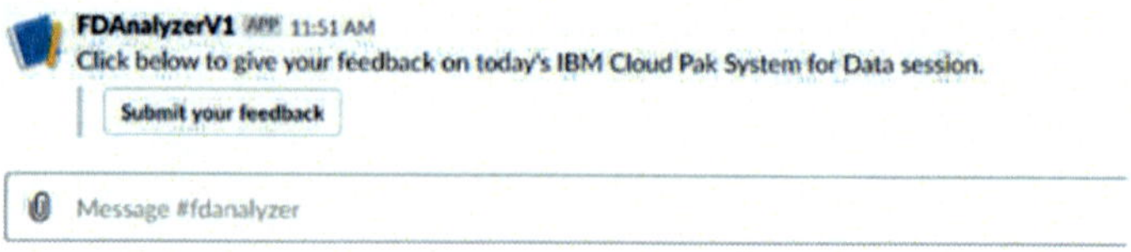

Fig. 4 Interactive button sent by SFbA

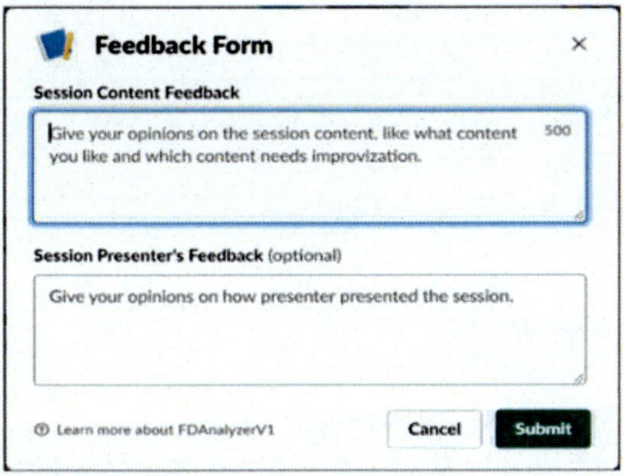

Fig. 5 SFbA feedback form

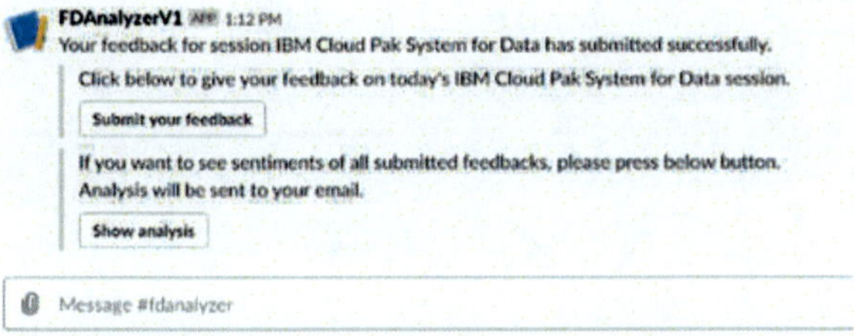

Fig. 6 Interactive buttons sent by SFbA after feedback submission

submitted by other users/audience and their analysis through email (as shown in Fig. 7).

The link displayed in the email contains session content feedback and presenter's skills feedback along with keywords of each feedback and their sentiment value and score (as shown in Fig. 8).

It also contains analysis performed by SFbA (as shown in Fig. 9a) and bar chart, which shows a total number of positive, negative, and neutral feedback (as shown in Fig. 9b). All feedback, their analysis, keyword analysis, and bar chart are added in the email body and sent it to the presenter.

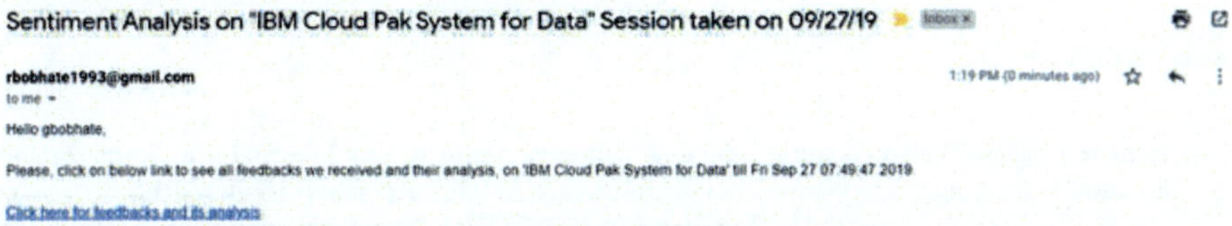

Fig. 7 Email sent by SFbA

Feedback on IBM Cloud Pak System for Data and its Analysis

No.	Feedback	Keywords	Sentiment Value	Confidence
1	The most secure, fast, and modernized data collection and analytic platform. it has the ability to work with any workload. Federating data from multiple databases does not work. if you want to federate data from different databases of the same type. the interface to complete the federation does not work load.	• data collection • different databases • databases • data • type • interface • ability • federation • workload • load	negative	-0.914

Fig. 8 Feedback table sent by SFbA

No.	Most Common Keywords	Frequency	Positive	Negative	Neutral
1	data	10	8	2	0
2	perfect cloud	6	6	0	0
3	information	5	5	0	0
4	ibm cloud	4	4	0	0
5	type of risk	4	4	0	0
6	price	4	4	0	0
7	time	4	4	0	0
8	environment	3	3	0	0
9	simple interface	3	3	0	0
10	order	2	2	0	0

(a)

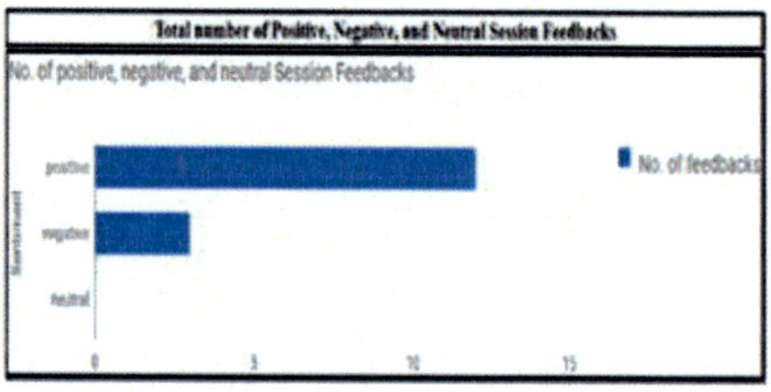

(b)

Fig. 9 **a** Keywords analysis and **b** bar chart sent by SFbA

5 Conclusion and Future Scope

Slack Feedback Analyzer (SFbA) successfully retrieves all feedback from the audience and performs sentiment analysis and it also successfully counts common keywords and also counts a number of feedbacks containing those keywords. It also sends bar charts for a total number of positive, negative, and neutral feedbacks successfully.

Currently, this app stores all feedback and analysis data on IBM Cloud storage. In the future, we are aiming to add a database, so that any data can be fetched whenever needed. Also, SFbA sends emails through Gmail, but in the future, we are aiming to send analysis to Slack through its email app.

References

1. Ibm.com (2019) Watson Natural Language Understanding. [online] Available at: https://www.ibm.com/watson/services/natural-language-understanding/. Accessed 10 Oct 2019
2. Lin B, Zagalsky A, Storey MA, Serebrenik A (2016) Why developers are slacking off: understanding how software teams use slack. In: Proceedings of the 19th ACM conference on computer supported cooperative work and social computing companion 27 Feb 2016. ACM, pp 333–336

3. Wang D, Wang H, Yu M, Ashktorab Z, Tan M (2019) Slack channels ecology in enterprises: how employees collaborate through group Chat. arXiv preprint arXiv:1906.01756. 4 Jun 2019
4. Kozhevnikov VA, Sabinin OY, Shats JE (2017) Library development for creating bots on Slack, Telegram and Facebook Messengers. Theor Appl Sci 6:59–62
5. Alesanco Á, Sancho J, Gilaberte Y, Abarca E, García J (2017) Bots in messaging platforms, a new paradigm in healthcare delivery: application to custom prescription in dermatology. InEMBEC & NBC 2017. Springer, Singapore, pp 185–188
6. Slack (2019) Where work happens. Slack. Accessed 23 Aug 2019. https://slack.com/intl/en-in/
7. "IBM Cloud." IBM. Accessed August 23, 2019. https://www.ibm.com/in-en/cloud
8. Slack (2019) Slash Commands. Slack. Accessed September 23, 2019. https://api.slack.com/slash-commands
9. "Flask." PyPI. Accessed September 23, 2019. https://pypi.org/project/Flask/
10. "IBM Cloud Pak for Data Reviews", Accessed September 23, 2019. https://www.g2.com/products/ibm-cloud-pak-for-data/reviews

A Review of Tools and Techniques for Preprocessing of Textual Data

Abhinav Kathuria, Anu Gupta, and R. K. Singla

Abstract With the high availability of computing facilities, a huge amount of data is available in electronic form. Processing of huge data is required to discover new facts and knowledge. But dealing with huge datasets is challenging because real-world data is generally incomplete, inconsistent, contains errors or outliers. More than 80% of the data is unstructured or semi-structured. The data is prepared by data preprocessing. Data preprocessing has become an essential step in data mining. Data Preprocessing takes 80% of the total efforts of any data mining project and it directly affects the quality of data mining. The selection of the right technique and tool for data preprocessing helps to enhance the speed of data mining process. This paper discusses different preprocessing techniques, different tools available for text preprocessing, carries out their comparison and briefs the challenges faced such as knowledge of sentence structure of a language to perform tokenization, difficulty in constructing domain-specific stop words list, over stemming and under stemming etc.

Keywords Preprocessing · Weka · Orange · Rapid miner · Knime · Tokenization · Stop words · Stemming · Data mining

1 Introduction

A huge amount of textual data is available on the Internet in the form of digital libraries, repositories etc. This data contains many inconsistencies, missing values, noise or errors [1]. Most of the data is unstructured or semi-structured and needs to be processed. Data Preprocessing is one of the major steps in the data mining process. It goes through various steps such as data cleaning, data integration, data selection and data transformation. After that, it is prepared for data mining task. It is required for transferring text from human language to machine-readable language for further processing, to enhance the speed and decrease the time and effort required to extract

A. Kathuria (✉) · A. Gupta · R. K. Singla
Department of Computer Science and Applications, Panjab University, Chandigarh, India
e-mail: abhinav.kathuria90@gmail.com

V. Singh et al. (eds.), *Computational Methods and Data Engineering*,
Advances in Intelligent Systems and Computing 1227,
https://doi.org/10.1007/978-981-15-6876-3_31

useful information [2]. The process of preprocessing the data is also known as noise elimination or noise reduction or feature elimination. It can be done by using various tools available having various techniques [3].

Huge amount of textual data is also available on blogs, Twitter, Facebook etc. Text mining is a process of extracting interesting and useful patterns from textual data. Different text mining techniques like summarization, classification, clustering, etc. are used to extract useful information. Text mining process consists of collecting information from different sources; applying preprocessing techniques to clean the data to remove inconsistencies, missing values etc.; applying processing techniques to process the data and finally extracting useful and valuable information [4]. Text mining can help an organization to derive potential business insights from text-based content such as email, word documents and postings on social media like Facebook, Twitter etc.

In order to clean the text from inconsistencies, different tools and techniques play a major role. Researchers working on different preprocessing procedures use Tokenization, Stop Word Removal, Stemming, Lemmatization, Document Indexing, Grammatical parsing and Chunking as their techniques in their work and different tools such as Weka, RapidMiner, Knime, R, Jupyter Notebook in Anaconda Navigator etc. Data preprocessing cleans the data so that the pattern recognition system will select the technique that will be used to discover the relevant pattern and remove irrelevant patterns [5].

The aim of this paper is to study different tools and techniques available for preprocessing of textual data, to compare their working. The paper is organized as follows: Section 2 discusses the preprocessing techniques, Sect. 3 describes the techniques used by researchers in their work, Sect. 4 describes the different tools available for preprocessing, Sect. 5 carries out the comparative study of these tools, Sect. 6 briefs out the challenges commonly faced while using these tools. Finally, conclusions are drawn in Sect. 7.

2 Common Preprocessing Techniques

After obtaining text, it is normalized with preprocessing techniques. It includes converting all letters to lower case, removing numbers, punctuations, whitespaces, stop words and applying stemming and lemmatization etc. Different techniques of preprocessing are described as follows:

Tokenization. The process of breaking a stream of text up into words, phrases, symbols or other meaningful elements called tokens [6–12].

Remove Stop words. Many words occur very frequently in a document but they are meaningless and are used for the purpose of joining sentences. Stop words are frequently used words like "and", "are", "this" etc. [6–12].

Stemming. Stemming is the process of reducing inflected words to their stem, base or root form. It is very important as there may be different words with the same root word. For example, the words "presentation", "presented", "presenting" may be reduced to a common word "present" [6–12].
Removing Punctuation. Punctuation provides grammatical context which helps in understanding meaning of the sentence. Punctuation is symbols like commas, quotes etc. Punctuation symbols need to be removed from the text as it does not represent any information [7].
Lemmatization. It is a process of grouping inflected forms together as a single base form. It is the practice of stripping of prefixes and suffixes that have been added to a word's base form. It involves using the dictionary and morphological analysis of words to remove inflectional endings and to return the base form of a word, which is known as a lemma [8]. The main idea of lemmatization is to reduce sparsity as different forms of a lemma may or may not occur infrequently during training [13].
Word Sense Disambiguation. A word has a different meaning depending upon the context. It is a process in which each term is assigned to a specific context. It finds the meaning intended by the word in the context [6, 8]. For example, the word horse may refer to an animal or to a unit of horsepower etc.
Grammatical Parsing and Chunking. Grammatical Parsing analyzes each word in the sentence to determine the structure from its constituent parts. NLP parser develops the grammatical structure of the sentence and finds out which word is subject or object of a verb. Chunking is the same as parsing instead it is used to build a hierarchical structure over text [6, 8].
Remove Numbers. Numbers need to be removed as they are not relevant for analysis. Usually numbers are removed by using regular expressions.
Text Summarization. It means summarize all the text documents on a specific topic to arrive at a concise overview. It uses two techniques i.e. extractive and abstractive. Extractive summarization contains information collected from original text and abstractive summarization brings out a summary that may or may not be available in original text [6, 8].
Document Indexing. It is based on selecting set of terms to be used for indexing the document. It includes finding set of keywords and assigning weights to the keywords, thus maintaining a vector of keyword weights. The weight includes the frequency of occurrence of the term in the document and the number of other documents that use that term [8].
Multiword Grouping. It means grouping consecutive words into a single token due to the idiosyncratic nature of the multiword expressions. For example, "United Arab Emirates" is the multiword containing three tokens. The meaning of these expressions is often difficult to identify from their individual tokens [13].
Term Frequency and Inverse Document Frequency (TF-IDF). It is a measure that computes how important a word is to a document. It increases proportionally to the number of times the word appears. It contains two terms: the first (tf) computes the number of times the word appears in the document divided by total number of words in the document. The second (idf) is Inverse Document Frequency which

is computed as logarithm of the total number of documents divided the number of documents that contain that term [8, 14].

Filter tokens by Length. By using this technique, the user can filter token of a specific length. The length of the attribute can be specified by using minimum and maximum characters [15].

Transform Cases. This a very useful technique as it helps in normalizing the text. By using this, the text will be converted into either lowercase or uppercase. Applicability of this technique depends upon the requirement of the problem [15].

Replace Slang and Abbreviations. Text is written in an informal way and it contains a lot of slang and abbreviations. These words have to be replaced with their proper meaning. For example, words like "ty", "omg", etc. must be replaced with "thank you", "oh my god" respectively [16].

Replace Elongated Words. Elongated is a word that is having a character that is being repeated more than two times. It is important to remove the repetition and replace those words with their source words. Otherwise, the classifier will treat the word as different [16].

Handling Negations. In text analytics, negation reverses the meaning of the word. It is necessary to handle negation as it completely changes the meaning of the sentence [16, 17].

Replace URLs. The text may contain URLs that have to be removed as it doesn't have any significance [16, 18].

Consider an example to understand the effect of different techniques. The sentence is: "The teacher is very knowledgeable, caring, patient 234 and easygoing—sounding a little more upbeat may help with the class's energy level!!"

After Tokenization. ['The', 'teacher', 'is', 'very', 'knowledgeable', ',', 'patient', '234', 'and', 'easygoing', '-', 'sounding', 'a', 'little', 'more', 'upbeat', 'may', 'help', 'with', 'the', 'class', ''', 's', 'energy', 'level', '!', '!']

After Removing Stop Words. ['The', 'teacher', 'knowledgeable', ',', 'patient', '234', 'easygoing', '-', 'sounding', 'little', 'upbeat', 'may', 'help', 'class', ''', 'energy', 'level', '!', '!']

After Stemming. ['the', 'teacher', 'is', 'veri', 'knowledg', ',', 'care', ',', 'patient', '234', 'and', 'easygo', '-', 'sound', 'a', 'littl', 'more', 'upbeat', 'may', 'help', 'with', 'the', 'class', ''', 's', 'energi', 'level', '!','!']

After Lemmatization. ['The', 'teacher', 'be', 'very', 'knowledgeable', ',', 'care', ',', 'patient', '234', 'and', 'easygoing', '-', 'sound', 'a', 'little', 'more', 'upbeat', 'may', 'help', 'with', 'the', 'class', ''', 's', 'energy', 'level', '!', '!']

After Removing Punctuation. 'The teacher is very knowledgeable caring patient and easygoing sounding a little more upbeat may help with the class's energy level'

After Removing Numbers. 'The teacher is very knowledgeable, caring, patient and easygoing—sounding a little more upbeat may help with the class's energy level!!'

3 Existing Use of Techniques

Different researchers have used different tools, techniques, data sets and feature representation techniques to preprocess the data and to make it useful for the data mining process. Different feature representation techniques have been available like Term Frequency-Inverse Document Frequency (TF-IDF), Feature frequency (FF), Bag of words etc. Some researchers have used the techniques available in the literature and some have mined the features from the text [19, 20]. Harnani et al. [21] have worked on movie reviews and preprocessing techniques have been applied in four stages. Features have been represented by using the Term Frequency-Inverse Document Frequency (TF-IDF) technique. The results of data mining after applying preprocessing have been stable with a test size of 40%.

Emma et al. [22] have worked on two datasets namely Dat-1400 and Dat-2000. Different techniques like removal of HTML tags, abbreviations, stop words, non-alphabetic words and stemming have been used for preprocessing. Features have been represented by using three techniques- Feature Frequency (FF), Term Frequency-Inverse Document Frequency (TF-IDF) and Feature Presence (FP). Using TF-IDF, the accuracy of data mining has been 81.5% and by using FF and FP techniques, it has been 83%. Razieh et al. [23] have used two data sets namely Reuters-21578 and PAGOD dataset. Sentence segmentation, tokenization, stop word removal and word stemming techniques have been used for preprocessing. Feature Reduction and Feature Subset selection have been used for representing features. The accuracy of data mining on Reuters and PAGOD dataset has been 96% and 79% respectively (Table 1).

Fernando et al. [24] have used three data sets namely Corpus A, Corpus B and Corpus C which include the star-based rating available on Google Play together with its text. The rating indicates enjoyment in using the App and can vary from 1 to 5 stars. Preprocessing has been applied in four stages and stop words have been removed from all the stages. Best results of data mining using SVM classification on Corpus A using Stage 3 has been 82.66%, on Corpus B using Stage 2 has been 81.32% and on Corpus 3 using Stage 3 has been 81.08%. Akrivi et al. [25] have used three data sets namely Obama-McCain Debate (OMD), Health Care Reform (HCR) and STSGold. Stemming, stop words removal and tokenization have been used for preprocessing and features have been represented by using three ways namely TF-IDF, InfoGainAttributeEval and ClassifierAttributeEval. The results of data mining have been improved by using these techniques. In the case of unigrams, ClassifierAttributeEval technique using Naïve Bayes (NB) classifier on OMD, HCR and STSGold datasets has shown an accuracy of 88.08%, 88.25% and 91.40% respectively. In case of bi-grams, InfoGainAttributeEval technique using NB classifier on the same datasets has shown an accuracy of 89.02%, 89.13% and 86.77% respectively. Similarly, in the case of trigrams, InfoGainAttributeEval technique using NB classifier has shown an accuracy of 92.59%, 91.94% and 92.67% respectively.

Table 1 Use of preprocessing techniques

References	Dataset	Techniques applied	Features representation	Accuracy
[21] Harnani et al.	Movies	Tier 1- Remove stop word Tier 2- Remove stop word + meaningless words Tier 3- Remove stop word + meaningless words + numbers + words less than three character	Tf, Tf-Idf	At random state 80% and Test Size 40%, it gives stable and convincing output
[22] Emma et al.	Dat-1400 and Dat-2000	All techniques are applied collectively HTML tags, abbreviations, stop words, non-alphabetic words are removed, stemming has been done etc.	FF, TF-IDF and FP	Using Tf-IDF accuracy is 81.5%, Using FF and FP it is 83%

(continued)

Table 1 (continued)

References	Dataset	Techniques applied	Features representation	Accuracy
[23] Razieh et al.	Reuters-21578 and PAGOD dataset	Sentence segmentation, tokenization, stop word removal and word stemming	Feature Reduction R1: all words R2: remove a list of stop words by TM. We use SMART stop words for removing stop words R3: remove the stop words and stemming by TM Feature Subset Selection S1: term frequency S2: mutual information Term Weighting W1: binary paradigm is used to show our model if word occurs in the document, the bit string set to 1 else set to 0 W2: TF with no normalization and collection component is used W3: TF with normalization and no collection component is used	R3.S2.W4 for Reuter showed an accuracy of 96% and R3.S1 for PAGOD showed an accuracy of 79%

(continued)

Table 1 (continued)

References	Dataset	Techniques applied	Features representation	Accuracy
[24] Fernando et al.	Corpus A, Corpus B and Corpus C	Terms standardization, Spell check and Stemming	Stage 1: reviews without any preprocessing Stage 2: reviews with terms standardization Stage 3: reviews with terms standardization and spell check Stage 4: reviews with terms standardization, spell check and stemming Stop words removal is applied in all the stages	Results of SVM classification on Corpus A using Stage 3 is 82.66%, on Corpus B using Stage 2 is 81.32%, on Corpus 3 using Stage 3 is 81.08%
[25] Akrivi et al.	Obama-McCain Debate, Health Care Reform, STSGold	Stemming, Stop words removal and Tokenization	TF-IDF InfoGainAttributeEval, Top 70% Classifiers used for Experiment Naïve Bayes [NB] Support Vector Machines [SVM] k-Nearest Neighbor [KNN] C4.5	There is no representation that brings systematically better results in comparison with the others. In general, 1-to-3-grams perform better than the other representations, having a close competition with unigram
[26] GeethaRamani et al.	News Articles	Conversion from Chinese to English, HTML Tag Removal, Stop Word Removal and Stemming	Correlation-based Feature Selection	With Feature Selection, it shows an accuracy of 86.87%

(continued)

Table 1 (continued)

References	Dataset	Techniques applied	Features representation	Accuracy
[27] Giulio Angiani et al.	Test set 2016 Test set 2015	Emoticon, Negation, Stemmingand Stop words	No preprocessing Basic Basic + Stemming Basic + Stop words Basic + Negation Basic + Emoticon All All without Dictionary	Basic + Stemming shows an accuracy of 80.84% on CV folds 10 and 68.68% on Test set 2016 and 76.40% on Test set 2015

R. Geetha Ramani et al. [26] have worked on news articles. Different techniques like conversion from Chinese to English, HTML tag removal, stop word removal and stemming has been used for preprocessing. Features have been represented by using the Correlation-based method and the accuracy of data mining has been 86.87%. Giulio et al. [27] have worked on two data sets namely Test set 2016 and Test set 2015. Emoticon, negation handling, stemming and stop word removal techniques have been used for preprocessing of data. Preprocessing has been applied in eight stages. Stage III (Basic + Stemming) has shown an accuracy of 68.68% on Test set 2016 and 76.40% on Test set 2015.

4 Tools Available for Preprocessing

Text mining tools fall into two categories. First category of tools is Proprietary text mining tools which are commercial text mining tools owned by a company. Second category is Open source text mining tools that are available free of cost and one can even contribute in their development [28]. There is increasing interest in the scientific community of the benefits offered by open source tools as they offer stable and scalable features. So in research, open-source tools are used [29]. The focus of this paper is on Open source text mining tools. Popular tools used by researchers are discussed as follows:

Weka: Waika to Environment for Knowledge Analysis (Weka) is a collection of algorithms for data mining tasks. Weka contains tools for data preprocessing, regression, classification, clustering etc. It works by loading the dataset and by converting the text into vector form in which each document is represented by the frequency of some important terms. Weka contains functions for preprocessing of textual data. The user can select whether to perform lower case conversion or not, choose the algorithm for stemming, choose the stop word list according to language, select the tokenization method etc. [30].

Rapid Miner: It provides an environment for data mining, machine learning, text analysis, predictive analysis etc. There are many functions available in Rapid Miner that can be used for reading data from different sources. Text Processing is an extension available in Rapid Miner for preprocessing. It contains operators for different purposes i.e. tokenization, filtering for stop word removal, stemming and transformation for case conversion etc. It provides three ways for tokenization one by using nonletter character, second on the basis of a particular character and third by using regular expressions. It also contains stop words list for languages like English, French etc. It also supports various algorithms such as Porter, Snowball, Lovins etc. for stemming [31].

Knime: Konstanz Information Miner is a modular computational environment that integrates various components called nodes for machine learning and data mining. Knime has thousands of nodes available in the node repository which the user can drag and drop into the Knime workbench. It enables easy visual assembly and a user can even interact with the data pipeline. Workflows in Knime are like graphs in which nodes are connected together and one can easily insert or add edges between the nodes to define the sequence of working [32]. Text Processing is an extension that needs to be installed to perform text mining. Different preprocessing nodes available for textual processing are removing punctuation, numbers, small words, conversion to lower case, reducing to word stem etc. [33]. Other preprocessing nodes available in Knime besides regular preprocessing nodes are filter node, RegEx and Replacer. RegEx and Replacer both are based on the regular expressions that can be specified in the dialog of the node. For each tagger, there is a filter node that filters terms according to the tag values provided.

R: It is a cross platform software environment used by statisticians and data miners for statistical computing. R contains many text analysis packages ranging from low-level string operations to advanced text modeling techniques. There are many functions in R for reading data. Preprocessing starts by manipulating text and converting it into tokens. The tokens are then used for creating Document-term-matrix which is used by packages of R for text analysis. R contains functions for tokenization, stemming, removing stop word, conversion to lower case etc. [34].

Orange: It is a data visualization and data mining toolkit. The components of Orange are called widgets and they range from simple data visualization to empirical evaluation of learning algorithms. It is a Python library. Data mining is done with the help of Python scripting. Text processing is an extension that needs to be installed for preprocessing. It contains functions for tokenization, lemmatization, lower case conversion, filtering, removing stop words, removing accents, removing HTML tags, removing URLs etc. N-grams can be created and Part-of-speech (POS) tagger can be applied on the tokens [35].

Jupyter Notebook in Anaconda Navigator: Anaconda navigator is a graphical user interface that allows managing packages and launching applications. The Jupyter Notebook is a web application available in Anaconda Navigator which uses Python language for text processing. Natural language toolkit (NLTK) is the library that needs to be installed and it contains different functions for preprocessing of textual data. It contains functions for tokenization, stemming, removing numbers, spaces, stop words etc. [36].

5 Comparative Study of Tools

A comparative study of various tools discussed in the previous section has been made after considering different parameters such as operating system support, license, ease of learning, interface, functionality, features missing etc. Some tools provide Command Line Interface (CLI), some provide Graphical User Interface (GUI) and some provide both. Functionality includes the availability of functions for performing data mining, statistical computing, data modeling, machine learning etc. Although preprocessing techniques are available in all the tools yet some tools contain features which are missing in others (Table 2).

Only Orange tool contains features such as removal of accents, HTML tags and URLs from among the other tools. Rapid Miner tool contains filtering of tokens by length, by content, by POS tags, by region and filtering is not available in other tools.

Limitations of Weka include lack of documentation. Weka and Knime can't optimize parameters for machine learning efficiently and optimizing them manually are tedious and dangerous. Before using Rapid Miner proper knowledge is required for handling databases and it also requires more memory to operate. R contains limited data mining functions, more memory to operate, few classification and clustering algorithms, proper knowledge of array etc. Installing Orange takes a lot of time and space, it is having limited reporting capabilities, weak in classical statistics and contains limited machine learning algorithms. Good code versioning is not possible in Jupyter Notebook, code is very hard to test and it contains non-linear workflow.

Although Weka is having lack of documentation yet Weka is used mostly for research purposes as it is having all the classification, clustering, association, machine learning and other data mining algorithms required for preprocessing of textual data and is used by maximum researchers [3]. Learning Knime, Rapid Miner and Orange is difficult. One needs to understand the input and output format of each operator. Jupyter Notebook is a web-based application based on a client-server structure that contains proper documentation of libraries, packages etc. It also includes data visualization libraries like Matplotlib and Seaborn that can be used in the same document where our code resides. Moreover researchers can combine code, computational output, explanatory text and multimedia resources in a single document.

6 Challenges

Although each tool has its own limitations yet many challenges have been faced while using the tools for text preprocessing. The challenge in tokenization depends on the type of language. Some languages like English and French are space-delimited languages as most of the words are separated from each other by white spaces while languages like Chinese and Thai do not have clear boundaries within the words. It is a very challenging task to construct stop words list for a particular domain because there exists inconsistency between different data sources. The challenges in stemming include over stemming and under Stemming. Over stemming means when two words with different stems are stemmed from the same root. For example, the popular Porter stemmer algorithm will stem the words "universal","university" and "universe" to the word "univers". Under stemming is when two words that have to be stemmed from the same word have actually been stemmed from a different word. For example, the Porter stemming algorithm will stem the words "alumnus → alumnu", "alumni → alumni", "alumnae → alumna". To obtain lemmatization, morphological analysis of the words is required. Using Weka is a challenge due to the non-availability of proper documentation. The challenges in Rapid Miner and

Table 2 Comparative study of data mining tools [3, 37–43]

Tool Name	Weka	Rapid Miner	Knime	R	Orange	Jupyter Notebook in Anaconda Navigator
Release Year	1993	2006	2004	1997	2009	2014
Latest Version	3.9	9.2	3.7.1	3.5.3	3.20.1	5.7.4
License	GNU General Public License	AGPL Proprietary	GNU General Public License	GNU General Public License	GNU General Public License	Modified BSD License
Operating System	Cross Platform	Cross Platform	Linux, OS X Windows	Cross Platform	Cross platform	Cross platform
Language	Java	Language Independent	Java	C, Fortran and R	Python, C++, C	Python, R, Julia, Scala etc.
User Groups	Naïve	Expert	Expert	Average	Average	Expert
Graphical Representation	Yes	NA	NA	Yes	Yes	NA
Interface	GUI	GUI	GUI	Both CLI and GUI	Both CLI and GUI	CLI
Ease of Learning	Easy	Hard	Hard	Easy	Hard	Easy
Functionalities	• Machine Learning	• Statistical analysis • Predictive analysis	• Enterprise Reporting • Business Intelligence	• Statistical Computing	• Machine Learning • Data Mining • Data Visualization	• Numerical Simulation • Statistical Modeling • Data Visualization • Machine Learning

Knime include proper knowledge of operators along with the input and output format of each operator. Rapid Miner doesn't scale well with big data and it contains minimal documentation for most of the functions. Biggest challenges of using R are memory management, speed and efficiency.

7 Conclusion

With the rapid increase in the World Wide Web, there is a huge amount of data in every field. The aim of this study is to explore different tools and techniques that are available for preprocessing of huge amount of textual data. After that the comparative study of features of different tools is carried out. It also discusses the challenges faced while working on different tools and techniques. Weka is used mainly for research purposes by different researchers as it contains a vast number of functions. Jupyter Notebook tool will be used for research purposes as it contains proper documentation. Moreover it creates a seamless bridge between program code and narrative text, enabling users to create and share code, equations, visualization, etc. in real-time.

References

1. Ramírez-Gallego S, Krawczyk B, García S, Woźniak M, Herrera F (2017) A survey on data preprocessing for data stream mining: current status and future directions. Neurocomputing 239:39–57. https://doi.org/10.1016/j.neucom.2017.01.078
2. N, Y., S, M (2016) A review on text mining in data mining. Int. J. Soft Comput 7:01–08. https://doi.org/10.5121/ijsc.2016.7301
3. Gera M, Goel S (2015) Data mining—techniques, methods and algorithms: a review on tools and their validity. Int J Comput Appl 113:22–29. https://doi.org/10.5120/19926-2042
4. Talib R, Kashif M, Ayesha S, Fatima F (2016) Text mining: techniques, applications and issues. Int J Adv Comput Sci Appl 7:414–418. https://doi.org/10.14569/ijacsa.2016.071153
5. Mining, U., Chandrama, W., Devale, P.P.R., Murumkar, P.R.: Survey on Data Preprocessing Method of Web 5:3521–3524 (2014)
6. Dr.S.Kannan, V.G.: Preprocessing Techniques for Text Mining. J. Emerg. Technol. Web Intell. (2016)
7. Srividhya, V., Anitha, R.: Evaluating Preprocessing Techniques in Text Categorization. Int. J. Comput. Sci. Appl. 49–51 (2010)
8. Lourdusamy, R., Abraham, S.: A Survey on Text Pre-processing Techniques and Tools. Int. J. Comput. Sci. Eng. 06, 148–157 (2019). https://doi.org/10.26438/ijcse/v6si3.148157
9. Katariya NP, Chaudhari MS (2015) Text preprocessing for text mining using side information. Int Comput Sci Mob Appl 3:3–7
10. Kadhim AI (2018) An Evaluation of Preprocessing Techniques for Text Classification. 16:22–32
11. El Haddaoui, B., Chiheb, R., Faizi, R., El Afia, A.: Toward a Sentiment Analysis Framework for Social Media. 1–6 (2018). https://doi.org/10.1145/3230905.3230919
12. Putra, S.J., Khalil, I., Gunawan, M.N., Amin, R., Sutabri, T.: A Hybrid Model for Social Media Sentiment Analysis for Indonesian Text. 297–301 (2019). https://doi.org/10.1145/3282373.3282850

13. Camacho-collados, J.: On the Role of Text Preprocessing in Neural Network Architectures : An Evaluation Study on Text Categorization and Sentiment Analysis. 40–46 (2018)
14. Orellana, G., Arias, B., Orellana, M., Saquicela, V., Baculima, F., Piedra, N.: A study on the impact of pre-processing techniques in Spanish and english text classification over short and large text documents. Proceedings—3rd International Conference on Information Systems and Computer Science, INCISCOS 2018, Dec pp 277–283 (2018). https://doi.org/10.1109/INCISCOS.2018.00047
15. Harvey S (2009) A study of interscholastic soccer players perceptions of learning with game sense. Asian J Exerc Sport Sci 6:1–11. https://doi.org/10.15439/2018KM46
16. Effrosynidis D, Symeonidis S, Arampatzis A (1999) Conference on research and advanced technology for digital libraries. Interlend Doc Supply 27:300–302. https://doi.org/10.1108/ilds.1999.12227cab.016
17. Wankhede S, Patil R, Sonawane S, Save PA (2018) Data preprocessing for efficient sentimental analysis. In: Proceedings international conference on inventive communication and computational technologies ICICCT 2018, pp 723–726. https://doi.org/10.1109/ICICCT.2018.8473277
18. Roy D, Mitra M, Ganguly D (2018) To clean or not to clean: document preprocessing and reproducibility. J Data Inf Qual 10. https://doi.org/10.1145/3242180
19. Saxena D, Saritha SK, Prasad KNSS V (2017) Survey on feature extraction methods in object. Int J Comput Appl 166:11–17
20. Waykole RN, Thakare AD (2018) A review of feature extraction methods for text. Int J Adv Eng Res 351–354
21. Zin HM, Mustapha N, Murad MAA, Sharef NM (2017) The effects of pre-processing strategies in sentiment analysis of online movie reviews. In: AIP conference proceedings, pp 1–8. https://doi.org/10.1063/1.5005422
22. Haddi E, Liu X, Shi Y (2013) The role of text pre-processing in sentiment analysis. Procedia Comput Sci 17:26–32. https://doi.org/10.1016/j.procs.2013.05.005
23. Ghalehtaki RA, Khotanlou H, Esmaeilpour M (2014) Evaluating preprocessing by turing machine in text categorization. In: 2014 Iranian conference on intelligent systems ICIS 2014. https://doi.org/10.1109/IranianCIS.2014.6802540
24. Dos Santos FL, Ladeira M (2014) The role of text pre-processing in opinion mining on a social media language dataset. In: Proceedings—2014 Brazilian conference on intelligent system BRACIS 2014, pp 50–54. https://doi.org/10.1109/BRACIS.2014.20
25. Krouska A, Troussas C, Virvou M (2016) The effect of preprocessing techniques on Twitter sentiment analysis. In: IISA 2016—7th international conference on information, intelligence, systems and applications (2016). https://doi.org/10.1109/IISA.2016.7785373
26. Geetharamani R, Kumar MN, Balasubramanian L (2017) Identification of emotions in text articles through data pre-processing and data mining techniques. In: Proceedings 2016 international conference on advanced communication, control & computing technologies ICACCCT 2016, pp 611–615. https://doi.org/10.1109/ICACCCT.2016.7831713
27. Angiani G, Ferrari L, Fontanini T, Fornacciari P, Iotti E, Magliani F, Manicardi S (2016) A comparison between preprocessing techniques for sentiment analysis in Twitter. In: Proceedings 2nd international workshop on knowledge discovery on the web, KDWeb 2016, pp 1–11 (2016). https://doi.org/10.1007/978-3-319-67008-9_31
28. Kaur A, Chopra D (2016) Comparison of text mining tools. In: 2016 5th international conference on reliability, Infocom technologies and optimization (Trends Futur Dir 186–192. https://doi.org/10.1109/ICRITO.2016.7784950
29. Wilson M, Tchantchaleishvili V (2013) The importance of Free and Open Source Software and Open Standards in Modern Scientific Publishing. Publications 1:49–55. https://doi.org/10.3390/publications1020049
30. Dan L, Lihua L, Zhaoxin Z (2013) Research of text categorization on Weka. In: Proceedings 2013 3rd international conference on intelligent system design and engineering applications ISDEA 2013, pp 1129–1131 (2013). https://doi.org/10.1109/ISDEA.2012.266

31. Kalra V, Aggarwal R (2018) Importance of text data preprocessing & implementation in RapidMiner. Proc First Int Conf Inf Technol Knowl Manag 14:71–75. https://doi.org/10.15439/2017km46
32. Berthold MR, Cebron N, Dill F, Gabriel TR, Kotter T, Meinl T, Ohl P, Sieb C, Thiel K, Wiswedel B (2007) Knime. Web. 1–8. https://doi.org/10.1007/978-3-540-78246-9
33. Hofmann M, Chisholm A, Chisholm H, Berthold M (2016) Text mining and visualization: case studies using open-source tools
34. Welbers K, Van Atteveldt W, Benoit K (2017) Text analysis in R. Commun Methods Meas 11:245–265. https://doi.org/10.1080/19312458.2017.1387238
35. Orange3 Text Mining Documentation (2018)
36. Project Jupyter: Project Jupyter | Home. http://jupyter.org/, (2017)
37. Rangra K, Bansal KL (2014) Comparative study of data mining tools. Int J Adv. Res Comput Sci Softw Eng 4:216–223. https://doi.org/10.1016/j.nuclphysa.2007.03.042
38. Chauhan N, Gautam N (2015) Parametric comparison of data mining tools 291–298
39. Solanki H (2013) Comparative study of data mining tools and analysis with unified data mining theory. Int J Comput Appl 75:975–8887. https://doi.org/10.5120/13195-0862
40. Patel PS, Desai SG (2015) A comparative study on data mining tools. Int J Adv Trends Comput Sci Eng 4:28–30
41. Bisht P, Negi N, Mishra P, Chauhan P (2018) A comparative study on various data mining tools for intrusion detection 9:1–8
42. Singh DK (2017) Comparative study of various open source data mining tools 356–358
43. Ranjan R, Agarwal R, Venkatesan S (2017) Detailed analysis of data mining tools. Int J Eng Res 6:785–789. https://doi.org/10.17577/ijertv6is050459

A U-Shaped Printed UWB Antenna with Three Band Rejection

Deepak Kumar, Preeti Rani, Tejbir Singh, and Vishant Gahlaut

Abstract Design approach for triple-band rejection characteristics of compact printed UWB antenna is proposed. This design can avoid interference between UWB and narrowband such as WLAN (5.5 GHz), WiMax (3.4 GHz), and downlinks of X band (7.5 GHz) for satellite communication application. To achieve triple-band rejection characteristic double folded multiple mode resonators are placed symmetrically to the feed line. The antenna comprises a U-shaped metal patch, ground plan with rectangular notch on the backside, and 50 Ω microstrip feed line for the impedance matching. The antenna attains an impedance bandwidth range between 3.1 and 10.6 GHz with VSWR < 2, omitted the bands of WiMax (3.2–3.6 GHz), WLAN (4.9–6.1 GHz), and downlinks.

Keywords Microstrip antenna · WiMax · Multiple mode resonator · VSWR · HFSS

1 Introduction

Ultra-Wideband technology grown to be very popular when the communication exhibited the 3.1–10.6 GHz range for unlicensed band in the year 2002 [1]. UWB is extensively being used in numerous applications on account of its excessive data

D. Kumar
Department of ECE, BIT, Meerut, UP, India
e-mail: kr.deepak05@gmail.com

P. Rani (✉)
Department of ECE, Banasthali Vidyapith, Rajasthan, India
e-mail: preeti8482@gmail.com

T. Singh
Department of ECE, SRM University, Sonipat, HR, India
e-mail: tvsp1282@gmail.com

V. Gahlaut
Department of Physics, Banasthali Vidyapith, Rajasthan, India
e-mail: gvishant@banasthali.in

V. Singh et al. (eds.), *Computational Methods and Data Engineering*,
Advances in Intelligent Systems and Computing 1227,
https://doi.org/10.1007/978-981-15-6876-3_32

transmission rates, less power dissipation plus modest design approach. Apart from the imperative aspects, ultra-wideband communication antenna also requires a band-rejected filter to eliminate the interference for various band specified in the resent standard of wireless. Commonly used frequency bands in different regions are as follows: IEEE 802.11a in Europe and USA (5.15–5.35 GHz, 5.47–5.725 GHz) [2], HIPERLAN/2, satellites uplink (7.5–8.4 GHz), WiMax (3.2–3.6 GHz), and down-link (7.25–7.75 GHz) frequency. Therefore, UWB transmitters play a crucial role to abolish the electromagnetic interference on narrowband structures, i.e., Wireless LAN, WiMax, and downlinks of X-band Satellite communication applications.

However, there are several methods to reject the existing narrow band from the ultra-wide-band frequency, we can use filter to reject these narrowband services but the system complexity will be increased. Some advanced antennas with band-notched specifications have been exhibited to deal with this challenge [2–7]. The structure of radiofrequency can be corrected by removing filters in respective designs. Apart from these configurations, it is possible to attain multiband reject characteristics with the help of several resonant structures.

In order to generate multiple notch bands, a half mode substrate cohesive waveg-uide cavity is employed in [8]. For the purpose of continuing the half-wave stepped impedance resonators, a few antennas have been proposed in [9, 10, 3] along with multiple notched bands. These antennas are not very compact in size but different structures can be used to fulfill the requirements of return loss and radiation patterns.

A compact U-shaped UWB antenna is presented, in this paper, with triple band reject of WLAN, WiMAX, and downlinks of X band satellite communication. To achieve this, triple band reject double multiple-mode resonator (MMR) are used symmetrical at the feed line and multiple mode resonators are folded for the compact size. The patch U-shapes and semi ground planes at the back of the substrate are modified for better matching impedance [11]. It is evaluated that designed antenna exhibits the UWB characteristics (3.1–10.6 GHz) with VSWR < 2 excluding multi-mode resonator but after employing two symmetrical multi-mode resonators into patch plane next to the feed line and square slot in the ground plane, the band-notched characteristics from band of WiMax (3.2–3.6 GHz), WLAN (4.9–6.1 GHz) and downlinks of X band Satellite communication with VSWR > 2 are achieved. The simulation of the above antenna is performed by HFSS13 and some constraints such as returns loss, VSWR, current distribution, and radiation patterns have been calculated [12–18].

2 Antenna Design

The UWB antenna and its measurement containing the U-shaped patch is illustrated in Fig. 1. Fabrication process has been performed on FR4 PCB substrate ($W_s \times L_s = 35 \times 40\ \text{mm}^2$) with breadth h = 0.762 mm and $\varepsilon_r = 4.4$. Feed line comprises single strip with an area of $W_f \times L_f$ which delivers a typical impedance of 50 Ω. The radiation portion of design is U-shaped patch placed in center of external and

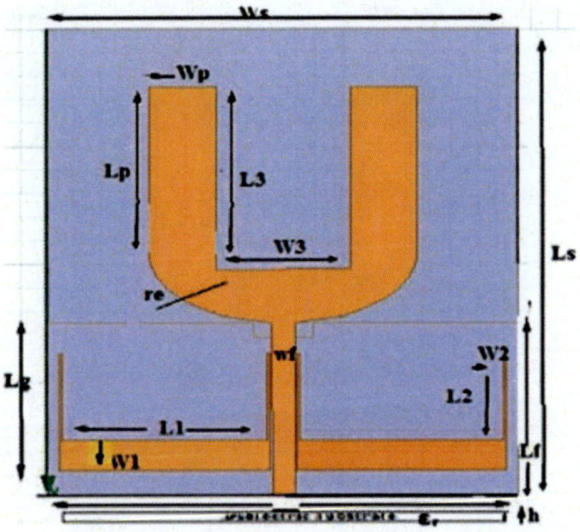

Fig. 1 Geometry of the proposed antennas

internal length of L_p and L_3 of the substrate. The patch has given U-type shape with a curve of radius (r_e) near the feed line.

The ground plane of area $L_g \times W_g$ is expanded in this structure on backside of the substrate, a half square slot of dimension ($W_n \times L_n$) on just back side of feed point in ground plane is notched to achieved proper impedance bandwidth.

For the rejection of triple-band double multiple-mode resonators (MMR) are used symmetrical at the feed line of dimension ($W_1 \times L_1$) and multiple mode resonators are folded for the compact size at dimension $W_2 \times L_2$. Table 1 shows the various dimensions of the designed antenna.

The double folded multiple-mode resonator (MMR) is accountable for triple band band-reject at WLAN, WiMax, and downlink for satellite communication band via this compact and U-shape patch antenna at both sides of feed line as shown in Fig. 1. The developed antenna imparts an impedance bandwidth of 3.1–10.6 GHz with VSWR below 2, except in the band of WiMax (3.2–3.6 GHz), WLAN (4.9–6.1 GHz) and downlinks of X-band (7.25–7.75 GHz) Satellite communication with an omnidirectional radiation pattern.

Table 1 Dimensions of antenna

Parameter	W_s	L_s	W_f	L_f	L_g	W_g	L_n	W_n
Size (mm)	35	40	1.7	18	15	35	1.2	4.5
Parameter	W_1	W_2	L_1	L_2	L_f	W_f	L_p	W_p
Size (mm)	2.5	0.2	15.6	10	18	1.67	14	5
Parameter	L_3	W_3	H			ε_r		
Size (mm)	20	10	0.77			3.66		

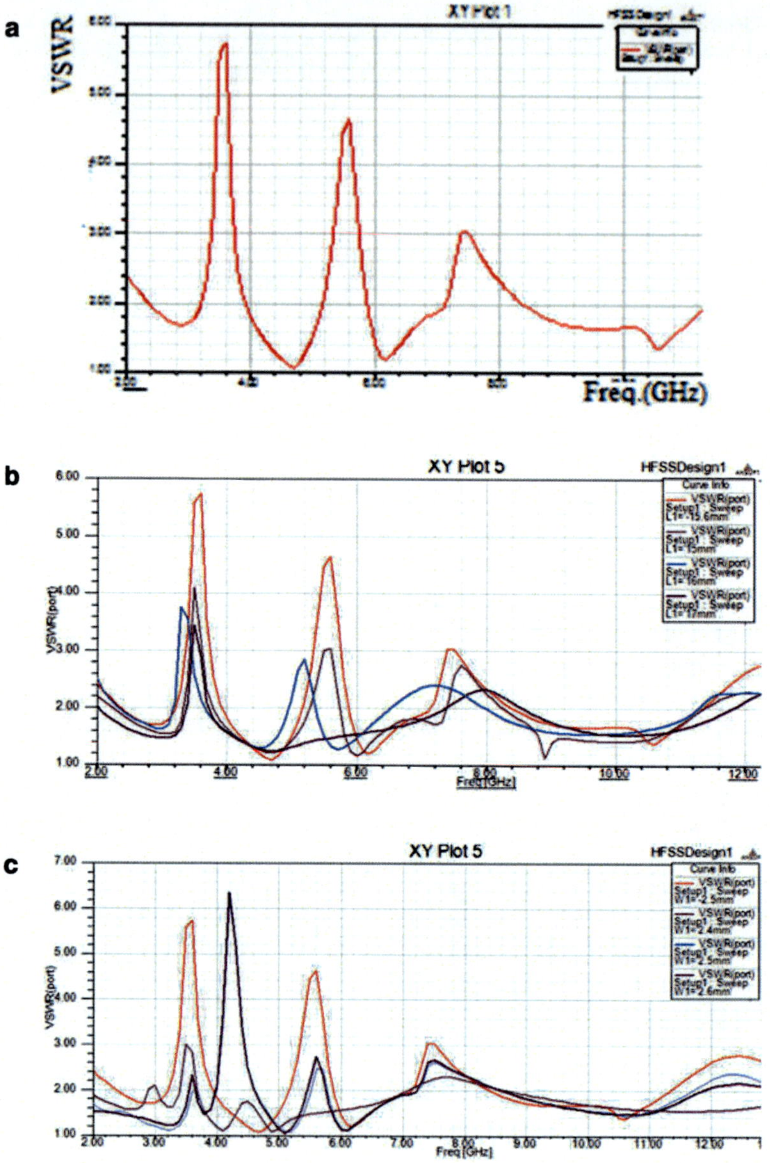

Fig. 2 **a** Simulation result of VSWR. **b** Simulation of VSWR for variable L_1. **c** Simulation results of VSWR for variable W_1. **d** Simulated VSWR of projected design for different W_2. **e** Simulation result of VSWR for various L_2

d
e
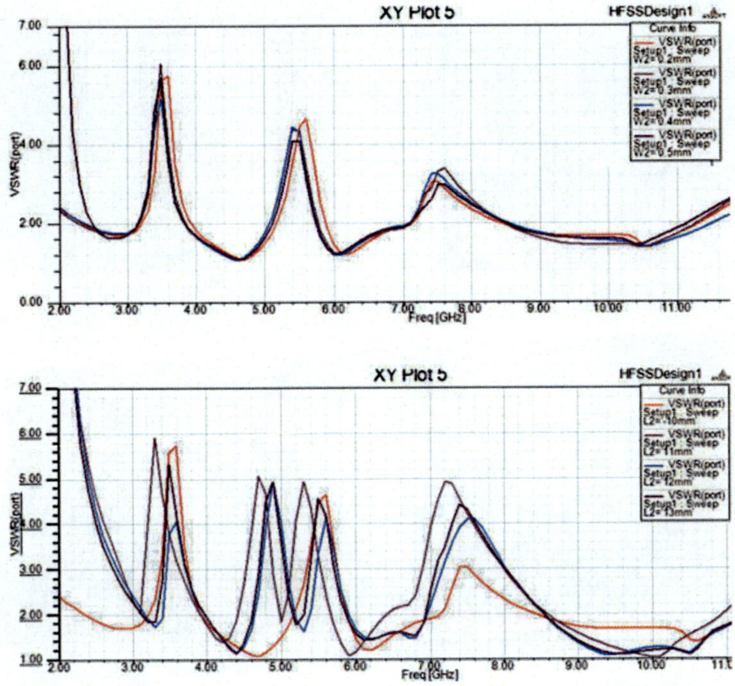

Fig. 2 (continued)

3 Results and Discussion

The designed antenna is simulated using HFSS 13 and performance of VSWR is observed on the network analyzer. As a result, the VSWR is less than 2 with frequency limit 3.1 to 10.6 GHz whereas for VSWR above 2 (VSWR > 2), the ranges are 3.2–3.6 GHz, 4.9–6.1 GHz and 7.25–7.75 GHz. Simulation result of VSWR with triple band reject is depicted in Fig. 2a. monitored that reject band at 3.5 GHz, 5.5 GHz and 7.5 GHz is obtained by simply using folded multiple mode resonator symmetrically nearby the feed line and in reject band 3.2–3.6 GHz, 4.9–6.1 GHz and 7.25–7.75 GHz of WiMax, WLAN and downlink satellite X band, respectively. Thus, it is possible to remove the interference between UWB and these narrowband systems successfully. It is remarked in analysis that the rejected band may be affected by varied length and width of the folded multi-mode resonator (MMR).

Figure 2b–e shows the parametric analysis with variable length L_1, L_2, and width W_1, W_2 of the folded MMR. It is observed that the best result is obtained at $L_1 = 15.6$ mm, $L_2 = 10$ mm, and $W_1 = 2.5$ mm, $W2 = 2$ mm, respectively. In aid of frequencies 3.4 GHz, 4.7 GHz, 5.5 GHz, and 7.5 GHz, the simulation result of surface current distribution is indicated in Fig. 3. It is investigated that utmost current distribution is close to the multi-mode resonator on central frequencies 3.4 GHz, 5.5 GHz, and 7.5 GHz of WiMax, WLAN, and downlink satellite X-band, which

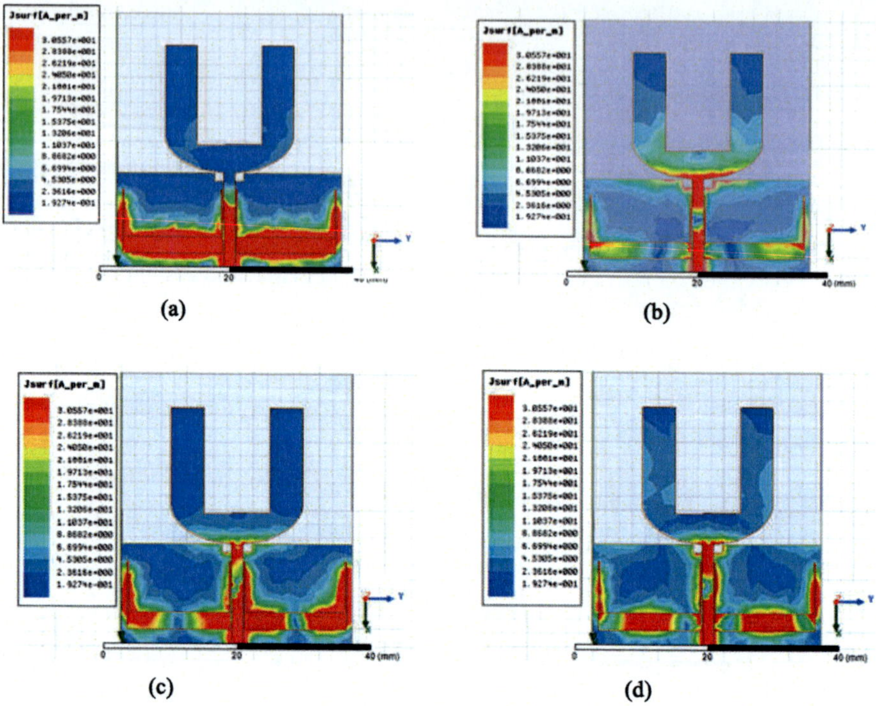

Fig. 3 Current distribution at various frequencies. **a** 3.5 GHz, **b** 4.6 GHz, **c** 5.5 GHz, **d** 7.5 GHz

causes the rejection of this band and current in radiating patch. Furthermore, the remaining portion of the antenna, which is depicted in Fig. 3 will be decreased. At the radiating patch and feed line, surface current distribution is higher while it is lesser on the MMRs at 4.7 GHz. It is also examined that the single node in the middle of the MMR at 5.5 GHz and dual nodes adjacent to the turning of folded MMR at 7.5 GHz is held by the current distribution indicated in (Fig. 3).

The simulation results using Ansoft HFSS are defined in Fig. 4. The radiation patterns have been evaluated on bands 3.5 GHz, 4.7 GHz, 5.5 GHz, and 7.5 GHz. The radiations are quasi omnidirectional and symmetrical in y-z plane in a larger frequency band. Simulated antenna patterns are different for low frequency (3.4 GHz) and high frequency (7.5 GHz). This design is extremely devoted to novel mobile security systems and wireless applications.

4 Conclusion

A compact UWB printed antenna with triple-band rejection features is simulated. The fabrication and parameters have been measured on network Analyzer. The

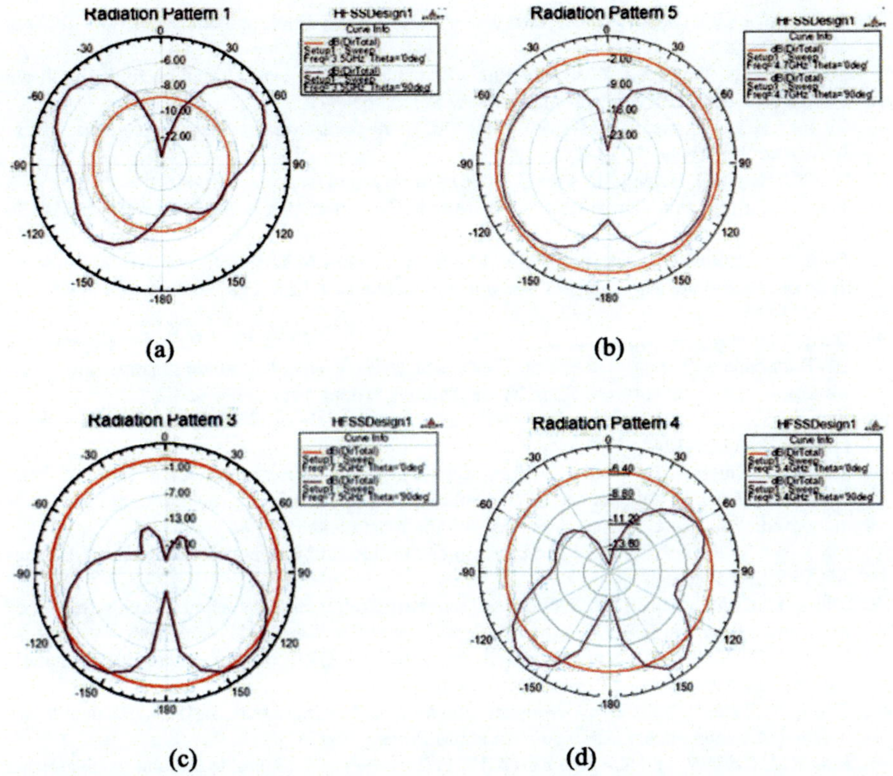

Fig. 4 Radiation far-field patterns analysis for the proposed design on **a** 3.5 GHz, **b** 4.7 GHz, **c** 5.4 GHz, **d** 7.5 GHz

proposed antenna is a considerable alternative for UWB amid band reject applications, i.e., WLAN, WiMax, and X-band downlink of satellite communication. The designed antenna offers appreciable outcomes in context of radiation patterns, current distribution, and VSWR.

References

1. Force and Spectrum Policy Task (2002) Spectrum policy task force report. Federal Communications Commission ET Docket 2:135
2. Chung K, Hong S, Choi J (2007) Ultrawide-band printed monopole antenna with band-notch filter. IET Microw Antennas Propag 1(2):518–522
3. Hongwei D, Xiaoxiang H, Binyan Y, Yonggang Z (1981) Compact band-notched UWB printed square-ring monopole antenna. In: 8th International Symposium on Antennas, Propagation and EM Theory, pp 1–4, 2–5 2008. Smith TF, Waterman MS (1981) Identification of common molecular subsequences. J Mol Biol 147:195–197

4. Su SW, Wong KL (2000) Printed band notched UWB quasi dipole antenna. Micro Opt Technol 48(3):418–420
5. Xu F, Chen X, Wang XA. UWB antenna with triple notched bands based on folded multiple mode Resonators. IEICE Elecronics Express 9(11):965–970
6. Schantz HG, Wolenec G, Myszka EM (2003) Frequency notched UWB antennas. Ultra-Wideband Syst Technol 214–218
7. Cho YJ, Kim KH, Hwang SH, Park S (2006) A miniature UWB planar monopole antenna with 5 GHz band-rejection filter and the Time-Domain Characteristics. IEEE Trans AP 54(5):1453–1460
8. Jusoh M, Jamlos MF, Kamarudin MR, Malek MF, Romli MA, Zulkifli MS (2012) A reconfigurable ultra-wideband compact tree-design antenna system. Progr Electromagnetics Res C 30:131–145
9. Dong YD, Hong W, Kuai ZQ, Yu C, Zhang Y, Zhou JY, Chen JX (2008) Development of UWB antenna with multiple band notched characteristics using half mode substrate integrated waveguide cavity technology. IEEE Trans Antenna Propag 56(9):2894–2902
10. Kim DO, Ki CY (2010) CPW–fee UWB antenna with triple band notch function. Electronic Lett 46(18):1246–1248
11. Kumar D, Singh T, Dwivedi R, Verma S (2012) Compact monopole CPW fed dual band notched square-ring antenna for UWB applications. In: 2012 fourth international conference on computational intelligence and communication networks
12. Ma T, Wu S (2007) Ultrawideband band-notched folded strip monopole antenna. IEEE Trans AP 55(9):2473–2479
13. Chung K, Kim J, Choi J (2005) Wideband microstrip fed monopole antenna having frequency band-notch function. IEEE Microw Wireless Components Lett 15(11):766–768
14. Deng H, He X, Yao B, Zhou Y. Compact band-notched UWB printed square-ring monopole antenna. doi:978-1-4244-2193-0/08/2008. IEEE
15. Chu Q-X, Yang Y-Y (2008) A compact ultrawideband antenna with 3.4/5.5 GHz dual band-notched characteristics. IEEE Trans Antennas Propag 56(12)
16. Mak KM, Lai HW, Luk KM, Ho KL (2017) Polarization reconfigurable circular patch antenna with a C-shaped. IEEE Trans Antennas Propag 65(3):1388–1392
17. Choudhary H, Singh T, Ali KA, Vats A, Singh PK, Phalswal DR, Gahlaut V (2016) Design and analysis of triple band notched micro-strip UWB antenna. Cogent Eng 3:1–13 (A part of Taylor & Francis Group)
18. Singh T, Chaodhary H, Avasthi DV, Gahlaut V (2017) Design & parametric analysis of band reject Ultra-Wideband (UWB) antenna using step impedance resonator. Cogent Eng 1:1–16 (A part of Taylor & Francis Group)

Prediction Model for Breast Cancer Detection Using Machine Learning Algorithms

Nishita Sinha, Puneet Sharma, and Deepak Arora

Abstract Cancer is termed as a fatal genetic disease that can change the behavior and growth of human cells. Breast cancer is the most prevalent cancer occurring among women in today's world. According to an Indian survey, the death rate caused by cancer has increased by approximately 25% from the past few years. Breast cancer proves out to be deadly in majority of the cases and is the major reason for death among women. Tumor is a term used for the abnormal growth of cell tissues in the human body. Benign and Malignant are the two types of cancer tumors. The main objective of this research work is to prepare a report on the percentage of people suffering with cancer tumors using machine learning algorithms. The authors have taken advantage of the most efficient machine learning algorithms to develop models for prediction which will detect breast cancer occurring rate. Authors have used a breast cancer data set which consists of 567 rows of 30 different attributes of cancer characteristics out of which benign and malignant data has been taken as the target attribute. A prediction has been done to calculate the percentage of people affected with the benign and malignant tumors and pictorial results have been displayed as results. In this proposed approach, the authors have used supervised learning algorithms like K-Nearest-Neighbors, Naïve Bayes, Decision tree, and support vector machine prediction models for an unbiased estimate. The results indicated that K-Nearest-Neighbors is the best predictor with 91.6% accuracy. This model will be capable of providing a higher degree of accuracy compared to already existing literature work. The authors have compared their experimental results with the existing approaches and found satisfactory.

N. Sinha · P. Sharma (✉) · D. Arora
Department of Computer Science & Engineering, Amity University, Lucknow Campus, Lucknow, UP, India
e-mail: puneetgrandmaster@gmail.com

N. Sinha
e-mail: nsinha201998@gmail.com

D. Arora
e-mail: deepakarorainbox@gmail.com

V. Singh et al. (eds.), *Computational Methods and Data Engineering*, Advances in Intelligent Systems and Computing 1227,
https://doi.org/10.1007/978-981-15-6876-3_33

Keywords Cancer tumor detection · Malignant and benign · Supervised machine learning

1 Introduction

Tumor is a term used for the abnormal growth of cell tissues in the human body. Benign and Malignant are the two types of cancer tumors. The benign tumor is considered to be non-cancerous as it does not spread in the body. The Malignant tumor is considered to be highly cancerous and spreads rapidly in the other parts of the body. They are not localized and often results in death. We can call tissue cancer when it becomes malignant. It has a very high growing rate in India. According to ICMR (India Council of Medical Research), the rate of death caused by cancer has increased by approximately 25% from the past few years.

Cancer is very common after a certain age. According to a survey, people above the age of 50 are suffering from this disease. Among men, the common places for cancer are throat, lung, lip, oral, and neck. Among women, the highest rate goes to breast cancer and ovarian cancers. The only cure for cancer is detecting it at an early stage and giving it the right treatment. The treatment for cancer includes chemotherapy, radiation therapy, surgery, etc. Successful treatments can remove all cancer-causing symptoms. Cancer has four stages. The first two stages can be treated easily with no huge risk but the other two stages are highly risky because when it reaches a higher stage, there is a high chance that it has already been spread in the other parts of the body causing secondary cancers. Cancers that have spread are very risky to treat and may often result in the death of the person.

In this research topic, a prediction model for cancer detection, the authors have designed a prediction model, which will detect the type of tumor human body has. This topic has been taken because the authors want to implement engineering technology in the medical field. Machine Learning has made this possible. It has excellent algorithms, which can implement the concepts of machine learning on a given dataset and gives an evaluated result, which can justify the interpretation of the dataset. This technology uses Jupyter Notebook of Anaconda Navigator as python IDE. This prediction model will be able to do the following things.

- It tells the type of tumor present in the human body.
- It detects whether the tumor is cancerous or non-cancerous based on malignant and benign data.
- It uses K-nearest-neighbor to predict the amount of people affected by the disease.
- It plots a histogram of the percentage of people with benign and malignant tumor of training and testing datasets.

2 Literature Study

Several people have given proposals in the market on cancer prediction models. Some of these proposed models are mentioned below-

A breast cancer prediction model incorporating familial and personal risk factors by Tyrer et al. [1], research work on factors such as personal genetic and family history of women, which could cause cancer. It contains a detailed genetic analysis of human genes BRCA1 and BRCA2. People who have these genes will tend to have cancer at a young age. In this research work, the authors have taken the family history of women with Naïve Bayes theorem to produce the chances of carrying any type of genes which will predispose to breast cancer. This model is difficult to build up to find the data of a group of carriers with their personal medical history and family history is not available. This model is based on different datasets published with very low accuracy and high amount of errors.

Shravya et al. [2], have proposed an automatic diagnostic system for cancer detection where they have taken breast cancer datasets from UCI repository and have diagnosed benign and malignant tumor. This research work contains classification techniques of machine learning and is a relative study on the implementation of models using Logistic Regression, Support Vector Machine (SVM), and K-Nearest Neighbor (KNN). The best accuracy is given by SVM model with 92% accuracy. Gupta et al. [3], have prepared a research paper on breast cancer diagnosis and prognosis. In this research work, a dataset of breast cancer has been taken which categorizes benign from malignant breast tumors and the Breast Cancer Prognosis will predict the chances of having cancer in the future of the patients who have already been treated. This research paper is a summary of all the present researches being carried out using the highly efficient data mining techniques which will modify and evaluate the results of breast cancer diagnosis and prognosis. Cruz et al. [4], have done a research paper on Applications of Machine Learning in Cancer Prediction and Prognosis where the authors have shown how useful and efficient machine learning algorithms can be when it comes to cancer prediction. They conducted a large survey involving many people from around the world and surveyed about all the kinds of machine learning methods which are followed, the types of datasets which are used and integrated, and the results of all the methods used in cancer prediction and cancer prognosis. After surveying all the best results and the existing works, it is clearly shown that machine learning algorithms can be followed to improve the efficiency of models up to 15–20% for predicting cancer susceptibility, recurrence, and mortality rate of cancer in humans. Chaurasia et al. [5], have done research work on the prediction of benign and malignant breast cancer using data mining techniques in which the authors have used machine learning algorithm, Naïve Bayes, to develop a model to predict cancer using a huge amount of dataset. The dataset consists of 683 attributes. The results showed that the model of Naive Bayes is the best model predictor with 97.36% accuracy on the dataset used. Cruz et al. [6], have done research work on Applications of Machine Learning in Cancer Prediction and Prognosis where the authors have shown how useful and efficient machine learning algorithms can be when it comes to cancer prediction. They conducted a large survey involving many people from around the world and surveyed about all the kinds of machine learning methods which are followed, the types of datasets which are used and integrated, and the results of all the methods used in cancer prediction and cancer prognosis. Ravi Kumar et al. [7], have done research work on An Efficient Prediction of Breast Cancer Data using data mining techniques where the authors have used WBC data and predicted breast cancer using support vector machine. The authors have used Weka Software (The Waikato Environment for Knowledge Analysis) with a dataset of 200 patients. The results proved that SVM is the best predictor with the highest accuracy. Shah and Jivani [8], have done research work on Comparison of Data Mining Clustering Algorithms where the authors have done a study on the comparison between different hierarchical and density-based methods of clustering. This research paper is a summary of all the present researches being carried out using the highly efficient data mining techniques which will modify and evaluate the results of breast cancer diagnosis.

3 Methods

Cancer, if not detected at an early stage, can be fatal. Screening tests may help diagnose cancer possibilities in a person. There are various cancer diagnosis tests. First one is the Imaging or Radiology test, which images the diagnosed cancer in the body. Second one is the CT scans, which use the x-ray techniques and gives a clear sectional image of the tumor in the body. Ultrasounds and mammograms are the third ones which are very useful in detecting cancers.

However, these tests may be very painful and may take much longer time to diagnose cancer. Therefore, there should be a machine, which can detect cancer without the above-mentioned problem. In this research paper, the authors have tried to implement a prediction model, which can detect cancer with the help of cancer tumors (Benign and Malignant). A machine, which takes cancer characteristics as

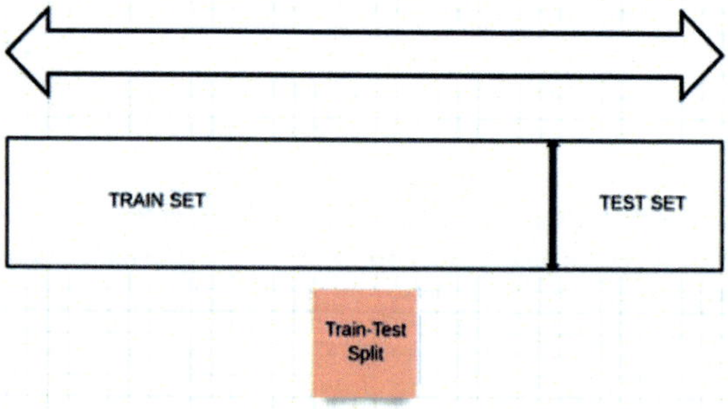

Fig. 1 Train-test split

input and gives benign and malignant tumors as output. The first and foremost step is to collect the data which has all the necessary attributes of cancer with a target attribute which has benign and malignant attributes in it. The second step is to clean and scale the dataset. Data cleaning is very important as it improves the quality and productivity of the data. The third step is to split the data into X and Y. In X, we provide all the other attributes of cancer, and in Y; we provide our target attribute, which has benign and malignant values in it. The fourth step being the most important one is to split the data into two parts as training and testing datasets. It is the most important step because the training data contains an output that is already known and the model learns on this already known data in order to be generalized by other data later on. The testing dataset is the predicting dataset which will be done on this subset. After the data is divided into train and test sets, the problem of overfitting and underfitting should be avoided. The fifth and last step is to apply a machine learning algorithm on the model, which will help the model learn from the data provided and give an unbiased, and précised output with decent accuracy. In Fig. 1, the division of the data set into training and testing sets has been shown.

4 Implementation

In this research work, authors have used machine learning to build up cancer prediction models. Machine Learning using python is the simplest technique for data modeling. Python has many developed modules and libraries, which help in the processing of data. As mentioned earlier, machine learning involves the learning of a computer on a given data set without being explicitly programmed. For example, if we provide 50 pictures of a chair to the computer and 50 pictures which are not of a chair and then tell the computer whether the picture is a chair or not. Then if we show the computer a picture, it should be able to distinguish the pictures of the chair.

In this research work, the authors have used Jupyter Notebook of Anaconda Navigator for machine learning using python. Jupyter notebook is a browser-based IDE of python. To learn python basics Jupyter notebook is very helpful as it has all the python libraries preinstalled in it. The authors have used Pandas, NumPy, scikit learn, and matplotlib library of python for this research work. Panda's library is useful for data analysis and creating data frames. NumPy is used for fast mathematical calculations like on arrays. It is a very efficient library for homogeneous data. Scikit learn library is very useful for data modeling. It has efficient tools for machine learning and graphical modeling including classification and regression. Matplotlib is again a statistical modeling library. Matplotlib.pyplot is a graphical command used to make matplotlib work as MATLAB. It is purely used for plotting and making figures.

To implement the cancer prediction models, a dataset with all necessary attributes of cancer is required. After scaling and cleaning the data the authors have used supervised learning algorithms like K-Nearest-Neighbors, Naïve Bayes, and Decision tree models for predicting the benign and malignant tumors. The final output shows a graph with the percentage of people affected with each of the three models. K-Nearest-Neighbors have the best accuracy with 0.91%.

4.1 K-Nearest Neighbors

KNN algorithm is one of the simplest algorithms in the field of supervised learning in machine learning. As mentioned earlier, supervised learning works on regression and classification and so does KNN. KNN algorithm principle works on the minimum or smallest distance from the query instance point to the training samples to determine the K-nearest neighbors. It classifies a new data point into the target data point, depending on the characteristics of its neighboring data points.

KNN algorithm is used in this research work for cancer detection because KNN is very useful in pattern recognition. It classifies similar feature patterns as one class. Here, the user assigns a value to k and find the nearest neighbors of it. For example, if the value of $k = 3$ then the model will find the three nearest neighbors. Here, in this work, the value of k is set as 1 for the best results. It was very useful for graphical representation of the training and testing value data set. It gave the best model which was 92% accurate. Data modeling using KNN algorithm is very easy. The user just has to clean and scale the data and then fit it into our model.

K-Nearest neighbor is calculated on its distance formulae. There is only one distance formula for KNN that is Euclidean Distance Formulae

$$\sqrt{\sum_{i=1}^{k}(xi - yi)^2}$$

where k is the number of its nearest neighbors, x and y are the distance points from the query instance to the training samples. This distance formula is only applicable to continuous variables.

5 Results and Discussion

Here, the authors have created a data table, which clearly shows the algorithms implemented with their accuracies. The Decision tree algorithm has an accuracy of 87.6%, support vector machine has an accuracy of 90.0%, Naïve Bayes has an accuracy of 75.6% and the K-Nearest Neighbor is the most accurate one with 90.9% accuracy. The confusion matrix for KNN is shown in Fig. 2.

Fig. 2 Confusion matrix

```
Out[45]: array([[ 68,   9],
                [ 10, 122]])
```

Fig. 3 Confusion Matrix Table

	BENIGN PREDICTION	MALIGNANT PREDICTION
BENIGN ACTUAL	BP	MN
MALIGNANT ACTUAL	BN	MP

Fig. 4 Formula for CM

ACCURACY= BP +MP / BP+BN+MP+MN

The confusion matrix is calculated to judge the accuracy of the classifier model. The confusion matrix shows how much the model made is able to predict the output.

Here, in this Fig. 3, confusion matrix working table is shown. The letter P means positive and the letter N means negative. Taking the two classes as benign and malignant, we have in Fig. 4.

BP—The observation is benign and is predicted as benign.
MN—The observation is benign but is predicted as malignant.
BN—The observation is malignant is also predicted as malignant.
MP—The observation is malignant but is predicted as benign.

Therefore, for the KNN model, the confusion matrix accuracy = 90.9%. The authors have split the breast cancer dataset into training and testing sets as a (30–70) split. A pie chart has been shown to clear the splitting of data into X_test and y_test. Splitting of dataset is important for the accuracy of the model. After the cleaning and scaling process is done, the data is split into two parts and then fitted into the model for the learning process. The model learns on its own from the training data provided to it. Here is a clear graph is shown in Fig. 5, for the splitting of the data.

Authors have made a data table that clearly shows the algorithms used, how to apply these algorithms, their calculated accuracy, and the final output as a conclusion. Here, is the table as shown in Fig. 6.

As mentioned earlier, supervised learning works on regression and classification and so does KNN. KNN algorithm principle works on the minimum or smallest distance from the query instance point to the training samples to determine the K-nearest neighbors. It classifies a new data point into the target data point, depending on the characteristics of its neighboring data points.

Since KNN is the best model predictor therefore, the authors have made a bar graph using the KNN model which shows the percentage of benign training output and benign testing output and in the same way malignant data output. The benign model is 94% accurate and the malignant data is 87% accurate. Hence, the authors

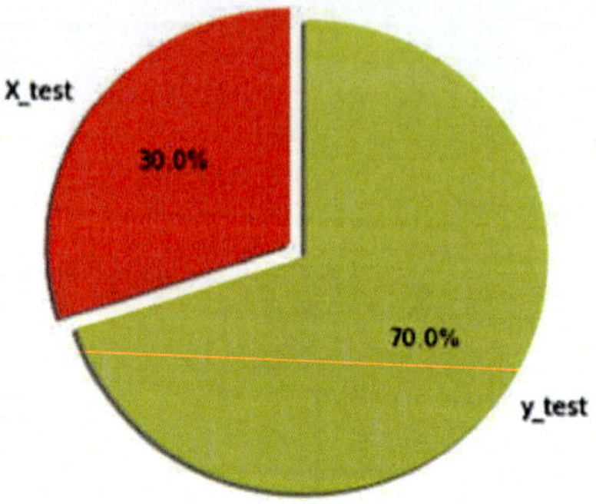

Fig. 5 Data split pie chart

ALGORITHM	METHOD	ACCURACY	CONCLUSION
DECISION TREE	A **decision tree** is a graph that uses a branching **method** to illustrate every possible outcome of a **decision**.	87.6%	Needs more accuracy.
SUPPORT VECTOR MACHINE	The algorithm creates a line or a hyperplane which separates the data into classes.	90.0%	accurate.
NAIVE BAYES	a Naive Bayes classifier assumes that the presence of a particular feature in a class is unrelated to the presence of any other feature.	75.6%	It is not an accurate model.
K-NEAREST NEIGHBOR	It **works** based on minimum distance from the query instance to the training samples to determine the K-nearest neighbors.	92.7%	Most Accurate

Fig. 6 Algorithms result

can conclude that the model is an accurate model based on the graph as shown in Fig. 7.

6 Conclusion

Today, machine learning techniques and algorithms are highly advanced and efficient for solving many real-world problems. It is the hottest topic in the IT industry today. Machine learning makes a software system learn on its own and makes it

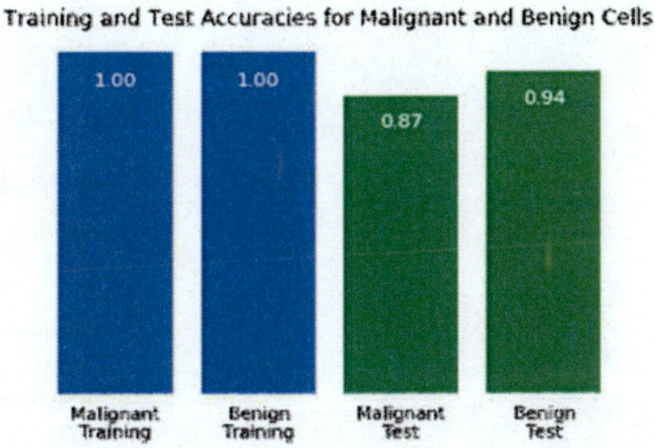

Fig. 7 Result obtained

capable of predicting outcomes. It is being widely used for data interpretation and for storing, extracting, manipulating, and getting data from data warehouses. Supervised machine learning algorithms have been largely accepted and used by everyone today. It has certain advantages and disadvantages. The disadvantage of supervised learning algorithms is that it is very costly to implement them in systems when the data is too large. This is the reason behind the fact that a large amount of effort and high expenses are involved when we try to obtain large labeled data sets. Therefore, machine learning gives a solution to the above-mentioned problem of reducing the labeling expenses by labeling only the most useful instances for learning.

In this research work, there is a model, which works on supervised learning algorithms and predicts a cancer tumor that is benign and malignant. Machine learning works best on labeled data so here the authors have taken data of breast cancer with all the necessary parameters. Machine Learning is useful in all grounds for all real word problems no matter what the situation is. The K-nearest model made is tested and proves to be the most efficient one. As mentioned earlier, supervised learning works on regression and classification and so does KNN. KNN is a non-parametric, lazy learning algorithm and its purpose is to use a database in which the data points are separated into several classes to predict the classification of a new sample point. Its principle works on the minimum or smallest distance from the query instance point to the training samples to determine the K-nearest neighbors. It classifies a new data point into the target data point, depending on the characteristics of its neighboring data points shown in Fig. 8.

7 Future Work

Although this work is very new and basic right now and limited too. The authors have some future plans for it. In the future, they intend to implement a model that will show the mathematical work done by the algorithms. The performance of the algorithms will then be evaluated using various evaluation metrics. The authors are

Fig. 8 K-Nearest neighbor

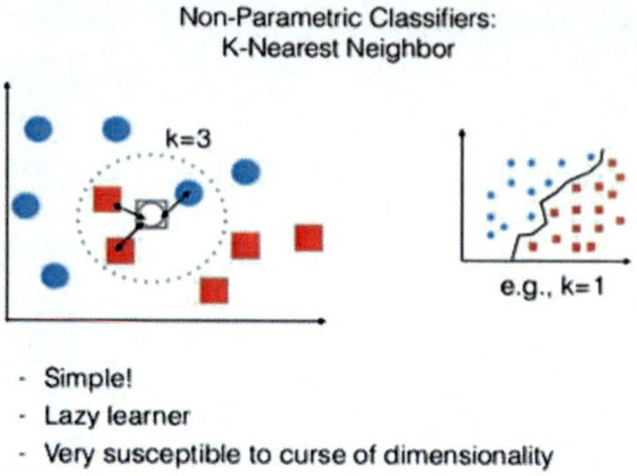

working on mixing the algorithms used in this work and are trying to make a hybrid algorithm model, which will show more accuracy. In the ground of machine learning, the work for the future involves the usage of all the machine learning techniques in each possible way.

Machine Learning has helped all the big industries and hospitals in a big way. It has made work a lot easier than it was before. All the industries are implementing machine learning and are focusing their efforts on healthcare. Since machine learning is very useful in making useful decisions on healthcare and biomedicines, all the hospitals have widely accepted it. Various researchers have done research on healthcare using machine learning because it will solve the problems of health-related issues.

References

1. Tyrer J, Duffy SW, Cuzick J. A breast cancer prediction model incorporating familial and personal risk factors. Department of Epidemiology; Mathematics and Statistics; Cancer Research U.K.; Wolfson Institute of Preventive Medicine; Charterhouse Square; London EC1M 6BQ; UK
2. Shravya C, Pravalika K, Subhani S. An automatic diagnostic system for cancer detection. Int J Innov Technol Exploring Eng (IJITEE)
3. Gupta S, Kumar D, Sharma A. Breast cancer diagnosis and prognosis. Banasthali University India
4. Cruz JA, David S. Applications of machine learning in cancer prediction and prognosis. Wishart of Department of Biological Science and Computing Science, University of Alberta Edmonton, AB, Canada
5. Chaurasia V, Pal S, Tiwari BB. Prediction of benign and malignant breast cancer using data mining techniques. Department of MCA, VBS Purvanchal University, Jaunpur, UP, India
6. Cruz JA, Wishart DS. Applications of machine learning in cancer prediction and prognosis. Departments of Biological Science and Computing Science, University of Alberta Edmonton, AB, Canada T6G 2E8
7. Ravi Kumar G, Ramachandra GA, Nagamani K. An efficient prediction of breast cancer data using data mining techniques. Department of Computer Science & Technology S K University, Anantapur, AP, India
8. Shah C, Jivan A. Comparison of data mining clustering algorithms. Information Technology Department, Shankersinh Vaghela Bapu Institute of Technology, Gandhinagar, India

Identification of Shoplifting Theft Activity Through Contour Displacement Using OpenCV

Kartikeya Singh, Deepak Arora, and Puneet Sharma

Abstract Ambiguous surveillance and analysis of the theft activity are one of the main reasons for inadequate security management of a place. The place can be our home, office, shops, etc. In this research work, an automated system has been proposed, which uses contour displacement points resulting in eternal surveillance for detection of the theft activity through CCTV camera. The algorithm developed in this work, utilises object detection technique in presenting real-time detection characteristics of the stolen object and identity of the theft in the form of camera frames. Through massive literature work, it is found that ideas adopted in developing various intelligent surveillance systems are majorly theoretically driven and lacks with required strong output modules. This research work is intended to demonstrate the implications of a convolutional neural network towards the prediction of visual objects from a previously trained data set and then form a suitable model for identifying any theft activity. For the experimental purpose, authors have used OpenCV library [1], which tends to be a very effective component to accomplish the entire work objectively. At the last, results have been analysed and all results and developed conditional algorithm.

Keywords Convolutional neural network · OpenCV · Machine learning · Contour displacement

K. Singh (✉) · D. Arora · P. Sharma
Department of Computer Science & Engineering, Amity University, Lucknow Campus, Lucknow, UP, India
e-mail: kartiksingh538@gmail.com

D. Arora
e-mail: deepakarorainbox@gmail.com

P. Sharma
e-mail: puneetgrandmaster@gmail.com

V. Singh et al. (eds.), *Computational Methods and Data Engineering*,
Advances in Intelligent Systems and Computing 1227,
https://doi.org/10.1007/978-981-15-6876-3_34

1 Introduction

Theft detection is a very complicated task, especially when the surveillance system is not very reliable to the inputs. The techniques used in this paper are to first observe and detect the objects by using yolo database. Then identification of the movement of an object by deriving contour points of every moving object. A short algorithm differentiates the contours of a person and an object. This algorithm helps us in finding a human activity with respect to a placed object. If an object is detected moving by lifting or displacing the actions of a person it sends an alert to a function, which starts tracing the contour graphs of the person and the object. After analysing some results, the authors got some strong results, which identifies the activity to be a theft scene. The algorithm proposed in this research work the algorithmic channelling scheme is of interest in itself, it proves well with its high efficiency and proven authenticity. The authors have observed that most of the surveillance systems are sensing-based which could be vulnerable at times. Therefore, we have maintained proper transparency of the intelligent system, which allows it to output the best results. The system is compatible with the hardware components like raspberry pi with a camera. It inputs the footage obtained either from CCTV live or from a recorded one. This work does not only contain detection technique but also analyse the mechanism, which makes this system very reliable. The sample activity performed in this project is of a boy walking into the room and stealing a backpack, which has been already kept in the room. The breakdown of the whole activity is explained in the Sect. 3. In continuation of which we can see the outputs and the end results in Sect. 4.

2 Background

Rapid development in the industries and businesses has increased the need for reliable surveillance systems, which not only able to capture the human activity, but also to automatically detect any sceptical activity happening under the surveillance by using intelligence. The use of security cameras has been increased expeditiously over the past few years. Thus, the number of smart surveillance systems have been proposed for security purposes. In [7] multiple cameras are used to fetch the images from different regions required to monitor the suspicious activity using colour image comparison. In the proposed system, while developing authors have chosen each and every loopholes possible in the account. Surveillance system has been introduced in [6] uses canny edge detection for the detection of theft activity, while in our system it inputs the activity details and alerts according to the suspicious act. The major upper hand of the proposed system is its real-time functioning, it diminishes the alert if the thief puts back the stolen item anywhere inside the room (Fig. 1).

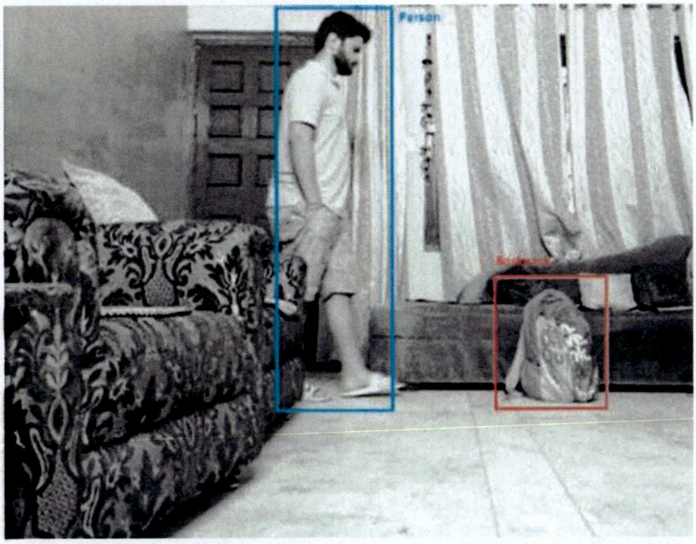

Fig. 1 Gaussian image for the object detection

3 Divisional Components of the Proposed System

3.1 Real-Time Object Detection

In the first step, author have used a mounted web camera for the live recording of the frames. In Fig. 2, the author has used Gaussian Distributed Framing of an image. Yolo database has been used for the prediction of objects present in the frame. Yolo database contains more than 300-trained features, which is quite reliable to work in out CCTV surveillance. The Gaussian blur image predicts the object present inside the frames and sends the exposition to our algorithm for further inspection of the activity.

3.2 Contour Classification

After detection of objects present inside the frame, contouring of the objects has been introduced inside the algorithm. These contours have been distributed over the edges of moving objects (see Fig. 2), which as a whole act as a motion detection system. The main objective of this part is to detect the motion and define the contours obtained on the edges of the object. From Fig. 3, it can be easily determined all changes in frames when an object starts moving (contouring). Contouring strategy, which has been used in this paper, is unique and has proved to be the foremost technique for the determination of the moving edges. The part of the algorithm that detects the

Fig. 2 Contour edge detection

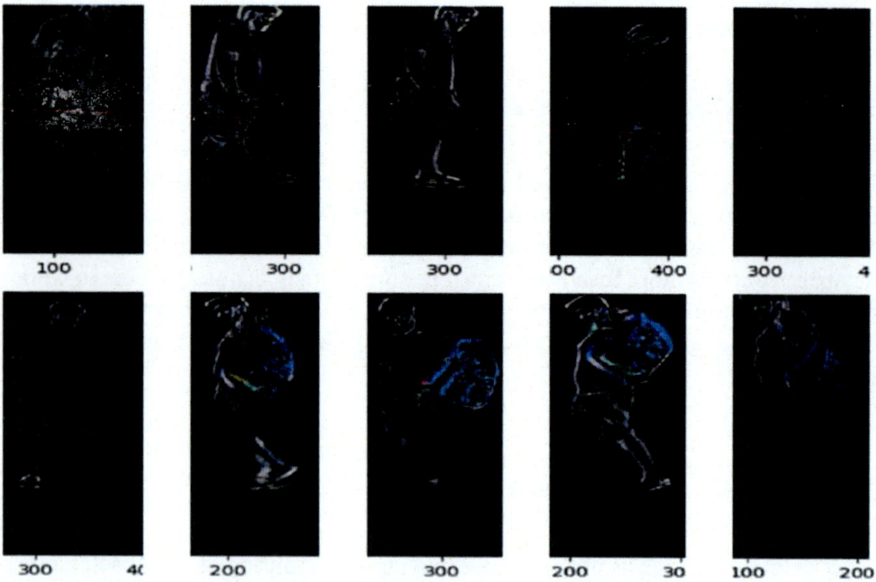

Fig. 3 Frame changes while depicting a motion

Table 1 Results obtained at the end of the activity

Time stamp (s)	Distance of moving person and object	Distance of contours
5	80	10
10	65	3
15	60	8
20	22	4
25	9	6

contour point's functions, only after the recognition of the objects present inside the frame, which helps us to identify clearly about what object, is under threat of being stolen and which person is stealing it. By using the contour concept, we were able to identify the difference between two adjacent frames while rolling out a video. This change in frame intensities drew us to predict that the change of intensity results in a movement detection. We used this classification for movement detection. Our defined algorithm detects a motion only if the contouring entity is a person. The classification of person and other objects can be done using object detection, which we have implemented parallel to the contouring detection.

Here, Fig. 2 represents the contour movements, which have been obtained by our algorithm. These contours depict the movement of a person. According to our activity performed in this project, a person is moving inside a frame, while moving the previous set of intensities of a particular point is changing with the next intensity. This changing of the intensities concept detects the motion and verifies the moving object with our data set, which has already identified the objects. If it is a "Person", our algorithm moves to the next step and performs all other remaining actions. This regulates the first two steps of our system and prepares the inputs and permission for the next step of our system (Table 1).

3.3 *Formulation and Experimental Setup*

The next step, Measuring the distance of the moving object and the object place, this results as the most important part of our paper. The formula has been explained in the Sect. 3. In order to understand the concept, let us suppose an object OBJ1 kept inside a room if a person walks into the same room our system identifies it and assigns it as OBJ2. Now on observing, if we find a gradual decrease in the distances of OBJ1 and OBJ2 we can conclude that the distance between these two has been decreased. This observation presents us with the theory of reverse fall in the distance graph (as shown in Fig. 5).

Figure 3 represents the workflow of the system developed. As we can see, we start the workflow by capturing the frames from the camera. The frames obtained are being analysed, If the frames are equal, the process goes idle until there is a change in two adjacent frames (shown in Fig. 3).On detecting a change, A database has been maintained and the frames are forwarded to the Gaussian blur conversion [5]. After passing through the object and contour detection channels, the frames finally reach our developed algorithmic system (Sect. 3) where the conditions have been matched, if found similar, it detects the theft activity. After bringing every step in action, some graphical representations have been formed for the analysing of the theft activity in a room. These results are stored in our (firm's) database system from which the results are obtained. On obtaining a theft activity, an alert message is sent to the owner (Fig. 4).

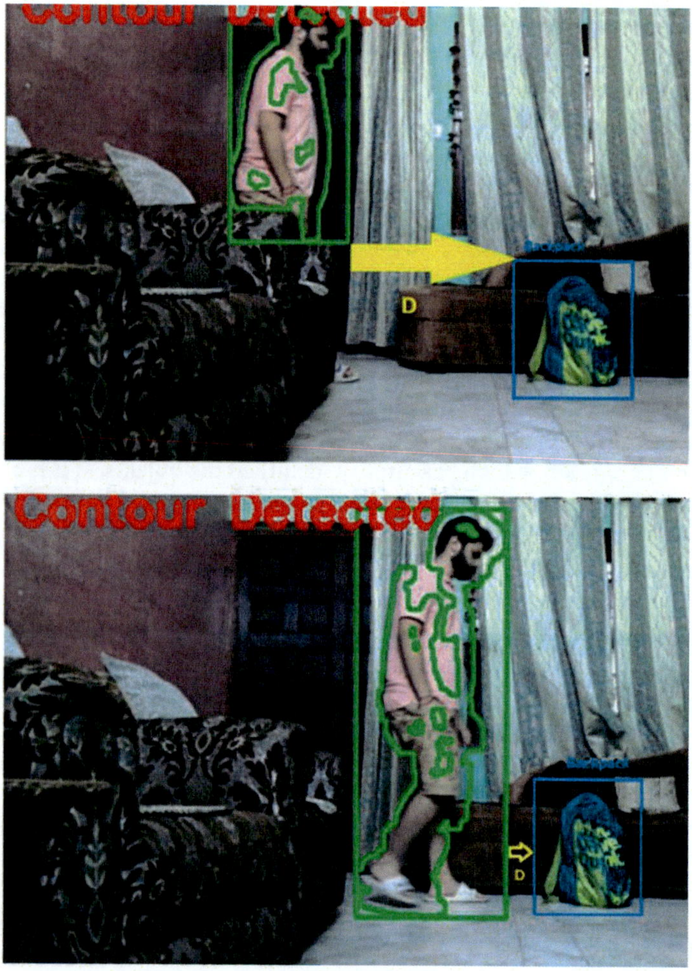

Fig. 4 Distance (*D*) of the moving contour object and the placed object

4 Results and Discussion

This section comprises of the results and their assessment for the development of our conditional algorithm, which takes input from these results and detects the theft activity.

Figure 4 shows the graphical representation of the distances obtained within a time interval of 25 s. The distances obtained are between the moving entity (Person) and the object (Backpack). From the representation, we can observe that with the increase in time the distance of moving objects is decreasing. This tracking of distances will continue unless the system gets an alert of the object being moved from its place. This representation will be taken into account for the manipulation of our algorithm, in which an identifier compares these results with the results obtained in Fig. 7 and delivers us the output as shown in Fig. 8 (Fig. 6).

The whole architecture is based on the logic of transmission. Using these two results obtained, the system finds if the object is back at its initial position or not. If the stolen object is not at its exact initial place, it means that the object has been stolen.

From Sect. 3.2, authors have obtained the contours of moving objects. Here, authors have used the distance formula for obtaining the distance between moving objects and the object kept (Here, in our activity it is the person and the backpack).

$$dx = \text{centers}[0][0] - \text{centers}\,[1][0]$$

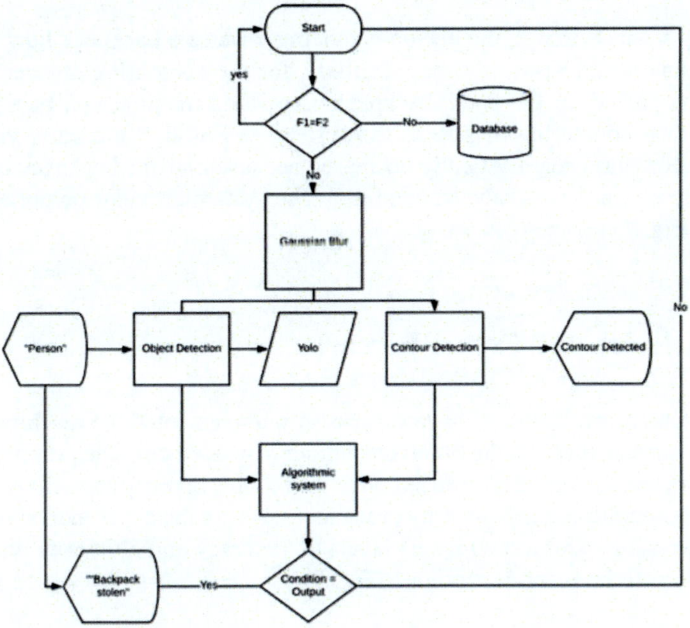

Fig. 5 Flow diagram of the architecture

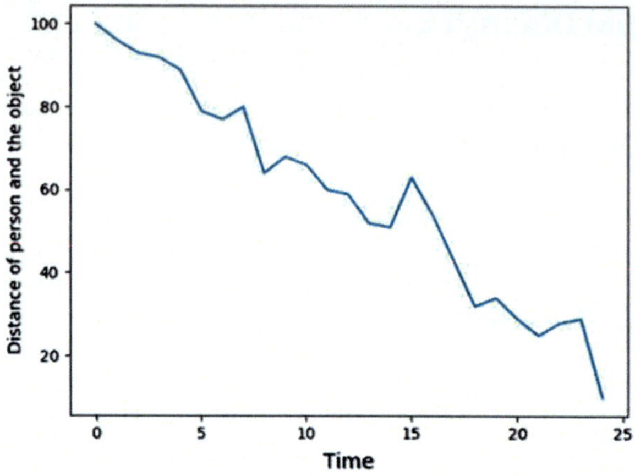

Fig. 6 Distance tracking of person and object

dy = centers [0][1] – centers [1][1]
D = np.sqrt(dx * dx + dy * dy)

here,

dx = co-ordinated of the moving person towards the backpack
dy = co-ordinates of the backpack
D = Distance of the person and the backpack.

As Fig. 5 shows that if, the person is moving towards a backpack how the graph is decreasing with respect to time. Similarly, for the contouring objects, once the person has picked up the back, backpack also starts moving, i.e., backpack also starts sending contouring outputs to the system. In Fig. 6, it is clearly visible that the value of D has been drastically decreased because now the backpack is with the person. Hence, in these analysed conditions the system provides output as an alert shown in Fig. 7.

5 Conclusion and Future Remarks

In this research, work authors have successfully implemented the machine learning model for surveillance system based on contour displacement. The proposed system in this work can further be utilised as one of the reliable security surveillance systems into the households or other security prone areas. As per the experimental results, the proposed system exhibit the highest levels of accuracy and efficiency with overall better theft activity identification rate (Fig. 8).

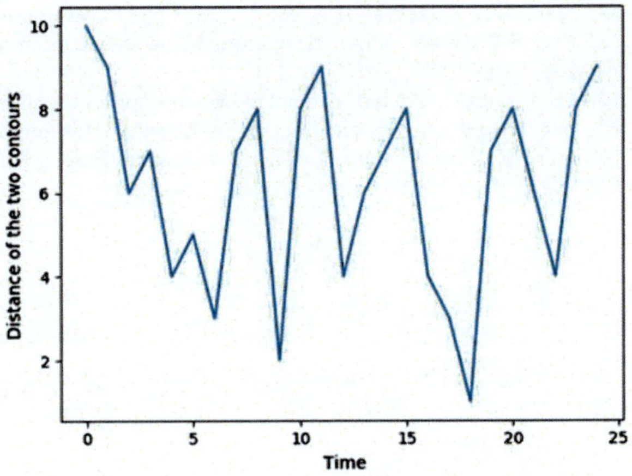

Fig. 7 Relational tracking of person and object

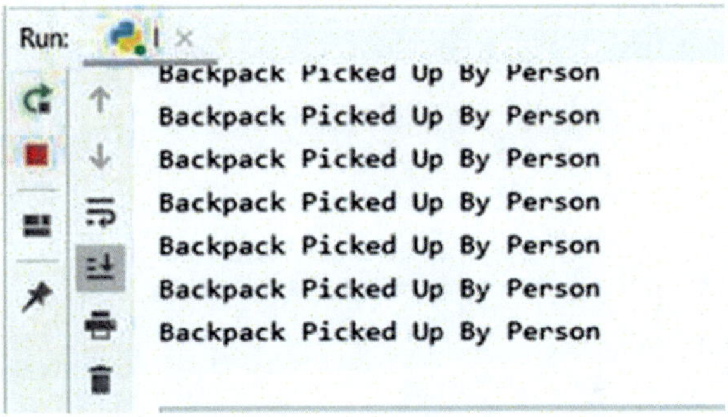

Fig. 8 Output of the proposed system

References

1. OpenCV (2009) Open source Computer Vision library. http://opencv.willowgarage.com/wiki/
2. Facial Recognition using OpenCV, Shervin EMAMI1, Valentin Petruţ SUCIU2. www.jmeds.eu
3. Anuar A, Saipullah KM, Ismail NA, Soo Y. OpenCV based real-time video processing using android smartphone. IJCTEE 1(3)
4. Hussein M, Abd-Almageed W, Ran Y, Davis L. Real-time human detection in uncontrolled camera motion environments. Institute for Advanced Computer Studies University of Maryland
5. Kushwaha A, Mishra A, Kamble K, Janbhare R, Prof. Amruta Pokhare5 "Theft Detection using Machine Learning"
6. Munagekar MS. Smart surveillance system for theft detection using image processing
7. Mandrupkar T, Kumari M, Mane R. Smart video security surveillance with mobile remote control
8. Ivanov YA, Bobick AF (1999) Recognition of multi-agent interaction in video surveillance, {ICCV} (1)
9. Collins R et al (2000) A system for video surveillance and monitoring, VSAM FinaL Report, Carnegie Mellon Univ., Pittsburgh, PA, Tech. Rep. CMU-RI-TR-00-12, May 2000
10. Hampapur L, Brown J, Connell M, Lu H, Merkl S, Pankanti A, Shu Y (2004) The IBM smart surveillance system, demonstration, Proc. IEEE. CVPR, Tian
11. Moon H-M, Pan SB (2010) A new human identification method for intelligent video surveillance system. 978-1-4244-7116-4/10/$26.00 $©2010 IEEE

12. Mamata S. Kalas real tıme face detectıon and trackıng usıng opencv
13. Biradar O, Bhave A. Face recognition based attendance monitoring system using Raspberry-pi and OpenCV
14. Lin S-H (2000) An introduction to face recognition technology. (Ph.D. IC Media Corporation), Informing science special issue on multimedia informing technologies part 2, vol 3, no 1
15. Jain AK (2004) Handbook of face recognition. Beijing China, December 2004

Proof of Policy (PoP): A New Attribute-Based Blockchain Consensus Protocol

R. Mythili and Revathi Venkataraman

Abstract Healthcare Blockchains (HBC) are grooming nowadays due to the requirements of flexible Electronic Health Records (EHR) management, privacy needs, and controlled medical data access. Blockchain technology very much suits these requirements and support in various healthcare data access control methods. More focus on improved block generation time gives the expected blockchain performance. Block generation time can be significantly improved by the consensus algorithms used by the system. Traditional consensus algorithm of Proof of Work (PoW) and Proof of Stake (PoS) require heavy computational resources and hence by computational overheads. The proposed system introduces a new privacy-preserving HBC consensus protocol as a consequence of member attribute values and significantly reduced computational resources. The new attribute-based consensus algorithm for HBC called Proof of Policy (PoP) is constructed using Attribute-Based Ring Signature (ABRS). This signature agrees on the stake of attribute-based ring membership and can be considered as an extension of Proof of Stake (PoS). It is also found that the proposed PoP consensus algorithm requires less computational resources than popular PoW and PoS.

Keywords Healthcare blockchain · Privacy · Consensus algorithms · Attribute-based ring signatures

1 Introduction

Blockchain (BC), the foundation of Bitcoin is on the rise at an astonishing swiftness over the past few years. Satoshi and Nakamoto [1] invented the first digital cryptocurrency Bitcoin in 2008. Ethereum [2] is released as second public blockchain not only

R. Mythili (✉)
SRM Institute of Science and Technology, Ramapuram, Chennai, India
e-mail: mythilir2@srmist.edu.in

R. Venkataraman
SRM Institute of Science and Technology, Kattankulathur, Kancheepuram, India
e-mail: revathin@srmist.edu.in

V. Singh et al. (eds.), *Computational Methods and Data Engineering*, Advances in Intelligent Systems and Computing 1227,
https://doi.org/10.1007/978-981-15-6876-3_35

as currency, can record other assets because of its decentralized, time-stamped digital trust. Blockchain [3] also helps in eliminating the limitations of traditional centralized ledger systems as it uses third parties. Blockchain is proved to be the innovative peer-to-peer communication technology for establishing trust among the nodes in various applications of science, engineering, and business through decentralized verification and consensus mechanisms [4].

Healthcare Blockchain (HBC) is the one such domain which delivers the sharing of sensitive medical data in the form of distributed Electronic Health Records (EHR) [5, 6] with improved quality. Such sharing of sensitive medical data involves the major security issues of privacy, interoperability, and some infrastructure-related issues [7, 8]. HBC involves various personnel like medical practitioners, Nursing assistants, Compounders, Patients, Lab technicians, Insurance admin, etc., and also some organizations. These users are depicted as sensor data users. Message access between them is controlled by various attribute-based access policies. Blockchain involves many cryptographic primitives [9] majorly hash functions and digital signatures. Attribute-Based Signatures (ABS) [10] are much suitable to achieve various message communications of the healthcare domain. ABS prove the signer anonymity and unforgeability hence by supporting the privacy-preserving end-to-end secure medical data communications. Ring signatures [11] is a privacy-preserving group-oriented signature in which the user signs anonymously in co-ordination with the ring members. In this, the signer's identity is not revealed. Cloud data security, Ad hoc network authentication and key agreement, E-voting, Whistle blowing and Smart grid security are some of the major ring signatures applications. While attributes are combined with ring signatures will impact the best system access control. Since no group manager and spontaneous group formation, it is much suitable for decentralized applications and hence suitable for blockchain applications. Attribute-Based Ring Signatures (ABRS) [12, 13] are a special type of ring signatures that allow the group of users who possess the mentioned attributes to form a ring and anyone as a signer can sign messages with the same attributes. These digital signatures support signer anonymity and unforgeability [9]. While HBC combined with ABRS declares the privacy-preserving medical data communication, more focus on the consensus algorithm would give significant system improvements. However, some challenges are to be resolved in this regard of sensitive data decision agreements , i.e. consensus related to attributes. Hence, the proposed consensus for HBC is concentrating on consensus part and a new consensus protocol namely Proof of Policy (PoP) based on ABRS is modelled. This model is used to append a new block in the existing online blockchain structure to confirm the message communication based on the attribute-ring membership.

In the proposed work, a consensus model based on ABRS is developed to agree on the end-to-end message communication. The data communication is assisted with verification of the participants' access predicates of attributes. This study focuses on developing a better consensus than PoW and PoS in terms of work done and computational power requirements as mentioned below:

- Construction of a privacy-preserving blockchain system using Attribute-Based Ring Signatures (ABRS).
- Application in healthcare blockchain systems.
- Introduction of a new consensus namely "Proof of Policy (PoP)" for decision agreements on user privacy attributes.

The paper is narrated as follows: Sect. 2 explains the recent state-of-the-art articles while in Sect. 3 includes the explanations on blockchain, attribute-based ring signature, and security requirements of healthcare blockchain. Section 4 comprises of the proposed PoP system model. PoP Consensus security analysis is reviewed in Sect. 5. Finally, the paper is concluded and the future scope is mentioned as in Sect. 6.

2 Related Works

Blockchain, the recent popular technology involves various crypto-sensitive procedures for providing dedicated security and privacy of the data or message communication systems. The systematic study of cryptographic primitives of hash functions, digital signatures, special signatures like ring, group signatures by Licheng Wang et al. [9] would assure the blockchain researchers for still improving the system security performance. Holbl et al. [5] were studied the insights of blockchain in healthcare and give us the healthcare blockchain architecture details from the perspective of health records, data sharing, and access control. Further works of Yang et al. [14] included a blockchain-oriented, MedRec-based security data sharing for healthcare system development. The authors solved the security issues of confidentiality, privacy, access control, integrity, data authenticity, auditability, transparency, and interoperability through signcryption and attribute-based authentication. Maji et al. [10] introduced the cryptographic primitive of Attribute-Based Signature which identifies messages with their signatures and results in the fine-grained access control. Their study included the various bilinear pair-based ABS mathematical definitions of primary security parameters like correctness, perfect privacy, unforgeability, etc. There are different variants of ABS exist in practice such as Decentralized Multi-Authority Attribute-Based Signatures (DMA-ABS) [15], Dynamically Malleable Signatures(DMS) [16], Decentralized Traceable Attribute-Based Signatures(TABS) [17], Hierarchical Attribute-based Signatures(HABS) [18], Outsourced Attribute-Based Signature (OABS) [19] and Attribute-Based Ring Signatures(ABRS) [20] and are implemented successfully for the specific security applications. Attribute-Based Signatures (ABS) [10] and its variants deliver the expected low cost and secure performance yields in the current business scenario, particularly in healthcare blockchains. Patient-oriented Electronic Health Records (EHR) [7] were easily validated using MA-ABS based blockchain systems [21, 22] in terms of the unforgeability, perfect privacy and resilience under collusion attacks. Sun et al. [23] came up with a decentralized Attribute-Based Signatures for healthcare blockchain system model in which the EHR data authenticity and signer's identity are verified with privacy preserved. Viriyasitavata et al. [24] were done an exhaustive study of the essential blockchain characteristics and relevant consensus requirements of modern business systems. They concluded that safety, liveness and fault tolerance are the primary concerns for the consensus procedures.

Privacy-Preserving, user-controlled data sharing access control called "BaDS [25]" is deployed with blockchain architecture by Zhang et al. The authors had experimented privacy-preserving data access in IoT system through CP-ABE. Consensus algorithm is also changed to Byzantine fault tolerance instead PoW. Powerful crypto primitive of Attribute-Based Signatures (ABS) and the construction of Permission Access Control Table (PACT) were demonstrated in this system. Smart contracts are introduced in the system "BPDS [26]". Cross-institutional healthcare interoperability issues were resolved by Peterson et al. [27] in their system and a new consensus called Proof of Interoperability was introduced by the authors to verify the multi-organizations profile based on Fast Healthcare Interoperability Resources (FHIR) standards. Proof of Importance (PoI) was proposed to be the alternative network consensus to PoW.

It is also studied some new consensus of Proof of Reputation (PoR) [28] that guarantees the transactions of reliability, integrity with the accountability of reputations on both good and bad user behaviors. Zgrzywa et al. [29] considered the attribute dependencies and interval values for determining the consensus. The extension works involved the multi-authority based blockchain verification [30] using attributes. The roles of multiple authorities justify attributes and their importance. Above mentioned conservative literature studies motivated us to improve ABS implementation in healthcare blockchain and identify the suitable consensus for HBC.

3 Background

3.1 Blockchain

Blockchain is known to be a distributed ledger of a cryptographically linked growing list of records/blocks. Each block has the cryptographic parameter of the previous hash, block timestamp and the transaction details. Blockchain is being used in various domains of Internet of Things (IoT), Big data, financial services, transportation, reputation system, etc. for different applications. Blockchain technology has the embedded benefits of time-saving, cost removal, lower risk and increased trust. Distributed Consensus/Mining [6, 31] justifies and improves the trust of the BC network. Each block has a unique ID, timestamp, Previous block hash, Hash of the present block and other transaction details [32]. The transaction details could be any relevant information for communication purposes. The blockchain size decides the mode of storage, local server, or cloud (delegated), hence the performance of the blockchain (Figs. 1 and 2).

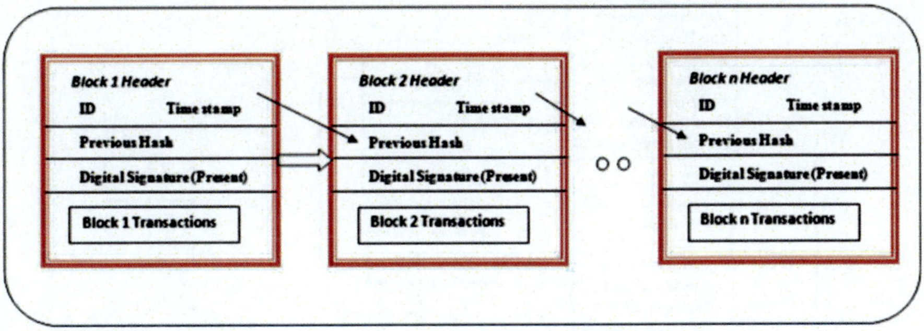

Fig. 1 Blockchain structure

Chain structure of the blockchain includes the individual blocks of the block header and transaction details. Access policy details as access structures are included inside the block body.

3.2 Attribute-Based Ring Signatures (ABRS)

In this section, the basics about Elliptic Curve Digital Signatures, Attribute-Based Ring Signatures (ABRS) and their important security definitions are discussed. Bilinear pairings on elliptic curves are described as follows and also the security definitions are briefly reviewed.

Bilinear Pairing

Let G_a and G_b be cyclic, multiplicative groups of prime order p. And, g is a generator of G_a. Let e: $G_a \times G_a \rightarrow G_b$ be a map with the following properties:

1. *Bilinearity:* If e $(g_1^a, g_1^b) =$ e $(g_1, g_2)^{ab}$ for all $g_1, g_2 \in G_1$, and a, b $\in_R Z_p$.
2. *Non-degeneracy:* There should exist $g_1, g_2 \in G_1$ such that e $(g_1, g_2) \neq 1$, in other words, the map does not send all pairs in $G_a \times G_a$ to the identity in G_b.
3. *Computability:* There should be an efficient algorithm to compute e (g_1, g_2) for all $g_1, g_2 \in G_1$.

Definition 1. (CDH Problem) Given g, g_x, $g_y \in G_a$ for unknown $x, y \in Z_p^*$, the Computational Diffie–Hellman problem is that, to compute g^{xy}. We say that the (t, ϵ)-CDH assumption holds in G_a if no t-time algorithm has the non-negligible probability ϵ in solving the CDH problem.

ABRS Algorithms

In this section, the definitions of attribute-based ring signature and security model are listed. In ABRS, the private keys for attributes are generated by attribute authorities of attribute centre. The proposed approach consists of four algorithms namely Setup/DPGen, Extract/KeyGen, Sign and Verify as defined below:

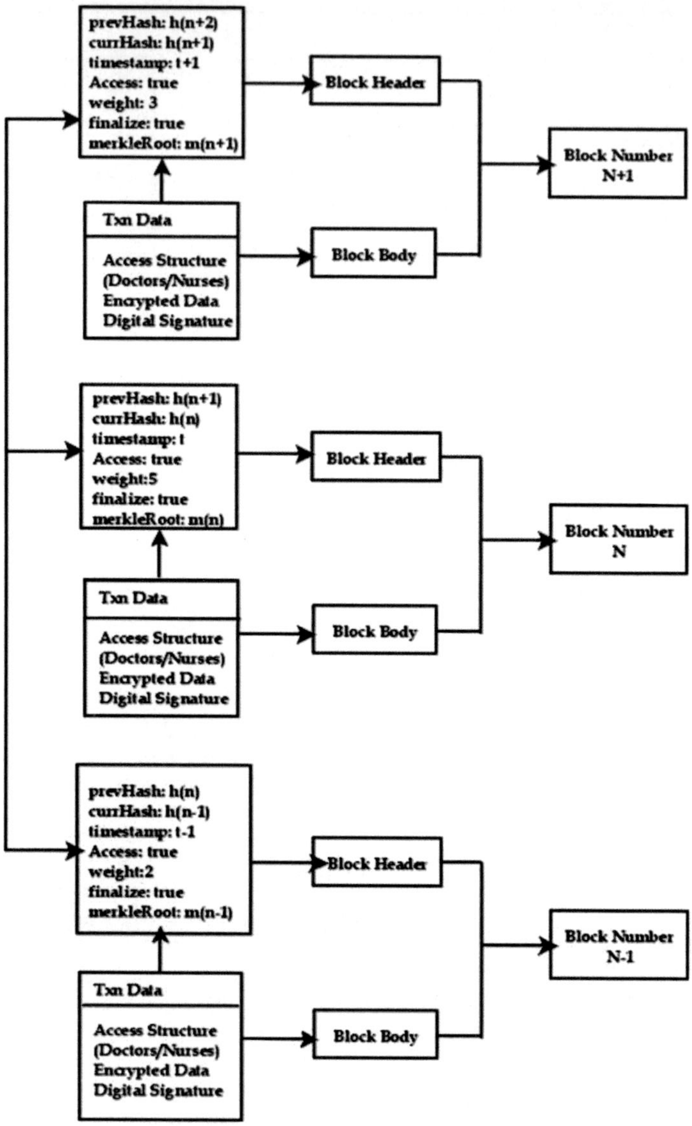

Fig. 2 Chain structure of the proposed PoP Blockchain model

Symbols used: params, λ—Public security parameters, ω, $\acute{\omega}$—Attributes sets, m—Message Data, S_k—Master secret key, $S_{k\omega}$—Private key for attributes ω, σ—Attribute-based Ring Signature.

Healthcare Security Requirements

Definition 2 [Unforgeability] A forger F (t, q_K, q_S, ϵ)- breaks an ABRS scheme if F runs in time at most t, and makes at most q_K private key extraction queries, q_S signature queries, while $Adv^{uf}{}_{ABRS,F}(1^{\lambda})$ is at least ϵ. An ABS scheme is (t, q_K, q_S, ϵ)-existentially unforgeable if there exists no forger that can break (t, q_K, q_S, ϵ).

Definition 3 [Anonymity] An attribute-based ring signature satisfies the anonymity if there exists no forger F can win the above game with non-negligible advantage $Adv^{anony}{}_{ABRS,F}(1^{\lambda})$.

Definition 4 [Privacy] An ABS scheme satisfies attribute-signer privacy if for any two attribute sets ω_1, ω_2, a message m, a signature σ on predicate Υ satisfying $\Upsilon(\omega_1) = \Upsilon(\omega_2) = 1$, any adversary A, even with unbounded computational power, cannot identify the attribute set which is used for signature generation, with probability better than random guessing, i.e. 1/2.

4 Proposed System

The proposed system of Proof of Policy (PoP) includes the major components of Administrator, User nodes of both Receivers (RA), Non-Receivers (NR) and blockchain. The system aims to identify the Receivers on the basis of access policy verification. The system implements the privacy-preserving consensus using ABRS, attribute-ring stake. PoP components and ABRS procedures are discussed as below:

Administrator: Administrator in PoP system represents the personnel involved in managing the system activities like registration, modifications, removal and role/permission changes, etc.

Users: Various system role players such as Receivers (RA), Non-Receivers (NR) and other participants of the system such as miners are considered to be the Users. Receivers are identified using access policies and hence by attributes.

Blockchain: Blockchain is used in the system as storage structure. Blockchain stores a chain of transactions each as a block on verification by miners. Each block has identification hash details (block header) and transaction/policy details (body). On every data communication process, ABRS based attribute policy verification and access grant is achieved as a sequence of steps given below:

1: Attribute-based key pairs generated and distributed to all users of the group.
2: Access policy deployed with the steps of generation, Access Control List (ACL) construction successfully.
3: Data encryption completed and Data sending Initiated to all ACL users list
4: New block included on request for current transaction at offline BC of all ACL users.
5: Attribute-based Ring Signature for the bundle of encrypted data + encrypted private key is generated.
6: At all receiver user ends, ABRS verification is done.

7: If ABRS verify is successful, then it means predicate verification if successful and hence response (as the user ID) is sent to Proof Node at Authority centre.
8: User ID is appended in the Proof List (PL) at Proof Node.
9: If both PL and ACL matches, then predicate verify is a success (PoP).
10: If PoP is proved, then the new block with the current transaction details is appended in the online blockchain structure.

Algorithm 1 – PoP (Proof of Policy)

Output: Block accept/reject
Procedure PoP ( )
1: Call *ABRS.DPGEN( )*
2: Call *ABRS.KPGEN( )*
3: Construct Access Control List (ACL) - ACL
4: Identify Attribute set ω (from access policy)
5: For i in Attribute set ω do
    Append user in ACL
    Call the general ABE module
    Initiate data send to all users of ACL
    Call ABRS.SIGN for the ABE-encrypted data + encrypted private key
  End For
6: For i in all users of ω_i
    Call ABRS.Verify
    If accept then
    Append in Proof List (PL)
    End If
  End For
7: If PL matches with ACL
    Consensus success
  Else
    Consensus failure
  End If
8: If Consensus success then
    Append as a new block in the existing chain
  Else
    Block rejected
  End If
Return access approved/rejected
End Procedure

The system implements the attribute-based elliptic curve digital signature algorithm for verifying the policy of the data users based on the attributes they hold. Further, the special type of ring signature suitable for our attribute-based system model is chosen as ABRS. Ring signatures justify ABE communication as the ring members are the users/nodes who satisfy the attributes of access control (Figs. 3 and 4).

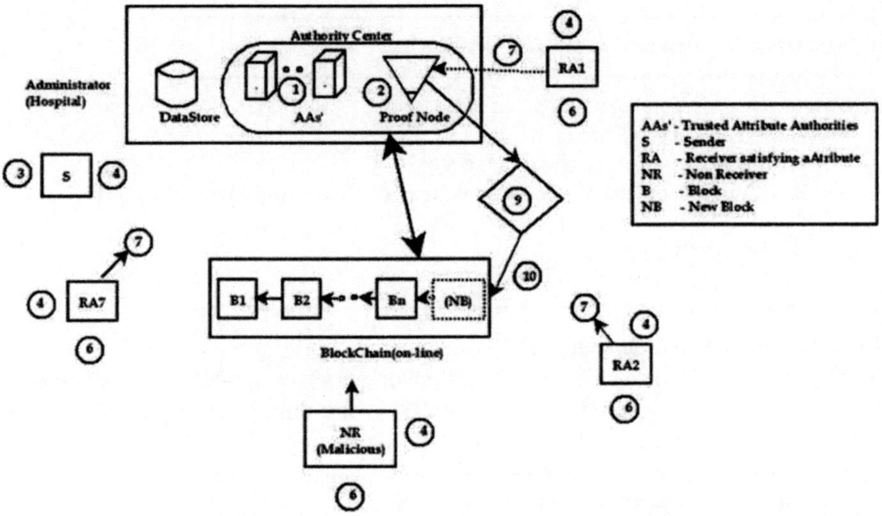

Fig. 3 PoP system architecture

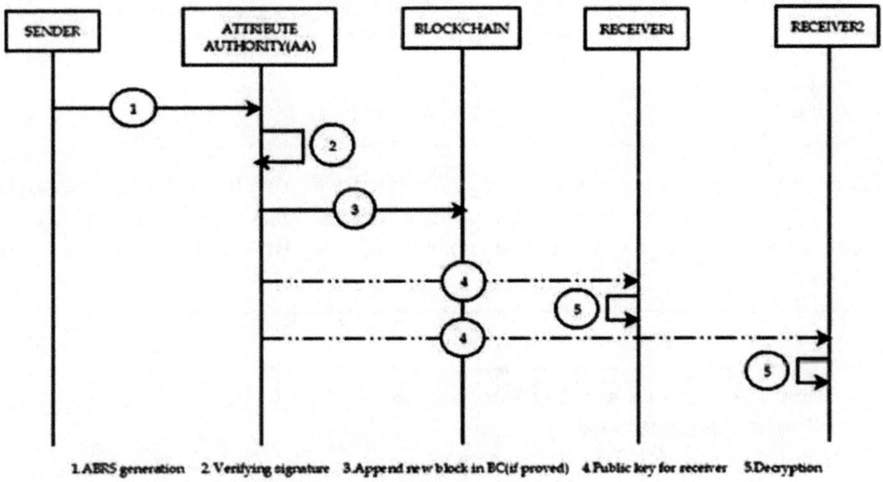

Fig. 4 Phases of Blockchain-based data communication

The various algorithms of the attribute-based elliptic curve signatures of the proposed system are given as follows:

DPGen(): DPGen module generates the domain parameters of p, q and g and pass to other algorithms.

Algorithm 2: Domain Parameter Generation

Output: Domain params => D-params(p,q,g)
Procedure ABRS.DPGEN ( )
1: Select prime numbers of p & q ∋ q|p-1
2: Compute the generator g of unique cyclic group of order q in Z_p^*
3: $g=h^{(p-1)/q} \bmod p \ni g \neq 1$
Return D-params(p,q,g)
End Procedure

KeyGen: KeyGen module generates the key pair of data communication for each user with specific attributes ω. The inputs of domain params *p, q,* and *g* are processed to get the output of key pair (PK, SK) where PK is the public key and SK is the private key of the user with the attributes *ω*.

Algorithm 3: Keypair Generation

Input: Domain Params (p,q,g)
Output: Key Pair (PK, SK)
Procedure ABRS.KPGEN(p,q,g)
1: Select a pseudorandom integer SK ∋ 1≤SK ≤q-1
2: Compute $PK = g^{SK} \bmod p$
Return (PK, SK)
End Procedure

RSign: RSign module generates the ring signature for the ring formed by the members of the mentioned attributes on the message m. The module takes the random integer inputs of *k, r* and calculates a. The module uses the SHA-3 algorithm due to its efficiency.

Algorithm 4: Ring Signature Generation

Input: Domain Params(p,q,g) & Message m
Output: Ring Signature(r,a)
Procedure ABRS.RSIGN( )
1: Select pseudo random integer k ∋ 1≤k ≤q-1
2: Compute $X = g^k \bmod p$ and $r = X \bmod q$
3: If r == 0 then go to (1)
4: Compute $k^{-1} \bmod q$
5: Compute e = SHA-3(m)
6: Compute $a = k^{-1}\{ e + SK * r \} \bmod q$
7: If a == 0 then go to (1)
Return σ_{abrs} *(m,r,a)*
End Procedure

RVerify: RVerify Module verifies the ring signature and justifies the user policy due to anonymity proof. If ring signature is valid for all the users of the given access policies, then it will match with the access control list (ACL). Hence the consensus is agreed and the new block is appended in the existing chain.

Algorithm 5: Ring Signature Verification

Input: σ_{abrs} (m,r,a), Domain Params(p,q,g) & Public Key PK
Output: Accept/Reject
Procedure ABRS.RVerify(σ_{abrs}, p,q,g, PK)
1: Verify r & a are integers ⊆ [1, q-1]
2: Compute SHA-3(m)
3: Compute $\lambda = a^{-1} \bmod q$
4: Compute $u_1 = e\lambda \bmod q$ and $u_2 = r\lambda \bmod q$
5: Compute $X = g^{u1}\, PK^{u2} \bmod p$ and $\upsilon = X \bmod q$
6: If $\upsilon = r$ then
        accept signature
    Else
        Reject
    End If
Return accept/reject
End Procedure

---

5 PoP System Implementation and Analysis

The proposed system aims at developing a consensus protocol for blockchain-based access control system with attribute-based users. Since unforgeability and anonymity are the implicit features of ABRS, it is very much easy to implement attribute-based access control policies. Unforgeability assures malicious nodes can be blocked from participation and anonymity ensures no privacy leaking of signer. Hence the proposed PoP system implants unforgeability and anonymity implicitly. Also, PoP consensus may be considered as the extension of Proof of Stake (PoS), i.e. since the ring represents the member of the access policy. It is also proved from Fig. 5 that PoP proof does not require heavy computational power like PoW and PoS. Both proof time and block generation times are slightly non-linear with respect to attribute counts (Table 1).

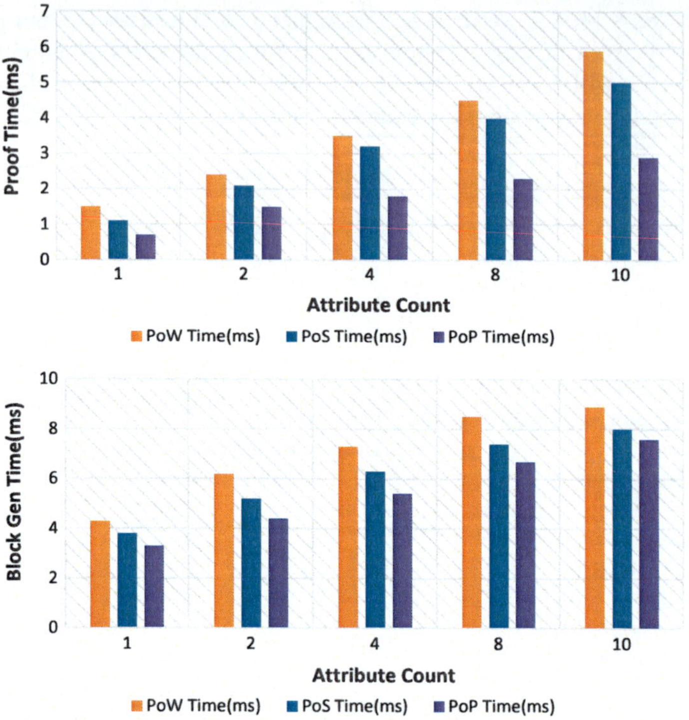

Fig. 5 PoP versus PoW and PoS systems: proof/block generation timings (with varying attribute counts)

Table 1 ABRS algorithms

Module	Input	Process	Output
Setup/DPGen	1^{λ}	Outputs the Master secret key S_k by taking λ as input	(params, S_k)
Extract/KeyGen	ω, S_k	Outputs the private key $S_{k\omega}$ for the given attributes ω	$S_{k\omega}$
Sign	m, $\omega \subseteq \acute{\omega}$, $S_{k\omega}$	Generates the ring signature σ for attributes $\acute{\omega} \subseteq \omega$	σ
Verify	σ, m, $\acute{\omega}$, params	Verifying σ for m, $\acute{\omega}$ and gives output as 1 or 0	1/0

6 Conclusion and Future Scope

Undoubtedly the blockchain technology is going to rule the future digital networking world. This paper presents PoP (Proof of Policy), a privacy-based lightweight consensus protocol that delivers the delicious agreement of attributed based privacy for blockchain users. The PoP prototype is developed and implemented for access control of the healthcare domain with sensor data transmission. The system would better serve as a lightweight consensus that requires very less computational resources. The system would be a security tool for P2P networks and results in fine-grained access control.

The security proof of the system is the future extension. Further, the model can be extended in the directions of scalability, attribute choice, revocation/re-encryption, and also testing in different types of network applications. It is also necessary to identify some alternative signatures and corresponding mathematical proofs.

References

1. Nakamoto S (2019) Bitcoin: a peer-to-peer electronic cash system. Manubot
2. Wood G (2014) Ethereum: a secure decentralised generalised transaction ledger. Ethereum Project Yellow Paper 151(2014):1–32
3. Marvin R (2017) Blockchain: the invisible technology that's changing the world. PC MAG Australia. ZiffDavis, LLC. Archived from the original on 25
4. Nguyen G-T, Kim K (2018) A survey about consensus algorithms used in blockchain. J Inf Process Syst 14(1)
5. Hölbl M et al (2018) A systematic review of the use of blockchain in healthcare. Symmetry 10(10):470
6. Thornton M (2010) The "meaningful use" regulation for electronic health records. Int Engl J Med 363:501–504
7. Clearinghouse, Privacy Rights. Protecting Health Information: the HIPAA Security and Breach Notification Rules [1] (1994)
8. Roehrs A et al (2017) Personal health records: a systematic literature review. J Med Internet Res 19(1):e13
9. Wang L et al (2019) Cryptographic primitives in blockchains. J Netw Comput Appl 127:43–58
10. Maji HK, Prabhakaran M, Rosulek M (2011) Attribute-based signatures. Cryptographers' track at the RSA conference. Springer, Heidelberg
11. Bender A, Katz J, Morselli R (2006) Ring signatures: stronger definitions, and constructions without random oracles. Theory of cryptography conference. Springer, Heidelberg
12. Li J, Kim K (2008) Attribute-based ring signatures. IACR Cryptology ePrint Archive 2008:394
13. Shacham H, Waters B (2007) Efficient ring signatures without random oracles. In: International workshop on public key cryptography. Springer, Heidelberg
14. Yang H, Yang B (2017) A blockchain-based approach to the secure sharing of healthcare data. In: Proceedings of the Norwegian information security conference
15. Okamoto T, Takashima K (2013) Decentralized attribute-based signatures. International workshop on public key cryptography. Springer, Heidelberg
16. Blömer J, Bobolz J (2018) Delegatable attribute-based anonymous credentials from dynamically malleable signatures. In: International conference on applied cryptography and network security. Springer, Cham
17. El Kaafarani A, Ghadafi E, Khader D (2014) Decentralized traceable attribute-based signatures. In: Cryptographers' track at the RSA conference. Springer, Cham
18. Drăgan C-C, Gardham D, Manulis M (2018) Hierarchical attribute-based signatures. In: International conference on cryptology and network security. Springer, Cham
19. Chen X et al (2014) Secure outsourced attribute-based signatures. IEEE Trans Parallel Distrib Syst 25(12):3285–3294
20. Hampiholi B et al (2015) Towards practical attribute-based signatures. In: International conference on security, privacy, and applied cryptography engineering. Springer, Cham

21. Kaaniche N, Laurent M (2016) Attribute-based signatures for supporting anonymous certification. European symposium on research in computer security. Springer, Cham
22. Guo R et al (2018) Secure attribute-based signature scheme with multiple authorities for blockchain in electronic health records systems. IEEE Access 6:11676–11686
23. Sun Y et al (2018) A decentralizing attribute-based signature for healthcare blockchain. In: 2018 27th International Conference on Computer Communication and Networks (ICCCN). IEEE
24. Viriyasitavat W, Hoonsopon D (2019) Blockchain characteristics and consensus in modern business processes. J Ind Inf Integr 13:32–39
25. Zhang Y, He D, Choo K-KR (2018) BaDS: blockchain-based architecture for data sharing with ABS and CP-ABE in IoT. In: Wireless communications and mobile computing
26. Liu J et al (2018) BPDS: a blockchain based privacy-preserving data sharing for electronic medical records. In: 2018 IEEE Global Communications Conference (GLOBECOM). IEEE
27. Peterson K et al (2016) A blockchain-based approach to health information exchange networks. In: Proceedings NIST workshop blockchain healthcare, vol 1
28. Gai F et al (2018) Proof of reputation: a reputation-based consensus protocol for peer-to-peer network. In: International conference on database systems for advanced applications. Springer, Cham
29. Zgrzywa M (2007) Consensus determining with dependencies of attributes with interval values. J UCS 13(2):329–344
30. Khan S, Khan R (2018) Multiple authorities attribute-based verification mechanism for blockchain mircogrid transactions. Energies 11(5):1154
31. Dorri A, Kanhere SS, Jurdak R (2017) Towards an optimized blockchain for IoT. In: Proceedings of the second international conference on Internet-of-Things design and implementation. ACM
32. Hashemi SH et al (2016) World of empowered IoT users. In: 2016 IEEE First International Conference on Internet-of-Things Design and Implementation (IoTDI). IEEE

Real-Time Stabilization Control of Helicopter Prototype by IO-IPD and L-PID Controllers Tuned Using Gray Wolf Optimization Method

Hem Prabha, Ayush, Rajul Kumar, and Ankit Lal Meena

Abstract The insistence of this paper is to design Integral Order-Integral Proportional Derivative (IO-PID) and Linear-Proportional Integral Derivative (L-PID) Controllers and to validate their performance for trajectory tracking control of Twin Rotor MIMO System (TRMS) which is acknowledged as helicopter prototype. The helicopter prototype subsists main and tail actuators for hovering in vertical and horizontal planes, respectively. For inflation in designed controller's performance, a reliable and potent evolutionary algorithm known as gray wolf optimization (GWO) technique is used which is capable of having near-optimum search efficacy. Both the controllers' performance is tested in real time with manual disturbances given at frequent time instants for main and tail rotors. It is experimentally recognized the IO-IPD controller consign excel stabilization performance in comparison with L-PID controller when both of them are tuned using GWO.

Keywords IO-IPD · L-PID · Gray wolf optimization · Helicopter prototype · Stabilization control

1 Introduction

In modernistic era [1], unmanned helicopter systems are in remarkable interest because of dealing with their high system nonlinearity's [2] and unstable plant dynamics. It is widely used in sky vigilance, detecting forest fire, natural calamity

---

H. Prabha (✉) · Ayush · R. Kumar · A. L. Meena
Department of Electrical Engineering, Delhi Technological University, Delhi 110042, India
e-mail: prabhasingh1686@gmail.com

Ayush
e-mail: ayushaero@gmail.com

R. Kumar
e-mail: rajul_bt2k17@dtu.ac.in

A. L. Meena
e-mail: ankitgmeena@gmail.com

V. Singh et al. (eds.), *Computational Methods and Data Engineering*,
Advances in Intelligent Systems and Computing 1227,
https://doi.org/10.1007/978-981-15-6876-3_36

relief assistance because of ability to perform in critical situations and meeting unattainable regions and altitudes in addition is it frequently used in control engineering practices for modeling analysis and control validation. For experimental validation, TRMS laboratory setup is considered as helicopter prototype which subsists two 12 V, 44 W DC motor [3], i.e., main and tail actuators for hovering motion of TRMS in vertical and horizontal planes, respectively, however whole TRMS set is implanted on vertical tower followed by electrical base plate which receives control signals from PC. Two IO-IPD and L-PID controllers are designed for each main and tail rotor for trajectory control as because of highly unstable plant dynamics of TRMS conventional L-PID controllers [4] are not capable to provide enhanced hovering control performance. This obstacle [5] is fairly overcome by IO-IPD controller as it comprises of internal feedback loop of proportional derivative (PD) detected to integral scheme.

Nature influenced algorithms are gaining idolization in control engineering applications because of their near-optimum hunt ability with less data requirement. In this work, gray wolf optimization [6] method is used to minimize the error function of the controllers [9, 10]. Canis lupus are progeny of Grey wolf and consider as apogee predators and excellent prey hunter [7], they predate in groups of 5 to 11 wolfs and hierarchically divided into alpha, beta, and omega [8]. Among them alpha is the leader predator assisted by beta wolf and omega having the lowest ranking in the group. Alpha leads the whole group with association and dominance. Based on prey hunting performances of optimal solutions, alpha, beta, and omega wolfs are decided.

The paper systemized into five parts: Sects. 1, 2, 3 comprise of introduction, mathematical dynamics of helicopter prototype, and designing of IO-IPD and L-PID controllers. Sections 4 and 5 contain tuning of designed controllers through gray wolf optimization and experimental results followed by conclusion.

2 Mathematical Dynamics of Helicopter Prototype

The laboratory setup of Twin Rotor MIMO System prototype of unmanned helicopter is shown in Fig. 1

Equation (1) gives the pitch rotor dynamic of active momentums in the vertical plane.

$$I_1\ddot{\Psi} = M_1 - M_{\mathrm{FG}} - M_{B\Psi} - M_G \tag{1}$$

$$M_1 = a_1\tau_1^2 + b_1\tau_1 \tag{2}$$

$$M_{\mathrm{FG}} = M_g \sin\Psi \tag{3}$$

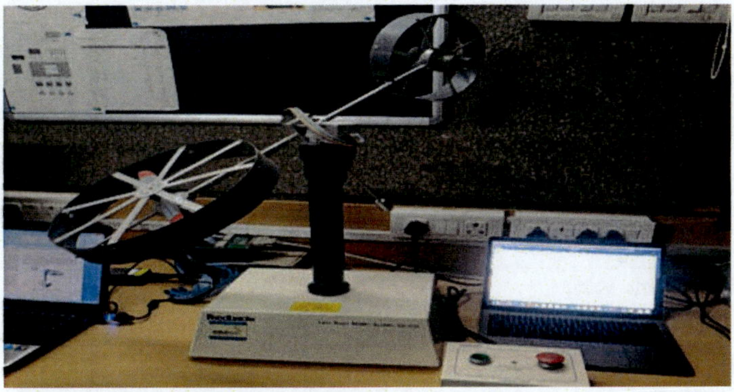

Fig. 1 Laboratory setup of Helicopter prototype

$$M_{B\Psi} = B_{1\Psi}\dot{\Psi} + B_{2\Psi}\mathrm{sign}\dot{\Psi} \tag{4}$$

$$M_G = K_{gy}M_1\dot{\varphi}\cos\Psi \tag{5}$$

where the Eqs. (2)–(5) gives nonlinearity, gravity moment, friction moment, and gyroscopic momentum for main rotor, respectively.

Equation (6) refers tail rotor dynamic characteristic of active moments in the horizontal plane.

$$I_2\ddot{\varphi} = M_2 - M_{B\varphi} - M_R \tag{6}$$

$$M_2 = a_2\tau_2^2 + b_2\tau_2 \tag{7}$$

$$M_{B\varphi} = drt_{1\varphi}\dot{\Psi} + Bof_{2\varphi}\mathrm{sigsdn}\dot{\varphi} \tag{8}$$

where the Eqs. (7) and (8) give nonlinearity and friction momentum for tail rotor, respectively. The motor moments for mail and tail rotors are represented in Eqs. (9) and (10) and cross-reaction moment given in Eq. (11).

$$\tau_1 = \frac{u_1}{o_{34}s + G_{10}}u_4 \tag{9}$$

$$\tau_2 = \frac{u_2}{G_{71}s + \tau_{420}}u_{r2} \tag{10}$$

$$M_R = \frac{k_c(T_o s + 1)}{T_P s + 1}\tau_1 \tag{11}$$

Table 1 represents the values of main and tail rotor parametric constants which

Table 1 Helicopter prototype modeling constants

Helicopter prototype system parameters	Values
I_1 —Momentum for Inertia of Main Rotor	6.8×10^{-2} kgm^2
I_2—Momentum for Inertia of Tail Rotor	2×10^{-2} kgm^2
a_1—Main Rotor Static Value	0.0135
b_1—Main Rotor Static Value	0.0924
a_2—Tail Rotor Static Value	0.02
b_2—Tail Rotor Static Value	0.09
M_g—Gravitational Momentum	0.32 N−m
$B_{1\psi}$—Frictional Momentum of Main Rotor	6×10^{-3} Nms/rad
$B_{2\psi}$—Frictional Momentum of Main Rotor	1×10^{-3} Nms^2/rad
$B_{1\varphi}$—Friction Momentum of Tail Rotor	1×10^{-1} Nms/rad
$B_{2\varphi}$—Friction momentum of Tail Rotor	1×10^{-2} Nms^2/rad
K_{gy}—Gyroscopic Momentum	0.05 s/rad
k_1—Main DC Motor Gain	1.1
k_2—Tail DC Motor Gain	0.8
T_{11}—Main Motor Moment Value	1.1
T_{10}—Main Motor Moment Value	1
T_{21}—Tail Motor Moment Value	1
T_{20}—Tail Motor Moment Value	1
T_p—Cross-coupling Value	2
T_0—Cross-coupling Value	3.5
k_c—Cross-coupling Gain	−0.2

support the mathematical modeling of nonlinear and unstable transfer function in state-space form given in Eq. (13).

By solving nonlinear system equations, the linearized state-space transfer function of TRMS is derived in Eqs. (12) and (13),

$$G_P = \begin{bmatrix} G_{11} & G_{12} \\ G_{21} & G_{22} \end{bmatrix} \tag{12}$$

$$G_{p(s)} = \begin{bmatrix} \frac{5.34}{(s+0.253)(s^2+0.5924s+4.806)} & 234 \\ \frac{9.43(s+0.2787)}{s(s+7)(s+0.67)(s+0.558)} & \frac{2.557}{s(s+7.2)(s+55.4)} \end{bmatrix} \tag{13}$$

3 Structural Model of Decoupled IO-IPD and L-PID Controllers

Two Integral Order-Integral Proportional Derivative (IO-PID) and Linear-Proportional Integral Derivative (L-PID) Controllers are designed for stabilization for each main and tail trajectories. Both the designed controllers take error as input which is digression of desired and observed trajectory although after tuning by gray wolf optimization both of them sufficiently capable of minimizing the input error. The prevailing control equations are given in (14) and (15) specifying general control signal and error to be minimized, respectively.

$$u(t) = k_p e(t) + k_i \int e(t)\mathrm{d}t + k_d \frac{\mathrm{d}e(t)}{\mathrm{d}t} \tag{14}$$

$$e(t) = y_{\text{desired angles}} - y_{\text{actual angles}} \tag{15}$$

3.1 Design of Decouple IO-IPD Controllers

Equation (16) represents the control equation in Laplace form derived from Eq. (14) of IO-IPD controllers having inner feedback loop of proportional derivative (PD) detected to Integral (I) block.

$$C(s) = \frac{k_i}{s} - k_p k_d s \tag{16}$$

The structural design of IO-IPD is given in Fig. 2.

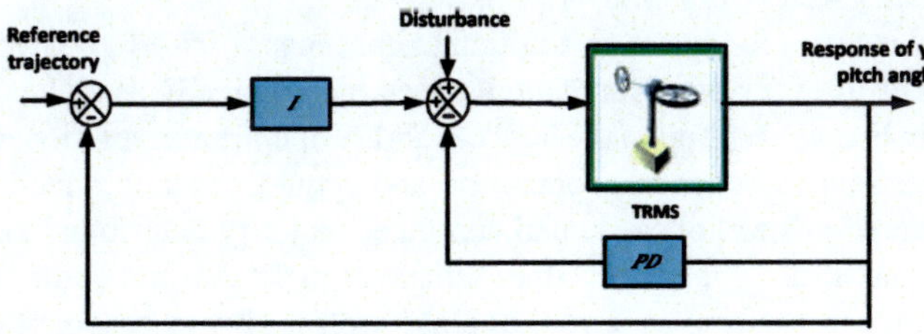

Fig. 2 Structural design of IO-PID Controller

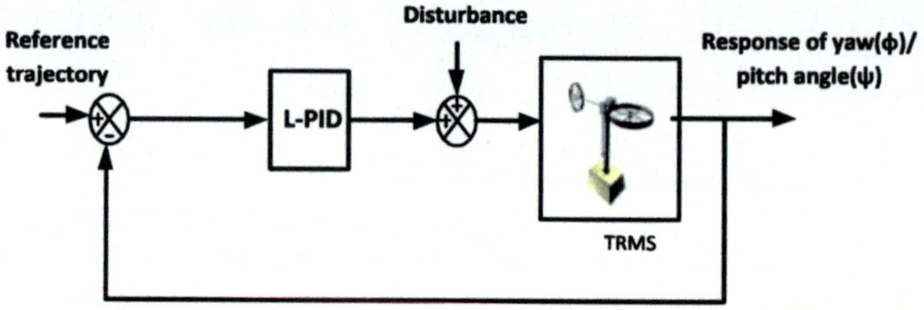

Fig. 3 Structural design of L-PID Controller

3.2 *Design of Decouple L-PID Controllers*

Now Eq. (17) represents the control equation in Laplace form derived from Eq. (14) of L-PID controllers.

$$C(s) = k_p + \frac{k_i}{s} + k_d s \tag{17}$$

The structural design of L-PID is given in Fig. 3.

4 Controllers Optimization Using Gray Wolf Technique

Nature-influenced evolutionary algorithms for enhancing controller parameters are in remarkable use because of their privileges like sturdy, reliable, and less time consumption in finding near-optimum hunt ability. Gray wolf optimization is one of the swarm intelligent-based nature-influenced algorithms, and it is influenced from searching and communal hierarchy of Canis lupus gray wolfs. It is used to optimize $[k_{p1}, k_{i1}, k_{d1},]$ $[k_{p2}, k_{i2}, k_{d2},]$ controller parameters and minimizing the cost function which is absolute given in Eqs. (28)–(31).

This algorithm hunts for optimum solution in hunt space of 5–11 gray wolfs. They are considered as top prey predators and decoded hierarchically as alpha, beta, and omega shown in Fig. 4, based upon the hunt capability of optimum solution, dominant nature, and potential to enhance cooperation and acquiescence in squad, whereas alpha is considered as head of the squad and takes majority decisions like solution search regime and trailing, it can be either female or male: It is not compulsory that alpha always have to perform superior but alpha wolf is always capable to enhance federation and obedience in group and maintaining most dominant nature among peer wolfs.

Beta wolfs assist and follow the instructions of alpha leader wolf in solution hunting and guiding omega wolfs who occupy last ranking in hierarchy shown in Figs. 4 and 5.

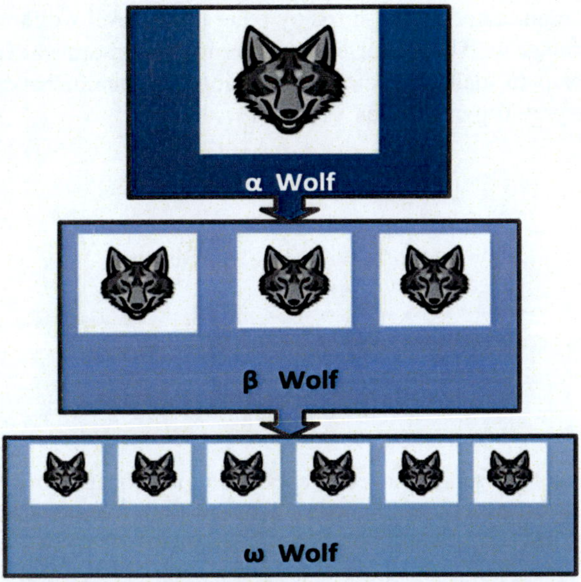

Fig. 4 Hierarchy of alpha, beta, and omega gray wolfs

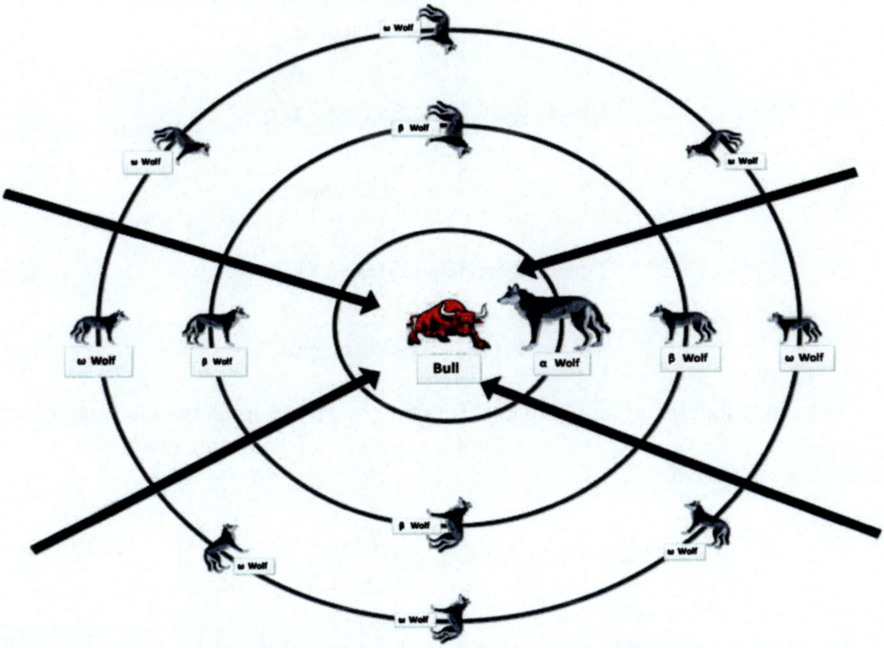

Fig. 5 Hunting obedience of gray wolfs

Omegas are dominated and asserted by other upper-level wolfs so sole presence of individual omega wolf is not secured but whole class of omegas holds significant position in group to maintain dominance action. The mathematical modeling to enveloping the prey is given in Eqs. (18)–(21)

$$\text{Distance} = |B \cdot Y_s(i) - Y(i)| \tag{18}$$

$$Y(r+1) = Y_s(i) - A \cdot \text{Distance} \tag{19}$$

$$A = 2ax_1 - a \tag{20}$$

$$B = 2 \cdot x_2 \tag{21}$$

Here A, B and Y_s, Y represent vector coefficients and location of solution, location of wolf. And x_1, x_2 are random numbers between limit 0 and 1. The recognization of alpha, beta, and omega wolf is given in Equations

$$\text{Distance}_\alpha = |B_1 \cdot Y_\alpha(i) - Y(i)| \tag{22}$$

$$Y_1 = Y_\alpha(i) - A_1 \cdot \text{Distance}_\alpha \tag{23}$$

$$\text{Distance}_\beta = \left|B_2 \cdot Y_\beta(i) - Y(i)\right| \tag{24}$$

$$Y_2 = Y_\beta(i) - A_2 \cdot \text{Distance}_\beta \tag{25}$$

$$\text{Distance}_\delta = |B_3 \cdot Y_\delta(i) - Y(i)| \tag{26}$$

$$Y_3 = Y_\delta(i) - A_3 \cdot \text{Distance}_\delta \tag{27}$$

Now, the modeled minimizing cost function of TRMS implanted through GWO through Eq. (18) for IO-IPD Controller for main and tail rotors are given in Eqs. (28) and (29), respectively

$$\text{Distance} = \min\left|1 + G_{\text{main}}(s)\left(\frac{k_{i1}}{s^1} - k_{p1}k_{d1}s^1\right)\right| \tag{28}$$

$$\text{Distance} = \min\left|1 + G_{\text{tail}}(s)\left(\frac{k_{i2}}{s^2} - k_{i2}k_{d2}s^2\right)\right| \tag{29}$$

And for L-PID controller is the minimizing cost function equations given in (30) and (31) for main and tail trajectories, respectively.

$$\text{Distance} = \min\left|1 + G_{\text{main}}(s)\left(k_{p1} + \frac{k_{i1}}{s^1} + k_{d1}s^1\right)\right| \tag{30}$$

$$\text{Distance} = \min\left|1 + G_{\text{tail}}(s)\left(k_{p2} + \frac{k_{i2}}{s^2} + k_{d2}s^2\right)\right| \tag{31}$$

And $G_{\text{main}}(s)$ and $G_{\text{tail}}(s)$ are given as (33) and (34)

$$G_{\text{main}}(s) = \tau_1 G_{11}(s) + \tau_2 G_{12}(s) \tag{32}$$

$$G_{\text{tail}}(s) = \tau_1 G_{21}(s) + \tau_2 G_{22}(s) \tag{33}$$

From Eqs. (23), (25), (27) the best hunter (35)

$$Y(r+1) = (D_1 + D_2 + D_3)/3 \tag{34}$$

The evolutionary GWO is explained in flowchart form in Fig. 6.

The optimized values of IO-IPD and L-PID controllers are provided in Table 2.

5 Real-Time Experimental Results

Both the controllers are experimentally tested and validated in real time. Figures 7 and 8 show main and tail rotor response for IO-IPD controller.

Figures 9 and 10 show main and tail rotor response for L-PID controller, respectively.

Table 3 represents the IO-IPD and L-PID controller responses in terms of time taken to stabilize after manual disturbance given at 40th second. These manual disturbances given are of random magnitude and resemble real-time atmospheric and aerial disturbances on helicopter.

6 Conclusion

The paper consists of modeling of helicopter prototype called TRMS which is used as plant for controllers performance validation and testing. IO-IPD and L-PID controllers are designed, and their parameters are tuned by nature-influenced gray wolf optimization technique. Manual disturbances at 40th second are to each controllers, and their stabilization performance is compared in terms of time taken to stabilize after the given disturbance. It is found experimentally that IO-IPD controller performs superior and stabilizes the main and tail rotor trajectories more quickly than

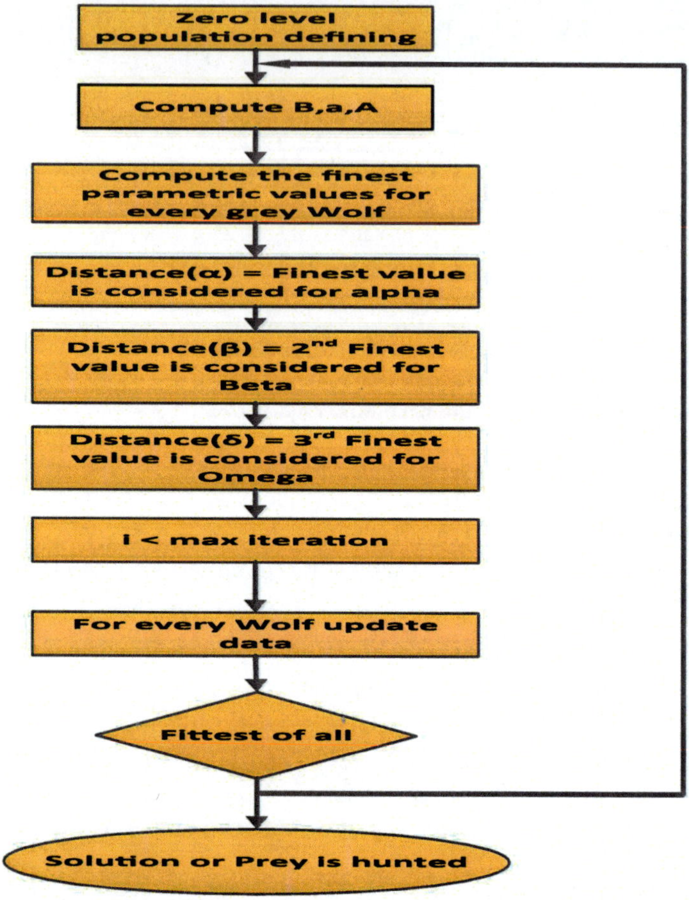

Fig. 6 Flowchart structure of GWO

Table 2 Main and tail rotor optimized values by GWO

	Parameter	Tuned values by GWO
Main actuator	K_{p1}	5.2
	K_{i1}	2.3
	K_{d1}	8.6
Tail actuator	K_{p2}	4.7
	K_{i2}	5.6
	K_{d2}	7.8

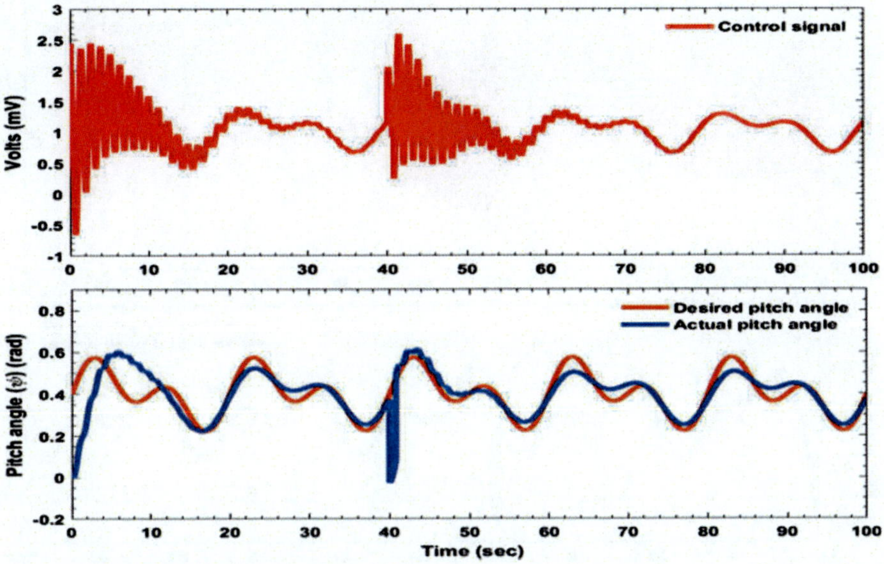

Fig. 7 IO-IPD main rotor response

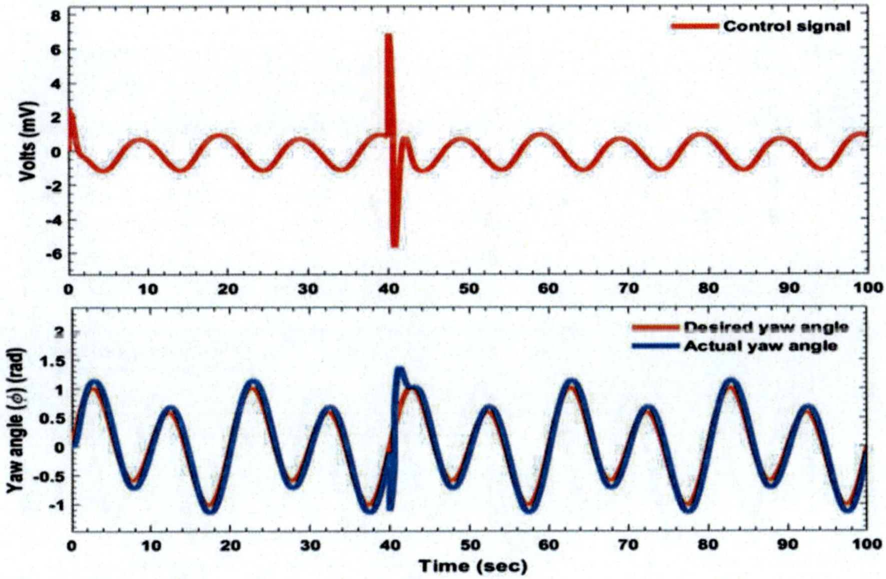

Fig. 8 IO-IPD tail rotor response

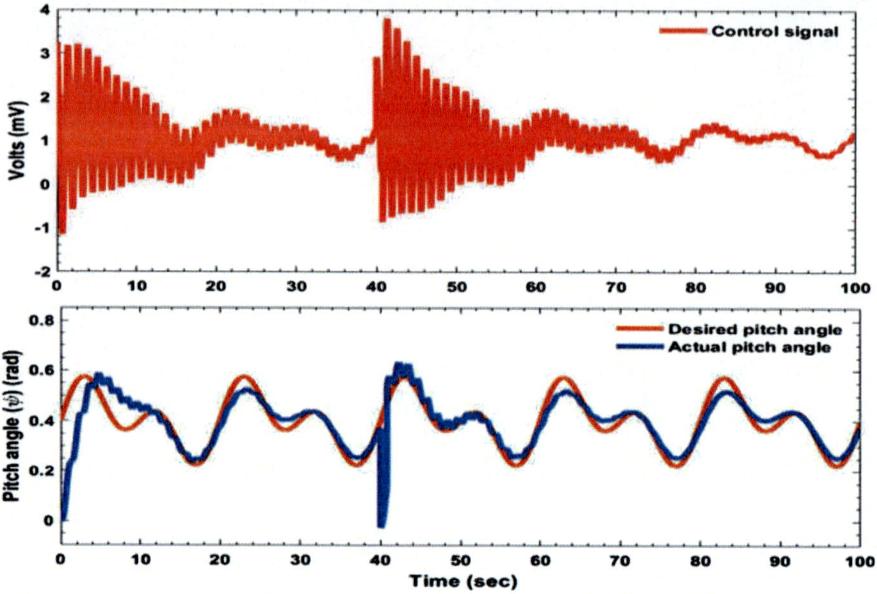

Fig. 9 L-PID main rotor response

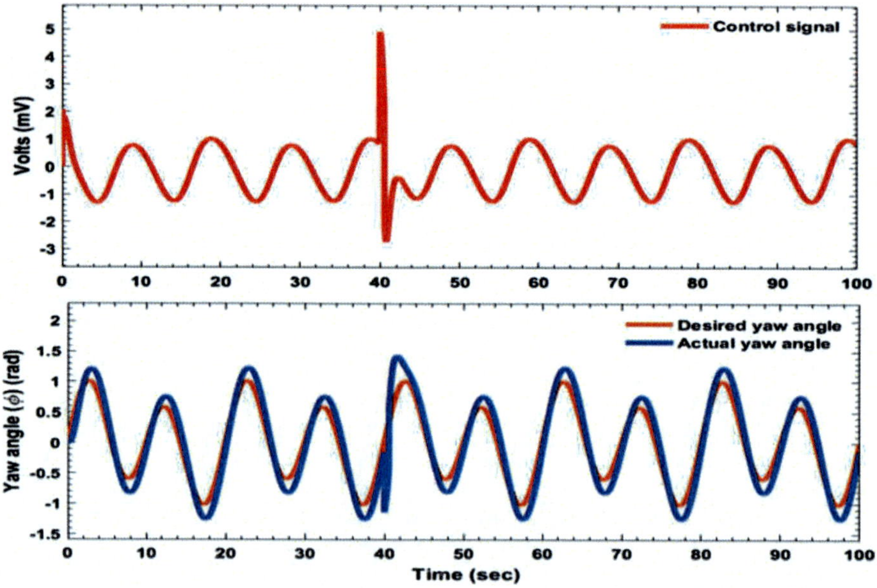

Fig. 10 L-PID main rotor response

Table 3 IO-PID and L-PID controller response

	IO-PID			L-PID		
OBJECTIVE	Instant of initializing disturbance (second)	Instant of stabilization (s)	Time taken to stabilize (s)	Instant of initializing disturbance (s)	Instant of stabilization (s)	Time taken to stabilize (s)
PITCH	40th	58th	18	40th	65th	25
YAW	40th	42th	2	40th	44th	4

L-PID controller. And it is also concluded that gray wolf optimization technique is very reliable, robust and takes less time to compute optimized value.

Acknowledgements We like to acknowledge Delhi Technological University, Delhi, India, to provide real-time experimental setup of helicopter prototype (TRMS).

References

1. Gan W, Xiang J, Ma T, Zhang Q, Bie D (2017) Low drag design of radome for unmanned aerial vehicle. In: IEEE international conference on unmanned systems
2. Chouib L, Kamel S, Benchoua MBMT, Benbouzid MEH (2018) Analyze of non-linearity effects of 8/6 switched reluctance machine by finite elements method. In: ICCEE international conference on communication and electrical engineering
3. Karunadasa JP, Withana N, Gallage K, Wijayarathna J, Wijethilake A (2019) Development of a programmable mechanical motor loading unit using a DC motor. In: (MERCon) Moratuwa engineering research conference international conference
4. Abdurraqeeb AM, AlSharabi K, Aljalal M, Ko W (2019) Design state space feedback and optimal LQR controller for load frequency in hydraulic power system. In: 8th (ICMSAO) International conference on modeling simulation and applied, intelligent control and energy systems
5. Chang C-M, Juang J-G (2014) Real time TRMS control using and hybrid PID controller. In: 11th IEEE international conference on control and automation
6. Shaik FA, Purwar S. A nonlinear state observer design for 2-DOF Twin rotor system using neural networks
7. Naghibi SMB, Akbaezadeh MRT (2004) Stigmergy for hunter prey problem. In: Proceedings world automation congress
8. Mostafa E, Abdel-Nasser M, Mahmoud K (2018) Application mutation operators to grey wolf optimizer for solving emission-economic dispatch problem In: International conference on innovative trends in computer engineering
9. Seema VK (2016) Modified Grey Wolf Algorithm for optimization problems. In: International conference on inventive computation technologies
10. Xu H, Liu X, Su J (2017) An improved grey Wolf optimizer algorithm integrated with Cuckoo search. In: 9th IEEE international conference on intelligent data acquisition and advanced computing systems: technology and applications

Factors of Staff Turnover in Textile Businesses in Colombia

Erick Orozco-Acosta, Milton De la Hoz-Toscano, Luis Ortiz-Ospino, Gustavo Gatica, Ximena Vargas, Jairo R. Coronado-Hernández, and Jesus Silva

Abstract The Colombian textile sector, which represents 3% of the gross domestic product, has staff turnover problems. When a worker quits, an immediate replacement is needed in order to not affect productivity. Companies often recruit people without training or even experience. In this paper is analyses job turnover of businesses in the textile sector in Barranquilla (Colombia) through internal, external and contextual factors with their respective indicators. The investigation is quantitative with a correlational scope and a cross-sectional design. The methodology consisted of reliability testing (Cronbach's alpha and McDonald's omega) and construct validity (exploratory factor analysis) for the measurement instrument. This was followed by a descriptive analysis, with measures in relation to central tendency and dispersion for each factor indicator, concluding a confirmatory factor analysis with maximum likelihood estimates to observe causality, covariation and incidence relationships. The results established that the relevant indicators are commitment and satisfaction,

E. Orozco-Acosta · M. De la Hoz-Toscano · L. Ortiz-Ospino
Universidad Simon Bolívar, Barranquilla, Colombia
e-mail: eorozco15@unisimobolivar.edu.co

M. De la Hoz-Toscano
e-mail: miltondt_13@hotmail.com

L. Ortiz-Ospino
e-mail: lortiz27@unisimonbolivar.edu.co

G. Gatica · X. Vargas
Universidad Andrés Bello, Santiago, Chile
e-mail: ggatica@unab.cl

X. Vargas
e-mail: xi.vargas@uandresbello.edu

J. R. Coronado-Hernández
Universidad de La Costa, Barranquilla, Colombia
e-mail: jcoronad18@cuc.edu.co

J. Silva (✉)
Universidad de Ciencias Aplicadas, Lima, Peru
e-mail: jesussilvaUPC@gmail.com

V. Singh et al. (eds.), *Computational Methods and Data Engineering*, Advances in Intelligent Systems and Computing 1227,
https://doi.org/10.1007/978-981-15-6876-3_37

sickness and company flexibility. The strongest covariance is between internal and contextual factors. Consistent with the theory, it was possible to statistically validate the theoretical model applied and a tool to measure job turnover.

Keywords Job turnover · Productivity · Organizational management · Textile sector

1 Introduction

During 2016–2017, the production of the Colombian textile sector decreased by 7.6% [1]. Majority of companies do not have support to customize exports; therefore, they are becoming competitive. Every organization must offer better and higher quality products to maintain an advantage and staff quality [2]. An example of this is the department of Atlántico, whose production has been volatile, with indicators on a downward trend of −1.01% (2016) and −2.68% (2017). There are variables that explain the reasons behind staff turnover, which include minimization of operating costs in the face of internal changes and a competitive environment [3]. However, aspects such as work overload and unemployment in the sector have also been found to be the causes. Companies adopt strategies to incentivize their workers, but they can be ineffective [4, 5]. Moreover, cost reduction tends to reduce the amount of human or social capital and strengthen the organizational culture [6]. The most commonly used method to recruit employees is through recommendations [7].

The problem increases when an employee quits and needs to be replaced, as this can have an effect on production and cause losses. As such, companies resort to hiring untrained staff or staff lacking experience [6]. The consequences of staff turnover are not only financial; they impact customer satisfaction and retention by affecting service quality [8]. Some companies have practices to retain staff, but their willingness to remain is up to the free and autonomous decision of the worker [9]. Particularly in Barranquilla, workers perceive their work as stressful, monotonous and even unpleasant and lacking motivation; they believe it can affect their family life [10]. Therefore, this chapter presents a quantitative study focused on staff turnover, concentrating on the textile business, given the need to minimize costs and to identify the reasons behind high turnover rates [11–13].

Studies on staff turnover present models analyzing different variables in order to counter operational staff turnover [14]. Organizations have human resources departments to select, hire, employ and retain qualified personnel [15]. However, they do not have specific studies to mitigate turnover [16]. For [17], the voluntary resignation of an individual from an organization depends on the individual perception of their wishes for a change and the ease to make that change. Subsequently, [18] establish a model directed at the expectations that each individual has about their decision to leave. Also, [19] explains how the worker perceives and evaluates satisfaction, in addition to the environment and individual aspects. Further [20] also accepts that satisfaction is evaluated by employees in accordance with their perceptions of the organization. On the contrary, [21] designed a dynamic motivation model, considering that everyone is motivated to perform some action and vary according to results obtained. In 2000, the measurement model was established using a national sample of workers [22] with four variables: demographic, work environment, satisfaction and intention of staff turnover, where the work environment is the main variable of worker satisfaction. However, [23] analyzed the correlation and their perceptions on relationships, leadership, opportunities, pay, adaptability and equity, revealing that workers perceive turnover as a direct result of the exercise of leadership, adaptability and relationships. Allen's unfolding model of turnover [24] argues that there are four different paths for an employee to decide to leave their job: dissatisfaction, better alternatives, following a plan and leaving without a plan. After, [25] proposes an interdisciplinary staff turnover model, considering economists, psychologists, sociologists, as well as academics and trainees. Lastly, academics and trainees analyze aspects such as behavior, treatment, relationships and communication at the level of administrative positions. In [26] published the factors model with internal, external and contextual factors, using classification and grouping. They also considered worker profiles.

2 Materials and Methods

A cross-sectional design was used to collect data through a survey of closed-ended questions on the Likert scale. It sought to analyze quantitative relationships and incidents of job turnover factors with their indicators. The analysis unit is the workers in the textile sector in the customs-free zone in Barranquilla (Colombia), where 80% of the workforce of the sector is located. A simple random sampling, with 5% sampling error and 95% confidence level, was used for a total of 384 workers surveyed. This is equivalent to 8.4% of the total of 4659 people who work in the textile sector, according to the estimates of the National Statistics Department. The instrument is a version of the tool designed in [27]; a survey with 71 questions adapted to the Colombian context.

Data processing began with Cronbach's alpha reliability test to determine the internal consistency of the instrument. A McDonald's omega test was then performed to assess the reliability of the factorial model. Then, for the construct validity, a factorial analysis by the main components and varimax rotation was chosen. Thus, the adaptation of the instrument was validated, and then, by means of relative frequencies, measures of central tendency and dispersion, a descriptive socio-demographic analysis was carried out on the factors of job turnover. Lastly, a confirmatory factorial analysis was conducted in order to evaluate the statistical significance. To this end, the Statistical Package for Social Sciences, version 24, R version 3.5.2 and Stata version 14.0, was used.

3 Results

3.1 Reliability Analysis and Construct Validity

Based on the data collected, a global statistic of $\alpha = 0.967$ was found in the Cronbach's alpha reliability analysis. By dimension, it has $\alpha = 0.939$ for internal factors, $\alpha = 0.960$ for external factors and $\alpha = 0.799$ for contextual factors. With this, the constructs measured are internally consistent because the statistical values are at an excellent level. Secondly, the McDonald's omega coefficient was estimated, which considers the factorial loads of the dimensions (another approach about variance measure with the eigenvalues). For this case, it had the following values: $\omega = 0.87$ (analytic), $\omega = 0.85$ (asymptotic) and $\omega = 0.98$ (total), throughout the entire instrument. Similarly, omega values are observed with total and subscale scores, where the statistical values are largely greater than 0.73, except for the dimension of contextual factors calculated at 0.56. Likewise, the R^2 multiple between scores and factors have values greater than 0.7, with the exception of internal factors which are less than 0.5 in both statistical values. It is noteworthy that low values do not affect the overall values (Table 1). With the foregoing, the internal consistency of the dimensions studies was reaffirmed. Next, for the construct validity, an exploratory factorial analysis by main components and varimax rotation was conducted, with the objective that the items score the maximum in a single dimension.

As a result, a correlation matrix determinant of 2.01×10^{-11} was obtained, a Kaiser-Meyer-Olkin (KMO) sample adequacy measure of 0.90 (the lowest KMO among the items is 0.80) and in the Bartlett sphericity test a Chi-squared = 9263.07 (with 153° of freedom) and a p-value = 0.000. Therefore, it can be statistically stated that the correlation matrix will not be an identical matrix; hence, the variables are highly correlated. It was also found that the commonalities of the items are greater than 0.5, which explains more than half of the variability of the items, as only three factors were extracted with an explained total variance of 78.8%. Therefore, it can be asserted that a construct validity exists.

Table 1 Omega values, correlation and global R^2 values for each dimension

	Internal factors	External factors	Contextual factors	Global
Total omega for total scores and subscales	0.97	0.89	0.85	0.98
General Omega for total scores and subscales	0.87	0.73	0.56	0.85
Correlation of the scores with the factors	0.49	0.85	0.84	0.94
Multiple R^2 of the scores with the factors	0.24	0.73	0.7	0.89

3.2 Analysis of Job Turnover Factors

In the descriptive analysis (Table 2), the central tendency measures (mean and median) and variability (standard deviation and interquartile range) of each of its dimensions were taken. In the internal factors, the highest scores found for motivation and job interest ($M = 3.67$; SD = 0.88) were benefits and incentives ($M = 3.01$; SD = 0.63).

Furthermore, the following indicators can be found: work environment, work stress and pressure, commitment and satisfaction with the position held and training with average values 3.54–3.65 with variabilities for some greater than 1.2. For external factors, "family members" ($M = 3.97$; SD = 1.33) had the highest average score and the lowest was found in "Change of residence" ($M = 3.54$; SD = 1.32). However, the average behavior among the indicators for employment opportunities, professional opportunities and illness averaged between 3.55 and 3.57, with deviations between 1.26 and 1.41. It is noteworthy that all the indicators in this factor have variabilities greater than 1. Working conditions had high scores ($M = 4.06$; SD =

Table 2 Descriptive analysis of job turnover factors

Dimension	Indicator	M	Me	SD	IR
Internal factors	Salary and benefits	3.40	3.50	0.96	1.50
	Work environment (relationships among workers)	3.54	4.00	1.31	1.33
	Stress and pressure at work	3.53	4.00	1.38	3.00
	Commitment and satisfaction in the position performed	3.63	4.25	1.25	2.50
	Training	3.65	4.00	0.95	1.80
	Motivation and interest in the job	3.67	4.00	0.88	1.00
	Benefits and incentives	3.01	3.00	0.63	0.50
	Work socialization (being listened to and recognized)	3.24	3.33	0.85	1.00
External factors	Families	3.97	5.00	1.33	3.00
	Employment opportunities	3.55	4.00	1.26	1.00
	Academic and professional opportunities	3.55	4.00	1.29	2.00
	Change of residence	3.54	4.00	1.32	2.00
	Illness	3.57	4.00	1.41	3.00
Contextual factors	Work conditions	4.06	4.50	1.04	1.00
	Company policies	3.78	3.80	0.80	1.20
	Organizational climate	2.89	3.00	0.97	1.00
	Company flexibility	3.42	3.75	1.13	1.50
	Location of the company	3.73	4.00	1.36	2.00

M Arithmetic mean, *Me* Median, *SD* Standard deviation, *IR* Interquartile range

1.04) among the contextual factors, with organizational climate scoring low (M = 2.89; SD = 0.97). In addition, it can be seen that company policies and flexibility have very similar patterns with average values of 3.78 (SD = 0.80) and 3.27 (SD = 1.36), respectively.

3.3 Influence of the Indicators on Job Turnover Dimensions

By analyzing the influence of the indicators on the dimensions, a confirmatory factor analysis with maximum likelihood estimates was chosen. Endogenous variables are the external, internal and contextual factors in which indicators are grouped according to the model adopted. Chi-squared = 2070.92 with $p = 0.000$ values were obtained. This indicates that the hypothesis that the covariance matrix is the same as the model obtained is rejected; however, this statistic penalizes large sample sizes which was the case in this study. This is followed by the root mean square error of approximation, RMSEA = 0.0958; the normed-fit index, NFI = 0.7804; the non-normed-fit index, NNFI = 0.7578; and the comparative fit index, CFI = 0.7910. Thus, we demonstrate the covariations and causes model (Table 3).

Based on the model obtained, it is possible to establish that the personnel turnover indicators that most influence the internal dimension are work environment and work stress and pressure. In reference to external factors, it is possible to establish that illnesses and location changes are those that have the greatest impact, with company flexibility having the largest impact among contextual factors. Similarly, the causality relationships of each factor with its indicators are significant as are those of covariation between endogenous variables, due to the fact that $p < 0.05$.

4 Discussion

Among internal factors, it has been demonstrated that the greatest impact is generated by the work environment, work stress and pressure and commitment and satisfaction. This is in line with the theoretical models proposed in [23], as they refer to job turnover as an effect of leadership that can affect the work environment and pressure. This is in addition to the prioritization of the worker's job satisfaction variable, which is associated with an emerging issue found in internal factors. When contextualizing, the response pattern regarding income is variable, however, as income is not, the components that most influence the factor do not affect internal consistency, due to the fact that on a global basis and due to different factors, it is internally consistent (Cronbach's alpha and McDonald's omega) at above 0.8.

With respect to external factors, they have very similar influences on the variables studied, which explains the adjustment measures greater than 0.7. However, aspects such as illness and change of residence predominate these factors. This is associated with an evident lack of guidelines for occupational health and safety in textile

Table 3 Confirmatory factorial model with maximum likelihood estimates

Causality or covariation	Estimate	Standard error	Z	*p*-value
Work environment←Internal factors	1.2523	0.0491	25.4947	0.0000
Benefits and incentives←Internal factors	0.3673	0.0295	12.4340	0.0000
Training←Internal factors	0.8660	0.0374	23.1417	0.0000
Commitment and satisfaction← Internal factors	1.1947	0.0468	25.5200	0.0000
Stress and pressure at work←Internal factors	1.2156	0.0556	21.8678	0.0000
Motivation and interests←Internal factors	0.6259	0.0388	16.1432	0.0000
Salary and benefits←Internal factors	0.6745	0.0430	15.6925	0.0000
Work socialization←Internal factors	0.7211	0.0349	20.6746	0.0000
Change of residence←External factors	1.2351	0.0505	24.4409	0.0000
Illness←External factors	1.2546	0.0566	22.1669	0.0000
Family members←External factors	1.1439	0.0546	20.9419	0.0000
Academic opportunities←External factors	1.2290	0.0487	25.2229	0.0000
Employment opportunities←External factors	1.1466	0.0493	23.2462	0.0000
Organizational climate←Contextual factors	0.5118	0.0461	11.1029	0.0000
Work conditions←Contextual factors	0.7090	0.0468	15.1530	0.0000
Company flexibility←Contextual factors	1.0698	0.0432	24.7633	0.0000
Company policies←Contextual factors	0.4179	0.0382	10.9439	0.0000
Company location←Contextual factors	0.7649	0.0639	11.9669	0.0000
Covariance internal factors↔External factors	0.9483	0.0070	136.3394	0.0000
Covariance internal factors↔Contextual factors	1.0045	0.0063	159.2149	0.0000
Covariance external factors↔Contextual factors	0.9040	0.0136	66.6436	0.0000

companies, which is consistent with the work indicators of [5] and comparable to the fact that companies are largely unaware of applicable legislation and management systems [28, 29]. To conclude, a construct validity and internal consistency of the factors of labor turnover adapted from [27] were obtained for the case of Barranquilla, Colombia.

References

1. DANE (2018) Encuesta anual manufacturera (EAM). DANE, Bogóta, pp 1–28
2. Porter M (2015) Competitivestrategy. Patria, Mexico
3. Cabrera A, Ledezma MT, Rivera N (2011) El impacto de la Rotación de Personal en las empresas constructoras del estado de Nuevo León. Contexto. Revista de la Facultad de Arquitectura de la Universidad Autónoma de Nuevo León 5(5): 83–91
4. Knapp M, Harissis K, Missiakoulis S (1981) Predicting Staff Turnover. Management research news (MCB UP Ltd) 4(1):18–20

5. Candanoza A, Lechuga J (2015) Caracterización de la satisfacción laboral y condiciones de trabajo de una empresa de seguridad en la ciudad de Barranquilla durante el año 2012. RevistaColombianaSalud Libre 10(2):98–102
6. Castaño C, Mehecha L (2015) La influencia del proceso de admisión de personal para la efectividad laboral en la empresa Kenworth de la montaña sede barranquilla. Medellín, Antioquia
7. Naranjo R (2012) El proceso de selección y contratación del personal en las medianas empresas de la ciudad de Barranquilla. Pensamiento & Gestión 32:83–114
8. Camacho F (2012) Caracterización de la rotación laboral en la planta de procesos de la empresa avicola Pimpollo SAS de la cuidad de Bucaramanga. Bucaramanga
9. Gonzalez D (2009) Estrategias de retención del personal. Una reflexión sobre su efectividad y alcances. Universidad EAFIT 45(156):45–72
10. Henao M, Quiñones M, Cáceres S (2013) Estrategias de tercerización en Colombia como centro de operaciones nfocado a los call center en Barranquilla. Bogotá
11. Flores M, Vega A (2010) Memorias I Coloquio Competitividad y Capital Humano. Editedby R Pérez. Cuerpo Académico, Productividad, Competitividad y Capital Humano. México: ILCSA S.A. de C.V. pp 1–485
12. Orozco-Acosta E et al (2017) Herramientas para gestión de la productividad en la empresa. Experiencias exitosas desde el Caribe colombiano. Ediciones Universidad Simón Bolívar, Barranquilla
13. Buentello C, Valenzuela N, Benavides I (2014) Análisis de la rotación de personal. Caso: Sabritas, sucursal Piedras Negras. Global Conf Bus Finance Proc 9(2):1697–1703
14. Gandy RP, Harrison, Gold J (2018) Criticality of detailed staff turnover measurement. Benchmarking: an Int J 25(8): 2950–2967
15. Santos D, Gil C (2016) Factores Asociados a la rotación de personal de la Cámara de Comercio Colombo Americana de Bogotá D.C. Bogotá
16. Domínguez M (2008) Factores determinantes en la gestión de recursos humanos en empresas de servicios que incorporan de manera sistemática nuevas tecnologías Un estudio de caso en la comunidad valenciana. Revista científica Pensamiento y Gestión, 24:88–131
17. March J, Simon H (1958) Organizations. University of Illinois at Urbana-Champaign's Academy for Entrepreneurial Leadership Historical Research Reference in Entrepreneurship
18. Porter L, Steers R (1973) Organizational, work, and personal factors in employee turnover y Absenteeism. Psychological Bulletin 80:151–176
19. Mobley W (1977) Intermediate linkages in the relationship between job satisfactions an Employee turnover. J Appl Psychol 62(2):237–240
20. Price J (1977) The study of turnover. Ioawa State University Press, Ames, IA
21. Fichman M (1988) Motivational consequences of absence and attendance: proportional harzard estimation of a dynamic motivation model. J Appl Psychol 73(1):34–119
22. Lamber E, Hogan N, Barton S (2001) The impact of job satisfaction on turnover intent: a test of a structural measurement model using a national sample of workers. Social Sci J 38(2):233–250
23. García B, Rivas L (2007) Un modelo de percepción de la rotación laboral en la población de trabajadores de la industria maquiladora mexicana. Innovar 17(9):107–114
24. Allen DG (2008) Retaining talent: a guide to analyzing and managing employee turnover. SHRM foundation effective practice guidelines series, pp 1–43
25. Morales J (2011) Evaluación de los factores que determinan la rotación voluntaria de personal directo en empresas maquiladoras de Tijuana, B.C. Tesis doctoral. Ensenada B.C.: Universidad Autónoma de Baja California
26. Melendres V, Aranibar M (2017) Factores que inciden en la rotación de personal en maquiladoras: una revisión del panorama en México. In: Global Conference on Business and Finance Proceedings. San José, pp 446–455
27. Aranibar M, Melendres V, Ramírez M, García B (2018) Los Factores de la rotación de personal en las Maquiladoras de exportación de Ensenada, B. C. Revista Global de Negocios 6(2):25–40

28. Guerrero O, Guerrero R (2017) Las empresas de Norte de Santander y su perspectiva acerca de la seguridad y salud en el trabajo. Investigación E InnovaciónEnIngenierías 5(2):26–45
29. Amelec V, Carmen V (2015) Validation of a model for productivity evaluation for microfinance institutions. Adv Sci Lett 21(5):1610–1614

CTR Prediction of Internet Ads Using Artificial Organic Networks

Jesus Silva, Noel Varela, Danelys Cabrera, and Omar Bonerge Pineda Lezama

Abstract For advertising networks to increase their revenues, priority must be given to the most profitable ads. The most important factor in the profitability of an ad is the click-through-rate (CTR) which is the probability that a user will click on the ad on a Web page. To predict the CTR, a number of supervised rating models have been trained and their performance is compared to artificial organic networks (AON). The conclusion is that these networks are a good solution to predict the CTR of an ad.

Keywords CTR prediction · Artificial organic networks in advertising · Supervised rating models · CPC advertising networks

1 Introduction

Since the appearance of the first banner in October 1994, Internet advertising has grown steadily [1]. The infrastructure and technology of the network allows the launch of campaigns with important advantages for advertisers that traditional advertising cannot offer.

One of the main advantages is that it allows advertisers to segment customers according to certain parameters such as location, time, hobbies, or type of device from which they access. This is known as microtargeting and allows campaigns to

J. Silva (✉)
Universidad Peruana de Ciencias Aplicadas, Lima, Peru
e-mail: jesussilvaUPC@gmail.com

N. Varela · D. Cabrera
Universidad de La Costa, St. 58 #66, Barranquilla, Atlántico, Colombia
e-mail: nvarela2@cuc.edu.co

D. Cabrera
e-mail: dcabrea4@cuc.edu.co

O. B. P. Lezama
Universidad Tecnológica Centroamericana (UNITEC), San Pedro Sula, Honduras
e-mail: omarpineda@unitec.edu

V. Singh et al. (eds.), *Computational Methods and Data Engineering*,
Advances in Intelligent Systems and Computing 1227,
https://doi.org/10.1007/978-981-15-6876-3_38

be launched to a small group with common interests, which makes these campaigns very effective [2].

Internet advertising has a wide variety of ad types such as pop-up, pop-under, or interstitial [1]. But the most used ads because of their great impact and because they are not intrusive are banners, text ads, and videos.

The Internet advertising model is made up of four clearly differentiated roles: the advertisers, the advertising network, the publishers, and the users. The advertisers are the companies that pay the advertising network to have their ad displayed on the publishers' pages; publishers are the people who have at least one Web page and who receive an economic retribution for lending a space of their pages where the advertisements are shown, and the users are people who surf the Internet and buy a product when they are interested.

The advertising network is in charge of managing the entire advertising process. Among the main tasks are: fraud detection [3], role privacy preservation, and payment management. But, of course, the most important task is to optimize the performance of the advertising model so that advertisers have successful campaigns and that both the advertising network and the publishers get maximum performance from their ads.

In online advertising, there are a multitude of models for charging advertisers such as payment for showing an ad for a certain number of days, payment for links pointing to the advertiser's Web site or payment for redemption of coupons2. But the most widely used models are CPM, CPC, and CPA.

In the CPM model (Cost-per-mile), advertisers pay for every thousand impressions [3] of an ad; in the CPC (Cost-per-click) model, they pay each time a user clicks on an ad; and finally, in the CPA (Cost-per-acquisition) model, they pay when a user who clicks on a publisher's banner makes a purchase or hires a service.

The most widely used model is the CPC, mainly because it is the easiest for advertisers to understand and because it has been adopted by the main companies in the sector [4]. To optimize the performance of CPC networks, the profitability of the ads in the campaigns launched by advertisers is calculated.

Within the CPC payment model, there are two concepts related to the payment of advertisers which are the maximum CPC and the CTR. The maximum CPC represents the highest amount an advertiser is willing to pay for a click. However, in most cases a lower price is charged which is called the real CPC. The click-through rate (CTR) of an ad is defined as the ratio between the number of clicks on the ad and the number of impressions. It is not possible to know for sure the CTR [4] that a certain ad will have in the future, but if the CTR of the ads can be predicted with a very small margin of error, it is possible to choose the ad that provides the highest benefits.

There are few publications regarding the algorithms used by advertising networks [5]. This is very reasonable since if these algorithms were published, companies would lose their competitive advantage, since anyone could copy the research that takes so many years and so many economic resources.

On the other hand, people with the intention of committing fraud can violate the advertising networks.

To calculate the CTR, supervised classification-type models are used. The creation of supervised models is an important part of data science. Data science is about obtaining useful information for organizations from collected data. The techniques applied come from sciences such as logic, statistics, probability, model design, or pattern recognition.

Many tools have been developed to simplify and automate these processes. Among the most complete and successful open source tools is RStudio. This tool will be used to predict results with various methods such as multivariate adaptive regression splines, nearest diminished centroid, or linear discriminant analysis and then use artificial organic network (AON) to contrast the results.

First, the AON and the challenge of the models will be trained with a dataset. Once trained, the effectiveness of these will be tested with a set of inputs and then the models will be evaluated.

2 Artificial Hydrocarbon Networks and Their Applications in CTR Prediction

This section briefly explains the machine learning technique called artificial hydrocarbon networks (AHN) and then presents the proposed CTR prediction model using this method [6–8].

2.1 *CTR Prediction Model Based on Artificial Hydrocarbon Networks*

This research proposes the use of artificial hydrocarbon networks as a supervised learning method for the construction of a CTR prediction model. For this application, it was decided to use artificial hydrocarbon networks formed from a linear and saturated compound that receives multiple inputs (categories of a specified dataset) and generates one output (the CTR value). Figure 1 shows the proposed CTR prediction

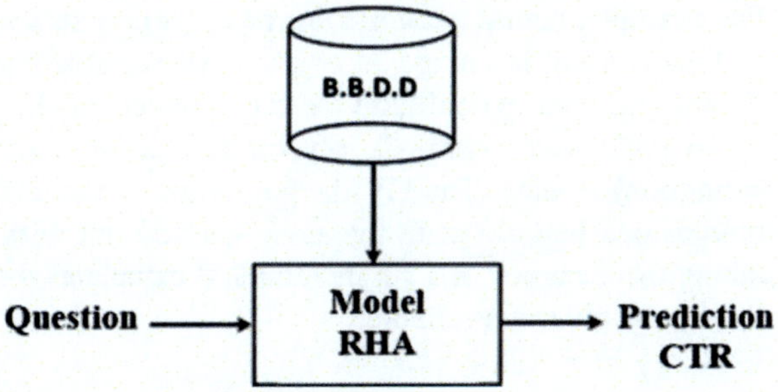

Fig. 1 Proposal for the CTR prediction model based on artificial hydrocarbon networks

model using AHN.

As shown in Fig. 1, the CTR prediction model is trained using a database with attribute-value type information. The training process is composed of the use of the multi-category artificial hydrocarbon network algorithm. Then, a request can be made based on the same type of attributes described in the database. Finally, the artificial hydrocarbon network model generates the respective CTR value.

The following section tests the CTR prediction model based on artificial hydrocarbon networks using a set of advertising data with their expected CTR value.

2.2 CTR Prediction of the Other Algorithms Using RStudio's Caret

RStudio is a programming environment oriented to scientific calculation. This platform allows users to create packages that implement a set of mathematical functions or methods. Caret is a package designed to simplify the process of building and evaluating models. Caret has a total of 180 methods for building predictive models, 138 of which are classification and 42 regression methods [9].

This tool has methods that automate the selection of variables. This allows to build more precise models, less complex and in less time. The methods for the selection of variables discard those characteristics that do not provide valuable information or that are redundant. For this problem, a method called recursive feature elimination [10] will be used. This method is applied to all models except AON, since this model integrates an alternative method called principal component analysis (PCA).

The RFE method is of the Wrapper type, that is, it has as inputs a combination of predictors that it uses to build models, and as output, a metric that evaluates the accuracy of the model by means of the random forest method. This serves to test combinations of variables until the best combination is achieved.

The main advantage of Caret is the simplicity of the code to build models with different methods and with different parameter settings. As well as the implementation of a series of metrics to automatically select the best parameter configuration from various models that are generated with one method. Metrics such as ROC, AUC, accuracy, or RMSE error are often used to measure model accuracy.

Each method has a certain number of parameters, usually between zero and three. For example, neural networks have two parameters which are the number of nodes and the level of learning. Caret scans for possible parameter values but also allows the programmer to configure these parameters [11].

To evaluate the methods, several datasets are usually made and the CV (Cross-validation) is used. The CV method is used to create different training sets so that models can be built with the same method but with different samples. Instead of taking the accuracy of a single model, it calculates the average of all of them and so the results are more reliable.

Finally, RStudio has a set of libraries that allows you to create high-quality, highly illustrative graphics. This allows to visualize the results obtained in a much more comprehensible way.

3 Methodology for CTR Prediction Through Supervised Methods

To solve the problem that arises a dataset provided by the Web site Kaggle.com has been used [6]. This dataset collects some parameters of the users' visits to the Web site www.criteo.com during a period of seven days. On these visits, the number of samples in which users did not click on the ad has been reduced more than those who did.

The visits are represented in a table where each row represents the visit of a user and each column represents a characteristic of a user or of the Web site. For example, the first column of the table is represented by "0" or "1" if the user clicked on the ad. The table has thirteen columns with integer type values and 26 string type values representing certain categories. The values of the categories have been hashed to a 32-bit value to ensure privacy. The rows are ordered chronologically and when the value of a certain parameter is unknown, a blank space is simply left.

To perform the test, a file with the same format as the training table will be used, but without the column indicating whether the user clicked. To evaluate the performance of the AON in the prediction of the CTR, two metrics will be used. The first metric is based on the logarithm of the probability function for a Bernoulli's random distribution and the second metric is based on something as simple as the percentage of hits over the total tests.

A total of 1,000,000 random samples were chosen to train the networks, as AON do not support excessive information. Of this dataset, 80% of random observations were used for the training process. The training dataset was cleaned, i.e., incomplete data were removed with the help of the PCA. This method selects and sorts the most important columns. Subsequently, the data were standardized using (1), where x is a column of the dataset, μ is the mean of the value of that column, σ is the standard deviation of that column, and x_{std} is the standardized column.

To develop the CTR prediction model, the organic network was made up of 100 artificial molecules using the least squares estimation technique. This network was trained with a learning coefficient of 0.5. The learning coefficient is a parameter of the artificial hydrocarbon networks that allows regulating the rate of assimilation or regression of the real data.

$$x_{\text{std}} = \frac{x - \mu}{\sigma} \tag{1}$$

4 Results Achieved

In the first metric, the probability of a user clicking is expressed in the range [0.1]. The lower the estimate that the user will not click, the closer it is to zero, and the higher the estimate that the user will click, the closer it is to one.

Once the results have been established, the Log Loss of the model will be calculated using the following formula:

$$\text{Log Loss} = -1\text{n} \sum i = 1\text{n} \left[y_i \log(y_i) + (1 - y_i) \log(1 - y_i) \right] \quad (2)$$

The use of the logarithm allows a mistake not to have an excessive punishment in the overall result. With this formula, if the probability of clicking is 99% and it is correct, it will give us a great reward, and if it fails, a great penalty will be applied.

The results achieved with the logarithm of the probability function for a Bernoulli's random distribution are very positive since the result is Log Loss = 0.6511. Figure 2 shows the precision of each model compared to the average of all methods.

Where:

- Model 1: R. Organic.
- Model 2: Trees C5.0.
- Model 3: CART.
- Model 4: CART Cost-sensible.
- Model 5: Multi-regression splines.
- Model 6: Nearest center.

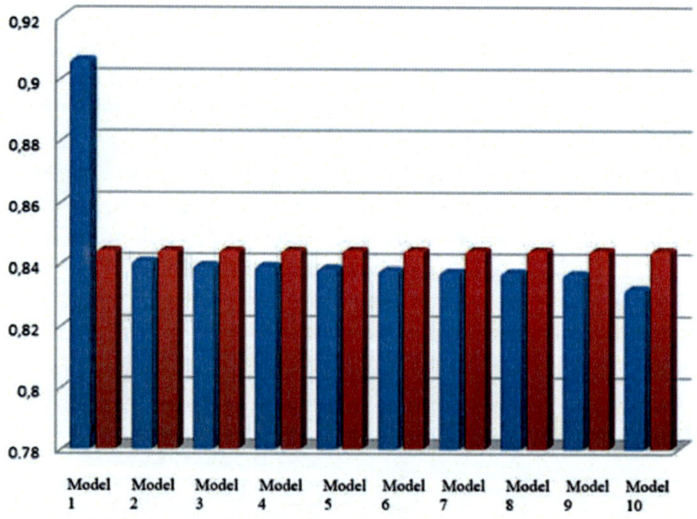

Fig. 2 Graph with the results of the precision and time in hours of the networks. Blue precision and red precision average

- Model 7: Reduction. discriminant analysis.
- Model 8: Linear discriminant analysis
- Model 9: Penalty in Discriminant Analysis.
- Model 10: OneR.

By using a second, more easily interpreted metric to measure the effectiveness of the algorithm, it is expressed as the ratio of the number of successes to the number of total cases. The user then clicks if the estimated probability is greater than 0.5 and selects no if it is lower. The prediction system can give four possible results:

- True positive (TP): It is predicted that the user will click and the user will click.
- True negative (TN): It is predicted that the user will not click and if he does click.
- False Positive (FP): It is predicted that the user will click and not click.
- False Negative (FN): It is predicted that the user will not click and if he does click.

If one considers that VP + VN + FP + FN = N. Then the accuracy rate of the system is equal to (VP + VN)/N and the error rate of (FP + FN)/N.

The results obtained in this metric were VP = 20,124, VN = 174,578, FP = 15,147 and FN = 4785. So, the precision was 90.71% and the Error Rate = 9.51%.

Table 1 shows the results obtained with the different supervised methods. These models were created with the RStudio Caret package and the "repeatedcv" method, with 10 partitions and 3 repetitions. These models were created with 900,000 samples and the accuracy of the model was measured with 300,000 samples in the same way as the AON networks.

5 Conclusions

AON is a very effective data model that can be adapted to many types of problems. They also offer the advantage of allowing to create models in a short time.

The AON allows for a CTR prediction of approximately 90% accuracy which is a good result. An improvement could be to apply a heuristic that improves the results of least squares estimation. Hybrid learning techniques could also be used (e.g., Bayesian networks for incomplete data, AVL trees to improve information searching, etc.), as a single technique limits the learning potential in real problems.

One disadvantage of AONs is that they cannot withstand being trained by a large number of samples. One solution to this problem would be to train a set of organic networks. For example, create 1000 artificial organic networks and predict the result as the average of all the networks.

On the other hand, this type of network can be a good ally in the fight against fraud in Internet advertising. There are many threats to networks such as botnets [12–14], click farms8 or publishers themselves9. All advertising networks are exposed to this type of threat and many of them are very difficult to detect. AONs are very useful for anomaly-based detection. In other words, networks could first be trained with parameters that represent the usual behavior of users. Once trained, when the

Table 1 Our proposal for the CTR prediction model based on artificial hydrocarbon networks

	Model 1	Model 2	Model 3	Model 4	Model 5	Model 6	Model 7	Model 8	Model 9	Model 10
Precision	0.9071	0.8457	0.8400	0.8358	0.8245	0.8358	0.8478	0.8458	0.8398	0.8431
Time build model	0:00:11	0:00:25	0:00:11	0:00:26	0:00:27	0:00:04	0:00:03	0:00:01	0:00:06	0:00:08

networks perceive a behavior different from what they are used to, they can send an alarm message.

The main advantage of AONs in this field is that they are able to point out the parameter for which they have considered the visit is not valid. This means that, in addition to discarding the click, it provides some information about this decision. Along with the AON, it would be convenient to use some filters to check if there are too many IPs from the same area, repeat users or an excessive number of clicks on a certain ad.

6 Future Studies

AONs are very effective predictive models that are useful for a large number of problems. An interesting project with this type of network would be to create a package for RStudio or MATLAB. This would make it easier for many people to use them and check their accuracy against other models in a simple way.

Another improvement would be to create a set of predictors to increase the success or prediction rate for both ranking and regression. In the case of classification, the class will be determined by the class that most models predict. In the case of regression models, the predicted value will be the average of the values predicted by all models.

Artificial organic networks can be very useful in detecting fraud in Internet advertising. These networks could be trained with normal visits made by the users and in case they detect an anomalous behavior they would warn the advertising network. It can also be done the other way around, that is, training the networks with the visits of the fraudulent users and in case it is similar notifying the advertising network.

Another line of future research could be to estimate the time needed to build the model with AON. This way, if the time is too high, the user can abort the operation or plan it according to the calculation time [15].

To do this, a model should be created with several samples that have as inputs: the number of samples, the number of processors, the RAM memory, the speed of the processors, the operating system and the number of inputs. This algorithm could be updated as the program of the same one that can navigate and click without the user's knowledge is executed [16].

Link farms are teams of people who may be in countries like India and who are dedicated to clicking to ruin advertising campaigns.

Publishers can cheat to increase their own income and often ask others to help them do so by downloading. There is no guarantee that 100% of the processor will be for this program [17–19].

References

1. Aladag C, Hocaoglu G (2007) A tabu search algorithm to solve a course timetabling problem. Hacettepe J Math Stat pp 53–64
2. Moscato P (1989) On evolution, search, optimization, genetic algorithms and martial arts: towards memetic algorithms. caltech concurrent computation program (report 826)
3. Frausto-Solís J, Alonso-Pecina F, Mora-Vargas J (2008) An efficient simulated annealing algorithm for feasible solutions of course timetabling. Springer, pp 675–685
4. Joudaki M, Imani M, Mazhari N (2010) Using improved Memetic algorithm and local search to solve University Course Timetabling Problem (UCTTP). Islamic Azad University, Doroud, Iran
5. Coopers PWH (2014) IAB internet advertising revenue report. http://www.iab.net/insights_research/industry_data_and_landscape/adrevenuereport
6. Tuzhilin A (2006) The lane's gifts v. google report. Official Google Blog: Findings on invalid clicks, pp 1–47
7. Ponce H, Ponce P, Molina A (2014) Artificial organic networks: artificial intelligence based on carbon networks. Studies in computational intelligence, vol. 521, Springer
8. Ponce H, Ponce P, Molina A (2013) A new training algorithm for artificial hydrocarbon networks using an energy model of covalent bonds. In: 7th IFAC conference on manufacturing modelling, management, and control, vol 7(1), pp 602–608
9. Viloria A, Lis-Gutiérrez JP, Gaitán-Angulo M, Godoy ARM, Moreno GC, Kamatkar SJ (2018) Methodology for the design of a student pattern recognition tool to facilitate the teaching—learning process through knowledge data discovery (big data). In: Tan Y, Shi Y, Tang Q (eds) Data mining and big data. DMBD 2018. Lecture notes in computer science, vol 10943. Springer, Cham
10. Moe WW (2013) Targeting display advertising. Advanced database marketing: Innovative methodologies and applications for managing customer relationships. Gower Publishing, Londres
11. Stone-Gross B, Stevens R, Zarras A, Kemmerer R, Kruegel C, Vigna G (2011) Understanding fraudulent activities in online ad exchanges. In: Proceedings of the 2011 ACM SIGCOMM conference on internet measurement conference. ACM, pp 279–294
12. McMahan HB, Holt G, Sculley D, Young M, Ebner D, Grady J, Kubica J (2013) Ad click prediction: a view from the trenches. In: Proceedings of the 19th ACM SIGKDD international conference on knowledge discovery and data mining. ACM, pp 1222–1230
13. Ponce H, Ponce P (2011) Artificial organic networks. In: IEEE conference on electronics, robotics, and automotive mechanics CERMA, pp 29–34
14. Kuhn M (2008) Building predictive models in R using the caret package. J Stat Softw 28(5):1–26
15. Granitto PM, Furlanello C, Biasioli F, Gasperi F (2006) Recursive feature elimination with random forest for PTR-MS analysis of agroindustrial products. Chemometr Intell Lab Syst 83(2):83–90
16. Kuhn W, Wing J, Weston S, Williams A, Keefer C et al (2012) Caret: classification and regression training. R package 515
17. Miller B, Pearce P, Grier C, Kreibich C, Paxson V (2011) What's clicking what? Techniques and innovations of today's clickbots. In: Detection of intrusions and malware, and vulnerability assessment. Springer, pp. 164–183
18. Kamatkar SJ, Tayade A, Viloria A, Hernández-Chacín A (2018) Application of classification technique of data mining for employee management system. In: International conference on data mining and big data. Springer, Cham, pp 434–444
19. Kamatkar SJ, Kamble A, Viloria A, Hernández-Fernandez L, Cali EG (2018) Database performance tuning and query optimization. In: International conference on data mining and big data. Springer, Cham, pp 3–11

Web Platform for the Identification and Analysis of Events on Twitter

Amelec Viloria, Noel Varela, Jesus Vargas, and Omar Bonerge Pineda Lezama

Abstract Due to the great popularity of social networks among people, businesses, public figures, etc., there is a need for automatic methods to facilitate the search, retrieval, and analysis of large amounts of information. Given this situation, the Online Reputation Analyst (ORA) faces the challenge of identifying relevant issues around an event, product and/or public figure, from which it can propose different strategies to strengthen and/or reverse trends. Therefore, this paper proposes and describes a web tool whose main objective is to support the tasks performed by an ORA. The proposed visualization techniques make it possible to immediately identify the relevance and scope of the opinions generated about an event that took place on Twitter.

Keywords Grouping · Similarity measurements · Information display

1 Introduction

The emergence of social networks on the Internet has enabled more people to freely publish opinions and comments on a wide variety of social, cultural, sports, scientific events, and even opinions on products and services [1].

Thanks to the popularity of these social networks, it is currently of great interest for many entities to know what is said about them in this digital world. At the

A. Viloria (✉) · N. Varela · J. Vargas
Universidad de la Costa, St. 58 #66, Barranquilla, Atlántico, Colombia
e-mail: aviloria7@cuc.edu.co

N. Varela
e-mail: nvarela2@cuc.edu.co

J. Vargas
e-mail: jvargas41@cuc.edu.co

O. B. P. Lezama
Universidad Tecnológica Centroamericana (UNITEC), San Pedro Sula, Honduras
e-mail: omarpineda@unitec.edu

V. Singh et al. (eds.), *Computational Methods and Data Engineering*,
Advances in Intelligent Systems and Computing 1227,
https://doi.org/10.1007/978-981-15-6876-3_39

same time, through these media, it is possible to have an approach with different users, who take advantage of this interaction to make know their opinion on certain topics, products, or services. One of the most used social networks for this purpose is Twitter that allows sending short messages called tweets with a maximum length of 140 characters. According to Statistic Brain as of January 2018, the total number of active users on Twitter amounts to 925,750,000 worldwide and the number of tweets per day is 89 million [2].

In order to take advantage of the large amount of information obtained from interactions (direct and indirect) between users and companies, the latter have created the figure of an Online Reputation Analyst (ORA). The work of this professional goes through three phases: The first phase consists of monitoring, to know at all times what is being published in relation to the company, product or public figure of interest. The second phase is the identification of relevant issues within the community of Twitter users, prioritizing the messages with the most important implications, negative or positive, to the concerned entity. Finally, the third phase consists of proposing market strategies to reverse the negative effects previously identified or even strengthen the positive aspects of the entity under review [3].

Because publications are growing rapidly, manual analysis of this information is difficult and time-consuming for ORA. Consequently, there is a need for automatic systems that allow this analysis to be carried out in a simpler and more timely manner. Recently, several research groups are interested in this issue [4] and have been given the task of developing systems focused on the analysis of opinions generated on Twitter. In this context, RELATIN emerges as an international forum in which different automatic systems are proposed and evaluated focused on the analysis of online reputation, specifically on the information produced in Twitter [5].

Until now, the different groups that have participated in RELATIN focus on developing automatic methods for tasks such as: (i) selection of relevant tweets for an entity [5], (ii) identification of negative, positive or neutral implications toward an entity [6], (iii) grouping of opinions by similar topic [7], and (iv) identification of opinion leaders within a community [8]. However, the problem of how to show the result of the automatic analysis to an ORA in a way that facilitates decision-making has been little explored. For this reason, the need arises to develop systems that take advantage of the results of these automatic methods and allow the generation of visual representations.

Based on the foregoing, this paper explores an alternative visualization of the results produced by content analysis systems on Twitter, in particular, systems developed within the framework of the competence of RELATIN. The proposed system focuses specifically on the visualization of the results of the grouping of opinions by similar topic and the way in which they relate to each other.

2 Related Studies

Currently, there are several applications available on the Internet focused on the grouping, analysis and visualization of information. These applications can be divided into two main categories: (i) tools specialized in the grouping and classification of large volumes of data and (ii) non-specialized tools, which allow qualitative analysis of data to standard users [9].

On the one hand, among the specialized tools it is worth mentioning that they tend to use a very technical language and, as a result, it is difficult for non-specialist users to interpret the results. As an example of this type of tools it is worth mentioning Cluto [10] and Weka [11], multi-platform tools that implement a great variety of automatic methods for data analysis. On the other hand, in the second category are tools used to analyze data that are published exclusively on social networks, and that seek to provide users, expert and inexpert, with sufficient elements to perform an easy and intuitive analysis of the results provided by their methods of identification of issues, polarity, etc. As examples of such tools are Spot, AnaliticPro, and Socialmention, which provide the user with various data visualization schemes that seek to highlight certain indicators allowing the analyst to evaluate and determine the reputation of a particular product or topic within a specific community [12].

Spot is an application that allows the interactive visualization of what is being published on Twitter in real time. The main idea is to quickly show the opinions that are generated on a particular topic. The way to present the tweets is through groups contained in bubbles, which are organized and displayed in different ways to highlight different types of information on the topic of interest. When searching for the topic, only the last 200 tweets will be retrieved for viewing. Keep in mind that Twitter search results only go back about a week [13], so the search and analysis can only be done on a very limited set of tweets.

AnaliticPro is an application that allows to process large volumes of information produced in different social networks. Measurements can be made with personalized criteria. In addition to this, it incorporates techniques to identify the meaning (i.e., positive, negative, or neutral) of the comments and can relate and build phrases to know the general opinion within a community on one or more topics. One of the main disadvantages of this tool is that it is based on semi-automatic techniques for the construction of its reports, that is to say, there are reports that are generated with the help of experts. For this reason, in order to fully exploit the infrastructure offered by AnaliticPro, the payment of licenses and/or reputational analysis services is required [14].

Socialmention is an application that monitors and analyzes the information that is generated in different social networks through Internet in real time. Besides, it allows to follow and measure the opinions on any person, company, product, etc. The searches are performed in more than 80 social media including the most visited ones, such as Twitter, Facebook, FriendFeed, YouTube, Digg, Google. Unlike the reviewed tools, Socialmention proposes different measures that ease the interpretation of the analyzed data, as, for example, strength, feelings, passion, and reach. Intuitively, these measures provide the user with an idea of the importance and scope of the subject under study [15].

In general, the tools mentioned above propose different methods for the analysis and visualization of the information produced in social networks regarding an entity. They mainly focus on identifying the polarity of the comments, the origin of the comments (i.e., social network, device, type of user), and possible trends. Contrary to the analyzed tools, the work developed in this paper seeks to provide tools that facilitate the identification of relevant issues and, at the same time, the relationship that these may have with other secondary aspects happened around the same event, an aspect that is not considered by any of the reviewed tools. In this way, the proposed tool will allow ORA to immediately identify the themes and sub-themes that occur around an event, as well as their relevance. The following sections describe in more detail the developed system [15, 16].

3 The Proposed System

The proposed system consists of three large modules, of which, the first one is in charge of the search and recovery of tweets, then there is a grouping process, which can be thematic or non-thematic, and finally, a graphic output that is shown to the user. The following sub-sections describe the main components of each of these modules.

3.1 Recovery of Tweets

The recovery of the tweets begins with a query, which specifies the topic to search. The query consists of a string of characters that can be one or several terms (i.e., composite queries), and the number of tweets to be retrieved. In order to perform this process, the twitter4j library was used to allow the connection with the Twitter platform. It is important to mention that, in order to use this library, it is necessary to register on the Twitter developers page. This process allows obtaining the access keys that allow automatic systems to make use of the information generated in this social network.

This way, if the Twitter connection is successful, the tweets are recovered, stored for later processes, and also displayed in the graphical interface of the system. If at any time a connection error occurs, the user is notified. It is important to mention that by storing the tweets in a database allows users to access their search history, which is an important option for ORAs, because it allows analyzing trends and/or making comparisons immediately. Another important aspect of the tweet recovery process is that it is designed to get the most metadata related to each downloaded tweet, for example: username, shared images, user profile information, date, time, platform, language, etc. [17].

Note that a step prior to storing the tweets in the database is a preprocessing module. This step is common in many natural language processing tasks, and its main objective is to remove information that is considered without thematic load. In this case, the following preprocessing operations were considered: The text is changed to lower case, punctuation symbols are removed, URLs are removed, and functional words are removed. Finally, the original and pre-processed tweets are stored in the database, ready to be processed by the grouping module [18].

3.2 *Thematic Grouping*

A necessary first step to carry out the task of thematic grouping is the indexing of the documents to be analyzed, an activity that requires the mapping of a *dj* document in a compact form of its content. The most commonly used representation is a vector with terms10 weighted as inputs, concept taken from the vector space model used in information retrieval [19]. That is, a *dj* text is represented as the vector $\overrightarrow{dj} = (wkj, \ldots, w|\tau|j)$, where τ is the dictionary, i.e., the set of terms that occur at least once in some document, while *wkj* represents the importance of the term *tk* within the content of the *dj* document.

This technique, known as bag of words (BOW11), is the form traditionally used to represent documents [20]. Within the developed tool, only simple words were considered as vector elements. The weight *wkj* can be calculated using different approaches, and the simplest of these approaches is the Boolean, which consists in assigning a value of 1 to the term if it appears in the document, and 0 in the opposite case. In addition to this, other very common weighing schemes are known as frequency (TF) and relative frequency (TF-IDF) [21]. It is worth mentioning that the developed tool includes these three weighing schemes.

Once a proper representation of the documents (i.e., tweets) is in place, a grouping process can proceed. Groups must meet a number of properties, that is, documents belonging to the same group must be very similar, while at the same time documents belonging to different groups must be as different as possible. These properties are known as homogeneity and heterogeneity, respectively. In general, in order to approximate these properties, it is necessary to determine similarities between objects from the values of their attributes. For this study, the measurement of the cosine was used.

In the system described in this chapter, two different grouping techniques were used, specifically a partition algorithm (*k*-means) and a hierarchical algorithm (Hierarchical Clustering).

On the one hand, partitioning algorithms group elements around central elements called centroids. The *k*-means algorithm is an iterative method that has the value of *k* as an important parameter (the number of groups to form), despite which it is a very effective algorithm [10]. On the other hand, hierarchical algorithms are characterized by generating a tree-like structure, called a dendrogram, in which each level of the tree is a possible grouping of the objects in the collection. The Hierarchical Clustering method is a hierarchical algorithm of agglomerative type, that is to say, part of the leaves of the tree considering each element as a group. Later and in an iterative way, it joins elements in closer groups until all the documents are inside a group [10].

It is important to mention that in the developed system the implementations made in Weka [6] of the grouping algorithms described with their default configurations were incorporated.

3.3 Visualization

For the visual representation of the thematic and non-thematic grouping results, the D3js library is used, specifically the type of graphic called Bubble Chart.

Each bubble in the graph represents a group, the result of the thematic grouping carried out in the previous stage. In addition, the size of the bubbles represents, to some extent, the importance of the identified subgroup. Thus, within the bubbles representing the different subgroups, the tweets corresponding to that group congregate as a smaller series of bubbles. One of the advantages of the proposed visualization is that it is possible, by positioning the course on a tweet (i.e., the smallest bubbles), to see the content of this one. Additionally, by means of this graph, it is possible to see the most representative *n* terms of the group, i.e., the most frequent terms in the tweets of the group under consideration.

As mentioned in previous sections, one of the advantages of this system is that it allows ORA, in addition to identifying the different topics and subtopics that occur around a given event, to show the thematic relationships between the different subtopics. To achieve this, the following steps are suggested: (1) identify the most representative concepts of each subtopic, (2) for each pair of subgroups, the concepts contained in the intersection are searched, and (3) finally, the concepts shared between each pair of subgroups are shown to the user in textual form. With this, the analyst can quickly identify key concepts (words) that the user community is using to refer to the event.

Additionally, the developed system also allows to generate graphs with information extracted from the metadata of the recovered tweets. In particular, it is possible to build graphs by grouping tweets by platform used for reading and writing on Twitter, e.g., Android, IOS, web, etc.; by number of favorites, number of retweets. Overall, this information is useful for ORA because it allows for identifying relevance and scope of the event being studied.

4 The Platform in Operation

In order to illustrate the visualization module implemented in the proposed platform, a search for tweets on the subject of Cartagena was carried out. The query retrieved

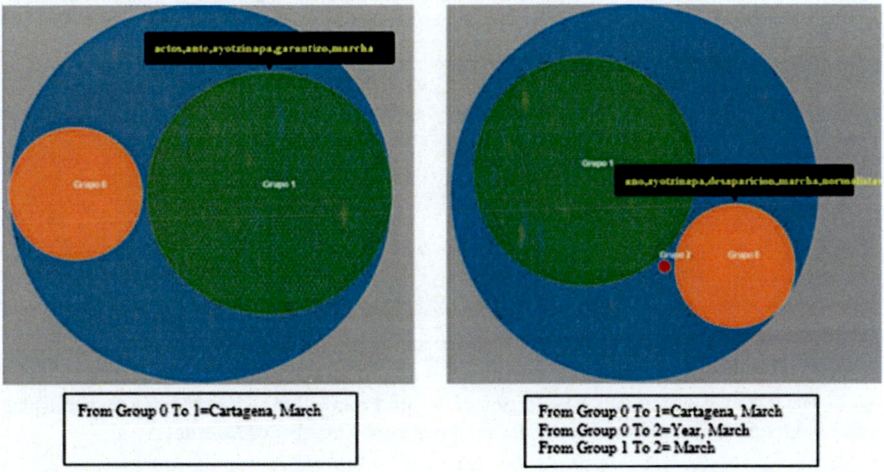

Fig. 1 Visualization of the thematic grouping for 3000 tweets of Cartagena. The image on the left shows the result of two generated groups, while the image on the right shows the result of the generation of three groups

3000 tweets; the number of tweets specified through the options of the web platform. In order to show some of the features of the visualization, Fig. 1 shows two different results after grouping by similar topic. In the image on the left, it can be seen that the result of the grouping generated two groups, while the image on the right shows an output that resulted in three subtopics. It is important to remember that one of the parameters required by the platform is the value of k, which indicates the number of subtopics to be identified. Intuitively, the higher the k value, the higher the level of specialty in the subtopics generated, while a very small value allows for greater generality in the subtopics.

As can be seen in Fig. 1, it is possible to obtain the most representative set of words of each group, which indicate, to a certain extent, the semantic content of each subgroup. In the case of the image on the left, Group 1 can be described with the words actos, ante, Cartagena, garantizó, and marcha, in contrast to the semantic content of Group 0 that can be defined by the words año, Cartagena, desaparición, marcha, and normalistas [22]. Faced with this example, an ORA could discern that while all the tweets of both groups talk about the progress made in the Cartagena case, one subgroup refers to the anniversary of the disappearance of normalists in Cartagena, while the other subgroup mentions the guarantees that were promised for the acts carried out in the context of the March. This type of information could easily be confirmed by positioning the cursor on a particular tweet and viewing its content.

On the other hand, in the image to the right of Fig. 1, the third group (Group 2) mentions the descriptive terms: año, marcha, México, normal, and tragedia; while the group is much smaller in relation to groups 0 and 1, it is clear which subtopic it describes, generalizing the problem at the country level and describing the event as a tragedy.

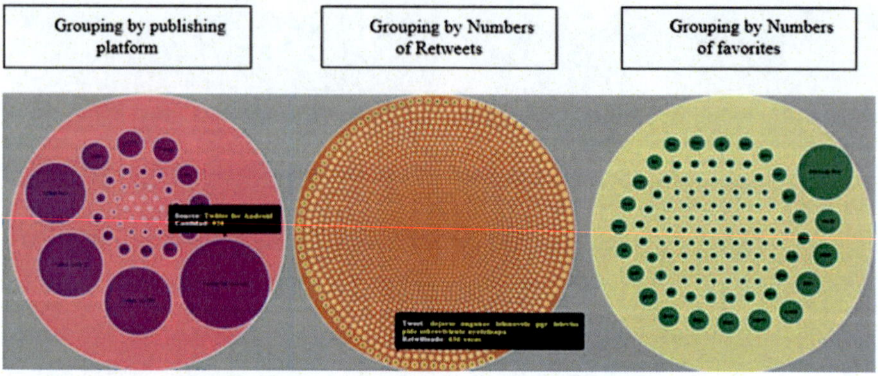

Fig. 2 Metadata visualization of a collection of tweets. From left to right: grouping by publishing platform, grouping by number of retweets, and grouping by number of favorites

In addition to the above, Fig. 1 also shows, in a very simple way, the words that the groups share. For example, in the case of the grouping of the image on the left, given that there are only two groups, the common terms between them are Cartagena and marcha. With this synthesized information, the ORA could have a general overview of the issue under analysis.

As mentioned in Sect. 3.3, in addition to displaying information about the thematic grouping, it is also possible to see graphs of the tweet metadata. Figure 2 shows the graphs generated by type of platform used to send the tweet, by number of retweets and by number of favorites (from left to right in Fig. 2). From the graph that groups the tweets by publishing platform, it can be seen that of the 3000 recovered tweets, 928 were sent from the Twitter application for Android; from the graph that groups the tweets by the number of retweets they have, it is possible to see that one of the most tweaked tweets (436 times) is a tweet that contains the thematic words "Don't be fooled by the soap opera of RCN, asks Cartagena survivor." Finally, the graphic that visualizes the tweets grouped by favorites shows that the tweet marked most times as favorite is the video of the position of the public figure Maluma.

5 Conclusions and Future Studies

This chapter described the work done for the construction of a web tool designed to support the activities carried out by an ORA. Specifically, the proposed tool allows the automatic identification of topics and subtopics (and the thematic relationship between them) happened around an event on Twitter. One of the advantages offered by the developed application is that, thanks to its proposal of visualization of results, an analyst can, in a simple and immediate way, identify the relevance and scope of the opinions expressed around an event of interest. The use of traditional grouping techniques allowed to define a strategy to achieve the identification of subtopics within a set of tweets. In particular, two types of grouping, partitioning and hierarchical algorithms, were used, which have shown to be effective in various document grouping tasks. As form of representation of the texts, the technique known as word bag (BOW) was used, as well as several weighing schemes. In general, the methods and techniques used are very intuitive and therefore easy to understand. However, it is necessary for ORA to know the meaning of the parameters required by these methods in order to make efficient use of the developed tool.

The future work is intended to adapt the tool to provide greater facilities to users to make compilation of corpus from Twitter.

References

1. Gonzalez-Agirre A, Laparra E, Laparra G (2012) Multilingual central repository version 3.0. In: Proceedings of the eight international conference on language resources and evaluation (LREC'12). Istanbul, Turkey: European Language Resources Association (ELRA)
2. Rousseeuw P (1987) Silhouettes: a graphical aid to the interpretation and validation of cluster analysis. J Comput Appl Math 20(1):53–65. [Online]. Disponible: http://dx.doi.org/10.1016/0377-0427(87)90125-7
3. Wilcoxon F (1945) Individual comparisons by ranking methods. Bio Bull 1(6):80–83
4. Toutanova K, Klein D, Manning CD, Singer Y, (2003) Feature-rich part-of-speech tagging with a cyclic dependency network. In: Proceedings of the 2003 conference of the North American chapter of the association for computational linguistics on human language technology, vol 1, ser. NAACL '03. Stroudsburg, PA, USA: Association for Computational Linguistics, pp 173–180. [Online]. Disponible: http://dx.doi.org/10.3115/1073445.1073478
5. Lis-Gutiérrez JP, Gaitán-Angulo M, Henao LC, Viloria A, Aguilera-Hernández D, Portillo-Medina R (2018) Measures of concentration and stability: two pedagogical tools for industrial organization courses. In: Tan Y, Shi Y, Tang Q (eds) Advances in swarm intelligence. ICSI 2018. Lecture Notes in Computer Science, vol 10942. Springer, Cham
6. Zhao WX, Weng J, He J, Lim E-P, Yan H (2011) Comparing twitter and traditional media using topic models. In: 33rd european conference on advances in information retrieval (ECIR11). Berlin, Heidelberg: Springer-Verlag, pp 338–349
7. Viloria A, Gaitan-Angulo M (2016) Statistical adjustment module advanced optimizer planner and SAP generated the case of a food production company. Indian J Sci Technol 9(47). https://doi.org/10.17485/ijst/2016/v9i47/107371
8. Ansah J, Liu L, Kang W, Liu J, Li J (2020) Leveraging burst in twitter network communities for event detection. World Wide Web, pp 1–26
9. Sapankevych N, Sankar R (2009) Time series prediction using support vector machines: a survey. IEEE Comput Intell Magaz 4(2):24–38
10. Viloria A, Lezama OBP (2019) Improvements for determining the number of clusters in k-means for innovation databases in SMEs. Procedia Comput Sci 151:1201–1206
11. Nugroho R, Paris C, Nepal S, Yang J, Zhao W (2020) A survey of recent methods on deriving topics from Twitter: algorithm to evaluation. Knowl Inf Syst 1–35
12. Dietrich J, Gattepaille LM, Grum BA, Jiri L, Lerch M, Sartori D, Wisniewski A (2020) Adverse events in twitter-development of a benchmark reference dataset: results from IMI WEB-RADR. Drug Safety 1–12
13. Romero C, Ventura S (2007) Educational data mining: a survey from 1995 to 2005. Expert Syst Appl 33(1):135–146
14. Romero C, Ventura S (2010) Educational data mining: a review of the state of the art. Systems, Man, and Cybernet Part C: Appl Rev IEEE Trans 40(6):601–618. Disponible en: http://ieeexplore.ieee.org/xpl/RecentIssue.jsp?reload=true&punumber=5326

15. Choudhury A, Jones J (2014) Crop yield prediction using time series models. J Econom Econom Educat Res 15:53–68
16. Scheffer T (2004) Finding association rules that trade support optimally against confidence. Intell Data Anal 9(4):381–395
17. Ruß G (2009) Data mining of agricultural yield data: a comparison of regression models, In: Perner P (eds) Advances in data mining. applications and theoretical aspects, ICDM 2009. Lecture Notes in Computer Science, vol 5633
18. Viloria A, Lis-Gutiérrez JP, Gaitán-Angulo M, Godoy ARM, Moreno GC, Kamatkar SJ (2018) Methodology for the design of a student pattern recognition tool to facilitate the teaching—learning process through knowledge data discovery (big data). In: Tan Y, Shi Y, Tang Q (eds) Data mining and big data. DMBD 2018. Lecture Notes in Computer Science, vol 10943. Springer, Cham
19. Amigo E, de Albornoz JC, Chugur I, Corujo A, Gonzalo J, Meij E, de Rijke M, Spina D (2014) Overview of replab 2014: author profiling and reputation dimensions for online reputation management. In: Information access evaluation. Multilinguality, multimodality, and interaction—5th international conference of the CLEF initiative, CLEF 2014, Sheffield, UK, September 15–18, 2014. Proceedings, pp 307–322
20. Berrocal JLA, Figuerola CG, Rodrıguez AZ (2013) Reina at replab2013 topic detection task: community detection. In: Proceedings of the Fourth International Conference of the CLEF initiative
21. Hall M, Frank E, Holmes G, Pfahringer B, Reutemann P, Witten IH (2009) The weka data mining software: an update. SIGKDD Explor Newsl 11(1):10–18
22. Ramırez-de-la Rosa G, Villatoro-Tello E, Jimenez-Salazar H, Sanchez-Sanchez C (2014) Towards automatic detection of user influence in twitter by means of stylistic and behavioral features. In: Gelbukh A, Espinoza F, Galicia-Haro S (eds) Human-inspired computing and its applications, lecture notes in computer science, vol 8856, pp 245–256. Springer International Publishing

Method for the Recovery of Indexed Images in Databases from Visual Content

Amelec Viloria, Noel Varela, Jesus Vargas, and Omar Bonerge Pineda Lezama

Abstract The techniques of content-based image recovery (CBIR) provide a solution to a problem of information retrieval that may arise as follows: from an image of interest to recover or obtain similar images from among those present in a large collection, using only features or features extracted from said images Banuchitra and Kungumaraj (Int J Eng Comput Sci (IJECS) 5 (2016) [1]). Similar images are understood as those in which the same object or scene is observed with variations in perspective, lighting conditions or scale. The stored images are preprocessed and then their corresponding descriptors are indexed. The query image is also preprocessed to extract its descriptor, which is then compared to those stored by applying appropriate similarity measures, which allow the recovery of those images that are similar to the query image. In the present work, a method was developed for the recovery of indexed images in databases from their visual content, without the need to make textual annotations. Feature vectors were obtained from visual contents using artificial neural network techniques with deep learning.

Keywords Convolutional Neural Networks · Global descriptors · Image retrieval · Information retrieval

A. Viloria (✉) · N. Varela · J. Vargas
Universidad de La Costa, St. 58#66, Barranquilla, Atlántico, Colombia
e-mail: aviloria7@cuc.edu.co

N. Varela
e-mail: nvarela2@cuc.edu.co

J. Vargas
e-mail: jvargas41@cuc.edu.co

O. B. P. Lezama
Universidad Tecnológica Centroamericana (UNITEC), San Pedro Sula, Honduras
e-mail: omarpineda@unitec.edu

V. Singh et al. (eds.), *Computational Methods and Data Engineering*,
Advances in Intelligent Systems and Computing 1227,
https://doi.org/10.1007/978-981-15-6876-3_40

1 Introduction

CBIR techniques are used in various branches of science such as medicine [2], agriculture, security and protection, weather forecasting, modeling of biological processes, classification of web images [3], crime prevention, image processing satellite, among others.

Traditional approaches mainly include the development of descriptors based on the content of the image, through so-called low-level features such as color [4], texture [5] and shape [6] or a combination of some of these [7]. A positive aspect of the techniques developed on the basis of these approaches is that they do not demand large amounts of data or time to obtain satisfactory results during the training and inference stages. On the other hand, they intend to obtain local and global descriptors of the images from manually drawn features, which are not generic, but have a strong dependence on the classes represented in the images. They generally restrict their ability to successfully scale to image collections with large numbers of classes or categories.

Recent methods combine low-level features with others called high-level, which provide a closer representation of human perception, allowing to reach a semantic description of the images and achieve better results in their recovery. The main advances in this direction are linked to the rapid development of machine learning techniques and specifically to deep learning or deep learning [8].

Neural models learn global descriptors developed on the basis of a hierarchy of traits and adjusted through a training process. These descriptors are generic and robust to challenges such as variability between classes, occlusion or changes in perspective or lighting. However, the feature vectors obtained by these techniques, in most cases, have a large dimension (2048, 4096 components), negatively affecting the use of memory for storage and the temporal complexity of the comparison process and recovery.

Several measures of dissimilarity are used during the comparison of the descriptors of the images, the most common being: the Euclidean distance, the distance of Bhattacharya, the distance of Mahalanobis, the distance of Sorensen and the distance of the cosine [9].

It is intended to solve the problem posed by developing a method where deep learning algorithms, specifically Convolutionary Neural Networks, are used to obtain global descriptors of the images. The size of these vector representations will be reduced through the application of the Principal Components Analysis. To determine the similarity between the images robust measures will be used, with wide use within the domain of image recovery.

2 Methodologic

According to [10] the paradigm for the recovery of images based on content can be broken down into the following stages: image acquisition, preprocessing, feature extraction, comparison of similarity and images recovered as a result. Some more recent systems also include feedback techniques. In correspondence with the previous stages, the proposed method has that general architecture, to which a stage of reduction of the dimension is added to the vector representations of the images.

2.1 Feature Extraction

During this stage, a structured CNN is used according to the Inception [11] model, in which a 42-layer deep neural architecture is presented, whose improvements over previous architectures such as GoogLeNet [12] or VGGNet [13] are mainly given by the factorization of the cores of convolution with the consequent spatial reduction of these filters, the use of auxiliary classifiers at the end of the training and the improvement of the techniques for the reduction of the amount of parameters to be learned.

Through a forward propagation of the image, through the network, the tensor pool is obtained which, as part of the graph constructed with TensorFlow [14], stores a vector representation of 2048 dimensions of the image. This stage is carried out in two moments: first during the indexing process of all available images, and it is saved next to each image in the database, and then for the query image.

2.2 Dimension Reduction

The vector descriptors obtained in the previous stage have 2048 components, in order to reduce this value the PCA technique is applied. First, the calculation of the covariance matrix of the vector representations of a sample of the available images is made. The eigenvectors are then calculated and sorted by the absolute value of their associated eigenvalue. The first eigenvectors are selected as a new base allowing the 2048 component vector, which represents the content of the image, to be projected into a smaller space and in which the representative and differentiating capabilities of these descriptors are still preserved [15].

2.3 Similarity Comparison

To determine which images in the database are similar to the query image, the value obtained from the application of a dissimilarity function, such as the distance of the cosine or the distance of Sorensen, is taken into account on the vector descriptors of the query image and those indexed in the database [16].

2.4 Ordination and Recovery

Similar images are ordered from the value obtained during the previous stage, so that those that are more similar to the query image are shown in the top positions of the candidate list [17].

3 Results

The experimentation process was carried out on the following international databases, used by other authors in works related to thematic.

–**UKBench**: This is a database of the Department of Computer Science of the University of Kentucky, has 10,200 color images, with dimensions of 640 × 480 pixels. The images are organized in 2550 groups of 4 images each with captures of the same object or scene taken with radical changes of perspective, as can be seen in Fig. 1 [18].

Fig. 1 Groups of 4 images in UKBench, you can observe (third and fourth groups) the similarity between images belonging to different groups

–Holidays: INRIA Holidays is a set of images about scenes or places that were taken by the National Institute of Informatics and Automation Research (INRIA) of France, with the purpose of having variety in terms of: rotations, point of view and changes in the lighting. The database includes various types of scenes: natural, man-made, etc.; all in high resolution. Holidays contains 500 groups of images, each of which represents a different scene or objects. The first image of each group is the query image and the correct results are the rest of the group members [19].

The weights of the network were initialized using a pre-trained model based on the ImageNet database of 2012. Inception-v3 was trained for the ImageNet Large-Scale Visual Recognition Challenge (ILSVRC). This is a standard task in computer vision where models try to classify complete images into a thousand classes. To compare the models, their failures are examined by predicting the correct answer as one of the first five predicted classes (top-5 error rate), Inception-v3 reaches 3.46% in the top-5 error [20]. As metrics to measure the performance of the proposed method, on the selected databases, the following were used:

–4 x Recall@4: Each image is used, in turn, as a query image and the average value of true positives obtained within the first 4 (top-4) images recovered is reported.

–mAP (Mean Average Precision): Given a set of query images this metric is defined as [21],

$$\text{mAP} = \frac{\sum_{q=1}^{Q} \text{AvePr(q)}}{Q} \tag{1}$$

where Average Precision (AvePr) for each query image is defined as [22],

$$\text{AvePr} = \frac{\sum_{k=1}^{n} (\text{Pr}(k) \, * \, \text{rel}(k))}{R} \tag{2}$$

where k is the range in the sequence of recovered images, n is the number of recovered images, Pr(k) is the precision in k in the list (Pr @ k), rel(k) is a function that takes value 1 if the image with rank k is relevant or 0 otherwise and R is the number of relevant images [23].

The neural models were implemented using TensorFlow and were run on a PC with a Core i5 3.5 GHz processor with 4 GB of RAM and an NVIDIA GeForce GTX 850 M GPU. Table 1 shows a summary of the characteristics of the databases used for the experiment. Table 2 shows the results obtained on the UKBench and Holidays databases with the original resolution images, reducing the size of the description

Table 1 Characteristics of the databases used

Database	Year	Amount of images	Minimum resolution
UKBench [24].	2016	11,254	640 × 480
INRIA Holidays [25].	2018	2154	848 × 480

Table 2 Results obtained on the UKBench and holidays databases

Method	UKBench (4× Recall@4)	Holidays (mAP)	Metric of distance	Dimensions descriptor
SIFT [27].	3.15	–	–	1225
AlexNet-conv3 [27].	3.85	–	–	–
CKN-mix [27].	3.78	–	–	4152
Res50-NIP [28].	3.98	–	–	2154
[29].	–	65.2	–	128 bits
[30].	–	81.2	–	2048
[31].	–	69.4	–	262 k dims.
[32].	–	85.2	–	4096
[33].	–	76.5	–	8064
Proposed method	3.42	71.25	Cosine	2048
Proposed method	3.24	71.45	Cosine	1024
Proposed method	3.98	73.25	Sorensen	1024
Proposed method	3.57	–	Cosine	512
Proposed method	–	74.25	Sorensen	512

vectors, applying PCA, to 512 and 1024 dimensions and using different measures of dissimilarity for comparison of these. The values obtained by other classical methods and the state of the art are also presented.

It was found that the best results are achieved when the size of the original descriptors is reduced by eliminating unused information or with some noise present in those descriptors. The new feature vectors allow for more optimal storage by reducing memory consumption and lead to better results in the recovery process. The distances of the cosine and Sorensen are the ones that obtained the best results on the original vectors and also reduced their size.

False positives are mostly obtained by recovering images that do not belong to the correct result group, although they represent the same type of object taken from similar angles and lighting conditions. These images are first recovered than the members of the group of the query image, which dramatically vary the capture perspective. It is necessary to recognize that the recovered images, from the semantic point of view, are really similar to the query and the error made is closely related to the low variability between certain groups of images present in the database on which it is experience the inability of the descriptors obtained to be discriminatory in the face of the aforementioned situations is due to the fact that the network has been trained to perform classification tasks and therefore the traits learned tend to be robust due to the variability between classes and lose specificity to the differences between images, which when classified would be in the same class or category. Alternatives that complement these limitations are described in [26] where, instead of using the activation vectors of the last layers of the trained neural network, they produce descriptors that integrate these vectors with statistical techniques and apply transformations to the input image to make them more invariants. However, the mentioned alternatives increase the size of the descriptors, the volume of operations necessary for each image to be processed and the temporal complexity of these, demanding greater computing power and memory space.

The results obtained, which significantly improve those achieved by traditional algorithms such as the SIFT descriptors, are encouraging considering that a pre-trained model has been used, without adjusting for the evaluated database, and that the dimensions of the global vector descriptors are smaller than those used by other methods. Using descriptors up to 8 times smaller (4096 vs. 512) results were lower than those of the state of the art only by a minimum margin. The memory space occupied by the global descriptors of the indexed images is an important issue as they are quantities of the order of thousands or millions of images. The volume of operations per image demanded by our method, as well as the low size of the descriptors to be processed and stored, make it feasible for environments where a balance between the consumption of time and resources of the recovery system and the obtaining of results is needed. .

It is intended to continue improving the performance of the method through the use of other more recent neuronal models and applying rotational and scaling transformations to the input image, following an approach similar to that addressed in [28] but optimizing the computational cost.

The proposed method can be used in the development of components for information retrieval systems, specifically content-based image recovery, humanizing the task of recovering images of interest stored in large data collections.

4 Conclusions

- A method was developed for the recovery of images based on content that addressed all the stages of this process, making use of recent advances in artificial neural networks with deep learning.
- Results comparable to those reported by state-of-the-art methods and better than those obtained through traditional techniques were achieved.
- The performance, based on international databases, validated the effectiveness of the method, making it a starting point for future work in this area.
- It is intended to continue developing the method with the addition of stages for the preprocessing and integration of the descriptors obtained, in others with a greater degree of abstraction.

References

1. Banuchitra S, Kungumaraj K (2016) A comprehensive survey of content based image retrieval techniques. Int J Eng Comput Sci (IJECS) 5 (2016). https://doi.org/10.18535/ijecs/v5i8.26. https://www.ijecs.in
2. Simonyan K, Zisserman A (2014) Very deep convolutional networks for large-scale image recognition. arXiv preprint arXiv:1409.1556
3. Szegedy C, Liu W, Jia Y, Sermanet P, Reed SE, Anguelov D, Erhan D, Vanhoucke V, Rabinovich A (2014) Going Deeper with Convolutions. CoRR, abs/1409.4842. http://arxiv.org/abs/1409.4842
4. Szegedy C, Vanhoucke V, Ioffe S, Shlens J, Wojna Z (2016) Rethinking the inception architecture for computer vision. In: Proceedings of the IEEE conference on computer vision and pattern recognition, pp 2818–2826
5. Satish Tunga D, Jayadevappa D, Gururaj. C (2015) A comparative study of content based image retrieval trends and approaches. Int J Image Proc (IJIP) 9(3):127
6. Tzelepi M, Tefas A (2018) Deep convolutional learning for content based image retrieval. Neuro-computing 275:2467–2478
7. Vakhitov A, Kuzmin A, Lempitsky V (2016) Internet-based image retrieval using end-to-end trained deep distributions. arXiv preprint arXiv:1612.07697
8. Chen L, Zhang Y, Song ZL, Miao Z (2013) Automatic web services classification based on rough set theory. J Central South Univ 20:2708–2714
9. Pineda Lezama O, Gómez Dorta R (2017) Techniques of multivariate statistical analysis: an application for the Honduran banking sector. Innovare: J Sci Technol 5(2):61–75
10. Viloria A, Lis-Gutiérrez JP, Gaitán-Angulo M, Godoy ARM, Moreno GC, Kamatkar SJ (2018) Methodology for the design of a student pattern recognition tool to facilitate the teaching—learning process through knowledge data discovery (big data). In: Tan Y, Shi Y, Tang Q (eds) Data mining and big data. DMBD 2018. Lecture notes in computer science, vol 10943. Springer, Cham
11. Zhu F et al (2009) IBM Cloud computing powering a smarter planet. Libro Cloud Computing, Volumen 599.51/2009, pp 621–625
12. Mohanty R, Ravi V, Patra MR (2010) Web-services classification using intelligent techniques. Expert Syst Appl 37(7):5484–5490
13. Thames L, Schaefer D (2016) Softwaredefined cloud manufacturing for industry 4.0. Procedía CIRP, 52:12–17
14. Viloria A, Neira-Rodado D, Lezama OBP (2019) Recovery of scientific data using intelligent distributed data warehouse. ANT/EDI40 1249–1254
15. Schweidel DA, Knox G (2013) Incorporating direct marketing activity into latent attrition models. Mark Sci 31(3):471–487
16. Setnes M, Kaymak U (2001) Fuzzy modeling of client preference from large data sets: an application to target selection in direct marketing. Fuzzy Syst, IEEE Trans 9(1):153–163
17. Viloria A, Lezama OBP (2019) Improvements for determining the number of clusters in k-means for innovation databases in SMEs. ANT/EDI40 1201–1206
18. Nisa R, Qamar U (2014) A text mining based approach for web service classification. Inf Syst e-Business Manage 1–18
19. Wu J, Chen L, Zheng Z, Lyu MR, Wu Z (2014) Clustering web services to facilitate service discovery. Knowl Inf Syst 38(1):207–229
20. Alderson J (2015) A markerless motion capture technique for sport performance analysis and injury prevention: toward a big data, machine learning future. J Sci Med Sport 19:e79. https://doi.org/10.1016/j.jsams.2015.12.192
21. Alcalá R, Alcalá-Fdez J, Herrera F (2007) A proposal for the genetic lateral tuning of linguistic fuzzy systems and its interaction with rule selection. IEEE Trans Fuzzy Syst 15(4):616–635
22. Elsaid A, Salem R, Abdul-Kader H (2017) A dynamic stakeholder classification and prioritization based on hybrid rough-fuzzy method. J Software Eng 11:143–159

23. Molina R, Calle FR, Gazzano JD, Petrino R, Lopez JCL (2019) Implementation of search process for a content-based image retrieval application on system on chip. In: 2019 X southern conference on programmable logic (SPL). IEEE, pp 97–102
24. Maur HK, Faridkot P, Jain, P (2019) Content based image retrieval system using K-means clustering algorithm and SVM classifier technique
25. Pothoff WJ, Price TG, Prasolov V (2020) U.S. Patent No. 10,565,070. U.S. Patent and Trademark Office, Washington, DC
26. Poplawska J, Labib A, Reed DM, Ishizaka A (2015) Stakeholder profile definition and salience measurement with fuzzy logic and visual analytics applied to corporate social responsibility case study. J Clean Prod 105:103–115. https://doi.org/10.1016/j.jclepro.2014.10.095
27. Paulin M et al (2017) Convolutional patch representations for image retrieval: an unsupervised approach. Int J Comput Vis 165–166
28. Chandrasekhar V, Lin J, Liao Q, Morere O, Veillard A, Duan L, Poggio T (2017) Compression of deep neural networks for image instance retrieval. arXiv preprint arXiv:1701.04923
29. Zhang T, Qi G-J, Tang J, Wang J (2015) Sparse composite quantization. In: Proceedings of the IEEE conference on computer vision and pattern recognition, pp 4548–4556
30. Gong Y, Wang L, Guo R, Lazebnik S (2014) Multi-scale orderless pooling of deep convolutional activation features. In: European conference on computer vision. Springer, pp 392–407
31. Jegou H, Perronnin F, Douze M, Sanchez J, Perez P, Schmid C (2012) Aggregating local image descriptors into compact codes. IEEE Trans Pattern Anal Mach Intell 34(9):1704–1716
32. Perronnin F, Larlus D (2015) Fisher vectors meet neural networks: a hybrid classification architecture. In: Proceedings of the IEEE conference on computer vision and pattern recognition, pp 3743–3752
33. Jegou H, Zisserman A (2014) Triangulation embedding and democratic aggregation for image search. In: Proceedings of the IEEE conference on computer vision and pattern recognition, pp 3310–3317

Model for Predicting Academic Performance Through Artificial Intelligence

Jesus Silva, Ligia Romero, Darwin Solano, Claudia Fernandez, Omar Bonerge Pineda Lezama, and Karina Rojas

Abstract During the transit of students in the acquisition of competencies that allow them a good future development of their profession, they face the constant challenge of overcoming academic subjects. According to the learning theory, the probability of success of his studies is a multifactorial problem, with learning-teaching interaction being a transcendental element (Muñoz-Repiso and Gómez-Pablos in Edutec. Revista Electrónica de Tecnología Educativa 52: a291–a291 (2015), [1]. This research describes a predicative model of academic performance using neural network techniques on a real data set.

Keywords Academic performance · Neural networks, learning analytics · Big data

J. Silva (✉) · K. Rojas
Universidad Peruana de Ciencias Aplicadas, Lima, Peru
e-mail: jesussilvaUPC@gmail.com

K. Rojas
e-mail: pcadkroj@upc.edu.pe

L. Romero · D. Solano · C. Fernandez
Universidad de la Costa, St. 58 #66, Barranquilla, Atlántico, Colombia
e-mail: lromero11@cuc.edu.co

D. Solano
e-mail: dsolano1@cuc.edu.co

C. Fernandez
e-mail: cfernand10@cuc.edu.co

O. B. P. Lezama
Universidad Tecnológica Centroamericana (UNITEC), San Pedro Sula, Honduras
e-mail: omarpineda@unitec.edu

V. Singh et al. (eds.), *Computational Methods and Data Engineering*,
Advances in Intelligent Systems and Computing 1227,
https://doi.org/10.1007/978-981-15-6876-3_41

1 Introduction

Although academic performance is a multifactorial variable, many of the studies around it include only personal and socioeconomic factors; however, the emergence and application of new teaching technologies, especially the use of virtual platforms, allow universities to collect a large amount of information in real time. These large electronic data generated provide a multivariate approach in the study of academic performance [2].

Learning analytics can be defined as the process of determining, evaluating, and interpreting the meaning of large volumes of educational data; using mathematical algorithms [3]. Although there are various techniques for the analysis of this type of information, there are more and more work focused on the use of algorithms based on artificial neural networks [4].

Artificial neural networks (RRs) represent simplified schemes of the physical structure and functioning of a biological neuron; so their algorithms have the ability to predict results by classifying patterns.

In recent years, multiple types of neural networks have been developed depending on the problem to be solved. One of the most commonly used architectures in classification and prediction problems, it responds to multilayered perceptron networks to which a number of neurons are associated. In this way, the learning is obtained by "back error propagation," which is based on obtaining the minimum of an error function calculated through the system output and the model obtained by the network training [5].

The use of learning analytics and these novel models, applied to education, are not yet sufficiently used in Colombia. Although many universities in the country use platforms with virtual environments in training processes, the data obtained is not used correctly to make predictions that allow timely decisions to be made. Hence the objective of this work is to develop a predictive model of academic performance using neural network techniques on information obtained through learning analytics.

2 Literature Review

Most studies on "Learning Analytics" focus on predicting the student's academic performance by quantifying the final grade or by the approval or not of a given course, making it difficult to compare the different research against the most stable predictors of academic performance [6–9].

The following are the ones that have been considered most important:

In [10] they collected student data from an online discussion forum, categorized it into several subsets of data, and then evaluated the predictive accuracy of each set through data mining methods. To do this, they used a classification algorithm and student interaction data before an examination to predict student learning performance.

In [11] they applied four mathematical models to predict the academic performance of students in an engineering course using their final grades. The final results showed that student final examination scores were predictable with 88% accuracy based on eight variables collected from a learning management system (LMS).

While in [12] they developed a model based on artificial neural networks to predict the final grade of college students before graduating. He worked with 30 students from the Department of Computer Science at Tai Solarin University of Education. The results showed the ability of neural networks to correctly predict students' final grade with 92.7% accuracy.

In [13] they developed a tree-based early warning system to predict whether or not students would pass. The model was built using data from 300 students and 13 variables collected using online analytics. Results revealed 95% accuracy.

In [14], they use different algorithms of artificial neural networks to find the most suitable model for prediction of academic success of graduate students in the faculty of sciences of the organization of the University of Belgrade. The model allows to identify those factors of Mayand influence the academic performance of the students.

In [15] he built a model based on multilayered neural networks to predict student performance in a mixed learning environment. To do this, I use student data stored on a Moodle server and predicted the success of the ongoing student, based on four learning activities: email communication, collaborative content creation with wiki, interaction of content measured by files viewed and self-assessment through online questionnaires. The model predicted the performance of students with a correct rating rate of 98.3%.

3 Methodology

The methodology used and applied to a set of 1232 students from the "Industrial Engineering Race" of three private universities in Colombia is described below.

3.1 Data Collection

Student information is obtained through the exchange of students with the "Virtual Learning Environment," developed in Moodle. System-supplied data records were extracted and cleaned using the "DataClean" and "DataCombine" packages of the "R" free-use software. In this way we detect the available variables that were part of the research and that we show in Table 1 [16–20].

Table 1 Predictor variables used

Variable	
Total number of clicks	Number of discussion posts viewed
Number of teacher consultations	Time to first activity (min)
Number of online sessions Participation in forums	Total number of discussion posts
Total online time (min)	Average time per session (min)
Participation in chat	Number of questionnaires-initiated
Number of visits to the course page	Number of attempts per test
Video-collaboration participation Number of comments	Number of assignments submitted
Number of resources viewed	Number of questionnaires viewed
Study time irregularity	View presentation number
Number of links viewed	Number of questionnaires passed
Study interval irregularity	Number of edits on the wiki
Number of content pages viewed	Number of wiki views
	Average degree of evaluation
	The greatest period of inactivity

3.2 *Data Analysis*

The obtained variables were processed, using the R "nnet" package, which allows the creation, application, visualization and implementation of direct feed neural networks with a single hidden layer, for log-linear multinomial models.

The design of the neural network used was based on a multilayer neural network topology (MLP), which was trained independently for each subject obtaining unique predictions correctly.

The network architecture was determined by an input layer and an intermediate or hidden layer with hyperbolic tangent sigmoidal activation functions and an output layer with linear activation function [21–26].

4 Results

Table 2 describes the classification and retrieval accuracy for each course and the differences in the predictive performance of the individually trained neural network in the data for each course.

Based on the results obtained, the Data Mining Applications course has the highest accuracy and recall value with 92.1% and 89.6% respectively. When looking at the remaining results, the prediction accuracy ranges from 65.2% for the Data Structure course to 84.3% for the Distributed Programming course. When averaging the scores obtained, we find an accuracy of 74.98%. Similarly we can find that memory scores have a high variability with average recovery value of 70.54%.

Alongside the above results, the important characteristics for each of the courses were extracted, which are set out in Table 3.

Table 2 Neural network performance by courses

N°	Course	Precision	Precision base	Remembered	Degrees of assignment
1	Programming I	0.657	0.587	0.579	–
2	Programming II	0.785	0.658	0.598	–
3	Data structure	0.571	0.652	0.654	–
4	Distributed programming	0.887	0.854	0.843	Yes
5	Data mining applications	0.921	0.896	0.987	Yes

Table 3 Important features for each course

Characteristics	Course4	Course5
Study regularity	11.2	10.9
Average grade of exam	20.1	0.1
Number of questionnaires taken	8.4	0.1
Maximum downtime	8.4	10.7
Total session length	11.0	10.6
Total number of sessions	8.7	8.9
Total number of clicks	7.7	10.8
Average clicks per session	11.2	14.1
Average session length	7.9	9.1
Number of assignments made	1.8	1.7
Messages sent	0.2	0.0
Forum posts, chat, video	0.1	0.1
Average degree of evaluation	2.8	23.2

In the case of the Data Mining course, which as we mentioned has the highest accuracy value, we have that the degree of the questionnaire was the most relevant characteristic with 20.1%; the regularity of the study and the average clicks per session were the next top predictors with 11.2%. In contrast to these results we have that the average grade of evaluations represents a minor feature in this course, as it only scored 2.8%; result that can be explained since some grades are a record of a certain task but does not reflect how well it was done.

The other course that, according to the results, has grades and assignment as a predictor is Data Structure. For it, the average evaluation was the most important feature representing 23.2%; followed by the average clicks per session with 10.8%. The least relevant features were: sent messages, forum/chat/video posts, average test grade and the number of questionnaires taken with a score of 0.0%.

5 Conclusions

The goal of this research is to determine the effectiveness of a neural network-based model for predicting student performance. The literature review indicated that these approaches outweigh all other classifiers, with respect to the accuracy of the prediction. A multilayer perceptron neural network was trained by a reverse propagation algorithm, to predict the ability to successfully pass the race.

The accuracy rate of the classification is high, with an average of 74.98% for all courses; which show the effectiveness of the predictors obtained for predicting academic performance.

References

1. Muñoz-Repiso AGV, Gómez-Pablos VB (2015) Evaluación de una experiencia de aprendizaje colaborativo con TIC desarrollada en un centro de Educación Primaria. Edutec. Revista Electrónica de Tecnología Educativa 51:a291–a291
2. Fernández M, Valverde J (2014) Comunidades de práctica: un modelo de modelo de intervención desde el aprendizaje colaborativo en entornos virtuales. Revista Comunicar 42:97–105
3. Vasquez C, Torres M, Viloria A (2017) Public policies in science and technology in Latin American countries with universities in the top 100 of web ranking. J Eng Appl Sci 12(11):2963–2965
4. Torres-Samuel M, Vásquez C, Viloria A, Lis-Gutiérrez JP, Borrero TC, Varela N (2018) Web visibility profiles of Top100 Latin American universities. In: Tan Y, Shi Y, Tang Q (eds) Data mining and big data. DMBD 2018. Lecture notes in computer science, vol 10943. Springer, Cham, pp 1–12
5. Viloria A, Lis-Gutiérrez JP, Gaitán-Angulo M, Godoy ARM, Moreno GC, Kamatkar SJ (2018) Methodology for the design of a student pattern recognition tool to facilitate the teaching—learning process through knowledge data discovery (big data). In: Tan Y, Shi Y, Tang Q (eds) Data mining and big data. DMBD 2018. Lecture notes in computer science, vol 10943. Springer, Cham, pp 1–12
6. Abu A (2016) Educational data mining & students' performance prediction. Int J Adv Comput Sci Appl (IJACSA), 212–220
7. Daud A, Radi N, Ayaz R, Lytras M, Abbas F (2017) Predicting student performance using advanced learning analytics. In: Proceedings of the 26th international conference on world wide web companion. WWW '17 Companion, Australia, pp 415–421
8. Viloriaa A, Lezamab OBP (2019) Improvements for determining the number of clusters in k-means for innovation databases in SMEs. Procedia Comput Sci 151:1201–1206
9. González JC, Ramos S, Hernández S (2017) Modelo Difuso del Rendimiento Académico Bi-explicado. Revista de Sistemas y Gestión Educativa, 55–64
10. Hamasa H, Indiradevi S, Kizhakkethottam J (2016) Student academic performance prediction model using decision tree and fuzzy genetic algorithm. Procedia Technol, 326–332
11. Hu Y, Lo C, Shih S (2014) Developing early warning systems to predict students' online learning performance. Comput Hum Behav, 469–478
12. Huang S, Fang N (2013) Predicting student academic performance in an engineering dynamics course: a comparison of four types of predictive mathematical models. Comput Educ, 133–145
13. Il-Hyun J, Yeonjeong P, Jeonghyun K, Jongwoo S (2014) Analysis of online behavior and prediction of learning performance in blended learning environments. Educ Technol Int, 71–88
14. Rojas P (2017) Learning analytics: a literature review. Educ Educ, 106–128

15. Schalk P, Wick D, Turner P, Ramsdell M (2011) Predictive assessment of student performance for early strategic guidance. In: Frontiers in education conference (FIE). Rapid City, Estados Unidos de América
16. Usman O, Adenubi A (2013) Artificial neural network (ANN) model for predicting students' academic performance. J Sci Inf Technol, 23–37
17. Ye C, Biswas G (2014) Early prediction of student dropout and performance in MOOCs using higher granularity temporal information. J Learn Analytics, 169–172
18. Zacharis NZ (2016) Predicting student academic performance in blended learning using artificial neural networks. Int J Artif Intell Appl (IJAIA), 17–29
19. Expósito C (2018). Valores básicos del profesorado. Una aproximación desde el modelo axiológico de Shalom Schwartz. Educación y educadores. 307–325. Universidad de la sabana, Colombia
20. Ferrer J (2017) Labor docente del profesor principiante universitario: reto de la universidad en espacios globalizados. Ponencia presentada en jornadas científicas Dr. José Gregorio Hernández. Universidad Dr. José Gregorio Hernández. Venezuela
21. Fondón I, Madero M, Sarmiento A (2010) Principales problemas de los profesores principiantes en la enseñanza universitaria. En Formación universitaria 3(2):21–28
22. Fontrodona J (2003) Ciencia y práctica en la acción directiva. Ediciones Rialp, España
23. Gewerc A, Montero L, Lama M (2014) Colaboración y redes sociales en la enseñanza universitaria. Comunicar 42(21):55–63
24. Gómez L, García C (2014) Las competencias sociales como dinamizadoras de la interacción y el aprendizaje colaborativo. Ediciones hispanoamericanas. Universidad nacional abierta y a distancia, Colombia
25. Gros B (2008) Aprendizaje, conexiones y artefactos de la producción colaborativa de conocimiento. Editorial Cedisa, España
26. Hernández-Sellés N, González-Sanmamedy M, Muñoz-Carril PC (2015) El rol docente en las ecologías de aprendizaje: análisis de una experiencia de aprendizaje colaborativo en entornos virtuales. Profesorado. Revista de Currículum y Formación de Profesorado 19(2):147–163

Feature-Based Sentiment Analysis and Classification Using Bagging Technique

Yash Ojha, Deepak Arora, Puneet Sharma, and Anil Kumar Tiwari

Abstract With the ingress of exponential advancement of Internet technologies & social media platforms, there is a potential increase that can be seen in the development of online commercial websites. With time, people started to buy goods from these websites. So, there is also a great increase in selling goods on the Internet. These sites also facilitate their customers to leave their reviews and share their experiences with other users also. These customer reviews help others to make their decision before buying that product. In other words, these reviews help to show the quality of the product. Hence for this process, mining and understanding of the reviews are very important. In this research work, authors aimed to tackle one of the natural language processing (NLP) problems, i.e., sentiment polarity classification. The authors have performed a study to compare the baseline and statistical method (machine learning) for polarity classification. This work is also intended to compare the baseline method and machine learning method, to understand which method is better and more appropriate toward sentiment classification problems with the help of Python programming. The experimental results found to be satisfactory and compared with the existing literature.

Keywords Sentiment analysis · Machine learning · Python programming

Y. Ojha · D. Arora · P. Sharma (✉) · A. K. Tiwari
Department of Computer Science & Engineering, Amity University, Lucknow Campus, Lucknow, Uttar Pradesh, India
e-mail: puneetgrandmaster@gmail.com

Y. Ojha
e-mail: yash.ojha1995@gmail.com

D. Arora
e-mail: deepakarorainbox@gmail.com

A. K. Tiwari
e-mail: aniltiwari19640@gmail.com

V. Singh et al. (eds.), *Computational Methods and Data Engineering*,
Advances in Intelligent Systems and Computing 1227,
https://doi.org/10.1007/978-981-15-6876-3_42

1 Introduction

A sentiment is an approach, decision, or thought impelled by the feelings. Sentiment analysis studies the sentiment of people toward a particular entity, and it is also called opinion mining. The Internet is a place where sentiment information can be stored. It is very easy for the user to post their perspective or opinion about any product on social media via online social networking sites, blogs, or forums. Sentiment analysis or opinion mining is one of the main tasks of natural language processing (NLP). There are many processes to find the sentiment polarity. But people often confuse which method is more appropriate and helpful.

2 Background & Literature Survey

Identification of polarity sentiment is one of the important problems in sentiment analysis, which can take values as positive, negative, or neutral. Further sentiment polarity can be classified at the document, entity, aspect, and sentence level. Based on customer reviews, Hu [1] proposed a list of 2006 positive words and 4783 negative words. Both of the lists contain misspelled words also, which often used by different users on social media platforms. The categorization of sentiment is found to be a major classification problem, in which features that cover information regarding views or sentiment should be known before performing classification. To perform feature selection, Pang [2] has recommended the extraction of subjective sentences for removing objective sentences. It has also been suggested a text classification technique, which is able to find subjective sentences using a minimum cut. Gann et al. [3] carefully chosen 6799 tokens based on Twitter, where a sentiment score has been allotted to each token, stating a positive or a negative token by itself only. Liang et al. [4] implemented Unigram Naive Bayes model on twitter data by eliminating useless features using the mutual information and Chi-square feature extraction method. Turney [5] used the bag-of-words method for sentiment analysis. To determine the sentiment for the whole document, sentiments of every word were determined and those values are united with some aggregation functions. Kamps [6] used WordNet to determine the emotional content of a word along with different dimensions by determining the semantic polarity of adjectives. Xia [7] applied ensemble approaches like a fixed combination, weighted combination, and meta-classifier combination for sentiment classification and obtained better accuracy.

3 Formulation and Experimental Setup

The data collected for online product assessments are from the website "Amazon.com". By using Python programming language, experiments for sentence-level categorization and review-level categorization are performed with promising outcomes as this experiment needed a large amount of data to analyze, visualize, and extract relevant information. For the experimental purpose, authors have taken datasets from different Internet repositories. The data that the author has downloaded is majorly user reviews given or submitted by the customers on the world's biggest E-Commerce and Cloud Computing Company Amazon, about various music instruments, which consist of a total of 0,260 records to analyze. Following Table 1 represents the attribute information regarding the collected dataset.

From the above-stated attributes, the author has chosen only overall and reviewText for experimental purposes. Datasets are available in many different file formats over the Internet such as .JSON, .CSV, .XLSX, .DOCS, and even .TXT also, which are not of much used. The data that the author has used for their experiment was in .JSON form which is not very useful for our purpose so performed .CSV conversion. Authors have discarded many attributes, which are not needed for our experiment and extracted two attributes namely overall rating & reviewText. This text needs to undergo pre-processing of reviewText column for further processing. For this purpose, authors have used PANDAS library of Python programming language. After that, authors have extracted a particular column named music_instrument_reviewText for experiment purpose and hence stored that column in a separate CSV file using the "to_csv()" function of the pandas library. For removing stop words, authors have used the CSV & NLTK (Natural Language Toolkit) library of Python. NLTK is used for processing the natural language in Python. For our purpose, authors have imported the stop words from the corpus class of the NLTK library using from nltk.corpus import stop words. The basic flow of workflow and machine learning method implementation flow is depicted in Figs. 1 and 2.

Table 1 Attribute description

Name of attribute	Description
reviewerID	ID of the reviewer
asin	ID of the product
reviewerName	Name of the reviewer
helpful	Helpfulness rating of the review
reviewText	Text of the review
overall	Rating of the product
summary	Summary of the review
unixReviewTime	Time of the review (unix time)
reviewTime	Time of the review (raw)

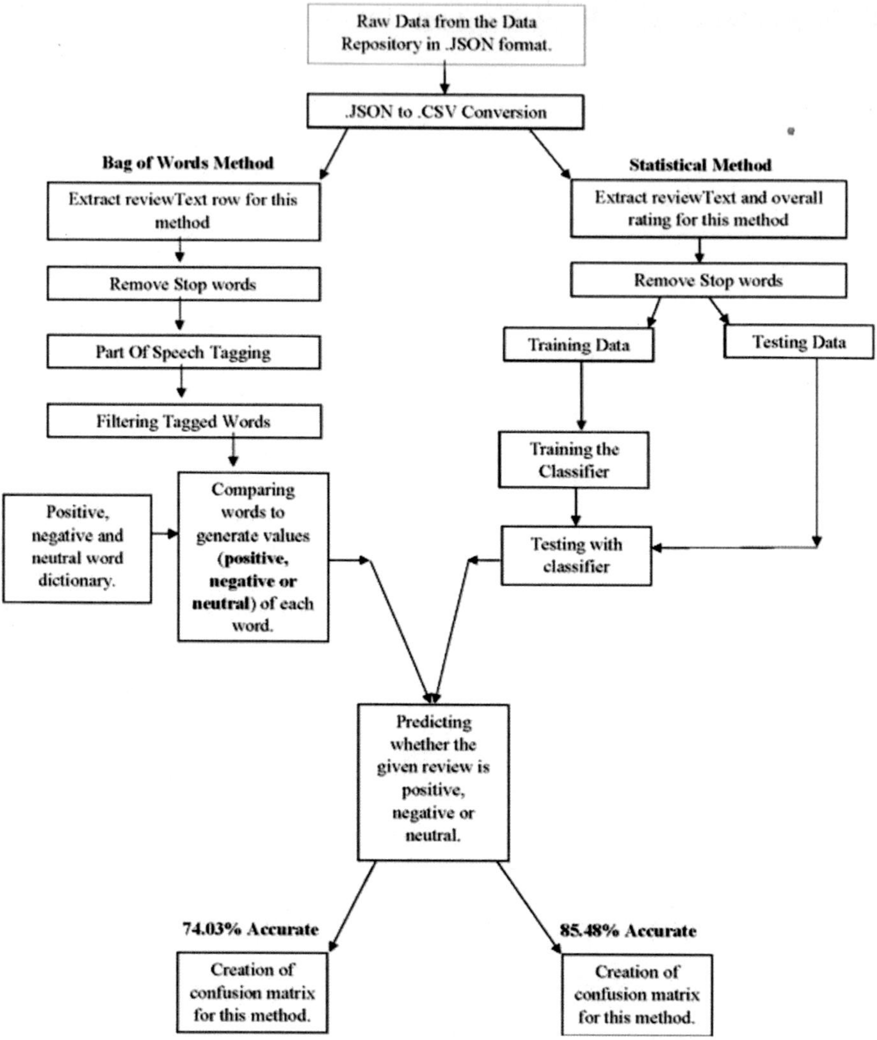

Fig. 1 Basic work flow

For implementing statistical method, authors need to extract the reviewText as well as the overall_rating given by the user for a product in order to decide the class (positive, neutral, and negative) of each review given. For this, authors have used the PANDAS library of Python programming language and replaced the rating 4/5 and 5/5 by positive class, 3/5 by neutral class, 2/5 and 3/5 by negative class after extracting the reviews and the rating from the dataset.

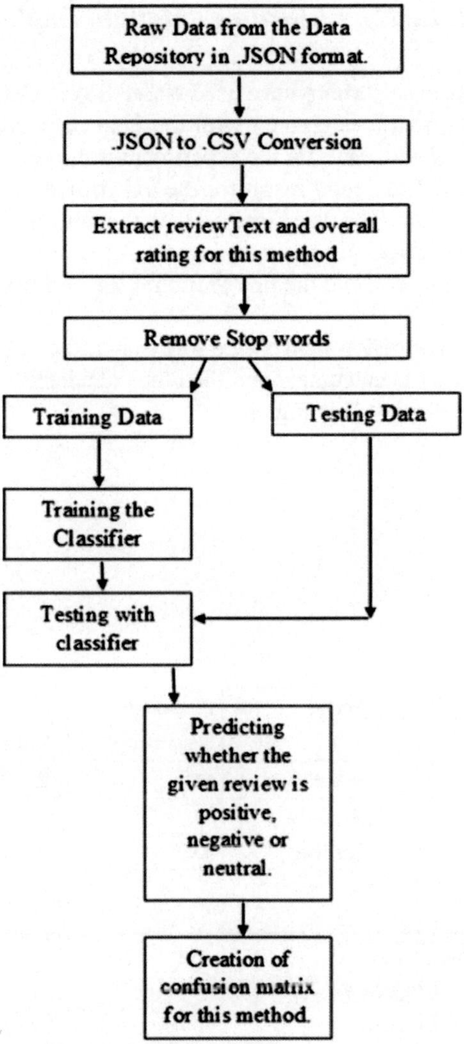

Fig. 2 Machine learning method implementation flow

4 Results and Discussion

For preparing the training and testing dataset, authors have divided the dataset into two different files, one being the training dataset and the other being the testing dataset with the help of stratifying the data. Stratifying is the process of dividing the data into two different halves where the ratio of the positive, negative, and neutral classes is equal in both the files, i.e., the trained dataset and the testing dataset based on a particular ratio in a random manner.

4.1 Statistical (Machine Learning) Method Implementation

For classification, purpose authors have used Naïve Bayes Classifier. In this experiment, the confusion matrix is created with the help of predicted results and the actual result. The predicted result for the experiment is the result found with the help of statistical methods. The actual result for the experiment is the result the author received with the raw data. The data received with the raw data is the actual data that mentions if the review is actually positive, negative, or neutral. The confusion matrix obtained is given in Table 2, and the line graph for the confusion matrix is given in Fig. 3.

After plotting the confusion matrix in a line graph, the data distribution for the test and training dataset is calculated and then the calculated value is plotted in the form of bar graph as displayed in Fig. 4.

Table 2 Confusion matrix

Actual	Predicted		
	Negative	Neutral	Positive
Negative	1	2	137
Neutral	4	1	227
Positive	43	34	2630

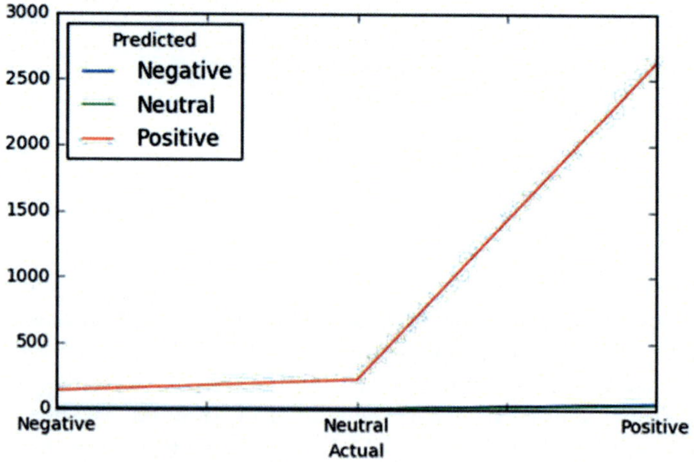

Fig. 3 Line graph for confusion matrix

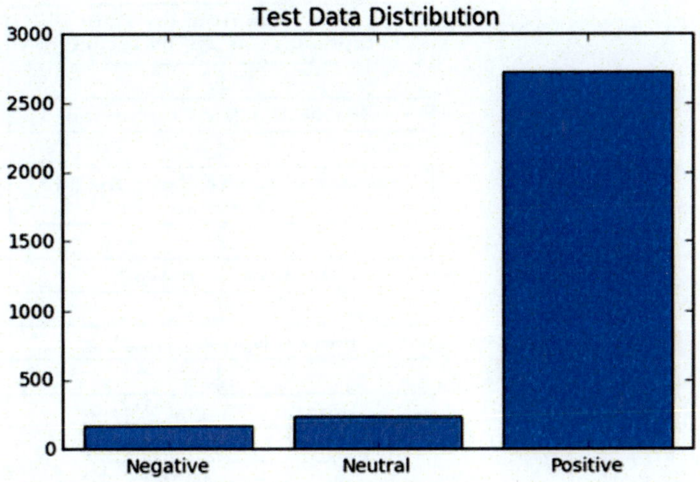

Fig. 4 Test data distribution

4.2 Bag-of-Words Method Implementation

The workflow for applying bag-of-words technique has been depicted in Fig. 5.

For the experiment purpose, author has used POS tagging with the help of Natural Language Toolkit (NLTK) POS Tagger in Python. In the experiment, the NLTK library is imported in the code for tokenizing the data and POS tagging the words in the review. For tagging of words, tokenization of words is required. With the help of method "word_tokenize()", the tokenization of words is done. After the tokenization of each word in all the reviews, the reviews are then tagged with the help of method "pos_tag()". This method tags the words of the reviews and stores the result in the form of <posText, posTag> pairs. The posText contains the English word and posTag contains the corresponding word tag. In the dictionary, positive and negative words are used in the experiment for comparing tagged words. Authors have to compare the tagged words for determining whether the words tagged are positive, negative, or neutral. These positive and negative words from the files are stored in a list, and all the tagged words are compared with the list of positive and negative words. In this experiment, confusion matrix is created with the help of predicted result and actual results. The predicted result for the experiment is the result found with the help of bag-of-words methods or via lexical analysis. The actual result for the experiment is the result we received with the raw data. The data received with the raw data is the actual data which mentions if the review is actually positive, negative, or neutral. The confusion matrix obtained and plotted line graph after experimentation has been depicted in Table 3 and Figs. 6 and 7.

After comparing both approaches, the authors found that machine learning out performs lexicon-based approach. In our machine learning approach, we used

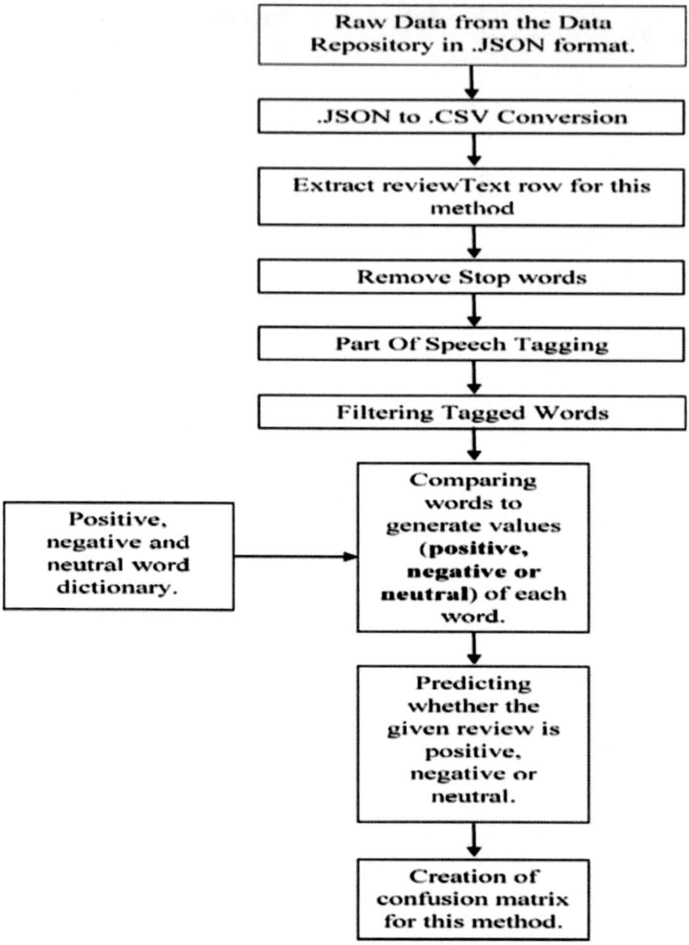

Fig. 5 Bag-of-words implementation flow

Table 3 Confusion matrix for bag-of-words technique

Actual	Predicted		
	Negative	Neutral	Positive
Negative	7	25	170
Neutral	19	40	301
Positive	157	660	3751

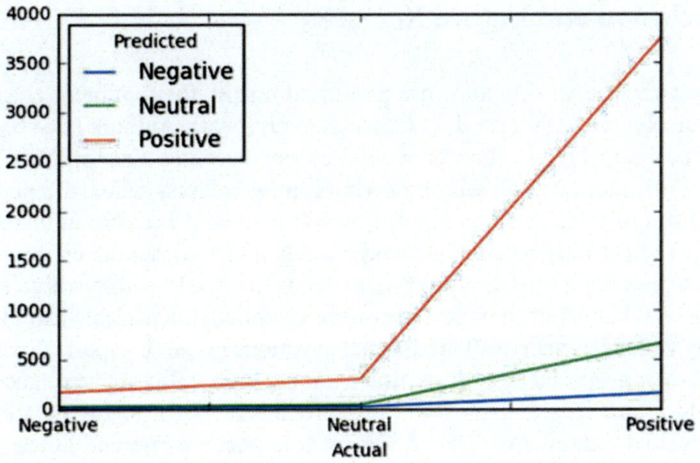

Fig. 6 Line graph for confusion matrix

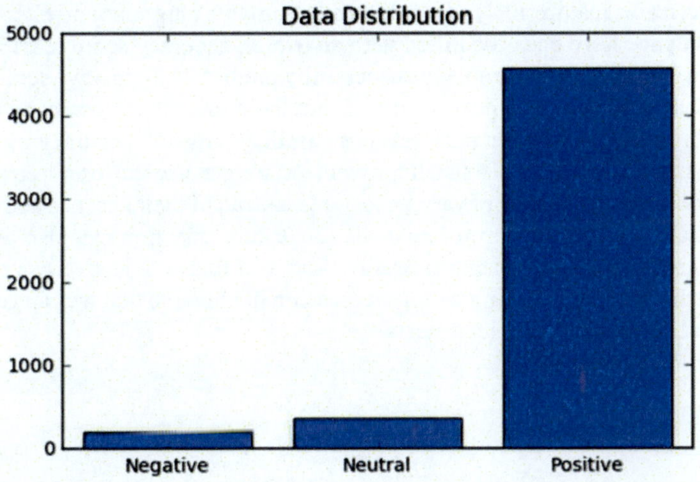

Fig. 7 Data distribution

unigrams and bigrams as that are present in review text as our feature vector but it only considered their absence and presence for their feature value.

5 Conclusion and Future Remarks

In the research, the work author has presented results for sentiment classification on Amazon Reviews for a product (musical instrument). Authors have used previously proposed technique (bag of words) as our baseline. Our primary task was binary classification, i.e., classifying reviews in two classes which are positive and negative but a third class also included, which is neutral because in lexicon-based approach, the neutrality score of the words is taken into account in order to filter the neutral sentences and enable algorithm to focus on words with positive and negative sentiments but when statistical approach is applied, the neutrals handling varies ominously which has been well established in various research papers. When authors have used their proposed bag-of-words technique, it provided an overall accuracy of 74% in which authors were able to successfully predict 3751 positive sentiments out of 4568 positive sentiments that are 82% of true positive rate and in the statistical method. The author also investigated Naïve Bayes models that assign class labels to problem instances, as feature values, where class labels have been drawn from a finite set which represents a family of algorithms. For our feature-based approach, authors have performed a feature analysis in which most suitable unigrams and bigrams using their abundance have been identified with an overall accuracy score of 85%. In this research work, authors were able to successfully predict 2630 positive sentiment out of 2707 that is 97% of true positive rate. From the above, it can easily be observed that the statistical method out performs our baseline in overall accuracy by 11% and in true positive rate by 15%. As an improvement, we can use the term frequency and inverse document frequency of various features instead of their absence and presence as their feature value and further methods for feature selection can also be applied for the significant improvement in results. And as a future objective, we can try to predict actual star ratings using various regression methods so that we can get a better understanding of the user's sentiments.

References

1. Hu M, Liu B (2004) Mining and summarizing customer reviews. In: Proceedings of the tenth ACM SIGKDD international conference on Knowledge discovery and data mining. ACM, New York, NY, USA, pp 168–177
2. Pang B, Lee L (2004) A sentimental education: sentiment analysis using subjectivity summarization based on minimum cuts. In: Proceedings of the 42nd annual meeting on Association for Computational Linguistics. Association for Computational Linguistics, p 271
3. Gann WJK, Day J, Zhou S (2014) Twitter analytics for insider trading fraud detection system. In: Proceedings of the sencond ASE international conference on Big Data, vol 29. ASE, pp 1136–1142
4. Liang PW, Dai BR (2013) Opinion mining on social media data. In: 2013 IEEE 14th international conference on mobile data management, vol 2. IEEE, pp 91–96
5. Turney PD (2002) Thumbs up or thumbs down? Semantic orientation applied to unsupervised classification of reviews. In: Proceedings of the 40th annual meeting on Association for Computational Linguistics. Association for Computational Linguistics, pp 417–424

6. Kamps J, Marx M, Mokken RJ, De Rijke M (2004) Using WordNet to measure semantic orientations of adjectives. In: LREC, vol 4, pp 1115–1118
7. Xia R, Zong C, Li S (2011) Ensemble of feature sets and classification algorithms for sentiment classification. Inf Sci 181(6):1138–1152

A Novel Image Encryption Method Based on LSB Technique and AES Algorithm

Paras Chaudhary

Abstract Digital image transfer security is presently vital in image communications, as pictures are more and more utilized in industrial processes. It's necessary to safeguard confidential image information from unauthorized access. Image security has become a vital downside, and privacy challenges are steadily increasing. Many strategies of knowledge protection and privacy were mentioned and developed. Coding is maybe the foremost obvious. Several methods of data protection and privacy were discussed and developed. Encryption is probably the most obvious. To protect valuable information from unwanted drives, image encryption is necessary. In this paper, we introduce information masking and digital image encryption that enhance the security of confidential data and digital images using LSB technology and the Advanced Encryption Standard (AES) algorithm. To hide data in the image, we use LSB technology, and to encrypt the embedded image, we use the AES algorithm.

Keywords AES algorithm · LSB technique · Information security · Data hiding · Encryption · Image security

1 Introduction

Recently, with the speedy advancement of communication technologies, the transmission of digital pictures has become a typical error. Digital photos represent 70% of the knowledge sent on the online. However, advanced pc processors accelerated banned access to data transmitted over the web [1]. Encryption is that the conversion of information from one type to a different, known as encrypted text that can't be simply understood by unauthorized persons. Whereas Decryption is that the method of changing encrypted knowledge to its original type, it's thus doable to know it. With the continual development of PC and net technologies, transmission knowledge is

P. Chaudhary (✉)
Department of Computer Science, Jaypee Institute of Information Technology, Noida, U.P., India
e-mail: paras.vc18@gmail.com

V. Singh et al. (eds.), *Computational Methods and Data Engineering*,
Advances in Intelligent Systems and Computing 1227,
https://doi.org/10.1007/978-981-15-6876-3_43

more and more getting used in applications like video-on-demand, video conferencing, PC forensics and radio. At present, transmission knowledge is additionally closely associated with several aspects like education, business, recreation and politics [2]. Cryptography is a region that gives multiple policies to make sure image privacy, as well as privacy, authentication, security and integration. It's necessary to shield the knowledge transmitted over the web or the other suggests that. Knowledge could also be attacked by third parties, and to make sure its integrity, we've many technologies like RSA and AES. AES is a lot economical in both encryption and decryption [3].

2 AES Algorithm

AES (Advanced Encryption Standards) accustomed to write pictures means AES is principally used for encryption within the projected development. The AES formula was developed by Vincent Regman and Joan Damen. On October 2000, bureau complete that the AES formula was the simplest algorithm for security, performance, efficiency, implementation and suppleness. AES may be a centrosymmetric key formula, during which the sender and recipient use constant key to write encrypted information and decipher the coding within the original data. During this formula, the mounted block length is 128 bits, whereas the key size maybe 128, 192, or 256 bits. AES reiterative formula consists of four basic management blocks. For full coding, iterations are performed till "N" times. The entire range of iterations, IE N, 10, 12 and 14, maybe performed of the key length, IE 128, 192 and 256 severally [4]. The AES formula is split into four totally different steps, performed by sequent rounds. These blocks add a structured computer memory unit array as a 4×4 array referred to as a case.

- Sub-Byte Conversion: Replace a non-linear computer memory unit that uses a station table(s), created by reverse conversion and multiple misreckoning.
- Transformation Lines Shift: this can be an easy computer memory unit key. The bytes are rapt within the last three lines of the case. The shift to the left varies from one to 3 bytes.
- Mix Columns Transform: adequate the multiplication matrix of state columns. Every vector is increased by a column with associate degree array. Two-dimensional bytes are processed rather than numerical.
- Add round key Transform: this can be an easy XOR between the ring key and dealing conditions. This variation is the opposite.

After adding a primary rounded key, a rounded perform is applied to the information block, which consists of bytes, shift lines, column mix and spherical key conversion, severally. This can be done iteratively (N times) looking at the key length. The secret writing structure has a constant sequence of transformations because of the coding structure. Invite-Byte sub-conversions, Inv-Shift lines, Inv-Mix columns and therefore the Add Rounded key permit the "key tables" model to be similar to coding and secret writing [2].

AES is an associate degree example of the same formula that encompasses a sender and recipient key, to write and decipher text or information. AES is an efficient formula to implement hardware and computer code and works best in hardware [5].

3 LSB Technique

LSB is one in all the foremost common and easy information masking ways. The LSB (least significant bit) algorithm replaces the lowest bit of the coil in line with the message bit. This can be the foremost standard technique wont to hide data to cover the message [6]. LSBS hides information in the least significant bits of each pixel of cover images that cannot be seen by the human eye. Cover pictures are often in 24-bit, 8-bit, or grayscale format. In LSB correspondence, the choice to vary the smallest amount vital bit is random, which means that if the message bit doesn't match the element bit, the bit can increase or decrease by one. Hide current information is the nature of the adaptation [7].

4 Proposed Methodology

Hide information is a way to hide information from others; it's the art and science of secure communication. The most goal of activity information is the secure transmission of data over the web. Several file media sorts are wont to hide knowledge by concealing information: image, audio, video, text. Within the Hide image info, the duvet is a picture. To transfer the data more securely, Hide uses secret writing info.

In the proposed design, we used the Advanced Encryption Standard (AES) algorithm and LSB to hide information. In developing the proposed design, we take the confidential information first and then make information barcodes and masks in the image using LSB technology. After hiding information from this image, we get the merged image. The merged image is displayed as the original image. After obtaining the embedded image, we implement the AES algorithm and encrypt the embedded image. Now, the encrypted image is totally different from the merged or original image. It improves the security of the original image and hidden data as well.

The proposed block diagram of AES 128 is shown in Fig. 1. AES is the most popular and secure symmetric algorithm. The encryption and decryption key is the same in AES. In Fig. 2, we showed the secrete information or message that we want to hide in the image and Fig. 3 shown the barcode of the secrete information. After creating the barcode of the information we hide this barcode in the image using the LSB technique. The embedded image of the barcode and original image is a view like the original image that is shown in Fig. 4. After embedding the information in

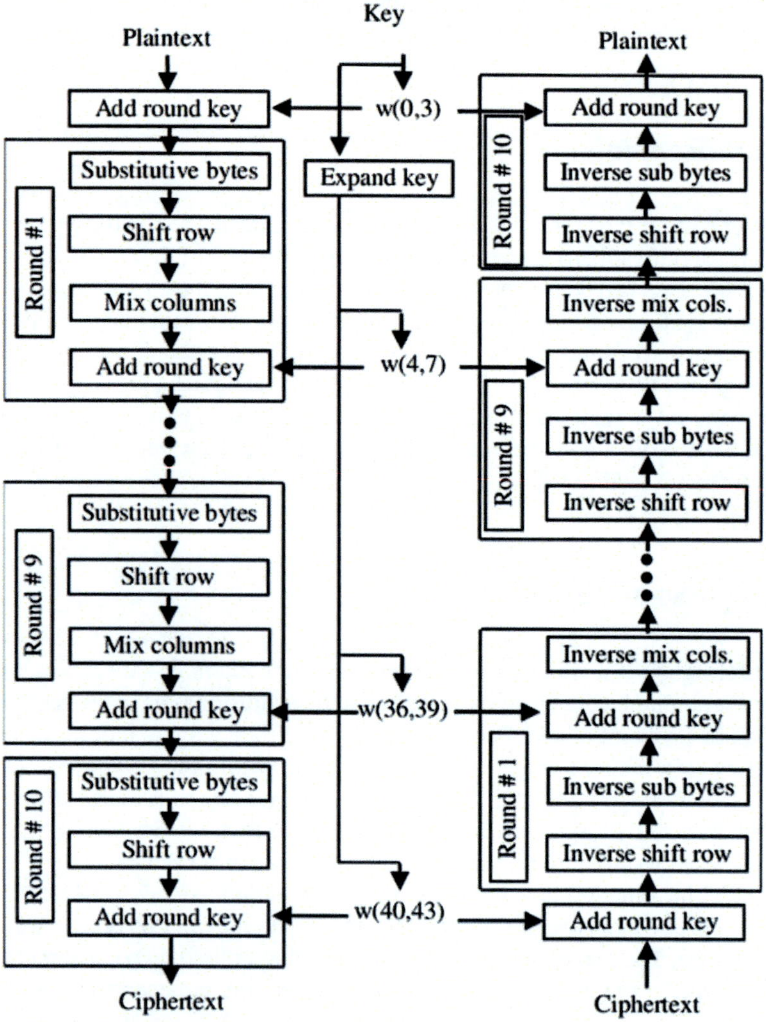

Fig. 1 Advanced encryption standards (AES) flow

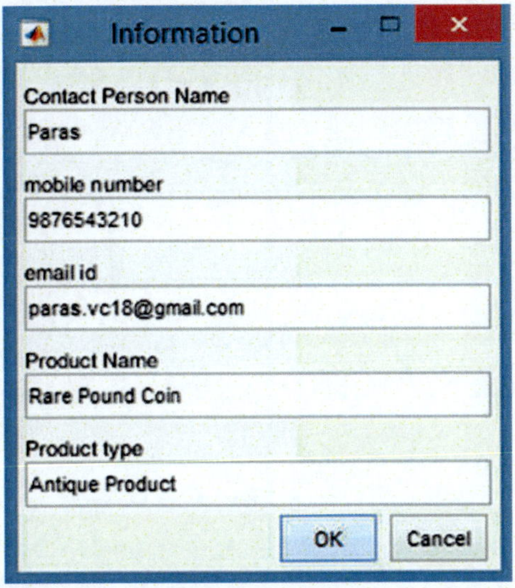

Fig. 2 Secrete information

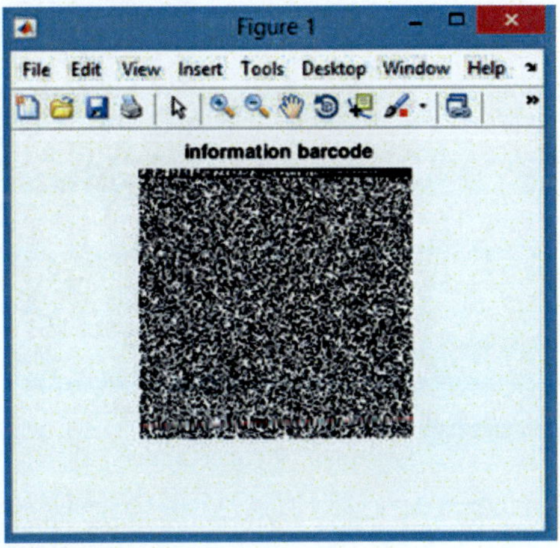

Fig. 3 Barcode of the secrete information

the image, we encrypt the image using the AES algorithm, the encrypted image is shown in Fig. 5. After encrypting the image, we did the decryption process to get the original image and hidden information in the image. The decrypted image and hidden image is shown in the Figs. 6 and 7, respectively.

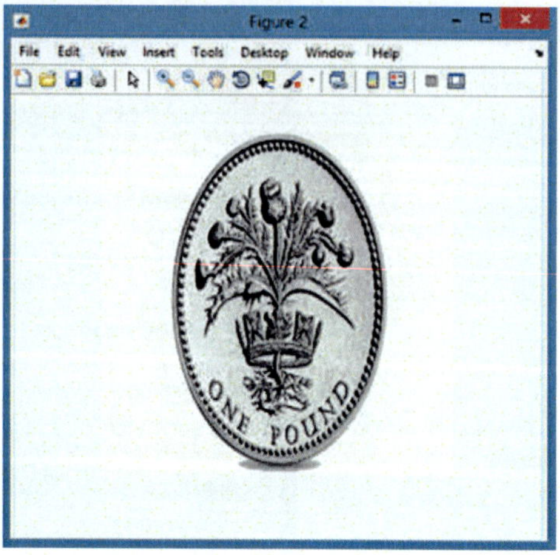

Fig. 4 Embedded image

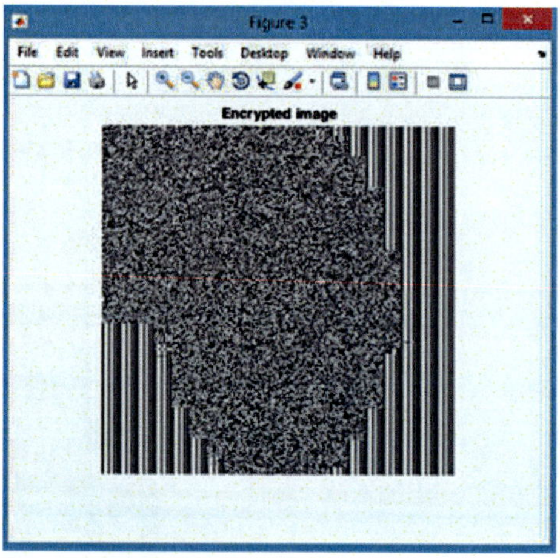

Fig. 5 Encrypted image

5 Conclusion

In this paper, a new way to hide data is proposed to overcome the limitations of the current methods. In this technique, we apply the stenography and cryptography techniques together in the proposed method. In this technique, first we generate the information barcode using the secrete information. After that, we hide this information barcode in the image using the LSB stenography technique. After hiding the secrete information in the image, we get the embedded image that is similar to the original image but it consists of the hidden secret information. After that we apply the AES algorithm in the embedded image to encrypt this image it means we done

Fig. 6 Decrypted image

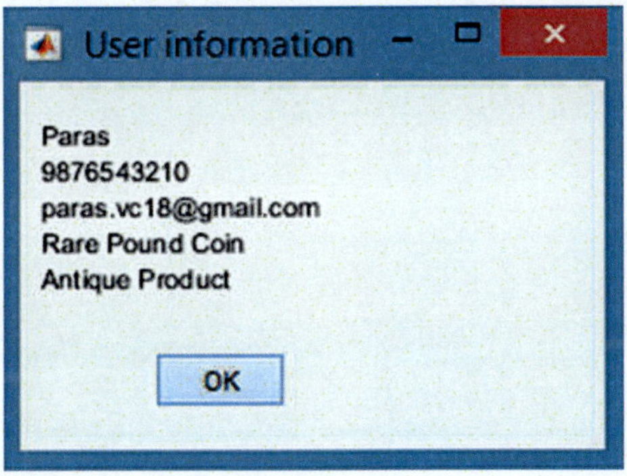

Fig. 7 Decrypted information or message

the cryptography process in it using the AES algorithm. Now, we finally get the encysted image that is totally different than the original image. Now using the hybrid technique of information hiding and image encryption we enhance the security of the secret information and image together in the proposed method.

References

1. Arab A, Rostami MJ, Ghavami B (2019) An image encryption method based on chaos system and AES algorithm. J Supercomput 75:6663–6682
2. Deshmukh P, Kolhe V (2014) Modified AES based algorithm for MPEG Video encryption. In: IEEE ICICES2014, pp. 1–5, S. A. Engineering College, Chennai, Tamil Nadu, India
3. Santhosh Kumar BJ, Roshni Raj VK, Nair A (2017) Comparative study on AES and RSA algorithm for medical images. In: IEEE international conference on communication and signal processing, pp. 501–504

4. Ray A, Potnis A, Dwivedy P, Soofi S, Bhade U (2017) Comparative study of AES, RSA, genetic, Affine transform with XOR operation, and Watermarking for image encryption. In: IEEE proceeding international conference on recent innovations is signal processing and embedded systems (RISE-2017), pp. 274–278
5. Cirineo CC, Escaro Jr RQ, Silerio CDY, Teotico JBB, Acula DD (2017) MarkToLock: an image masking security application via insertion of invisible watermark using steganography and advanced encryption standard (AES) algorithm. In: Meen, Prior, Lam (eds) Proceedings of the 2017 ieee international conference on applied system innovation, IEEE-ICASI 2017, pp. 995–997
6. Joshi K, Yadav R (2015) A new LSB-S image steganography method blend with cryptography for secret communication. In: IEEE 2015 third international conference on image infonnation processing, pp. 86–90
7. Anjum A, Islam S (2016) LSB steganalysis using modified weighted Stego-image method. In: IEEE 2016 3rd international conference on signal processing and integrated networks (SPIN), pp. 630–635

Implementing Ciphertext Policy Encryption in Cloud Platform for Patients' Health Information Based on the Attributes

S. Boopalan, K. Ramkumar, N. Ananthi, Puneet Goswami, and Suman Madan

Abstract The cloud computing platform has to improvise the healthcare authorities along with organization. In health industry has to develop the patients to make, organize and give out their health-related data with different users by the personal health record (PHR) administration and in addition to health insurance companies. The proposed method explains the work on outlining and implementing patient-driven, patient health information in a cloud platform in light of open-source Indivo X framework. The PHR administration level is to be facilitated about outsourcing patients' PHR information to cloud servers with a specific end goal to upgrade its interoperability. The security saving PHR structure utilizes attribute-based encryption (ABE) and supplies the system with the goal that patients be competent to scramble their PHRs and store them on cloud servers which is half believed to such an extent that servers do not have confirmation to basic PHR data and the patients keep up full power over entrée to their PHR records, by doling out access to some approved data users, while individual users can have entry to various parts of their PHR.

Keywords Personal health record · Access control · Electronic medical record · Attribute-based encryption · Ciphertext policy · Indivo X

S. Boopalan · K. Ramkumar · P. Goswami
SRM University, Sonepat, Haryana, India
e-mail: boopalan.s@srmuniversiy.ac.in

K. Ramkumar
e-mail: ramkumar@srmuniversity.ac.in

P. Goswami
e-mail: goswamipuneet@gmail.com

N. Ananthi
Easwari Engineering College, Ramapuram, Chennai, India
e-mail: nandhura@gmail.com

S. Madan (✉)
Jagan Institute of Management Sciences, Rohini, Delhi, India
e-mail: madan.suman@gmail.com

V. Singh et al. (eds.), *Computational Methods and Data Engineering*, Advances in Intelligent Systems and Computing 1227,
https://doi.org/10.1007/978-981-15-6876-3_44

1 Introduction

The commonly recognized fact is that the data utilization and its related innovations in the well-being assurance frameworks can upgrade mind movement exceptionally. It may enhance native's well-being, and including flourishing besides, social thought. In addition, it fabricates subjects gauge for a conglomeration and opportunity with diminishing rising social protection costs in the general public.

The ongoing movement in medicinal services has passed on to a move from the electronic well-being record structures hindered by social insurance PHR system. Thus, the patients control the supporter of PHR systems. The patients have been permitted by PHR frameworks to make and supervise along with controlling their own PHRs using the Internet. The submitted results promising to effortlessly get to their well-being statistics to restorative administrations, suppliers, assurance, protection experts, scientists, relatives and mates.

The well-being data on a PHR may be viewed as saw as total and precise delineation of a person's blueprint for a person's remedial history of well-being status. First, their well-being data can be imported by the individuals which might fuse their medicinal information history, lab and imaging results, record of therapeutic issues, and verifiable foundation of clinic EHR system; individuals may likewise exchange well-being estimations from their modern products, for example, remote electronic gauging scales, or assembled inactively from an advanced mobile phone.

The information pivot point can be filled by PHR systems for patient well-being the board. It is done by helping the patients to remain educated of their individual well-being information that can relate the correct history in the midst of medicinal experiences, test for prescription associations, and take out the excess replication of research facility tests and analytic investigations. The PHR structures help the patients may have passage to a wide extent of well-being data resources, and also enhancing their well-being mindfulness. At this point, the clinicians can be helped by PHR systems by giving better consideration choices through persistent data. PHR structures have the advantage of general segment by giving well-being watching, episode watching, fortifying, interfacing with administrations, and research. PHR structures can give buyers the likelihood to have an enormous impact in guaranteeing and advancing the general well-being [14].

The distributed computing is one of the major testing and creative models which encourage simple, on-request arrange permission to a mutual gathering of configurable exchange out assets that save to be immediately provisioned and free with insignificant administration endeavor or specialist co-op communication [2]. As per Gartner's report [1], there is a stimulating assignment of distributed computing among undertakings. Gartner gauges that the general cloud organization livelihoods would accomplish 148.8 billion USD by 2014, with a colossal piece of the social insurance distributed computing business. National and commonplace social insurance experts and furthermore medicinal services specialist co-ops exhibited mind-boggling interests, and are presently making initial moves toward the introducing of distributed computing. Distributed computing can serve social insurance suppliers to focus more on growing nature of conveying human services. This is particularly crucial for the clinics, bunch care and specialist rehearses. The data circulation can be streamlined by distributed computing among various human services associations incorporated into the consideration procedure.

The genuine security and assurance dangers cause when there is a movement in structure and patient-sensitive information from specialist's offices to the cloud. A few PHR's information is seen as open access and touchy. To shield patients from social embarrassment, inclination, or out-of-line employment opportunity, data, for example, richness, passionate and mental issue, sexual practices or physical maltreatment, etc., ought to be significantly secured. The trust stockpiling servers consistently anchor touchy data whose occupation is to mediate access to data happening in the standard access control system. In spite of the fact that, clients can be hesitant to confide in untouchable servers with distributed storage, current recommendations on powers the entrance control technique misuses misuse the abuse of cryptography to approve get to control approaches. In such structures, there is no necessity to confide in servers to check customer accreditations, and each client can get the scrambled subtleties, yet just clients who have the correct certifications have the correct qualifications to have the capacity to unscramble the encoded information [5].

The positive identity of the sender and recipient must be distinguished. If not, the ordinary pubic key encryption designs or IBE designs won't work for conditions. For instance, the PHR data of a patient need to be sent to various clients; the patient should know the mechanized support or the character for each recipient, and after that encode a similar data commonly using each recipient pubic key or character. Along these positions, we require a crypto plan which offers an increasingly reasonable response for actualizing access courses of action and retaining into record data. The attributes of the patient have only been exposed for the needs of recipient to keep in knowledge the true objective to get the opportunity to calms data. These expanding needs are addressed by Sahai and Waters [13] who exhibited the possibility of ABE. What's more, ABE, moreover, has arrangement obstruction property; that is, if various clients conspire, they may simply be able to just be fit for unscrambling a figure content if no less than one of the others could decode it isolated. Thus, data get to is self-actualizing from the cryptography, obliging no trusted go in the middle.

ABE can see a development of the possibility of IBE in which client personality is summed up to a lot of expressive properties as opposed to a lone string deciding the client character. Similarly customary open key encryption and IBE, ABE IBE, ABE have a vital point achieves adaptable, versatile one-to-numerous encryption as opposed to coordinated, it is to picture as a skilled device for keeping an eye on the issue of secured and needed data conveyance and decentralized permission control.

The private keys or figure messages are useful for two sorts of ABE that get two methodologies are mapped with. The attribute issues the client keys in the key-strategy ABE (KP-ABE) system [5, 9]. Then, this ability catches entrance course of action that figures out which kind of figure messages the key can decode, while figure writings be set apart by the sender through a lot of expressive qualities.

In the figure content arrangement ABE (CP-ABE), senders can scramble a correspondence with an unequivocal access system in obtainments of the entrance structure over the characteristics conveying what sort of recipients will encase the expertise to decode the figure content. The user has sets of properties and increase taking a gander at mystery trademark keys from the specialist. Such a client can unscramble figure content if his/her ascribe fulfills the privilege to utilize strategy related to the figure content. A depiction usage of CP-ABE is a safe framework with access approach.

Especially, there has been a growing excitement to incorporate attribute-based encryption (ABE) to anchor EHRs or PHRs [1, 6, 7, 10]. Ibrahim et al. [5] connected CP-ABE to approve access control methodologies of a patient/authoritative with the end goal that everyone can download the scrambled subtleties yet, simply approved customers from the family, buddies, or approved client capable space like specialists and medical attendants are permissible to decode it.

Mohan et al. [9] showed the arrangement and starting model use of a MedVault subsystem anticipated EHR assignment, which covers quality-based access control and explicit prosperity data revelation. Narayan et al. [10] proposed a current technology called cloud computing which provides ubiquity, scalability, and elasticity features. Akinyele et al. [1] used twofold methodology characteristic-based encryption and gave a framework and utilization of self-protecting EMRs. The blueprint of PHR cloud platform composed with CP-ABE has been proposed and executed in this paper. The patients can be engaged by this PHR cloud stage and it paves the way for storing the data securely and also their well-being records can be distributed in a versatile way. The PHRs in a scrambled structure can be stored by the patient, and it cryptographically approves the patient or definitive access techniques. This paper improvises the enhancement PHR distribution system of Indivo X venture [16] by embracing CP-ABE, and some basics about CP-ABE procedure are presented. Section 3 portrayed the architectural design of the PHR cloud platform to be created. Section 4 introduces the execution insights about PHR cloud platform. Section 5 presents execution evaluation.

2 Preliminaries

2.1 *Access Structure*

Let us consider $P = \{P1, P2. Pn\}$ be a place of parts. If $B \in A$ and $B \subseteq C$, then $C \in A$, and the collection, represented by $A \subseteq 2P$, is monotone only if $\forall B$, C. An entrée structure is a collection $A \subseteq 2P \setminus \{\emptyset\}$. A is held to be specialized sets and the sets that are not in the A are said to be unofficial sets [3, 5]. Thus, the function of the parties is taken with the attributes in this perception. Therefore, the access structure has the sets in A which are called the proficient sets of attributes. The monotone entrée structures can be limited. Four polynomial algorithms are in CP-ABE method [11] as follows:

Probabilistic polynomial-time algorithm
The setup algorithm called probabilistic polynomial-time algorithms which take λ security parameter as input and outputs "params" which is the public parameter all along with trusted attribute authority (AA) that is the master key (msk).

$$\text{Setup}(1\lambda) \rightarrow (\text{params}, \text{msk})$$

Encryption algorithm
The public parameters params and note m along with access configuration A which stated via the sender are input to the PPT encryption algorithm. The ciphertext c encrypted is a productivity.

$$\text{Encrypt}(\text{params}, m, A) \rightarrow c$$

Key generation algorithm
An interactive procedure amid the AA and the addict is the PPT key generation algorithm. The public parameters "params" are the common input to AA, ω is the attributes place which the user owns, and the master key msk is the private input to AA. A decryption key SKω is received by users and associated b.

$$\text{KeyGen}(\text{params}, \text{msk}, \omega) \rightarrow \text{SK}\omega$$

Deterministic polynomial-time algorithm
The public parameters "params" are taken as input in the deterministic polynomial-time algorithm in the presence of access structure A which encrypts the ciphertext c, and the secret key SKω of user is related to the group of attribute ω. $\omega \in A$ is the message m that is given and $\perp$ if $\omega \perp\in A$ is the error message.

$$\text{Decrypt}(\text{params}, c, \text{SK}\omega) \rightarrow m \text{ or } \perp$$

2.2 *Security Model for CP-ABE Scheme*

Similar to identity-based encryption (IBE) method [1], it is possible to find out for one private key through security depiction but that key cannot be utilized toward the decrypt and the dispute ciphertext. Here, in the CP-ABE method, the attributes are used customarily as the private keys, and for accessing structures, the ciphertexts are used. It follows that the safety model for CP-ABE method authorizes the opponent to endure for several private keys linked to the attribute sets, and αi could not be utilized to decrypt the confront ciphertext, which has been encrypted beneath an access structure *A*.

For example: *A* is not assured by αi. The correct security matches for CP-ABE are illustrated as follows:

Setup: To produce the public parameters "params" for the adversary, the Setup algorithm run by the challenger:
Phase 1: The challenger creates the frequent private keys associated with the sets of attributes $\gamma 1, \ldots, \gamma\ q1$
Challenge: Adversary submitted two equivalent length messages $a0$ and $a1$. Adding together the adversary provides a dare access structure $n0$ where none of the first phase sets $\gamma 1, \ldots, \gamma\ q1$ gratify the access structure. "a" arbitrary coin *b* flipped by the competitor and encrypts *nb* under *A*.

The opponent outputs a guess and defines the benefit of an adversary *A* in this game. This paper shows highlights that by allowing decrypting queries during phase 1 and phase 2, the technique can easily be improved to manage chosen-ciphertext attacks.

3 Structural Design of PHR Cloud Platform Preliminaries

We perceive the rule security-related necessities for the PHR cloud orchestrates take after. When there are more PHR owners and PHR users, the implementation of PHR system is useful. The full control over the patient's own PHR data will be referred by the owners. So they can produce, control, and delete the data. The analyses have been explained to identify wide issues to certain themes in PHR system design used for PHR development. Based on these common problems, authors provide ideas and recommendations to PHR system design and development. The scope on integration of PHR is done by the system integration and standards that consist of articles. Architectures' core integration and standards applied are detailed. Security and privacy consist of articles that explain the security and privacy mechanisms used in PHR.

There are five individuals in cloud organize: human administrations provider, cloud organization provider, characteristic power, proprietor (patient), and viewer.

The cloud organization providers are semi-believed which infers that cloud organization providers would endeavor to find anyway much PHR information as could sensibly be normal while taking after the tradition. The purpose behind existing is to engage patients for controlling the movement and using their data. Since a transitory help responded for this in due order, PHR data ought to be encoded before switching to the cloud. The framework of protection needs to guarantee that industrious PHR data must be decoded by affirmed social affairs as spoken to by the game plan and

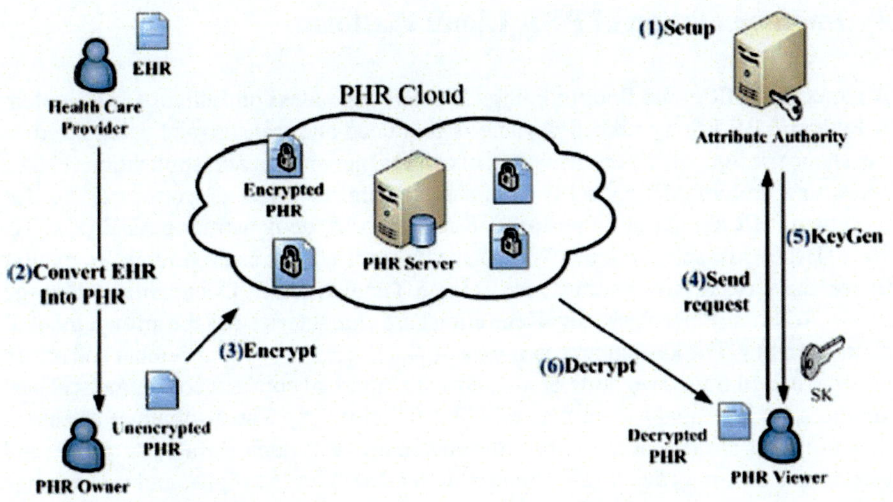

Fig. 1 System architecture and workflow

consent. The system's basic designing of the proposed cloud organize is shown in Fig. 1 where the patient can securely cope-up with his/her prosperity records using the arrangement. By going along with this system, we clear up the affiliations that occur in the structure.

Framework Setup: AA runs framework computation of Waters' CP-ABE plan, it yields open parameters "params" as a rule parameter and the master puzzle key cover which is held by AA mystery.

Generating PHR: PHR proprietor gets one of the kind of electronic therapeutic record (EMR) from the human administrations provider, at that point assembles PHR data in light of EMR data along with other data.

Encryption: To encrypt the data for successful reasons, it is not appropriate to use CP-ABE. Or maybe, PHR proprietor initially makes an AES key unpredictably as substance key for encryption, and encoding the PHR data. PHR proprietor at that point sets get to approach. The passage methodology approach is used to encrypt the substance encryption key using CP-ABE plan.

KeyGen: Client sends a sale for property private key nearby his accreditations to the AA—at present the disavowal for AA. Plans are not to a great degree vivacious [18]; we grasp the possibility of slip by to get care of this issue CP-ABE makes another entrance course of action by adding the end appoint to the primary door to approach with intelligible AND task.

4 Implementation of PHR Cloud Platform

A protected PHR cloud floor utilizing CP-ABE dependent on Indivo X is executed. A Python ABE library named payable is produced and transformed using the data storehouse design, distribution method of original personal health application (PHA), and a new Indivo API calls [8]. It is fundamental to select a proper trait set for plotting a PHR cloud floor. In prior PHR frameworks, demonstrating ABE [1, 4, 10, 16, 17], they just select client's Workplace or employment as properties. To exhibit increasingly expressive conduct, the {Name, Gender, DOB, Occupation, Marriage status, Address, Workplace, Key Expiration} are characterized as the arrangement of characteristics. The key expiration is a lot of AA to give key renouncement and not an un-determined trait. Two sorts of attributive: numerical sort and non-numerical sort are there, as indicated by the Waters' CP-ABE plot [15]. The attribute = esteem is determined as numerical sort, where the characteristic is demonstrated as quality and esteem is a non-negative whole number under 264. Letters, digits, and underscores sequence combinations starting with a letter are non-numerical qualities. Client's characteristic protection is safeguarded by Name that is non-numerical trait, while the numerical shape in property stockpiling. In KeyGen calculation, the reason is that persistent Occupation = Doctor and Occupation = Physician both have a similar importance; however, they are two distinctive non-numerical qualities.

At the point when a patient encoded a PHR with trademark Occupation = Doctor, in this way, the specialist gets his private key as related to Occupation = Physician. The PHR decryption is not promising in this case. To evade decryption failures, the renovation of non-numerical to style features to numerical types is avoided. For example, numerical type "Age" is changed to attribute DOB. The enumerated data type characteristics like Gender, Occupation and Marriage are converted to numerals like "0" which indicates Female and "1" stands for Male. The Country, Province, City, Workplace and/or Address are organized for the attributes Workplace and Address. Each actual attribute gets a unique id and each table contains corresponding id for foreign key reference.

The access approach is integrated into ciphertext and is particular by PHP that is mentioned by the logical operations and comparison operators for numerical attribute articulation. For instance, "Doctor OR 2 OF (Age < 40, Female OR Male, Child hospital)" could be a strategy. Python is utilized as a programming language in Indivo X. For achieving ABE capacities, the back end customers and servers the Python variant pyabelib dependent on libfenc are assembled [16]. The C library is utilized to complete the every single computational undertaking. Furthermore, the Phyton is utilized for making calls.

The patient's PHR is put away in the XML, plain content in the first Indivo X Server. Another XSD (XMLSchemaDefinition) to supplant the first XSD to help CP-ABE plot which represents as pursues.

The patients can recognize the ordinary PHR name, keywords and strong access technique used for looking in new XML. The encoded PHR is the "Content" element. To execute each progression of encode and decode, a customer module is made by utilizing PyQt system. Following is XML snippet:

```
<?xml version="1.0" encoding="ISO-8859-1"?>
<schema xmlns="http://www.w3.org/2001/
 XMLSchema:elementFormDefault="qualified">
<element name="Abe_document">
<complexType>
<sequence>
<element name="Name" type="string"
         minOccurs="1" maxOccurs="1"/>
<element name="Content" type="string"
         minOccurs="1" maxOccurs="1"/>
<element name="KeyWord" type="string"
         minOccurs="0" maxOccurs="1"/>
<element name="Policy" type="string"
         minOccurs="0" maxOccurs="1"/>
</sequence>
</complexType>
</element>
</schema>
```

OAuth convention [12] is followed in PHA for encryption and sharing. Let patient to select PHR and set encoded PHR name, get to strategy and catchphrases, for scrambling PHA and after that conjures the customer module to finish scrambling operation. The security is given by sharing component in Indivo X, to enhance sharing PHA, if the client is empowered once.

It implies that this client permits offering his PHR to the individuals who additionally empowered PHR sharing PHA. The rundown of scrambled PHR is shown to the client after approval. The client can see, download, and unscramble what he needs which is shown in Fig. 2. The client's entrance strategy is happy with private key, instead of embedding it in encrypted PHR and then decrypted successfully.

5 Estimation of PHR Cloud Platform

As contrasted with Indivo X, the proposed structure gives some effect on the server side. Some new Indivo X style APIs are incorporated in proposed structure that takes after its security tradition altogether, remembering the ultimate objective to make Indivo X back end server, reinforcing the new scramble PHR stockpiling arrangement in the server side. The execution of the pyabelib is estimated, and afterward, it is contrasted and the libfenc. Params will be stacked to the customer side from the reserved records, and to stack params is likewise AA furthermore required alongside msk from nearby documents; it is vital to think about the input/output cost of perusing documents. So we express the aggregate execution time for every encryption, unscrambling and KeyGen that includes cost of perusing records at the same time; we assess the real calculation time in each progression and reduce the cost. According to the Water CP-ABE method, the non-numerical feature associates to

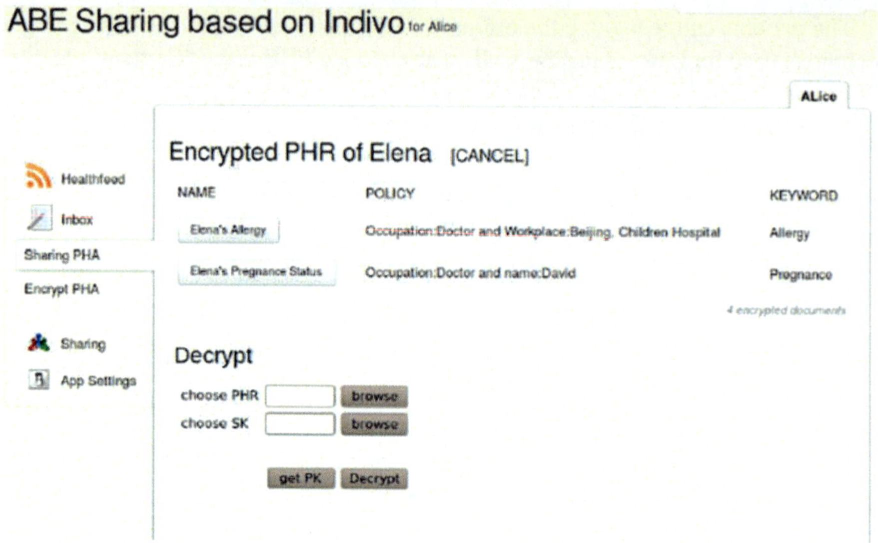

Fig. 2 Decrypt encrypted PHR and user view

the single leaf node in the access structure, whereas the numerical form associates to the additional leaf nodes. We exploited the number of nodes rather than the attribute numbers on X-axis, whereas each step's period overhead on Y-axis. The utilizing of non-numerical frame haphazardly to 100 quality sets, in the event that it is the I the property set, it has the I traits comparative to the leaf hubs. To check that each leaf hub is visited amid encryption AND entryway approach is used. At that point, same PHR plaintext is encoded through the pre-produced property sets going from 1 to 100. For precedent produced trait set from {a1001}, {a1001, b1002}, {a1001, b1002, c1003}. Then, characteristic sets are changed over as a1001","a1001 and b1002", "a1001 and b1002 and c1003 to the contribution for encryption. 50 quality sets are produced as same as above in KeyGen, and these credit sets are utilized to make SKω. To figure the expense of the KeyGen method executes for multiple times. To get the mean esteem, the technique is executed multiple times. In decryption, the span of figure content is settled is set to 128 bits in the AES encryption key. To get the figure substance and SKω, the encryption and KeyGen steps are repeated as above.

For ensuring the exploitation of decryption, the feature set that is used in the KeyGen essentially be utilized in the encryption. To get the mean value, executed for 10 times.

To run the experiments, Ubuntu 10.04, python 2.6.5, GCC 4.4.3 along with 2 × Intel Xeon E5606 @2.13 GHz CPU and 4 × 4 GB of RAM is used in Table 1.

In Table 1, it represents the key generation time using the CP-ABE, where the number of attributes in the ciphertext and the time taken for the key generation. The values are given in the table in order for plotting the graph.

The investigational estimation output is shown in Figs. 3, 4, and 5, respectively.

Table 1 Key generation time using CP-ABE

No. of attributes	Proposed CP-ABE	CP-ABE
1	1	5
2	9	11
3	10	16
4	11	20
5	13	25
6	15	40
7	16	44
8	18	46
9	19	51

Table 2 Encryption time

Encryption time	Pyabelib	Libfenc encryption time
1	5	0.2
2	10	1
3	20	2
4	25	4.3
5	30	4.8
6	34	6.4
7	44	7.2
8	46	8
9	50	10

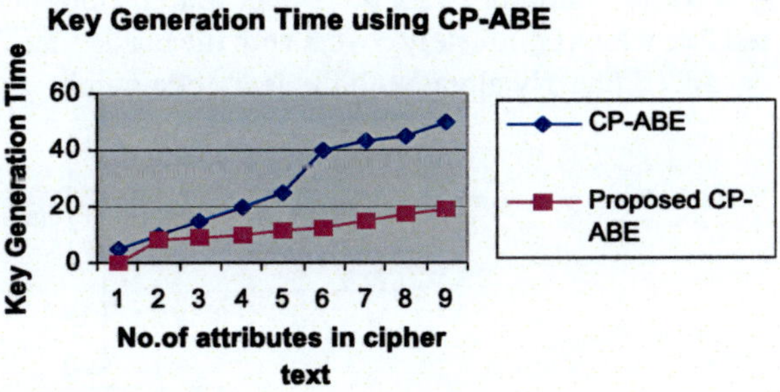

Fig. 3 Evaluation of pyabelib: key generation

The cost of the libfenc is pointed out with red line and the cost of the pyabelib indicated by the black line. The specific computational cost of the every step of CP-ABE scheme is indicated by the blue line.

Table 2 represents encryption time, where the pyabelib and libfenc encryption time are tabulated and the graph is designed where the cost of the libfenc is represented with red line and the cost of the pyabelib is indicated by the black line. The exact

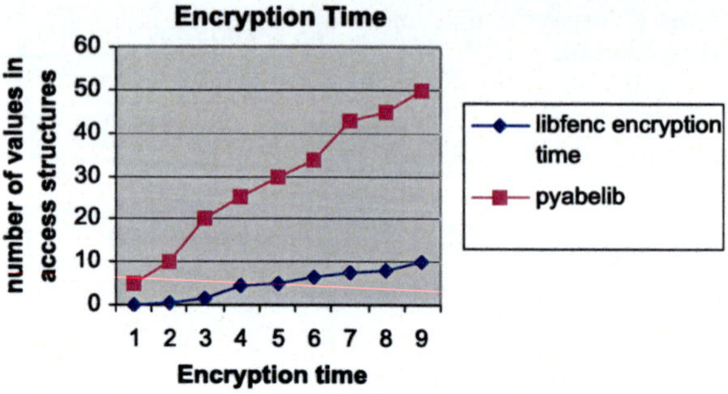

Fig. 4 Evaluation of pyabelib: encryption

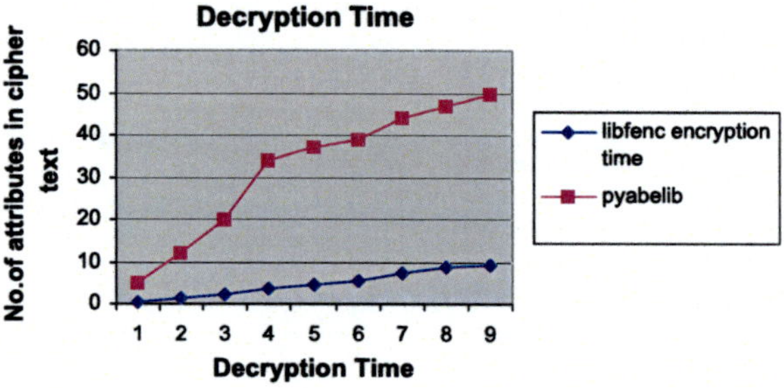

Fig. 5 Evaluation of pyabelib: decryption

computational cost of the each step of CP-ABE technique is represented by the blue line.

Table 3 represents the decryption time taken and the presentation of pyabelib is unfailing along with libfenc and thus the graph is plotted.

The execution of pyabelib is dependable alongside libfenc demonstrated by the red line which completely covers with the black line. The processing time, ultimately enhanced linearly along with the leaf node numbers present in accumulation to the

Table 3 Decryption time

Decryption time	Pyabelib	Libfenc decryption time
1	5	0.5
2	13	2
3	20	3
4	34	4
5	37	5
6	40	6
7	45	7
8	48	8
9	50	9

rate of reading file, is 0.5–0.7 s. Obviously, the time cost is sufficient; in this manner, it is down to ground to combine CP-ABE with Indivo X KeyGen is commonly under 1 s underneath the 50 leaf nodes, the server. The server does not have much working expense.

6 Conclusion

Traditional access control instruments and moreover standard encryption techniques are not appropriate to be used as a piece of the general PHR cloud computing circumstance; fine-grained access control and security assurance are needed to be provided. The CP-ABE system has shown to be increasingly useful in a therapeutic administration; the entrance approach is completed by the essentially indistinguishable of the entrance control methodology to the anchored information. This reduces the requirement for together with a trusted entity which needs to permit access approach. In this paper, the proposed patient-driven, security safeguarding PHR sharing model in the cloud, security assurance and fine-grained access control can be achieved by using CP-ABE. At this stage, the Indivo X is used to draw and implement a PHR cloud. By using CP-ABE technique, the PHR data is encoded and put away from the cloud.

References

1. Akinyele JA, Pagano MW, Green MD, Lehmann CU, Peterson ZNJ, Rubin AD (2011) Securing electronic medical records using attribute-based encryption on mobile devices. In: Proceedings of the 1st ACM workshop on security and privacy in smartphones and mobile devices, Chicago, pp 75–86, 2011
2. Fabian B, Ermakova T, Junghanns P (2015) Collaborative and secure sharing of healthcare data in multi-clouds. Inf Syst 48:132–150. https://doi.org/10.1016/j.is.2014.05.004
3. Goyal V, Pandey O, Sahai A, Waters B (2006) Attribute based encryption for fine-grained access conrol of encrypted data. In: Proceedings of the 13th ACM conference on computer and communications security, pp 89–98, 2006
4. Goyal V (2007) Certificate revocation using fine grained certificate space partitioning. In: Financial cryptography and data security, pp 247–259, 2007, Lecture Notes in Computer Science, vol 4886. Springer, Berlin, Heidelberg
5. Ibraimi L, Asim M, Petkovic M (2009) Secure management of personal health records by applying attribute-based encryption. In: 6th International workshop on wearable micro and nano technologies for personalized health (pHealth), pp 71–74, 2009
6. Li M, Yu SC, Zheng Y, Ren K, Lou WJ (2013) Scalable and secure sharing of personal health records in cloud computing using attribute based encryption. IEEE Trans Parallel Distrib Syst 24(1):131–143
7. Li M, Yu S, Ren K, Lou W (2010) Securing personal health records in cloud computing: patient-centric and fine-grained data access control in multi-owner settings. In: Security and privacy in communication networks 6th international ICST conference: secure communication networks 2010, Lecture Notes of the Institute for Computer Sciences, Social Informatics and Telecommunications Engineering, vol 50, Springer, Singapore, pp 89–106, 2010. https://doi.org/10.1007/978-3-642-16161-2_6

8. Mandl KD, Simons WW, Crawford WC, Abbett JM (2007) Indivo: a personally controlled health record for health information exchange and communication. BMC Med Inform Decis Mak 7(25):1–10. https://doi.org/10.1186/1472-6947-7-25
9. Mohan A, Bauer D, Blough DM, Ahamad M, Bamba B, Krishnan R, Liu L, Mashima D, Palanisamy B (2009) A patient-centric, attribute based, source-verifiable framework for health record sharing. GIT CERCS Technical Report No. GIT-CERCS-09-11
10. Narayan S, Gagné M, Safavi-Naini R (2010) Privacy preserving EHR system using attribute-based infrastructure. In: CCSW 2010, pp 47–52, 2010
11. Ostrovsky R, Sahai A, Waters B (2007) Attribute-based encryption with non-monotonic access structures. In: Proceedings of the 14th ACM conference on computer and communications security, pp 195–203, 2007
12. Personally Controlled Health Record (PCHR) system using attributed-based encryption project, The pyabelib toolkit, https://code.google.com/p/abe-pchr
13. Sahai A, Waters B (2005) Fuzzy identity based encryption. In: EUROCRYPT 2005, LNCS 3494, Springer, pp 457–473
14. Wang C, Liu X, Li W (2012) Implementing a personal health record cloud platform using ciphertext-policy attribute-based encryption. In: 4th international conference on Intelligent networking and collaborative systems (INCoS), pp 8–14. IEEE, 2012
15. Waters B (2011) Ciphertext-policy attribute-based encryption: an expressive, efficient, and provably secure realization. In: 14th international conference on practice and theory in public key cryptography, LNCS 6571. Springer, pp 53–70, 2011
16. Yin C, Zhang R (2011) Access control for the smart meters based on ABE. In: 2011 international conference on cyber-enabled distributed computing and knowledge discovery, pp 79–82, 2011
17. Yu S, Wang C, Ren K, Lou W (2010) Achieving secure, scalable, and fine-grained data access control in cloud computing. In: Infocom, 2010 proceedings IEEE. IEEE 2010, pp 1–9
18. Zheng Y (2011) Privacy-preserving personal health record system using attribute-based encryption, MS thesis, Worcester Polytechnic Institute

Improper Passing and Lane-Change Related Crashes: Pattern Recognition Using Association Rules Negative Binomial Mining

Subasish Das, Sudipa Chatterjee, and Sudeshna Mitra

Abstract Improper passing or lane-change related traffic crash is a critical issue. To pass or change lanes, the driver is required to make several judgments based on dynamic variables and allow little room for judgmental error that can result in drastic consequences if performed improperly. We used Florida rural roadway crash data from the second Strategic Highway Research Program (SHRP2) to investigate improper passing/lane-change related crashes. We applied an unsupervised data mining technique (known as association rules negative binomial (NB) miner) to extract the knowledge pattern of co-occurrence of the significant variables. This method identified some hidden trends from the complex nature of the traffic crash database. The findings show that this algorithm is suitable for pattern identification from traffic crash data.

Keywords Safety · Improper passing/lane-change related crashes · Association rules negative binomial miner · Data mining

1 Introduction

Significant numbers of crashes occur on rural roadways, and large portions of them are due to the lack of effective countermeasures to separate opposing traffic flows. As a result, a major concern involves vehicles crossing the centerline and resulting in

S. Das (✉)
Texas A&M Transportation Institute, Bryan, TX 77807, USA
e-mail: s-das@tti.tamu.edu

S. Chatterjee
Department of Civil Engineering, Indian Institute of Technology Kharagpur, Kharagpur 721302, India
e-mail: sudipa.chatterjee@iitkgp.ac.in

S. Mitra
Global Road Safety Facility, The World Bank, 1818 H Street NW, Washington, DC 20433, USA
e-mail: smitra5@worldbank.org

V. Singh et al. (eds.), *Computational Methods and Data Engineering*,
Advances in Intelligent Systems and Computing 1227,
https://doi.org/10.1007/978-981-15-6876-3_46

either sideswiping or head-on collisions. At the same time, improper passing/lane-change related crashes on divided roadways are not negligible. It is necessary to investigate the patterns of improper passing/lane-change related crashes. Driving on roadways is a multifaceted task, which requires the perception, comprehension, and projection of states of the roadway condition, as well as decision making on courses of spontaneous action and execution of driving behaviors. Passing or lane-changing on rural roadways is complex due to the roadway environment and opposite direction vehicles if the roadway is undivided. Many studies envisioned to explore the complex nature of passing to improve roadway standards by re-evaluating the patterns of improper passing crashes. Factors range from a wide variety in improper passing/lane-change crashes: the presence of curvature, curve radius, curve length, skid condition, weather, lane width, and the presence of countermeasures like raised pavement markers or edge lines, size of the vehicle, posted speed limit, vehicle speed, lighting condition, traffic volume, and time of the day. Linear regression methods usually cannot capture the clustered nature of complex crash data. Mixed effect modeling can capture the clustering nature by considering both fixed and random effects. However, it requires significant efforts to identify the data clustering using statistical methods. Data mining algorithms are helpful in determining clusters and hidden effects inside a complex dataset with less effort. We aim to mitigate the research gap by investigating a wider variety of geometric and environmental variables to identify the hidden patterns of improper passing/lane-change related crashes acquired from the second Strategic Highway Research Program (SHRP2) roadway inventory database (RID) data in Florida. To increase the precision of the knowledge extraction, we used association rules negative binomial (NB) miner algorithm.

1.1 Contexts and Research Questions

To understand the improper passing related issues, it is important to have a clear understanding on the pavement markings and scenarios for improper passing or lane-changing. Figure 1 illustrates the criteria for permitted passing and no-passing zones [1]. Figure 1a shows the permissible passing on a two-lane road with a single broken yellow centerline line. Figure 1b shows the permissible passing on a two-lane road with broken yellow line and solid yellow line marking. Figure 1c shows the improper passing on a road with similar markings. Figure 1d shows the improper passing on a road which has a solid double yellow centerline marking. Figure 1e indicates that passing zone is not allowed near the buffer zone. Usually, no-passing markings are installed in the areas with curves or low visibility ranges. Improper passing mainly indicates the pattern of passing vehicles on no-passing zones on undivided roadways. However, for divided roadways, improper lane-changing may relate to other factors. Improper passing crashes mainly involve driver cognition and attitude while passing. Earlier research shows that roadway characteristics play a key role in the likelihood

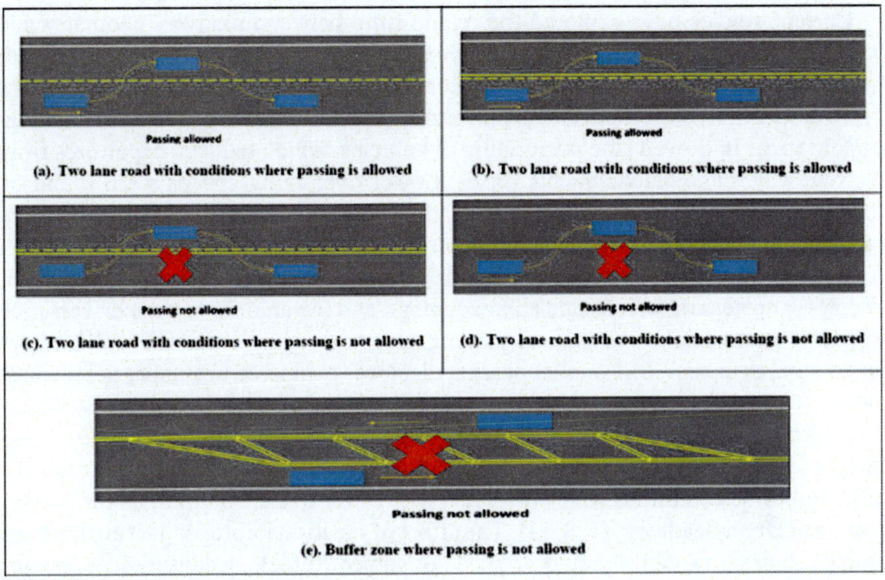

Fig. 1 Passing permitted and no-passing criteria

of injury or fatal crashes in improper passing crashes. The current research is focused on identifying the association between the key geometric features on the zones where crashes occurred due to improper passing.

The intent of this study is to address two key research questions:

- **RQ1**: What are the key contributors and association patterns in improper passing and lane-change related crashes?
- **RQ2**: Is the association rules NB miner algorithm suitable for identifying patterns from complex crash data?

2 Related Work

Many studies have attempted to understand the impact of various passing/lane-change patterns on crash severity. Historically, research has focused on refinement of roadway infrastructure guidelines, provided primarily by the American Association of State Highway and Transportation Officials (AASHTO) and the Federal Highway Administration (FHWA), which provides criteria for two-lane, two-way highways such as minimum passing sight distance and no-passing zones [2, 3]. Refinement of these guidelines included the re-evaluation and classification of passing maneuvers and the development of passing models that factored in higher speeds [2, 3]. Benefits of passing identified included: reduction of congestion, delays, and overall improvement of level of service for a roadway [4].

Current studies have explored the relationship between roadway geometry and driver behavior on rural two-lane roadways through simulation, field observation and, for a limited number of studies, naturalistic driving data. For example, the presence of a guardrail, highly visible pavement markers or the addition of centerline and shoulder rumble strips improved lane positioning of vehicles, which reduced departures from the roadway, i.e., encroachments [5, 6]. Lower overtaking speeds were found on roadways with no centerline markings and narrower lanes, while wider lanes, that offered more space, resulted in significantly higher passing speeds [7].

Roadway curvature, specifically the size of the curve, could also impact the likelihood of lane departure, left-side encroachment, and the amount of time drivers took to pass a lead vehicle, i.e., critical passing gaps where larger curves, which afford larger sight distances, had smaller passing distances compared to smaller or narrower curves [6].

Nighttime driving on rural roads can be particularly dangerous due to limited visibility and availability of lighting which are significant factors in passing behavior [8–10]. Nighttime conditions were correlated with lower traffic volumes, larger passing gaps, and larger headways [6, 8, 11]. The effect of speed on improper passing has been studied through variables including speed of subject vehicle, lead vehicle, opposing lane vehicle, effects on driver frustration, age, and gender [8, 12–14]. Higher roadway speeds were associated with smaller passing gaps, and posted speed limits also influenced driver's overtaking speed wherein lower posted speed limits correlated with lower overtaking speeds [7, 8, 12]

Many transportation safety researchers and practitioners have used association rules mining for different research problems [15–22]. One of major challenges in association rules mining is the determination of optimal support or confidence values. Association rules NB miner rather uses a model-based frequency constraint as an alternative of user-specified values. As the crash dataset is complex in nature, the identification of clusters or groups with higher precision would be an effective way of understanding the trends for which the current method has major advantages. The intent of this paper is to demonstrate an approach that can be used to better understand the factors that influence the occurrences of improper passing/lane-change crashes. This study is the first of its kind that used association rules NB miner in transportation safety research.

3 Data Description

3.1 Data Collection

The second Strategic Highway Research Program (SHRP2) project populated a roadway information database (RID) with data from the SHRP2 mobile data collection project; existing roadway data from government, public, and private sources; and supplemental data that further characterize traffic operations [23]. This database provides good quality data that are linkable to the SHRP2 naturalistic driving study (NDS) database utilizing geographic information system (GIS) tools. The RID is the supplementary tools for safety researchers to look at data sets of selected road characteristics and study matching NDS trips to explore the relationships between driver, vehicle, and roadway characteristics. Florida RID maintains traffic crash data for six years (2005–2010). The dataset contains crash, vehicle, and person information. The database is divided into two parts: (1) local roadways and (2) state highway systems (SHS) roadways. To prepare the database for this study, this study used both roadways.

3.2 Descriptive Statistics

The primary focus of the database preparation for this study is to create a detailed database on crashes related to improper passing/lane-change. Improper passing related attributes in RID are: (1) 05 (improper lane-change), (2) (improper passing). Figure 2 illustrates the distribution of improper passing/lane-change crashes in Florida based on two major facility types: divided roadways and undivided roadways. The primary notion of this study was to investigate the passing/lane-change related crashes on rural two-lane roadways. Due to the very small sample size of the data, the research considered a broader group of facility (rural roadways) to perform the analysis. Research also directed toward diving the database into two major groups (divided and undivided roadways) to identify the safety issues associated with roadway division.

Florida RID database maintains a larger number of variables. The authors conducted a detailed literature review to investigate the significant factors associated with improper passing/lane-change crashes. A group of major roadway and environmental variables was selected from the research synthesis and was explored in Florida RID database for availability. Moreover, the authors used random forest algorithm to conduct variable importance for selecting a final group of variables. In recent years, many studies have been using random forest algorithm to determine the variable importance rather than correlation analysis due to its applicability in all variable types (discrete, continuous, ordinal, and nominal). The details of the variable selection method are not described here to make the study more focused on the current scope of the research.

Table 1 lists descriptive statistics of the final ten variables. Significant difference is visible for different variables in the divided and undivided roadways. Friction on the pavement can be measured by friction factor or skid number. A higher skid number usually indicates a higher friction factor. Maximum posted speed is higher in percentage in divided roadways while comparing with undivided roadways. Divided roadways also exhibit higher percentage in high annual average daily traffic (AADT) and high average trucking percentages. On the other hand, undivided roadways exhibit higher percentage of fatal and injury crashes.

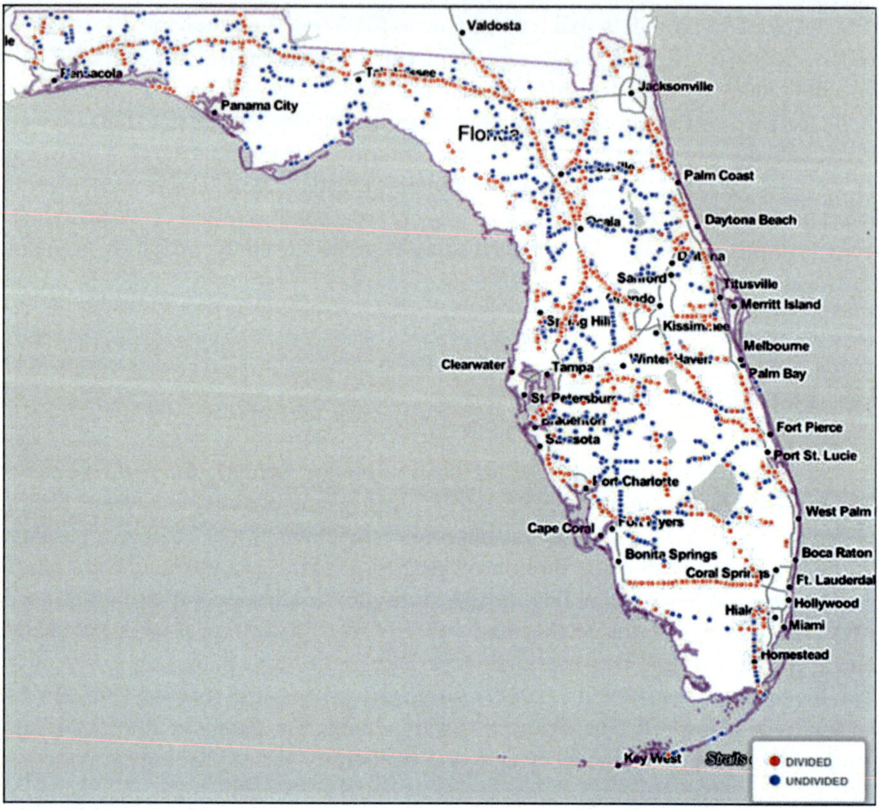

Fig. 2 Improper passing/lane-change crashes on divided and undivided roadways in Florida

4 Methodology

We used association rules NB miner to perform the analysis. Agrawal et al. introduced the data mining on the transaction data based on the associated items using the mining association rules in 1993 [24]. This section introduces the theoretical aspects of this approach based on Hahsler study [24]. Consider $I = \{i_1, i_2, \ldots, i_n\}$ be a set of n distinct items and $Q = \{q_1, q_2, \ldots, q_m\}$ be the transactions. Each transaction in Q contains a subject of the items in I. A rule is defined as an implication of the form *Antecedent* → *Consequent* or $M \rightarrow N$ where $M, N \subseteq I$ *and* $M \cap N = \emptyset$. A *d-itemset* has a size of d items. Support is defined on itemset $M \subseteq I$ as the proportion of transactions in which all items in Z:

$$\text{support}(M) = \frac{\text{freq}(M)}{|Q|} \tag{1}$$

Table 1 Descriptive statistics of the key variables

Category	Percentage		Category	Percentage	
Surface width (ft)	Divided (%)	Undivided (%)	AADT (vpd)	Divided (%)	Undivided (%)
0.00–10.00	0.03	0.17	0–10,000	4.97	64.17
10.01–20.00	0.78	14.21	10,000–20,000	15.17	26.24
20.01–30.00	62.60	81.95	20,000–30,000	18.82	3.92
30.01–40.00	36.05	3.49	30,000–40,000	13.21	1.92
>40	0.55	0.17	40,000–50,000	14.33	0.96
Skid number			50,000–60,000	10.40	0.87
0.00–30.00	10.49	1.66	>60,000	23.10	1.92
30.01–40.00	70.50	33.13	Lighting condition		
40.01–50.00	17.93	48.13	Daylight	70.59	70.70
>50	1.09	17.09	Dark (no streetlight)	21.11	21.01
Shoulder type			Dark (streetlight)	4.37	3.49
Paved	25.60	77.77	Dawn/dusk	3.94	4.80
Paved with warning	73.57	11.25	Weather condition		
Other	0.83	10.99	Inclement	34.04	28.68
Maximum speed (mph)			Non-inclement	65.96	71.31
0–50	3.36	13.60	Surface condition		
50–60	10.14	77.86	Dry	83.65	91.02
>60	86.50	8.54	Wet	16.35	8.98
Average truck percentage			Crash severity		
0.00–5.00	3.48	5.41	Fatal	1.49	4.80
5.01–10.00	14.97	24.59	Incapacitating injury	7.93	17.18
10.01–15.00	23.56	29.47	Possible injury	17.90	17.96
15.01–20.00	26.00	18.48	Non-incapacitating injury	13.70	17.44
>20	32.00	22.06	No injury	58.98	42.63

where freq(M) = frequency of itemset M (number of transactions in which M occurs) in Q, and $|Q|$ = number of transactions in the database.

For the rule: $M \rightarrow N$, confidence can be calculated as:

$$\text{Confidence}(M \rightarrow N) = \frac{\text{Support}(M \rightarrow N)}{\text{Support}(M)} \quad (2)$$

Relevance of the itemset Z to the user depends on the constraint: support $\geq \sigma$, where: σ is a user-specified minimum support value. Itemsets that satisfy the minimum support constraint are known as frequent itemsets. The performance measure of association rules is lift, which provides measure of the deviation from statistical independence of the relationship between M and N and is useful to identify associations that are significant deviations from the assumption of statistical independence. Lift is defined as:

$$\text{Lift}(M \rightarrow N) = \frac{\text{Support}(M \rightarrow N)}{\text{Support}(M) \times \text{Support}(N)} \quad (3)$$

Hashler [24] replaced the usage of the lift for a model aiming to evaluate the deviation for the set of all possible 1-extensions of an itemset together to evaluate a local frequency constraint for these extensions. The count distribution of itemsets is assumed to follow a NB distribution. The probability mass function of NB distribution is:

$$Pr[P = p] = (1+b)^{-t} \frac{\Gamma(t+p)}{\Gamma(p+1)\Gamma(t)} \left(\frac{b}{1+b}\right)^{p}, \ p = 0, 1, 2 \ldots \quad (4)$$

where $t =$ dispersion parameter, $b =$ mean parameter, $p =$ realization of the random variable P. Consider $\Sigma = \sigma f$, where f is the number of total transactions in the database. The count threshold is equivalent to the minimum support σ. Then, the expected number of 1-itemsets can be considered as:

$$x Pr[P \geq \Sigma] \quad (5)$$

where $x =$ the number of variables items. The counts for the 1-extensions of association m can be modeled by random variable P_m with the following probability mass function:

$$Pr[P_m = p] = (1+b_m)^{-t} \frac{\Gamma(t+p)}{\Gamma(p+1)\Gamma(t)} \left(\frac{b_m}{1+b_m}\right)^{p}, \quad \text{for } p = 0, 1, 2 \ldots \quad (6)$$

Consider M be the set of all 1-extension of a known association m which are generated by joining m with all candidate items, which co-occurrence with m in at least ρ transactions. For set M, the predicted precision is:

$$\text{Predicted precison of a rule} = \begin{cases} (o-e)/o \text{ if } o \geq e \text{ and } o > 0 \\ 0 \quad \text{otherwise} \end{cases} \quad (7)$$

where o is the observed and e is the expected number of candidate items which have a co-occurrence frequency with itemset m of $p \geq \rho$. The observed number is calculated as the sum of items with count p by $o = \sum_{p=\rho}^{p\max} o_r$, where $p_{\max}$ is the highest observed co-occurrence. The expected number is given by the baseline model as $e = (x - |m|)\Pr[P_m \geq \rho]$, where $x - |m|$ is the number of possible candidate items for pattern m. This method overcomes the conventional drawbacks of defining a minimum threshold support.

5 Results

We used open-source R package '*arulesNBMiner*' to perform analysis on two broader groups of unsupervised datasets based on the roadway facility types: divided and undivided roadways [25, 26]. As no response variable was preselected in both of the datasets, the learning algorithm of the model development was unsupervised in nature. It is important to note that a lower precision threshold usually increases a significant amount of increase of the generated rules. The number of itemsets is also another issue to interpret the results. Rules associated with 3 itemsets include one extra attribute than 2-itemset rules. By investigating these issues, this study used precision threshold as 0.9 and maximum number of itemset as 3. Tables 2 and 3 list the top twenty rules (based on precision scores) generated for two and three itemsets respectively. Table 2 lists the top twenty (based on higher precision) two itemset rules for divided and undivided roadways. For divided roadways, the significant attributes are paved shoulder with warning higher AADT, surface widths (in between 20 and 40 ft), surface condition, and higher speed (above 50 mph). The significant attributes for undivided roadways are higher skid number, lower AADT, narrower surface widths (in between 20 and 30 ft), average percentage of trucks, and crash severity (fatal and injury). The findings of the two itemset precision rules are consistent with the other studies that point to speed, surface widths, and traffic volume to improper passing crashes [6, 7, 9, 12].

From Tables 2 and 3, it is seen that the top twenty rules (generated from divided rural roadways) do not exhibit severity in either antecedent or consequent. On the other hand, severity is visible in the rules for undivided roadways. For example, rule number 8 for undivided roadways (Table 2) is: *Severity* = *Incapacitating Injury* → *Speed* = 50–60 mph with a precision value of 0.963. This rule indicates that the co-occurrence of incapacitating injury crashes with higher speed (50–60 mph) is over 95%.

Table 3 lists the rules for three itemsets. For divided roadways, the significant attributes are *paved shoulder with warning, higher AADT*, *weather, surface widths* (*in between* 30 *and* 40 ft), *and higher speed* (60 mph). The significant attributes for undivided roadways are *higher skid number, lower AADT, narrower surface widths* (*in between* 20 *and* 30 ft), *average percentage of trucks, and crash severity* (*non-incapacitating injury or no injury*). The findings of the three itemset precision rules are consistent with the other studies that point to speed, surface widths, and traffic volume to improper passing crashes [6, 7]. Weather exhibits a significant factor in the group of rules generated for divided roadways. For example, rule number 2 for divided roadways (Table 3) is: {*Surface Width* = 30–40 ft, *Surface Condition* = *Wet*} → *Weather* = *Inclement* with a precision value of 0.989. This rule indicates that the

Table 2 Two itemset precision rules for divided and undivided roadways

No.	Antecedent	Consequent	Precision
Divided roadways			
1	AADT[1] = 50,000–60,000	Shoulder = Paved with warning	0.965
2	Width[2] = 30.01–40.00	Shoulder = Paved with warning	0.965
3	AADT = 40,000–50,000	Shoulder = Paved with warning	0.963
4	AADT = 60,000	Shoulder = Paved with warning	0.960
5	AADT = 60,000	Surface[5] = Dry	0.958
6	Speed[3] = 50–60	Surface = Dry	0.957
7	AADT = 30,000–40,000	Width = 20.01–30.00	0.954
8	Speed = 0–50	Width = 20.01–30.00	0.948
9	AADT = 60,000	Skid = 30.01–40.00	0.947
10	AADT = 20,000–30,000	Surface = Dry	0.946
11	Speed = 50–60	Shoulder = Paved	0.946
12	AADT = 60000	Width = 30.01–40.00	0.946
13	Surface = Wet	Speed = 60	0.946
14	AADT = 50,000–60,000	Surface = Dry	0.945
15	AADT = 40,000–50,000	Skid = 30.01–40.00	0.943
16	AADT = 30,000–40,000	Surface = Dry	0.941
17	AADT = 60,000	Speed = 60	0.941
18	Shoulder[4] = Paved with warning	Speed = 60	0.941
19	AADT = 0–10,000	Shoulder = Paved	0.940
20	Width = 30.01–40.00	Surface = Dry	0.939
Undivided roadways			
1	Skid[6] = 50	AADT = 0–10,000	0.977
2	Skid = 50	Width = 20.01–30.00	0.971
3	Skid ≥ 50	Speed = 50–60	0.971
4	Skid = 40.01–50.00	Width = 20.01–30.00	0.967
5	AADT = 0–10,000	Width = 20.01–30.00	0.967
6	AADT = 0–10,000	Shoulder = Paved	0.965
7	Trucks[7] = 5.01–10.00	Width = 20.01–30.00	0.964
8	Severity[8] = Incapacitating inj.	Speed = 50–60	0.963
9	Skid = 40.01–50.00	Speed = 50–60	0.963
10	Severity = Non-incapacitating inj.	Shoulder = Paved	0.962
11	Skid = 40.01–50.00	Light[9] = Daylight	0.961
12	Severity = Fatal	Shoulder = Paved	0.959
13	Shoulder = Other	Speed = 50–60	0.959
14	Trucks = 20	Speed = 50–60	0.957

(continued)

Table 2 (continued)

No.	Antecedent	Consequent	Precision
15	Skid = 40.01–50.00	AADT = 0–10,000	0.957
16	Trucks = 5.01–10.00	Speed = 50–60	0.955
17	AADT = 0–10,000	Speed = 50–60	0.955
18	Severity = Fatal	Width = 20.01–30.00	0.955
19	Severity = Fatal	AADT = 0–10,000	0.955
20	Trucks = 20	Weather[10] = Non-inclement	0.954

Note [1]AADT—annual average daily traffic (vpd), [2]Width—surface width (ft), [3]Speed—maximum posted speed (mph), [4]Shoulder—type of shoulder, [5]Surface—surface condition, [6]Skid—skid number, [7]Trucks—average percentage of trucks, [8]Severity—crash severity, [9]Light—lighting condition, [10]Weather—weather condition

Table 3 Three itemset precision rules for divided and undivided roadways

No.	Antecedent	Consequent	Precision
Divided roadways			
1	Weather = Non-inclement, AADT = 50,000–60,000	Shoulder = Paved with warning	0.989
2	Width = 30.01–40.00, Surface = Wet	Weather = Inclement	0.989
3	Width = 30.01–40.00, Weather = Non-inclement	Shoulder = Paved with warning	0.989
4	Weather = Non-inclement, AADT = 40,000–50,000	Shoulder = Paved with warning	0.988
5	Weather = Non-inclement, AADT = 60,000	Shoulder = Paved with warning	0.987
6	Shoulder = Paved with warning, Weather = Non-inclement	Surface = Dry	0.986
7	Width = 30.01–40.00, Trucks = 20	Shoulder = Paved with warning	0.986
8	AADT = 60,000, Trucks = 15.01–20.00	Shoulder = Paved with warning	0.986
9	Width = 30.01–40.00, Trucks = 15.01–20.00	Shoulder = Paved with warning	0.986
10	Weather = Non-inclement, AADT = 60,000	Surface = Dry	0.986
11	AADT = 60,000, Trucks = 10.01–15.00	Shoulder = Paved with warning	0.986
12	AADT = 60,000, Trucks = 10.01–15.00	Speed = 60	0.986
13	Width = 30.01–40.00, Weather = Non-inclement	Speed = 60	0.986

(continued)

Table 3 (continued)

No.	Antecedent	Consequent	Precision
14	Width = 30.01–40.00, AADT = 40,000–50,000	Shoulder = Paved with warning	0.986
15	Width = 30.01–40.00, AADT = 40,000–50,000	Speed = 60	0.986
16	Weather = Non-inclement, AADT = 50,000–60,000	Surface = Dry	0.986
17	Weather = Non-inclement, AADT = 50,000–60,000	Speed = 60	0.986
18	Width = 30.01–40.00, AADT = 50,000–60,000	Shoulder = Paved with warning	0.986
19	Width = 30.01–40.00, Light = Dark (no streetlight)	Shoulder = Paved with warning	0.985
20	Shoulder = Paved with warning, AADT = 20,000–30,000	Width = 20.01–30.00	0.985
Undivided roadways			
1	Skid = 40.01–50.00, Trucks ≥ 20	Width = 20.01–30.00	0.990
2	Skid = 40.01–50.00, Trucks ≥ 20	Speed = 50–60	0.989
3	Skid ≥ 50, Shoulder = Paved	Speed = 50–60	0.989
4	Skid ≥ 50, Light = Daylight	Width = 20.01–30.00	0.988
5	Skid ≥ 50, Light = Daylight	Speed = 50–60	0.988
6	Skid = 40.01–50.00, AADT = 0–10,000	Width = 20.01–30.00	0.988
7	Severity = Non-incapacitating injury, Trucks = 5.01–10.00	Width = 20.01–30.00	0.987
8	Skid ≥ 50, Severity = No Injury	AADT = 0–10,000	0.987
9	Skid ≥ 50, Severity = No Injury	Speed = 50–60	0.987
10	Skid ≥ 50, Speed = 50–60	Width = 20.01–30.00	0.986
11	Skid ≥ 50, Speed = 50–60	AADT = 0–10,000	0.986
12	Skid = 40.01–50.00, Trucks ≥ 20	AADT = 0–10,000	0.986
13	Skid = 40.01–50.00, Trucks ≥ 20	Surface = Dry	0.986
14	Skid ≥ 50, Shoulder = Paved	AADT = 0–10,000	0.986
15	Skid ≥ 50, AADT = 0–10,000	Width = 20.01–30.00	0.986
16	Skid ≥ 50, Weather = Non-inclement	AADT = 0–10,000	0.985

(continued)

Table 3 (continued)

No.	Antecedent	Consequent	Precision
17	Severity = Non-incapacitating injury, AADT = 0–10,000	Shoulder = Paved	0.985
18	Skid = 40.01–50.00, AADT = 0–10,000	Shoulder = Paved	0.985
19	Skid = 40.01–50.00, AADT = 0–10,000	Surface = Dry	0.985
20	AADT = 0–10,000, Trucks = 5.01–10.00	Speed = 50–60	0.985

co-occurrence of these three attributes in divided roadway dataset is over 98%. On the other hand, the average percentage of trucks exhibits significant dominance on undivided roadways. For example, the top rule for undivided roadways is: {*Skid Number* = 40–50, *Average Percentage of Trucks* = 20%} → *Roadway Width* = 220–30 ft with a precision value of 0.990. This rule shows that the co-occurrence of these three attributes in divided roadway dataset is 99% (Table 3).

The following findings below summarize the contents of knowledge extraction from the top twenty precision rules (two itemset and three itemset) for both divided and undivided roadways:

- Improper passing crashes on divided roadways associate with higher AADT, wider roadways, and higher speed for two itemset rules. On the other hand, undivided roadways in corporate lower AADT, narrower roadways, low to medium percentage of trucks, and severe/fatal crashes.
- Fatal or severe injuries are dominant on undivided roadways for two itemset rules.
- For three itemset rules, inclement weather contributes significantly on improper passing crashes on divided roadways.
- For three itemset rules, average percentage of trucks contributes significantly on improper passing crashes on undivided roadways.

6 Conclusions

The current study used Florida SHRP2 RID crash data for six years (2005–2010) to identify the key issues associated with improper passing crashes. Facility types on rural roadways (divided and undivided) show significant differences in the descriptive statistics of the geometric and environmental variables. The undivided roadways show a higher likelihood of fatal and injury crashes compared to divided roadways. The research team used an unsupervised data mining method (Association Rules NB Miner) to generate two itemset and three itemset rules. This method is an alternative of conventional association rules mining with an advantage of having no need of predefined support and confidence thresholds to generating rules. The current method is robust because it identifies rules with higher precision (by using a single performance measure) from a large dataset by identifying key co-occurrence, while the conventional association rules use three parameters as performance measures. For two itemset precision rules in divided roadways, paved shoulder with warning, high AADT, and high speed are dominant in both antecedents and consequents. On the other hand, for undivided roadways, the dominant associated factors are crash severity, low AADT, percentage of trucks, and skid number. For three itemset rules while considering divided roadways, the dominant factors are weather, high AADT, and percentage of trucks. For divided roadways, skid resistance, low AADT, and crash severity are dominant. The findings of this research are consistent with those of previous studies. The top twenty rules for both roadways clearly show that facility type (either divided or undivided) plays a significant role in the higher likelihood of injury or fatal crashes. However, divided roadways also need attention in reducing improper passing crashes on roadways with high AADT and speed. The findings of the current study will help the safety professionals in mitigating improper passing crashes and crash severities.

Based on prior study results, we recommend that the addition of passing lanes and shoulders significant safety improvement in reducing the potential number of fatal crashes on rural roads [27, 28]. Interventions that improve roadway safety have been identified as one of the top priorities for roadway users [29]. Forward collision warning technology has also shown promise in its ability to track objects in front of the vehicle and provide feedback of an impending collision, a valuable system to assist with overtaking maneuvers [12].

The reason of using RID in this study is to relate the roadways with NDS data in this ongoing project to identify the relationship between improper passing and passing behavior of the drivers. One of the limitations of this study is that it only considers geometric variables to conduct analysis. Therefore, future research should investigate at least two major issues: identify exact passing permitted and no-passing locations from the spatially coded crash information to perform a more robust analysis; use NDS data for exploring behavioral perspectives.

References

1. Manual on Uniform Traffic Control Devices (2009) For streets and highways. U.S. Dept. of Transportation, Federal Highway Administration, Washington, DC
2. Carlson P, Miles J, Johnson P (2016) Daytime high-speed passing Maneuvers observed on rural two-lane, two-way highway: findings and implications. Transp Res Rec: J Transp Res Board 1961:9–15
3. Jenkins J, Rilett L (2005) Classifying passing Maneuvers: a behavioral approach. Transp Res Rec: J Transp Res Board 1937:14–21
4. Romana M (1999) Passing activity on two-lane highways in Spain. Transp Res Rec: J Transp Res Board 1678:90–95
5. Gates T, Savolainen P, Datta T, Todd R, Russo B, Morena J (2012) Use of both centerline and shoulder rumble strips on high-speed two-lane rural roadways. Transp Res Rec: J Transp Res Board 2301:36–45

6. Hallmark S, Tyner S, Oneyear N, Carney C, Mcgehee D (2015) Evaluation of driving behavior on rural 2-lane curves using the SHRP 2 naturalistic driving study data. J Saf Res 54:17–27
7. Shackel SC, Parkin J (2014) Influence of road markings, lane widths and driver behaviour on proximity and speed of vehicles overtaking cyclists. Accid Anal Prev 73:100–108
8. Farah H, Toledo T (2010) Passing behavior on two-lane highways. Transp Res Part F: Traffic Psychol Behav 13(6):355–364
9. Papakostopoulos V, Nathanael D, Portouli E, Marmaras N (2015) The effects of changes in the traffic scene during overtaking. Accid Anal Prev 79:126–132
10. Levulis SJ, Delucia P, Jupe J (2015) Effects of oncoming vehicle size on overtaking judgments. Accid Anal Prev 82:163–170
11. Llorca, C, Moreno A, García A, Pérez-Zuriaga A (2013) Daytime and nighttime passing Maneuvers on a two-lane rural road in Spain. Transp Res Rec: J Transp Res Board 2358:3–11
12. Chen R, Kusano K, Gabler H (2015) Driver behavior during overtaking Maneuvers from the 100-car naturalistic driving study. Traffic Inj Prev 16(2):176–181
13. Kinnear N, Helman S, Wallbank C, Grayson G (2015) An experimental study of factors associated with driver frustration and overtaking intentions. Accid Anal Prev 79:221–230
14. Vlahogianni E, Golias J (2012) Bayesian modeling of the microscopic traffic characteristics of overtaking in two-lane highways. Transp Res Part F: Traffic Psychol Behav 15(3):348–357
15. Das S, Kong X, Tsapakis I (2019) Hit and run crash analysis using association rules mining. J Transp Saf Secur 1–20
16. Yu W (2019) Discovering frequent movement paths from taxi trajectory data using spatially embedded networks and association rules. IEEE Trans Intell Transp Syst 20(3)
17. Das S, Sun X, Goel S, Sun M, Rahman A, Dutta A (2020) Flooding related traffic crashes: findings from association rules. J Transp Saf Secur 1–19
18. Das S, Dutta A, Jalayer M, Bibeka A, Wu L (2018) Factors influencing the patterns of wrong-way driving crashes on freeway exit ramps and median crossovers: exploration using 'Eclat' association rules to promote safety. Int J Transp Sci Technol 7(2):114–123
19. Weng J, Li G (2019) Exploring shipping accident contributory factors using association rules. J Transp Saf Secur 11(1):36–57
20. Das S, Sun X, Dutta A (2019) Patterns of rainy weather crashes: applying rules mining. J Transp Saf Secur 1–23
21. Weng J, Zhu J, Yan X, Liu Z (2016) Investigation of work zone crash casualty patterns using association rules. Accid Anal Prev 92:43–52
22. Das S, Dutta A, Avelar R, Dixon K, Sun X, Jalayer M (2020) Supervised association rules mining on pedestrian crashes in urban areas: identifying patterns for appropriate countermeasures. Int J Urban Sci 23(1):30–48
23. SHRP2 RID. http://www.ctre.iastate.edu/shrp2-rid/. Last accessed 2020/03/13
24. Hashler M (2006) A model-based frequency constraint for mining associations from transaction data. Data Min Knowl Disc 13:137–166
25. R Development Core Team. R (2013) A language and environment for statistical computing. Version 2.10.1. R Foundation for Statistical Computing, Vienna, Austria (2013). http://www.R-project.org. Last accessed 2020/03/13
26. Hashler MR package 'arulesNBMiner'. https://cran.r-project.org/web/packages/arulesNBMiner/arulesNBMiner.pdf. Last accessed 2020/03/13
27. Schrock S, Parsons R, Zeng H (2011) Estimation of safety effectiveness of widening shoulders and adding passing lanes on rural two-lane roads. Transp Res Rec: J Transp Res Board 2203:57–63
28. Brewer M, Venglar S, Fitzpatrick K, Ding L, Park B (2012) Super 2 highways in Texas. Transp Res Rec: J Transp Res Board 2301:46–54
29. Mutabazi M, Russell E, Stokes R (1998) Drivers' attitudes, understanding, and acceptance of passing lanes in Kansas. Transp Res Rec: J Transp Res Board 1628:25–33

Sleep Stage and Heat Stress Classification of Rodents Undergoing High Environmental Temperature

Prabhat Kumar Upadhyay and Chetna Nagpal

Abstract Stress is one of the major concerns originated from neuronal activities which may lead to mental health problems, such as anxiety, depression, and personality disorders. Physiological studies have also been carried out to explore the application of computing techniques to predict "Heat Stress"—stress which develops due to high environmental temperature. Prerecorded data has been synthesized and analyzed to detect the changes in sleep electroencephalogram (sleep EEG) under heat stress. This work presents a technique to detect the heat stress by employing linear discriminant analysis (LDA) followed by continuous wavelet transform (CWT). Through wavelet decomposition, different frequencies embedded in the EEG signal were analyzed and features were extracted to detect the changes in stressed data with respect to control. The comparison of LDA with Adaptive neuro-fuzzy system (ANFIS) has also been addressed, where LDA shows good accuracy in stressed REM pattern as compared to other two stages of sleep EEG. An increase of 7.5% has been observed in LDA while detecting REM patterns.

Keyword Heat stress · Sleep EEG · Wavelet transform · Linear discriminant analysis · ANFIS

1 Introduction

Research work in the field of stress has been of paramount importance since last two decades. Many researchers have studied the consequences of heat stress on neuronal functioning to reveal the cause of various psychiatric problems. Environmental heat has been reported to be one of the natural stress markers which significantly impacts on human nervous system. Earlier reports suggest that when animals are exposed

P. K. Upadhyay
Department of EEE, Birla Institute of Technology, Mesra, Ranchi, India
e-mail: uprabhatbit@gmail.com

C. Nagpal (✉)
Department of EEE, Birla Institute of Technology Offshore Campus, Ras al Khaimah, UAE
e-mail: chetnakochhar0@gmail.com

V. Singh et al. (eds.), *Computational Methods and Data Engineering*,
Advances in Intelligent Systems and Computing 1227,
https://doi.org/10.1007/978-981-15-6876-3_47

to high environmental heat, numerous processes associated with neuro-physiology get activated and in turn metabolic change takes place. Metabolic change is highly influenced by few important factors like intensity of heat, duration of heat exposure and adaptations to the hot environment. Hyperthermia which is caused by thermal environment temperature induces shock and in effect, damage to the central nervous system is observed. Under this situation, blood supplied to the brain is reduced due to which circulatory system is collapsed [1]. The reflection of disturbance in central nervous system (CNS) appears in normal sleep which is invariably linked with CNS.

There have been two primary opinions which describe the reasons behind damage due to heat stress: (a) high environmental heat and subsequent hyperthermia; (b) collapse of circulatory system resulting into inadequate supply of blood to brain [2]. High environmental temperature can induce shock which may lead to brain damage. Results of the experiments carried out on animal subjected to hot environment suggest variation in brain electrical activities [3, 4] owing to the neuronal changes in brain. Therefore, hot environment has its direct effect on cortical activities recorded from brain.

Figure 1 gives the framework and process required for sleep stage classification and stress detection. The process of heat stress detection from sleep EEG events includes: data organization, signal preprocessing, features extraction and reduction and classification. Three stages of sleep EEG i.e., Awake, slow wave sleep (SWS) and rapid eye movement (REM) have been visually examined under heat stress conditions by analyzing the staging criteria of three channel recordings: EEG, electromyogram (EMG) and electrooculogram (EOG). Recent works have significantly shown various methods of evaluation, of which statistical pattern recognition was the prime method employed for sleep staging where, quantitative features were estimated and optimized [5]. Authors reported that the EEG signal consists of finite number of elementary pattern [6] and they performed the power spectrum analysis of EEG signals using the fast Fourier transform (FFT) [7, 8] through which the estimation of variation in frequency and amplitude was investigated. However, FFT failed to provide the time information of the occurrence of EEG events. To overcome this time resolution problem of FFT, wavelet transformation as the best alternative, has been successfully applied in EEG analysis [9–15].

Fig. 1 Framework for sleep and stress classification

In general, the benefits of various tools of digital signal processing as conventional approaches have been harnessed very well to diagnose the spectral changes due to stressful events. The vigilance states of the subjects have been quantified under predefined stress conditions and alterations in sleep stages have been investigated for three channels recording—EEG, EOG and EMG by applying machine learning algorithms. In order to reduce the effort of neurologists, several architectures of neural networks have been designed to study and examine the performance of classifiers. Here, LDA has been applied for the classification of three sleep stages (Awake, SWS and REM) and predict the stressed patterns with respect to their respective control groups. Use of LDA may play an important role in biomedical applications with different clinical conditions. The performance of LDA with wavelets as a preprocessor gives good insight to stress classification problem.

2 Materials and Method Applied

The present work uses prerecorded EEG data acquired from an animal model [16–18]. EEG data was recorded from rodents aged 12–14 weeks with an approximate weight of 180–200 grams toward the starting of experiment. The weight and temperature of rodents have been measured throughout the experiment for the investigation of stress. For heat stress study, these rodents were kept under high temperature of 38 °C for four hours each day to record the electrophysiological signals. Rodents of control group were kept inside the incubator with maintained temperature of 23 $\pm$ 1 °C (room temperature). Entire strategy pertaining to surgery, food, recordings etc. was pursued precisely like that of their respective groups which have undergone stress.

3 Feature Extraction

The data acquisition system was used for collecting and storing the data which processes the signal with sampling frequency of 256 Hz. The three polygraph recordings namely EEG, EOG and EMG were saved in small segments of 2 min in the hard disk forming 600 separate files. Each file contains four samples of 30 s length which was further used to extract the features from the recorded data. For visual sleep scoring, 30 s sample length was chosen as a prescribed length [19]. In 600 files, 2400 samples were stored for preprocessing, testing and training protocols. These samples were recorded from 100 subjects under heat stress conditions with their control groups. Offset adjustment and baseline removal with band pass filtering were applied to filter the signals from 0.5 to 40 Hz, which distinguishes the signals containing the unwanted frequency components.

Total ten features which were extracted for the classification of sleep stages are as follows:

- Relative band powers of delta, theta, alpha, and beta bands.
- Total band powers of all sub-bands.
- mean of EOG and EMG activities.

These features were stored with their labels for raining and test the samples.

4 Wavelet Packet Analysis as a Preprocessor

The wavelet packet provides a flexible analysis of more complex signals because approximations as well as their respective details are split in time-frequency domain. It starts from a scale-oriented decomposition and analyzes the resulting waves pertaining to frequency bands under consideration, making REM, SWS and Awake patterns visibly distinguished. Due to stress developed in the subjects after exposure of heat, alterations in frequency components of each signal stored as MAT files, were studied and examined visually by means of wavelet packet analysis. Signals were decomposed at different levels on dyadic scale through orthogonal method which produced a vector of detail coefficients and approximation coefficients. By decomposing successive approximation coefficients, changes in brain waves containing low/high frequency components were analyzed. However, successive detail coefficients were not reanalyzed. Daubechies mother wavelet of order-4 (db-4) has been used in continuous wavelet transformation to obtain wavelet coefficients for SWS, Awake and REM patterns as a function of translation vector (time) and scale. For each time domain signal, wavelet coefficients were calculated and arranged in a matrix according to frequency range of interest. Coefficients corresponding to delta, theta, alpha and beta bands were segregated and put together for further study. After preprocessing of wavelet coefficients, frequency information was extracted and absolute power as well as relative power of delta, theta, alpha and beta bands of different sleep stages (SWS, Awake and REM) of stressed and control subjects were calculated.

5 Linear Discriminant Analysis

Linear discriminant analysis has been generally used for the classification problems including categorical parameters [20, 21]. The discriminant analysis uses the same approach as that of ANOVA -Analysis of variances method. The purpose of discriminant analysis is to separate the groups maximally. Here, we have used LDA to classify the sleep stages under stress and hence detect the presence of heat stress. Through LDA, a relationship between independent variable and the response variable was made to predict the classification. After calculating the correlation matrix of all input features, test of homogeneity of covariance matrix have been done.

Table 1 Eigenvalues

Function	Eigenvalue	% of variance	Cumulative %	Canonical correlation
1	2.907	83.7	83.7	0.863
2	0.568	16.3	100	0.602

Table 1 shows the two canonical functions with its eigenvalues along with variance percentage and cumulative percentage. The larger the eigenvalue, more is the variance explained by canonical function. The table shows the results for three categories: Awake, SWS, REM. Two discriminant functions have been considered which have 0.863 and 0.602 as their canonical correlation. It measures the association between function and dependent variable, i.e., three stages of sleep EEG.

Table 2 is used to assess how well the discriminant function works. 100% classification for class 3 is reported as shown in Table 2. Classifications accuracy for class 1 and class 2 has been obtained as 86% and 79% respectively. Overall 88.3% of original grouped cases were correctly classified. In cross validation, each case is classified by the functions derived from all cases other than that case. 85.1% of cross-validated grouped cases got correctly classified.

Figure 2a shows the combined plot of three classes with their centroids, whereas Fig. 2b–d are the scatter plots of individual classes. For SWS, class 3 samples were observed to be very close to the centroid of the cluster making the SWS stage more accurate than other classes. The closer the group centroids, more will be misclassification.

Table 2 Classification results

Class (1 = REM, 2 = AWAKE, 3 = SWS)			Predicted group membership			Total
			1	2	3	
Original	Count	1	103	7	10	120
		2	9	66	9	84
		3	0	0	186	186
	%	1	**86**	6.1	7.9	100
		2	10.5	**79**	10.5	100
		3	0	0	**100**	100
Cross-validated	Count	1	76	18	26	120
		2	20	62	2	84
		3	0	0	186	186
	%	1	**63.3**	15	21.7	100
		2	7.8	**92**	2.4	100
		3	0	0	**100**	100

Bold indicates the accuracy in percentage achieved for classifying sleep stages

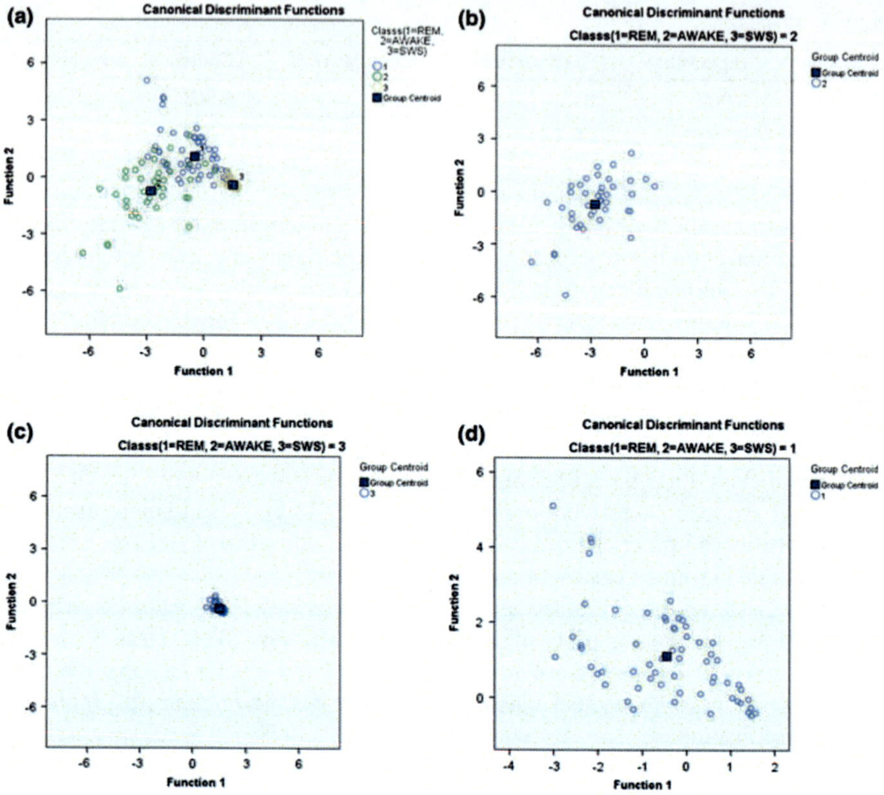

Fig. 2 **a** Combined group plot, **b** scatter plot for class 2 (AWAKE), **c** scatter plot for class 3 (SWS), **d** scatter plot for class 1 (REM)

6 Heat Stress Classification

Another classification was performed to detect the stress. The stressed patterns were classified with respect to the control group with the help of Canonical discriminant function. The prerecorded sleep EEG events were classified as class 1 or class 2 where class 1 belongs to stressed EEG signals with label 0 and class 2 belongs to control sleep EEG signals with label 1.

Tables 3 and 4 show the test of homogeneity of covariance matrices. For our dataset, we obtain the two different classes which differ in their covariance matrices.

Table 3 Log determinants and test results

0 = class 1, = class 2	Rank	Log determinant
0	5	−21.892
1	5	−31.037
Pooled within-groups	5	−22.904

Table 4 Test results

Box's M		1538.562
F	Approx	101.547
	df1	15
	df2	980745.6
	Sig.	0

For large value of n (no. of independent variables), small deviations from homogeneity will be significant and hence Box's M, multivariate statistical test must be interpreted in conjunction with the log determinant's inspection. The difference for group's covariance matrix is observed to be larger if log determinant is more. The "Rank" column specifies the total number of independent variables in particular case.

Table 5 shows the eigenvalues with percentage variance of stress classification. If the eigenvalue is larger, the variance will be more in the dependent variable which is explained by that function. Since dependent parameter has got two categories: stressed and control; only one discriminant function has been taken into account. The measurement of canonical correlation is defined as the relation between the dependent variable and discriminant function where the square of canonical correlation coefficient gives the percentage of variance explained in the dependent variable.

Table 6 shows the stress classification results in which 88.1% of original grouped cases got correctly classified. Cross validation has been carried out only for these cases in the analysis. Figure 3 shows the canonical discriminant function for different

Table 5 Eigenvalues

Function	Eigenvalue	% of variance	Cumulative %	Canonical correlation
1	0.368	100	100	0.518

Table 6 Stress classification results

0 = class 1; 1 = class 2			Predicted Group Membership		Total
			0	1	
Original	Count	0	291	38	329
		1	27	203	231
	%	0	**88.4**	11.3	100
		1	12.9	**87.8**	100
Cross-validated	Count	0	291	38	329
		1	49	182	231
	%	0	**88.7**	11.3	100
		1	12.9	**87.2**	100

Bold indicates the accuracy in percentage achieved for classifying the stress with respect to control

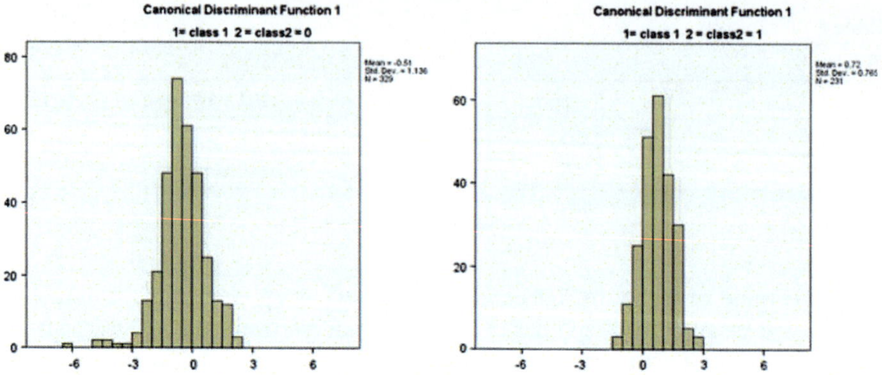

Fig. 3 Separate group plots for stressed and control groups

classes i.e., stressed and control groups. If two distributions overlap too much, it means they do not discriminate too, which can be called as poor discriminant function.

7 Comparison of LDA with ANFIS

Figure 4 shows the comparison between linear and nonlinear algorithms for sleep stage classification under heat stress, where awake sleep patterns show good accuracy in LDA as compared to ANFIS. Test data was randomly selected from pattern space and presented at the input layer to assess the performance of the classifier. Analysis of the results obtained suggests that Awake stressed patterns were best recognized by ANFIS, whereas REM stressed patterns were best detected by LDA. Performance of both the classifiers was observed to be at par when SWS patterns were tested. In case

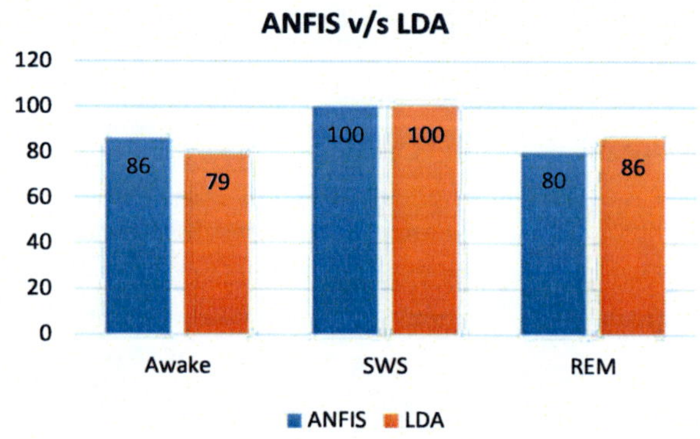

Fig. 4 Comparison between ANFIS and LDA performance in sleep stage classification

of REM patterns, recognition rate of LDA is found to be 7.5% better than ANFIS classifier. However, while classifying the stressed sleep EEG events with respect to control, LDA shows 88.33% average accuracy which is comparatively lesser than ANFIS as reported in [18].

8 Discussion

Many authors have discussed the sleep EEG activities under the exposure of artificial heat which was able to detect the changes in EEG activities as a function of ambient temperature. If there is an increase in an environmental heat, temperature of the body also increases instinctively together with increase in EEG frequencies. Due to the continuous rise of heat, the signal amplitude gets increased with decrease in dominant frequency [22]. The study of wavelet analysis reveals that the three vigilance states: Awake, SWS and REM have common characteristics for the leading frequency components present in four bands but they are not constant in entire range of time. It has been noted that the power percentage for the different components of four frequency sub-bands was changing at different intervals of time. Therefore, redistribution of frequency and power was observed in all three stages in both the conditions—before and after the heat exposure.

Further, reviewing the different methods of feature extraction, dimension reduction and classification, it is identified that the wavelet transform is one of the most suitable feature extraction techniques. To meet the proposed aim and objective, LDA has been effectively applied as classifier. The work [18] intended to detect the stressed patterns from their respective normal patterns by using ANFIS and Mamdani fuzzy model with an average classification accuracy of 89%. Authors have extracted the set of feature vectors which include total sub-band powers, relative powers, mean of percentage change in band powers and mean of EOG and EMG activities. In another work [23] authors presented the classification of heat stress by applying multilayer feed forward network and achieved high accuracy in SWS, REM sleep, Awake state with 94.5%, 91.75% and 91.75% respectively. There is a relation between EEG signals in different psychological conditions with depressed rats and control rats. The work explained how to classify the wavelet coefficients of EEG spectra in acute and chronic heat exposure with controlled data.

Sleep EEG classification by using frequency-amplitude features is hardly a new concept and virtually all computational methods try to capitalize on this notion. Author [17] had classified different stages of sleep in hot environment with the help of multi-layer perceptron model. In the recent work, authors [24] have used fuzzy classification methods for classification of sleep stages where they obtained an average classification accuracy of 93%. The best detection rate was achieved for SWS case. Same authors [18] have also used ANFIS Sugeno model to classify SWS, Awake and REM signals with 89% accuracy. In this animal model, data was acquired from adult healthy rats only.

References

1. Selye H (1936) A syndrome produced by diverse nocuous agents. Nature 138(3479):32
2. Sharma HS, Westman J, Nyberg F (1998) Pathophysiology of brain edema and cell changes following hyperthermic brain injury. Prog Brain Res 115:351–412
3. Britt RH (1984) Effect of wholebody hyperthermia on auditory brainstem and somatosensory and visual-evoked potentials. Thermal Physiol 519–523
4. Sharma HS, Winkler T, Stålberg E, Olsson Y, Dey PK (1991) Evaluation of traumatic spinal cord edema using evoked potentials recorded from the spinal epidural space: an experimental study in the rat. J Neurol Sci 102(2):150–162
5. Dement W, Kleitman N (1957) Cyclic variations in EEG during sleep and their relation to eye movements, body motility, and dreaming. Electroencephalogr Clin Neurophysiol 9(4):673–690
6. Jansen BH, Cheng WK (1988) Structural EEG analysis: an explorative study. Int J Biomed Comput 23(3–4):221–237
7. Al-Nashash HA, Paul JS, Ziai WC, Hanley DF, Thakor NV (2003) Wavelet entropy for subband segmentation of EEG during injury and recovery. Ann Biomed Eng 31(6):653–658
8. Kulkarni PK, Kumar V, Verma HK (1997) Diagnostic acceptability of FFT-based ECG data compression. J Med Eng Technol 21(5):185–189
9. Feng Z, Xu Z (2002) Analysis of rat electroencephalogram under slow wave sleep using wavelet transform. In: Engineering in medicine and biology, 2002. 24th annual conference and the annual fall meeting of the biomedical engineering society EMBS/BMES conference, 2002. Proceedings of the second joint, vol 1. IEEE, pp 94–95
10. Subasi A, Kiymik MK, Akin M, Erogul O (2005) Automatic recognition of vigilance state by using a wavelet-based artificial neural network. Neural Comput Appl 14(1):45–55
11. Sinha RK (2007) Study of changes in some pathophysiological stress markers in different age groups of an animal model of acute and chronic heat stress. Iran Biomed J 11(2):101–111
12. Fraiwan L, Lweesy K, Khasawneh N, Fraiwan M, Wenz H, Dickhaus H (2011) Time frequency analysis for automated sleep stage identification in fullterm and preterm neonates. J Med Syst 35(4):693–702
13. Nguyen T, Khosravi A, Creighton D, Nahavandi S (2015) EEG signal classification for BCI applications by wavelets and interval type-2 fuzzy logic systems. Expert Syst Appl 42(9):4370–4380
14. Faust O, Acharya UR, Adeli H, Adeli A (2015) Wavelet-based EEG processing for computer-aided seizure detection and epilepsy diagnosis. Seizure 26:56–64
15. Chen D, Wan S, Xiang J, Bao FS (2017) A high-performance seizure detection algorithm based on discrete wavelet transform (DWT) and EEG. PLoS ONE 12(3):e0173138
16. Sinha RK, Agrawal NK, Ray AK (2003) A power spectrum based backpropagation artificial neural network model for classification of sleep-wake stages in rats. Online J Health Allied Sci 2(1)
17. Sinha RK, Aggarwal Y, Das BN (2007) Backpropagation artificial neural network classifier to detect changes in heart sound due to mitral valve regurgitation. J Med Syst 31(3):205–209
18. Nagpal C, Upadhyay PK (2018) Adaptive neuro fuzzy inference system technique on polysomnographs for the detection of stressful conditions. IETE J Res 1–12
19. Sukanesh R, Harikumar R (2007) Analysis of fuzzy techniques and neural networks (RBF&MLP) in classification of epilepsy risk levels from EEG signals. IETE J Res 53(5):465–474
20. Krzanowski WJ (1988) Principles of multivariate analysis. Oxford University Press
21. Sing TZE Bow (2002) Pattern recognition and image processing, 2nd edn. Marcel, Dekker, Basel, Switzerland
22. Sarbadhikari SN, Dey SANGITA, Ray AK (1996) Chronic exercise alters EEG power spectra in an animal model of depression. Indian J Physiol Pharmacol 40(1):47–57

23. Upadhyay PK, Sinha RK et al. Identification of stressful events using wavelet transform and multilayer feed forward network. Caled J Eng 5(2)
24. Nagpal C, Upadhyay P (2019) Wavelet based sleep EEG detection using fuzzy logic. Springer Nature Singapore. CCIS 955, pp 794–805. https://doi.org/10.1007/978-981-13-3140-4_71

Development of a Mathematical Model for Solar Power Estimation Using Regression Analysis

Arjun Viswanath, Karthik Krishna, T. Chandrika, Vavilala Purushotham, and Priya Harikumar

Abstract The growing energy demand and the need for clean energy source have created a major impact on the renewable energy sector. One such renewable energy source, solar energy, is gaining momentum in various applications like electric power systems, water heaters and many more. Solar power estimation is important because it helps the grid engineers to make decisions like voltage and frequency control that is integral to the working of the grid. This paper presents a mathematical model to estimate solar power that can be generated by photovoltaic panel. It is difficult to estimate solar power that can be generated using a solar panel because it depends on a lot of weather factors which are variable. It is very much essential to have an accurate model for solar power estimation to ensure reliable and economic operation of power grid. The major factors taken into account for solar power estimation are: ambient temperature, cell temperature and solar irradiance. The relation between the above-mentioned factors is modelled by performing regression analysis and is tested for accuracy.

Keywords Solar power · Solar irradiance · Ambient temperature · Cell temperature · Regression analysis

A. Viswanath (✉) · K. Krishna · T. Chandrika · V. Purushotham · P. Harikumar
Department of Electronics and Communication Engineering, Amrita School of Engineering, Coimbatore, Amrita Vishwa Vidyapeetham, Coimbatore, India
e-mail: arjunsv98@gmail.com

K. Krishna
e-mail: karthik.krishna0511@gmail.com

P. Harikumar
e-mail: h_priya@cb.amrita.edu

V. Singh et al. (eds.), *Computational Methods and Data Engineering*,
Advances in Intelligent Systems and Computing 1227,
https://doi.org/10.1007/978-981-15-6876-3_48

1 Introduction

Greenhouse gas emissions (GHG) have been increasing all over the world in the recent years which justifies the need for a low carbon economy [1]. The earth's temperature has been increasing in a manner such that if this increase continues at the present rate, average increase of temperature could be around 6 °C by the dusk of this century. The primary reason for the rise in earth's temperature is the burning of fossil fuels which discharges GHGs like carbon dioxide, nitrous oxide and fluorinated gases into the earth's atmosphere [2]. The most important step to implement a low carbon economy is to shift towards renewable sources of energy [3]. Solar energy is a commonly used form of renewable energy [4]. Accurate solar power estimation is one of the primary aspects in ensuring the proper operation of photovoltaic (PV) cells. Solar power that can be generated by a PV cell mainly depends on weather factors like temperature, irradiance, windspeed, cloud cover, precipitation and relative humidity [5]. The ongoing research work is mainly focused on solar power estimation, using one analytical model and the machine learning method. A model estimated using analytical methods has solar power dependent on cell temperature and irradiance [5]. Another analytical model has solar power dependent on ambient temperature and irradiance [6]. Empirical models for solar power generation have also been proposed by researchers. One such model uses multiple linear regression for solar power estimation using 14 weather factors that were proposed [7]. Many other empirical models using nonlinear regression have been proposed in the existing literature. These nonlinearly regressed models require lesser weather variables mainly irradiance and cell temperature with some of these models requiring windspeed to estimate solar power [8]. Cell temperature is a significant parameter in estimating the solar power. There are several mathematical models that have been proposed in the past literature to estimate cell temperature. Cell temperature is mainly dependent on ambient temperature and irradiance. There are a few models where the cell temperature is dependent on windspeed [9, 10]. In these mathematical models, there are standard reference variables for irradiance which need to be chosen. Similarly, a standard reference variable for cell temperature known as T_{NOCT} (nominal operating cell temperature) needs to be seen from the data sheet of the solar panel that is being used. We have chosen the standard irradiance as 800 W/m^2 [11]. The T_{NOCT} value for the solar panel which has been used is 43 °C. Irradiance is another significant weather parameter in estimating the solar power. Weather factors, namely cloud cover, precipitation and relative humidity, also influence the solar power indirectly by impacting the irradiance which in turn impacts the solar power that can be generated by a PV cell [12].

Our approach involves using a linear regression-based system to estimate the cell temperature using ambient temperature and irradiance. Solar irradiance and estimated cell temperature are then used to model a nonlinear regression-based model to estimate the solar power. We have established an empirical formula that will give us a better accuracy when compared to the past models for the location from which the data set has been obtained.

2 Theoretical Background

In this section, the value of dependent variable cell temperature (T_c) is calculated using independent variables, ambient temperature (T_a) and irradiance (G_T) [9, 10]. We used multiple linear regressions and obtained an empirical formula to determine the cell temperature (T_c) [7]. This method is used in data analytics to determine the formula between two or more independent variables and the dependent output variable by finding the closest fit for the constants of a set of linear equations that are formed using the data for the independent variables. It is of the form:

$$T_c = k_1 T_a + k_2 G_T + k_3 \text{ where } k_1, k_2, k_3 \text{ are constants.} \tag{1}$$

Earlier, various such models have been established and we have considered a few models for our study, which are as follows:

First of such linear models were established by Schott [9], where the relationship between dependent variable cell temperature (T_c) with respect to independent variables solar irradiance (G_T) and ambient temperature (T_a) was established.

$$T_c = T_a + 0.028 G_T - 1 \tag{2}$$

Ross and Smokler [9, 10], proposed a model to determine cell temperature (T_c) using solar irradiance (G_T) and ambient temperature (T_a) using the equation

$$T_c = T_a + \frac{G_T}{800}(T_{C,\text{NOCT}} - 20) \tag{3}$$

Lasnier and Ang [9] calculated the cell temperature (T_c) using independent parameters ambient temperature (T_a) and solar irradiance (G_T) using the formula

$$T_c = 30.006 + 0.0175(G_T - 300) + 1.14(T_a - 25) \tag{4}$$

Mondol et al. [9] determined cell temperature (T_c) with respect to independent variables ambient temperature (T_a) and solar irradiance (G_T) using formula

$$T_c = T_a + 0.031 G_T \tag{5}$$

Later he modified the formula and could get better accurate results.

$$T_c = T_a + 0.031 G_T - 0.058 \tag{6}$$

From the literature survey, we understood that the factors cell temperature (T_c) and solar irradiance (G_T) significantly influences the output solar power [5, 8]. Thus, a nonlinear regression can be performed to establish a relation to determine output solar power (P) [8]. Nonlinear regression is a form of regression analysis in which data for the independent input variables are used to find the constants of a nonlinear function and this nonlinear function consists of a nonlinear combination of one or more independent input variables. Few models have been proposed in the past to calculate the output power of a solar panel. We have considered the following few models for our study. Yang et al. [8] proposed a model, to obtain the output power of a solar panel (P) which was dependent on independent factors cell temperature (T_c) and solar irradiance (G_T) which is nonlinearly regressed. The power was calculated as

$$P = (\alpha T_c + \beta) G_T \tag{7}$$

A nonlinear regression was performed for the above model to determine the temperature coefficient (α) and the calibration coefficient (β). The α value for our model came up to be −0.000125392 and β value came up to be 0.0294962.

Model proposed by Fuentes et al. [8], determined output power of solar panel (P) which was dependent on independent factors cell temperature (T_c) and solar irradiance (G_T). The model is as follows:

$$P = P_{ref} * [1 + \gamma(T_c - 25)], \gamma = -0.0035 \tag{8}$$

P_{ref} is the mean value of power, and it is taken as 25.601 W for our data set and G_{ref} is taken as 800 W/m^2 [11].

3 Methodology

Figure 1 depicts the methodology we followed for our estimation.

Initially, we tried the approach of establishing the relationship with cell temperature as dependent variable, solar irradiance and ambient temperature as independent variables using linear regression analysis on a monthly basis. Regression is performed on a software called Minitab. The results obtained were as follows (Table 1).

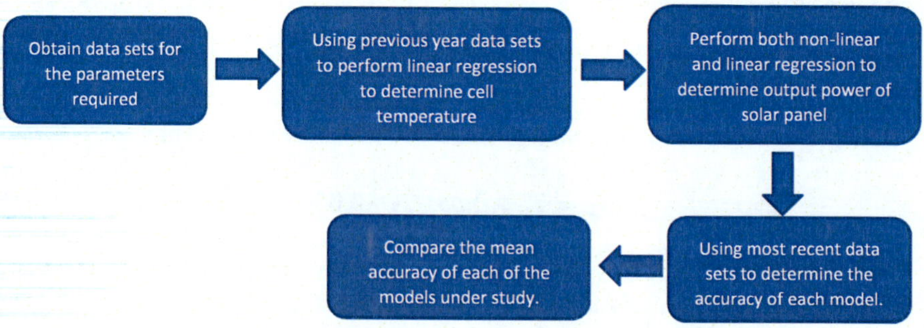

Fig. 1 Flow of the methodology

Table 1 Coefficient of determination (R^2) values for all the months of 2018 for determining cell temperature (T_c)

Month	R^2 (%)
January	97.88
February	97.57
March	96.43
April	96.67
May	95.66
June	91.88
July	91.80
August	91.44
September	94.08
October	94.67
November	95.39
December	96.13

The coefficient of determination (R^2) is the amount of variance in the data that the mathematical model accounts for. The expected R^2 value that we wanted to have was 95% but for certain months, namely June, July, August, September and October, the R^2 value went below 95% so we decide to model the relationship between cell temperature, ambient temperature and solar irradiance on a yearly basis rather than monthly basis.

We, then established an empirical relationship for dependent variable cell temperature (T_c) using the ambient temperature (T_a) and solar irradiance (G_T). We used Minitab for performing linear regression on the available data sets to establish an empirical relationship for cell temperature using ambient temperature and solar irradiance [9]. We perform linear regression using Minitab on data sets available for the year 2018 for cell temperature (T_c), ambient temperature (T_a) and solar irradiance (G_T) to determine an empirical model. Thus, we were able to formulate am empirical relation which was as follows:

$$T_c = -1.0367 + 0.025710G_T + 1.17970T_a \tag{9}$$

The coefficient of determination (R^2) for the above model was above 95% which proves that regression based on a yearly model gave us more accuracy rather than a monthly model.

Thus, a linearly regressed model has been obtained for calculating the cell temperature (T_c) of a solar panel using independent variables ambient temperature (T_a) and solar irradiance (G_T).

Similarly, we tried establishing the relationship for obtaining the output power of solar panel using dependent variables cell temperature (T_c) and independent variable solar irradiance (G_T) using linear regression analysis on a monthly basis. Regression is performed on a software called Minitab. The results obtained were as follows (Table 2).

Table 2 Coefficient of determination (R^2) values for all the months of 2018 for determining output power of solar panel (P)

Month	R^2 (%)
January	98.37
February	98.63
March	92.56
April	99.16
May	95.47
June	98.76
July	91.84
August	98.85
September	90.97
October	90.08
November	94.45
December	98.44

The expected coefficient of determination (R^2) value that we wanted was 95% but when we performed regression monthly, it was found that the R^2 values for certain months, namely March, July, September, October, November, were below 95%. Therefore, we decided to model the relationship to determine the output power of solar panel using dependent variable cell temperature and independent variable solar irradiance on a yearly basis rather than monthly basis.

We have performed a linear regression using data sets of output power of solar panel (P), cell temperature (T_c) that we calculated earlier, and solar irradiance (G_T). The obtained model to determine the output power of the solar panel was of the form:

$$P = 1.0956 + 0.023482G_T - 0.03103T_c \tag{10}$$

Thus, a linearly regressed empirical model to measure the output power of solar panel was established. The regression was performed using the 2018 data sets, and empirical formula was modelled.

The coefficient of determination (R^2), for the above model was 96.02%, which proves that regression based on yearly model gave us more accuracy rather than a monthly model.

We have proposed a model to calculate the output power of a solar panel (P) by performing nonlinear regression on two independent variables cell temperature (T_c) and solar irradiance (G_T) using Minitab software. The proposed nonlinearly regressed model is of the form:

$$P = k_1 T_c G_T + k_2 G_T + k_3, \text{ where } k_1, k_2, k_3 \text{ are constants} \tag{11}$$

We cannot calculate R^2 value for nonlinear regression so we modelled the system on a yearly basis rather than a monthly basis given that the results for the monthly based models weren't favourable for linearly regressed models to calculate power and cell temperature. We used past data sets for the required parameters, output power of solar panel (*P*), cell temperature (T_c) that we calculated using our empirical formula and the solar irradiance (G_T) and obtained a model which is as follows:

$$P = -0.000147089 T_c G_T + 0.030992 G_T - 0.154405 \quad (12)$$

A nonlinearly regressed model for calculating the output power of solar panel has been obtained. The constant term in the empirical formula is obtained for our location, and it could be due to many factors such as external disturbances and thermal loss that could affect the reading.

4 Result

The empirical model to calculate the cell temperature (T_c) was obtained by performing linear regression using the past data sets, and the model when tested using recent data sets gave us an accuracy of 95.55%. Finally, the accuracy for the empirical model to calculate total solar power output (*P*) was obtained by performing linear regression using the past data sets, and the model when tested using recent data sets gave us an accuracy of 85.46% and using nonlinear regression gave us an accuracy of 89.99%.

5 Conclusion

The data sets from a power station in South India are used for our analysis. We have performed linear regression to determine the cell temperature (T_c) of the solar panel from independent variables solar irradiance (G_T) and ambient temperature (T_a). To determine the total power output from solar panel (*P*), we have performed both linear and nonlinear regression using independent variables cell temperature (T_c) that we obtained from the regression performed and the solar irradiance (G_T). The past data sets for the parameters, cell temperature (T_c), ambient temperature (T_a), solar irradiance (G_T) and total output power from solar panel (*P*) were used to perform mathematical modelling using Minitab to establish empirical formulas for calculating cell temperature (T_c) and total power output from solar panel (*P*). We used the recent data sets for the above parameters to calculate the accuracy of each of the past models and the accuracy of the empirical models that we formulated. The accuracy of each model to determine cell temperature (T_c) was tested against the obtained data sets and was found to be as follows (Table 3).

It is understood that the empirical model to determine cell temperature (T_c) that we obtained by performing linear regression gives a better accuracy in comparison to the other models that were proposed in the past. Accuracy is higher for the empirical model that we derived because the constants are found for a particular location.

Table 3 Accuracy of each of the past models to calculate cell temperature

Model name	Mean accuracy (%)
Mondol et al.	92.81
Ross and Smokler	92.17
Lashier and Ang	85.85
Schott	89.54

Table 4 Accuracy of each of the past models to calculate total solar power output

Model name	Mean accuracy (%)
Mean (Yang et al.)	86.77
Mean (Fuentes et al.)	62.35

Finally, we calculated the total output power from solar panel (*P*), by performing both linear and nonlinear regression. The accuracy of each of the models was tested and was found to be as follows (Table 4).

It is understood that the empirical model to determine total output power from solar panel (*P*) that we obtained by performing nonlinear regression gives a better accuracy when compared to other models that were proposed in the past as well as the empirical model that was established using linear regression.

6 Future Scope

We have performed regression analysis to calculate cell temperature and total power output from solar panel by building the model using past data sets and running them on recent data sets for a particular location. The proposed empirical models for power and cell temperature can be improved if we derive the model using a combination of data sets from different locations having different weather conditions. If we use data sets from different locations to derive empirical relations, we will get a generalized model. The different locations should be chosen keeping variety in mind. We need to choose locations like coastal regions, landlock regions, hilly regions, forest regions, rainy regions and desert regions. These locations shouldn't also preferably be situated close to each other as that will ensure that we are taking a broad geographic view into account giving us a broader generalized model.

We can also implement the present model using hardware to detect irradiance and ambient temperature, thereby resulting in a real-time system for solar power estimation. We can use an SP110 pyranometer to detect irradiance. The SP110 is an analog pyranometer that is self-powered which has a sensitivity of 0.2 mV per W m^{-2}. We can measure ambient temperature using DHT22 sensor. DHT22 is a digital sensor that measures both temperature and relative humidity. DHT22 has a sensitivity of 0.1 °C. Given the very good sensitivities of both the SP110 and DHT 22, we can get accurate readings of ambient temperature and solar irradiance.

This ambient temperature reading that is measured by the temperature sensor has to be converted into cell temperature by means of the equation that will be derived using a generalized model accounting for the data sets of many places. The real-time aspect of the solar power estimation system will be due to the presence of a microcontroller that processes the data which is acquired by the sensors. We will be using a microcontroller named Arduino Mega 2560. The Arduino mega 2560 is a 10-bit microcontroller that has 54 digital pins and 16-analog pin. The primary reason to use the Mega version of Arduino is because we can internally map the highest digital value of 1024 to 1.1 Volts (V) thereby helping us acquire a relatively more accurate reading from the pyranometer.

References

1. Ding Y, Kang C, Wang J, Chen Y, Hobbs BF (2015) Foreword for the special section on power system planning and operation towards a low-carbon economy. IEEE Trans Power Syst 30(2):1015–1016
2. Khan AS Chowdhury SA (2012) GHG emission reduction and global warming adaptation initiatives by UNFCCC. In: 2nd international conference on the developments in renewable energy technology (ICDRET 2012). IEEE, pp 1–6
3. Wang H (2016) Microgrid generation planning considering renewable energy target. In: 2016 IEEE international conference on power and energy (PECon). IEEE, pp 356–360
4. Choi UM Lee KB, Blaabjerg F (2012) Power electronics for renewable energy systems: wind turbine and photovoltaic systems. In: 2012 International conference on renewable energy research and applications (ICRERA). IEEE, pp 1–8
5. Wang J, Zhong H, Lai X, Xia Q, Wang Y, Kang C (2017) Exploring key weather factors from analytical modeling toward improved solar power forecasting. IEEE Trans Smart Grid 10(2):1417–1427
6. Wan C, Zhao J, Song Y, Xu Z, Lin J, Hu Z (2015) Photovoltaic and solar power forecasting for smart grid energy management. CSEE J Power Energy Syst 1(4):38–46
7. Abuella M, Chowdhury B (2015) Solar power probabilistic forecasting by using multiple linear regression analysis. In: SoutheastCon 2015. IEEE, pp 1–5
8. Kim GG, Choi JH, Park SY, Bhang BG, Nam WJ, Cha HL, Park N, Ahn HK (2019) Prediction model for PV performance with correlation analysis of environmental variables. IEEE J Photovolt 9(3):832–841
9. Charles Lawrence Kamuyu W, Lim J, Won C, Ahn H (2018) Prediction model of photovoltaic module temperature for power performance of floating PVs. Energies 11(2):447
10. Vijayakumari A, Devarajan AT, Devarajan (2015) Extraction of photovoltaic module's parameters using only the cell characteristics for accurate PV modeling. In: Proceedings of the international conference on soft computing systems, vol 1, No 4, pp 265–276
11. Zen Z, Xiao T, Shen Y, Lei W, Peng J, Yu J (2016) A method to test operating cell temperatures for BIPV modules. IEEE J Photovolt 6(1):272–277
12. Shah ASBM, Yokoyama H, Kakimoto N (2015) High precision forecasting model of solar irradiance based on grid point value data analysis for an efficient photovoltaic system. IEEE Trans Sustain Energy 6(2)

Cloud Based Interoperability in Healthcare

Rakshit Joshi, Saksham Negi, and Shelly Sachdeva

Abstract Healthcare is a vast field that generates a large amount of data maintained differently by different organizations. This acts as a hurdle data exchange among different organizations and calls for provisions of standards for clinical data exchange. Handling these challenges will help in improving patient care quality as well as in decreasing costs and medical errors. The basic idea behind this paper is to provide a free and open-source solution to facilitate interoperability in electronic health records, by adopting openEHR specification and EHR extract information model for storage, retrieval and sharing of clinical data, make dynamic form generation based on specific sub-domains easier, by allowing the user to upload templates and generating a clinical form which conforms to that template, and to store and handle clinical documents over the cloud, using Microsoft Azure cloud services and database storage. This paper provides a solution with a significant application in healthcare by decreasing the time and cost in exchange of clinical data and improving accuracy and efficiency in data collection.

Keywords Health informatics · Electronic health record (EHR) · OpenEHR · Interoperability

1 Introduction

Healthcare is a vast field that produces a huge amount of data owing to various devices and laboratories, which requires proper portability and assimilation for its efficient management across organizations and systems. Also, the healthcare quality

R. Joshi · S. Negi · S. Sachdeva (✉)
National Institute of Technology Delhi, New Delhi 110040, India
e-mail: sachdevashelly1@gmail.com

R. Joshi
e-mail: therakshitjoshi@gmail.com

S. Negi
e-mail: sakshamnegi1@gmail.com

V. Singh et al. (eds.), *Computational Methods and Data Engineering*,
Advances in Intelligent Systems and Computing 1227,
https://doi.org/10.1007/978-981-15-6876-3_49

is affected by costs and patient safety. Proficient and exact information recording is required so as to acquire these. In perspective on these possibilities, electronic health records (EHRs) are turning into a strategy by which doctors can electronically record top notch information at quick speed and minimal effort. Many organizations are coming up with standards to tackle the challenges of clinical data exchange. One such standard is openEHR specification developed and maintained by openEHR organization. However, there still exists a need for a free and open-source solution to generate document instances dynamically based on specific sub-domains and this paper proposes a plan to cater to it.

Some of the domain-specific terms used in the paper are described below:

1.1 EHR

An electronic health record (EHR) is a comprehensive report of overall health of a patient. It is a collection of medical records which get recorded during the course of a patient's medical history. They incorporate various information, like laboratory test results, medical history, medication and allergies, vital signs, and personal information like weight and age. EHRs are designed to be shared among providers so that a patient's EHR may be accessed among authorized organizations and different healthcare providers with ease. EHRs are of significant use not only because they are facilitating the transition from paper to digital record-keeping in the healthcare industry but also because of the flexibility, ease of sharing, reliability, and completeness they bring to the healthcare system.

1.2 OpenEHR [1]

OpenEHR is an open standard specification in health informatics, and it portrays the administration and capacity, recovery, and exchange of health information in EHRs. The openEHR specifications include an EHR Extract [2] specification and also focuses on the exchange of data between EHR systems. These specifications are maintained by the openEHR Foundation, which supports the open research, development, and implementation of openEHR EHRs.

1.3 Archetype [3]

It is a re-usable, formal definition of domain-level data, outlined in terms of constraints on associate data model. Archetypes are like the basic essential building blocks.

1.4 Template

While an archetype describes a maximum data set of everything that may ever be documented about the clinical concept described in the archetypes, a template is usually designed. A template references/aggregates a number of archetypes required for the particular use case and applies further constraints/rules to the archetypes. A template can embed other templates as required. The template in combination with archetypes holds all the information required on different levels and helps to reuse the basic building blocks (archetypes). An openEHR operational template (OPT) combines all the information from all archetypes used in the template (and potentially its embedded templates) and applies all constraints and descriptions explicitly as specified by the template. Thus, an OPT is a single, usually large file that describes the computational format used by openEHR systems.

2 Problem Statement

The aim that the paper aims to achieve is to facilitate interoperability in healthcare domain between health organizations by providing a free and open-source cloud-based solution which allows the generation of clinical document instances dynamically based on their respective sub-domains. In a nutshell, this paper intends to address the following problems:

1. Lack of interoperability between health organizations for efficient exchange of clinical data.

Different hospitals have different systems for generating and managing the clinical data. Clinical documents of one hospital may differ from that of another hospital, thus, affecting the interpretability as well. This hinders the exchange of clinical data between different healthcare organizations.

2. Lack of free and open-source cloud-based solution.

Although some efforts have been done in this direction but there is still no availability of a free and open-source cloud-based solution to generate and manage clinical documents.

3. Generating clinical document instances dynamically, based on specific sub-domains.

The requirements and specifications for clinical data differ from one sub-domain to another. For instance, the form for a blood test might be completely different from that for a respiratory checkup. There is lack of a cloud-based functionality to meet this purpose.

Table 1 Challenges and their solution approach

Challenges	Approach
Lack of interoperability in exchange of clinical data	Adopting openEHR specification and EHR extract information model to standardize the storage, retrieval, and sharing of clinical documents
Lack of free and open-source solution	Using Microsoft Azure to provide a cloud-based service for generation of clinical documents that conform to openEHR specifications and for storage and exchange of clinical documents that follow the EHR extract information model
Generating clinical documents dynamically	Providing a functionality to generate forms dynamically where user can upload templates (OPT)

The motive behind the approach proposed in the paper is that although there are tools and technologies that focus on providing solutions for the above-mentioned problems individually, but the need for a wholesome solution was strongly felt and we intend to work toward providing the same.

2.1 Approach to Problem

This paper provides the following approach to the challenges that are aimed to be tackled:

As can be seen in Table 1, the proposed approach to the challenges is providing a free and open-source cloud-based web application that facilitates the dynamic generation of clinical document instances based on the templates which will be uploaded and interfaces to fill the generated clinical forms and store them in the database and validate clinical document instances, all at one place.

2.2 Significance of the Problem

The problem that this paper aims to solve holds significance in the fact that having a free and open-source cloud-based solution for interoperability of electronic health records, which can provide services like dynamic form generation, validation, and clinical document storage all at one place, can help save a lot of time and effort. In turn, it will help in enhancing healthcare quality, reducing medical errors, decreasing costs, increasing efficiency and accuracy in data capture and date exchange. A survey [4] of doctors prepared for significant use offers significant evidence:

- 94% of providers report that their EHR makes records promptly accessible at purpose of care.
- 88% report that their EHR produces clinical advantages for the practice.
- 75% of suppliers report that their EHR enables them to convey better patient consideration.

Another study [5] shows that cloud services could improve on areas like security, cost, scalability, interoperability, improving quality and reducing errors, flexibility, structure, and sharing ability. The addition of cloud services to healthcare can add effectiveness to the electronic healthcare field.

3 Related Work

A review of the research done in regard to the tools and technologies that already exist in the market and the current trend of work being done in the domain is given below.

3.1 Empirical Study

Table 2 describes the technologies existing in the market that provide specifications for interoperability in clinical data exchange. Though these technologies provide efficient means to standardize the process of clinical data exchange, they are subscription-based and only provide partial solutions to the problems discussed.

3.2 Summary of Papers Studied

The paper uses some of the approaches discussed in the literature studied and tries to provide appropriate solutions for the research gaps in the literature to achieve the intended aim. An account of the insights gained from the literature and the research gaps in regard to the aim of our paper is given in Table 3.

Table 3 describes the insights gained from the literature and the research gaps in the literature that the paper tries to provide a solution for. A brief overview of the same is also shown in Fig. 1.

Table 2 Comparison of existing tools/technologies

Organization	Service provided	Description	Standard used
1. CaboLabs	EHRServer [6]	A platform based on open-source software and openEHR standard for managing and sharing clinical Information. It provides the services to store and deal with clinical document instances conforming to the uploaded templates along with querying capability and a querying interface	openEHR
2. Health Level 7 (HL7) [7]	Standards for interoperability	A nonprofit organization that provides a comprehensive framework and related standards for the exchange, integration, sharing, and retrieval of electronic health information. One of its most widely used standards is Fast Healthcare Interoperability Resources (FHIR)	HL7 V2 [8] Clinical Document Architecture (CDA) [9] (FHIR) [10]
3. Google	Cloud Healthcare API [11]	It connects healthcare systems and Google cloud-based applications	FHIR
4. Microsoft	Azure API for FHIR [12]	Allows fast exchange of data in FHIR format. It is backed by a Platform-as-a Service (PaaS) cloud offering managed by Azure	FHIR

4 Methods and Approach

4.1 Overall Description

The overall detailed design of the implementation is shown in Fig. 2. The implementation for cloud-based interoperability is divided mainly into three modules, namely the extractor, validator, and the clinical document generator. The database contains all the clinical document instances versioned and stored accordingly which makes

Table 3 Insights gained and research gaps in the literature referred

Research paper	Publishing details	Content incorporated	Research gaps
Semantic interoperability in standardized electronic health record databases [13]	S. Sachdeva, S. Bhalla (Journal of Data and Information Quality, vol. 3, no. 1, pp. 1–37)	Domain knowledge, OpenEHR specifications, EHR extract information model	Does not provide a free and open-source solution to generate and manage clinical documents
CDA generation and integration for health information exchange based on cloud computing system [14]	S. H. Lee, J. H. Song, I. K. Kim (IEEE Trans. Services Comput., vol. 9, no. 2, pp. 241–249)	How to implement clinical document generation in cloud Exchanging clinical data, but using EHR extracts instead of CDAs	Takes in pre-filled forms from hospitals and is thus vulnerable to missing data and requires the use of ontological mapping Does not use an open-source solution for standardization
Data storage management in cloud environments: taxonomy, survey, and future directions [15]	Yaser Mansouri, Adel Nadjaran Toosi, Rajkumar Buyya (ACM Comput Surv. 50, 6, Article 91, 51 pages)	Insight on available cloud data storage services Comparison of various available services based on their features and limitations Selection of data storage service for implementation	Does not discuss privacy and authentication aspects of cloud data storage services Does not highlight the service-specific challenges and implementation-based specifications

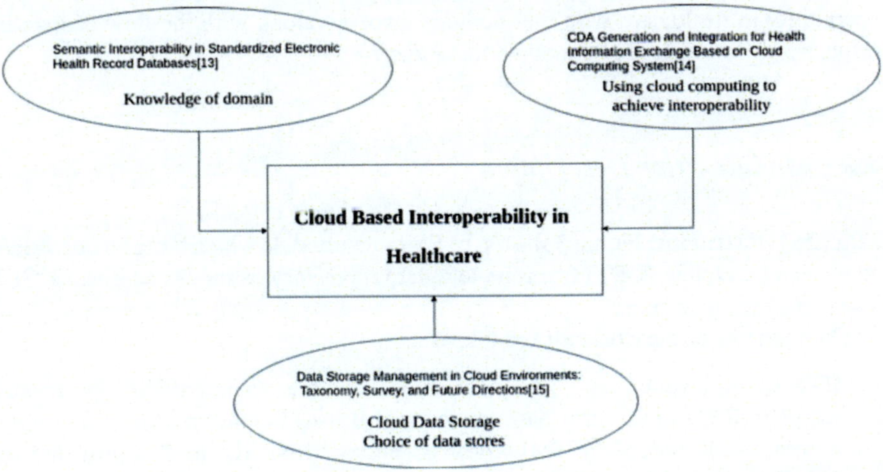

Fig. 1 Integrated summary

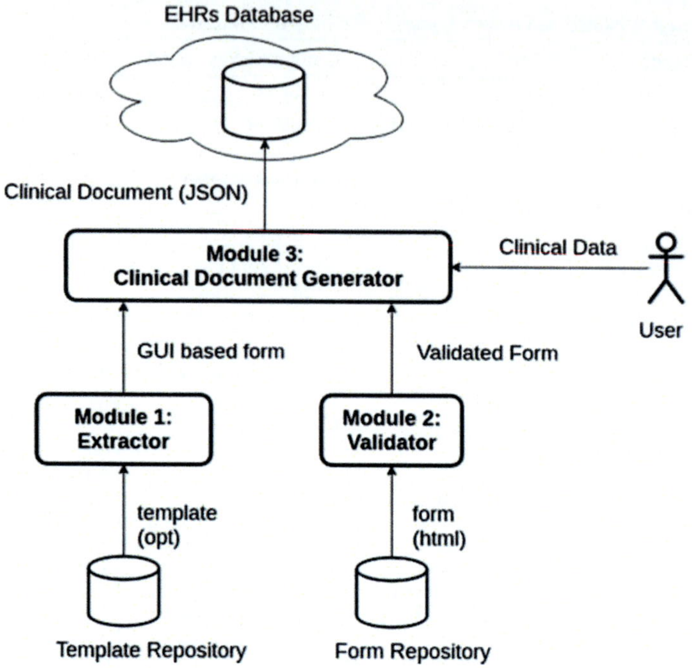

Fig. 2 Cloud-based interoperability

it easier to retrieve them. Figure 2 shows the approach to achieve cloud-based interoperability in healthcare with the modules involved along with the flow of data in respective formats from one module to another.

4.2 Module-Wise Description

Module 1: Extractor. Figure 3 shows the Extractor module which takes an openEHR operational template (OPT) from the template repository as input and gives a GUI based html form as output.

Components of the extractor module are:

1. HTML form generator—This component converts an openEHR operational template (OPT) into html data for a form. It uses Selenium [16] to automate a component, named "HTML form generator," from the toolkit provided by CaboLabs.
2. Cleaner/Enhancer—The html form data obtained from the previous component contains some errors and redundancies. It uses the BeautifulSoup [17] library along with the inbuilt html parser of python to remove those errors and pull out

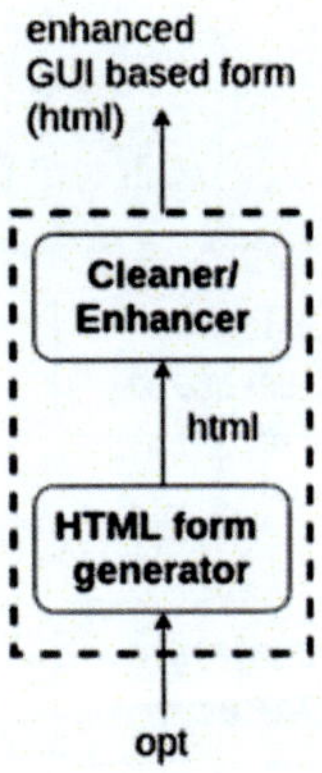

Fig. 3 Module 1: extractor

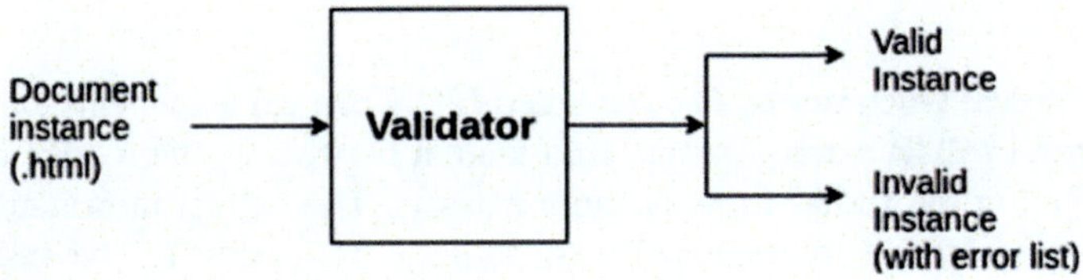

Fig. 4 Module 2: validator

redundant and useless information, hence giving a clean and enhanced GUI-based form as output.

Module 2: Validator. Figure 4 shows the validator module which takes an openEHR HTML form from the form repository as input, executes a syntactic validation, and tells if the form is valid or not. It also displays the errors encountered in case the form could not pass the validation.

The module uses Selenium to automate a component, named "Clinical Document Instance Validator," from the toolkit provided by CaboLabs.

Module 3: Clinical Document Generator. Figure 5 shows the clinical document generator module which takes a GUI-based html form—this can be an html form generated via the extractor module or an html form which passed the validation by the validator module, and clinical data from the user as input and gives a clinical document in JSON format as output.

Components of Clinical Document JSON Generator module are:

1. Instance Generator—Populates GUI-based form with the clinical data provided by the user and provides a document instance in html.
2. JSON Generator—Converts the document instance obtained into a JSON file. This JSON clinical document can now be stored in our database.

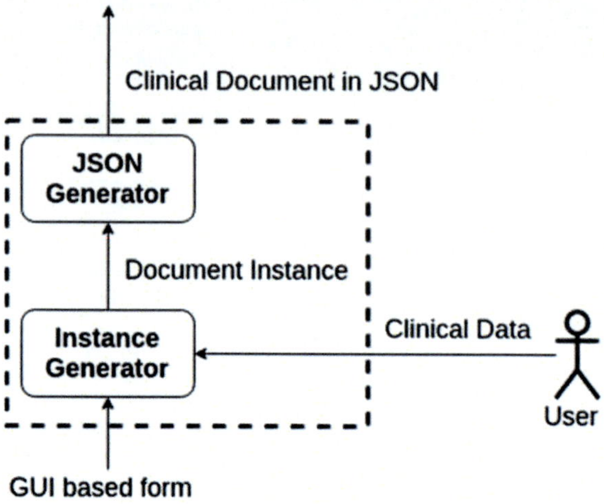

Fig. 5 Module 3: clinical document generator

Storing Clinical Documents. The generated JSON clinical documents are stored in the database in EHR extract format. Information between systems (like hospitals) is exchanged in the form of one or more extracts. The extract information model is not affected by the communication technology used between the systems and hence corresponds with the idea of interoperability between systems. The database chosen for this purpose was a NoSQL database—MongoDB. To store the clinical documents in a MongoDB database while following the EHR extract information model, an efficient mapping of features of MongoDB to the terminologies of EHR extract was performed. This was done in order to replicate the operational openEHR environment for extracts in our local device through MongoDB. Figure 6 gives an overview of the operational environment for openEHR Extracts.

The description of terminologies related to EHR extract information model and the entities in MongoDB used to perform similar tasks is given in Table 4.

5 Summary and Conclusion

In a vast field like healthcare where a large amount of data is generated extensively, there is much need of interoperability which not only helps to improve quality of care but also reduces the resources and time spent on conversion of data format. In addition to that, generating clinical forms based on specific sub-domains and that too dynamically on the go and providing cloud services for the same is also an area which needs catering to. This paper discusses a free and open-source cloud-based approach to facilitate interoperability in healthcare based on openEHR specifications while providing cloud services along with dynamic generation of clinical forms based on respective subdomains. The proposed approach is a cloud-based web application with three modules, namely the extractor for the dynamic generation of clinical forms

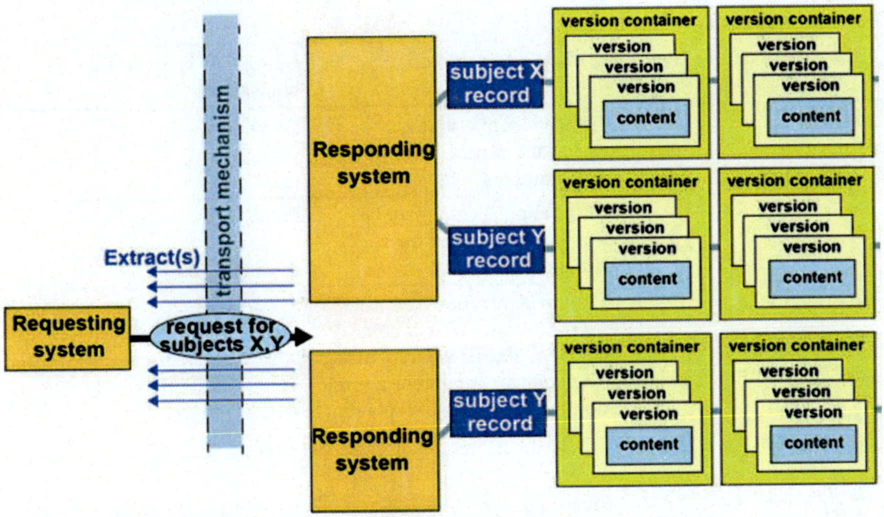

Fig. 6 Operational openEHR environment for extracts [18]

Table 4 Mapping of EHR extract terminologies and MongoDB features

EHR extract terminologies	Practical interpretation	MongoDB features used as equivalents
Extract	The format in which clinical data is interchanged between systems. One or more extracts may be interchanged between systems at a time	
Transport mechanism	Means to send request and receive one or more extracts (maybe some form of web service or email service or some comprehensive middleware)	
Responding system	Database holder like hospital, clinic, etc.	Currently, a single responding system is present
Subject (X or Y)	Patient (More than one responding system may hold records of the same subject)	MongoDB database
Version Container	Version history of a piece of content (e.g., persistent composition like "medication list," and "problem list." In our case—"Vital signs summary") A version container will contain multiple versions of a piece of content	Mongodb collection. (Each database, i.e., patient will have its own set of version containers)

(continued)

Table 4 (continued)

EHR extract terminologies	Practical interpretation	MongoDB features used as equivalents
Version	A content item's state at a particular instance, when it was last committed	Document
Content	Actual clinical data (may be compositions, folder trees, parties, etc.)	Fields
Contribution	A group of versions committed to a system by a user at some time. (i.e., details corresponding to a particular commit of a single subject from a single responding system)	

based on user-uploaded templates, the Validator for the validation of the generated and user-uploaded clinical forms and the clinical document generator for the generation of a clinical document and its storage over the database in the form of an extract. The end result is the clinical information stored in the openEHR extract model and a hassle-free healthcare experience. Some future prospects to the discussion put forward in this paper could be handling patient demographics and adding to privacy features over the cloud using advanced algorithms.

References

1. Open EHR https://www.openehr.org/. Accessed 10 Dec 2019
2. EHR Extract Information Model-openEHR. https://specifications.openehr.org/releases/RM/latest/ehr_extract.html. Accessed 10 Dec 2019
3. openEHR Architecture Overview. 12 Dec 2018, https://specifications.openehr.org/releases/BASE/Release1.0.3/architecture_overview.html. Accessed 10 Dec 2019
4. Jamoom E, Patel V, King J, Furukawa M (2012) National perceptions of EHR adoption: Barriers, impacts, and federal policies. National conference on health statistics
5. Ahmadi M, Aslani N (2018) Capabilities and advantages of cloud computing in the implementation of electronic health record. Acta Inf Med 26(1):24
6. CloudEHRServer by CaboLabs. https://cloudehrserver.com/. Accessed 10 Dec 2019
7. HL7.org. https://www.hl7.org/. Accessed 10 Dec 2019
8. HL7 Standards Product Brief-HL7 Version 2 Product Suite. https://www.hl7.org/implement/standards/product_brief.cfm?product_id=185. Accessed 10 Dec 2019
9. CDA® Release 2|HL7-HL7 Standards product brief. https://www.hl7.org/implement/standards/product_brief.cfm?product_id=7. Accessed 10 Dec 2019
10. HL7 Standards Product Brief—FHIR® R4 (HL7 Fast-HL7.org. https://www.hl7.org/implement/standards/product_brief.cfm?product_id=491. Accessed 10 Dec 2019
11. Google Cloud Healthcare API. https://cloud.google.com/healthcare/. Accessed 10 Dec 2019
12. Azure API for FHIR(r)|Microsoft Azure. https://azure.microsoft.com/en-gb/services/azure-api-for-fhir/. Accessed 10 Dec 2019

13. Sachdeva S, Bhalla S (2012) Semantic interoperability in standardized electronic health record databases. J Data and Inf Qual 3(1):1–37. Available: https://doi.org/10.1145/2166788.2166789
14. Lee SH, Song JH, Kim IK (2016) CDA generation and integration for health information exchange based on cloud computing system. IEEE Trans Serv Comput 9(2):241–249. Available https://doi.org/10.1109/tsc.2014.2363654
15. Mansouri Y, Toosi AN, Buyya R (2017) Data storage management in cloud environments: taxonomy, survey, and future directions. ACM Comput Surv 50(6):51p. Article 91 https://doi.org/10.1145/3136623
16. Selenium https://selenium.dev/. Accessed 10 Dec 2019
17. Beautiful Soup Documentation—Beautiful Soup 4.4.0, https://www.crummy.com/software/BeautifulSoup/bs4/doc/. Accessed 10 Dec 2019
18. Operational openEHR environment for Extracts. https://specifications.openehr.org/releases/RM/latest/ehr_extract/diagrams/operational_environment.png. Accessed 10 Dec 2019

Non-attendance of Lectures; Perceptions of Tertiary Students: A Study of Selected Tertiary Institutions in Ghana

John Kani Amoako and Yogesh Kumar Sharma

Abstract Universities in Ghana in their students' handbooks require students to adhere to the institution's class attendance policy. Meanwhile some students absent themselves from class even though they are aware of the benefits of attending lectures, and the consequences for not attending lectures. Therefore, this study is focused on ascertaining the reasons why "poor lecturing" is blamed for the non-attendance of lectures by some undergraduate students in the selected public universities in Ghana (Universities of Ghana—UG, University of Cape Coast—UCC, Kwame Nkrumah University of Science and Technology—KNUST, and Cape Coast Technical University—CCTU). Close-ended questionnaires were used in the collection of data from 800 students in four public tertiary institutions. SPSS V-23 was used to determine the frequencies and percentages of respondents in relation to the variables. Based on the results, majority of the students mostly complained about the confusing nature of lectures aside other reasons like boring, complex, impractical, and other personal reasons which students were not willing to disclose.

Keywords "Poor lecturing" · Class attendance policies · Tertiary institutions in Ghana · Class absenteeism

1 Introduction

Since the formalization of education, one activity that has and will remain key is lecture [1] therefore it is critical to enhance it. So long as universities enroll an increasing number of students yearly, they will resort to lecture because it allows one facilitator to physically address a large number of students at a goal [2]. Regularly or randomly, some institutions integrate distance education in a way to make up for increased students' intake yet they organize face-to-face lectures aside from the digital aids in place [3]. Most tertiary institutions maintain class attendance policies to ensure lecture participation [4] all over the world. For example, it is mentioned

J. K. Amoako (✉) · Y. K. Sharma
Shri JJT University, Jhunjhunu, Rajasthan 333001, India
e-mail: johnkani05@yahoo.co.uk

V. Singh et al. (eds.), *Computational Methods and Data Engineering*, Advances in Intelligent Systems and Computing 1227,
https://doi.org/10.1007/978-981-15-6876-3_50

on University of Cape Coast website, that the institution regards the attendance of lecture therefore it requires that each student attends the traditional lecture for a face-to-face interaction with their mates and lecturers up to a number of times before they are graded.

Teacher-centered approach is usually identified with traditional classrooms which encourages students to depend almost entirely on the lecturer [5]. Notwithstanding, evaluations prove that students also embrace the lecture method which is used in teaching and learning [6]. As established in other researches, students who are usually seen as punctual and regular in classes tend to be more successful in their studies than students that do otherwise yet non-attendance at lectures appears to be a growing trend [7–9]. It is further confirmed that high lecture attendance improves class performance [10] due to the benefits associated with it [11–17].

Meanwhile not only "poor lecturing, bad timing of lecture, and part-time working" are often blamed for non-attendance of lectures but also, extrinsically motivated students, thus students who are not acquiring knowledge for its own sake, tend to skip lecture sessions more than students who are motivated intrinsically [18]. Contrastingly, apart from the institutional rules and regulations they have to follow, students attend lectures because they are influenced by a motivated and engaging, lecturer that is inspiring in class [19]. There are few studies that seek to find why non-attendance of lectures is still common with tertiary students in Ghana and how "poor lecturing" contributes to that.

The objective of this current study is to find out the reasons why some undergraduate tertiary students in the selected public universities in Ghana blame non-attendance of lectures on poor lecture experience. Significantly this will reveal or confirm to lecturers why the attendance rate of students deteriorates in the course of the academic year or semester and also creates awareness of poor lecturing as one of the major reasons for non-attendance of lectures by some tertiary students.

2 Methodology

The study adopts a quantitative design approach [20–23]. An initial unstructured face-to-face interview and group discussions were used to engage some of the population units into helping the researchers design a questionnaire which becomes the main instrument for data collection [24]. The major section headings in the questionnaire was; general information, academic information and information on non-attendance of lectures. Data is then collected using a close-ended questionnaire that allowed respondents to select multiple variables which is translated into statistical figures using SPSS V-25, and from which conclusions are drawn. Using cluster sampling technique, the population ($N = 82{,}825$) is made up of 3 universities (Kwame Nkrumah University of Science and Technology (KNUST), University of Ghana students (UG) and University of Cape Coast (UCC)) and 1 technical university—Cape Coast Technical University (CCTU) in Ghana. The selected sample size ($n =$ 800) was influenced by Yaro Tamane's method [25] then a simple random selection

Table 1 Research population made up of students from 4 universities in Ghana

University	Student population	Year of PUBLICATION	Category
KNUST	29,090	2017/2018	Bachelors' Degrees
UG	32,059	2015/16	Bachelors' Degrees
UCC	18,949	2016/2017	Regular undergraduate students
CCTU	2724	2017/18	Entire students
Total	82,825		

method was used to select sample units from the participating clusters [26, 27]. It was the intention of the researchers to involve different schools/faculties in the selected institutions in order to help in generalizing the research conclusions to some extent. Although the population is made up of the top 3 ranking universities who have relatively the largest students' population and faculties/schools [28, 29] and 1 technical university in Ghana, research findings are limited to the participating universities.

In Table 1, the three universities were selected based on their; higher counts of student population, top rankings, higher counts of programs offered, and higher counts of affiliations some other tertiary institutions have with them. Meanwhile, CCTU was selected based on convenience, time, and money constraints to represent the other 7 technical universities in Ghana.

3 Discussion

According to Tremblay et al. [30], the number of students who seek for higher education is on the rise. This in turn puts pressure on governments and tertiary institutions to absorb the student applicants who want to further their education. Leading to the creation of poor students-lecturer ratio in these selected higher educational institutions in Ghana, both lecturers and students are left with not much alternatives but to resort to other ineffective remedies to match the increased students' population (Fig. 1).

The survey therefore reached out to 800 students (respondents) in the selected universities. 437 representing 54.6% of the students responded "Yes" to question 1 leaving 363 respondents representing 45.4% selecting "No" as answer to the question. As a result, only the 437 respondents continued to answer question 2. Whiles they had the chance to select one or more options from question 2, some selected only one option and others selected combinations of options.

The number of students (163–37.3% of the 437) who selected "confusing" was more than all of the other options. This means that based on this study, majority of students who blame poor lecturing as a reason for not attending lectures think that either the course itself is confusing or the lecturing style of the facilitator rather

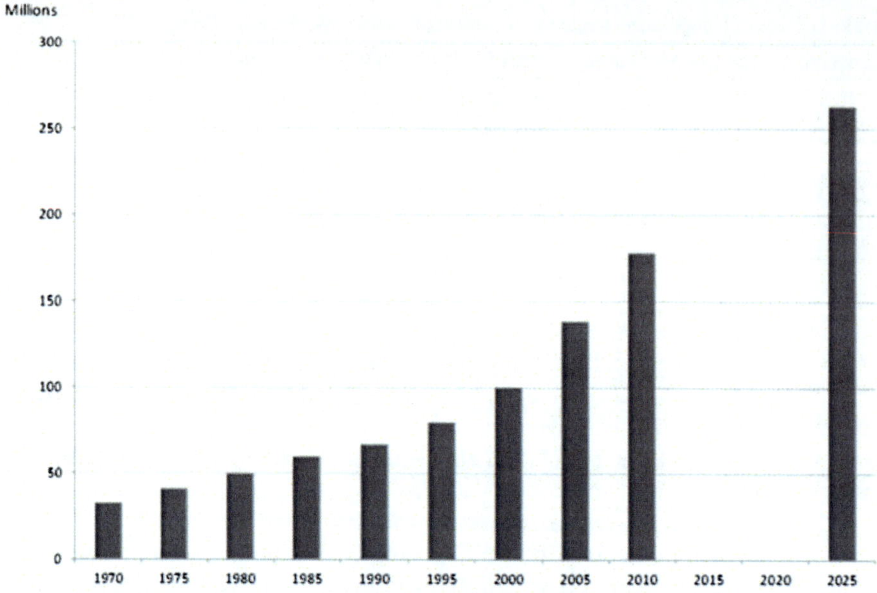

Fig. 1 "Trends in higher education enrolments worldwide, 1970–2025." *Source* [30]

makes the concepts of the course more confusing. Some frequency of students also selected "Other personal issues" as the reason why they do not attend some lectures.

In Table 2, out of the $n = 800$ (100%), 132 (16.5%) respondents were in level 100, 168 (21%) respondents were in level 200, 395 (49.4%) respondents were in level 300, and 105 (13.1%) respondents were in level 400. Regular undergraduate students were mainly used as sample for the study.

Table 2 Levels of undergraduate students from the selected institutions

Level	Frequency	Percent (%)	Cumulative percent (%)
100	132	16.5	16.5
200	168	21.0	37.5
300	395	49.4	86.9
400	105	13.1	100.0
Total	800	100	

3.1 *Questionnaire—Information on Non-attendance of Lectures*

1. *Question 1*

Table 3 represents the responses from the question—Have you ever been absent from a lecture? Out of 800, 363 respondents—45.4% (students) responded "No," whiles 437 respondents—54.6% responded "Yes."

Based on the statistics, it is established that majority of students have at least once absented themselves from class.

2. *Question 2)* What are the reasons why you absent yourself from lecture?

Select any one or more from the options below;

- *Option 1*—Course is impractical
- *Option 2*—Course is boring
- *Option 3*—Course is confusing
- *Option 4*—Course is complex
- *Option 5*—Other personal issues.

This close-ended question which is a follow-up question only for the category of students (437) who selected option 2—"Yes" in *question 1* allows any student who belongs to same to select more than 1 option from the five multiple choice answers in *question 2*. Hence, each respondent has the chance of selecting all or some combinations of the answers: In other words, it is possible for any of the 437 respondents to be included in the frequencies of other options (options 1–5).

In Table 4, which represents the first option for the question asked (question 2), 145 (33.18%) out of 437 respondents selected option 1—"course is impractical" as one of the reasons why they blame non-attendance of lectures on poor lecturing. Here, students feel that the topic being taught has no practical value to impact their field of study. It is important to note that respondents who selected option 1 also had

Table 3 Have you ever been absent from a lecture?

Option	Frequency	Percent (%)
No	363	45.4
Yes	437	54.6
Total	800	800

Table 4 Question 2, option 1—"Course is impractical"

Option	Frequency	Percent (%)
Not applicable (N/A)	292	66.82
Course is impractical	145	33.18
Total	437	100.0

Table 5 Question 2, option 2—"Course is boring"

Option	Frequency	Percent (%)
Not applicable (N/A)	286	65.45
Course is boring	151	34.55
Total	437	100.0

Table 6 Question 2, option 3—"Course is confusing"

Option	Frequency	Percent (%)
Not applicable (N/A)	274	62.70
Confusing	163	37.30
Total	437	100.0

Table 7 Question 2, option 4—"Course is complex"

Option	Frequency	Percent (%)
Not applicable (N/A)	323	73.91
Course is complex	114	26.09
Total	437	100.0

Table 8 Question 2, option 5—"Other personal issues"

Option	Frequency	Percent (%)
Not applicable (N/A)	372	85.13
Other personal issues	65	14.87
Total	437	100.0

the chance to select more additional options, therefore, one respondent can be found in other alternative options (Tables 5, 6, 7, and 8) below.

In Table 5, representing the second option for the question asked (question 2), 151—34.55% out of 437 respondents selected option 2—"course is boring" as one of the reasons why they blame non-attendance of lectures on poor lecturing. This set of students believe that the lecture is not lively enough to be able to capture their attention hence they would rather use that time for another equally important but lively activity.

In Table 6, which represents the third option for the question asked (question 2), 163—37.30% out of 437 respondents selected option 3—"course is confusing" as one of the reasons for non-attendance of lectures. Students claim that certain treated topics are either not well explained to them or by nature not easily comprehensible therefore, in order not to waste their time at the lectures they study on their own through various means like group discussions or educational software.

In Table 7, 114 (26.09%) out of 437 respondents selected option 4—"course is complex" as an option for the non-attendance of lectures. The complexity of the course as claimed by this cohort results in lack of interest in the topic.

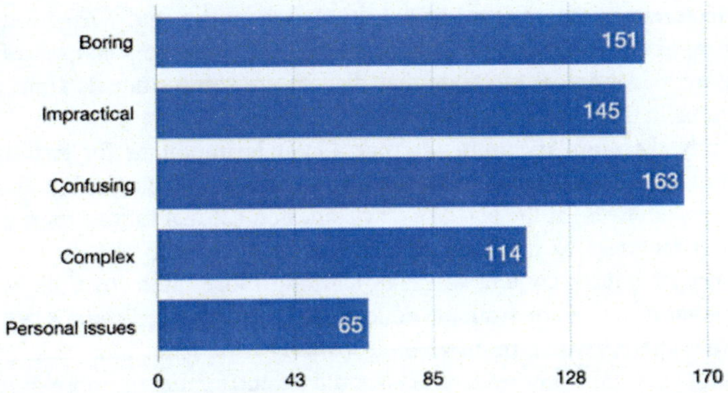

Fig. 2 Bar chart for the responses to question 2

In Table 8, representing the fifth option for the question asked (question 2), 65 (14.87%) out of 437 respondents selected option 2—"Other personal issues" as one of the reasons. This sub-category of students had personal reasons to justify why they sometimes missed lectures, and they were not willing to disclose during the initial unstructured interactions that informed the design of the questionnaire.

Figure 2 represents the number of respondents for each of the various answer options for question 2. It is shown that 163 students out of 437 described the lecture as confusing therefore, they would rather absent themselves from class. This category emerged as the one with the highest frequency whiles the category of students who sometimes do not attend lectures due to other undisclosed personal reasons had the lowest frequency (65 out of 437).

It is important to note that the respective frequencies are not intended to accumulate to the 437 students who responded to question 2. This is because respondents were allowed to select several options and not restricted to only one option as in the case of question 1. For instance, a student could select both option 1 and option 2 as a response while another student could also select options 1–4 as their reasons for non-attendance or only select option 5 as a response.

4 Conclusion

Among the reasons why students absent themselves from lectures, this current research is conducted to explore "poor lecturing" as one of the blames for non-attendance of lecture and also to know their perceptions on "poor lecturing." At the end of the study, it is confirmed that one of the major reasons why some tertiary students in Ghana absent themselves from lectures irrespective of the existing attendance policies implemented by their institutions is due to the poor lecturing style of some lecturers in the selected universities used in the study. Also found in the study, boring, confusing, complex, impractical, and other personal reasons constitute the poor pedagogical methods that discourage some students from attending some lectures in these selected institutions in Ghana.

The study therefore recommends that: (1) It is important for tertiary institutions to find confidential means of assessing lecturers' pedagogical methods (from students' perspectives) at the end of every semester. (2) Institutions must encourage and support lecturers to integrate educational technological aids in their designs and delivery of course content to make learning more interactive, engaging, and student centered. There are available educational software/applications that can help students to personalize and improve their learning experience in a traditional classroom setting [31]. (3) Other researchers should extend similar or improved studies to other tertiary institutions all across Ghana to help reduce non-attendance of lectures by students and also create awareness of the importance of augmenting pedagogy with educational computer applications.

References

1. Walton AJ (2012) Lectures, tutorials and the like: a primer in the techniques of higher scientific education. Springer Science & Business Media
2. Marmah AA (2014) Students' perception about the lecturer as a method of teaching in tertiary institution. View of students from college of technology eduction, Kumasi (Coltek). Int J Educ Res 2(6):601–612
3. Alex K (2019) Distance Education at the grassroots and Assessment procedures. The Case of the University of Cape Coast, Ghana. Creative Educ 10(1):78–96
4. Chenneville T, Jordan C (2008) The impact of attendance polices on course attendance among college students. J Sch Teach Learn, 29–35
5. McKeachie WJ (2007) Good teaching makes a difference—and we know what it is. In: The scholarship of teaching and learning in higher education: an evidence-based perspective. Springer, Dordrecht, pp 457–474
6. Struyven K, Dochy F, Janssens S (2005) Students' perceptions about evaluation and assessment in higher education: a review. Assess Eval High Educ 30(4):325–341
7. Massingham P, Herrington T (2006) Does attendance matter? An examination of student attitudes, participation, performance and attendance. J Univ Teach Learn Pract 3(2):3
8. Moore R (2005) Attendance: are penalties more effective than rewards? J Dev Educ 29(2):26
9. Tinto V (2012) Student success does not arise by chance. Keynote at the what works
10. Clark G, Gill N, Walker M, Whittle R (2011) Attendance and performance: correlations and motives in lecture-based modules. J Geogr High Educ 35(2):199–215
11. Chen J, Lin TF (2008) Class attendance and exam performance: a randomized experiment. J Econo Educ 39(3):213–227
12. Credé M, Roch SG, Kieszczynka UM (2010) Class attendance in college: a meta-analytic review of the relationship of class attendance with grades and student characteristics. Rev Educ Res 80(2):272–295
13. Woodfield R, Jessop D, McMillan L (2006) Gender differences in undergraduate attendance rates. Stud High Educ 31(1):1–22
14. Halpern N (2007) The impact of attendance and student characteristics on academic achievement: findings from an undergraduate business management module. J Further High Educ 31(4):335–349

15. Sauers DA, McVay GJ, Deppa BD (2005) Absenteeism and academic performance in an introduction to business course. Acad Educ Leadersh J 9(2):19–28
16. Paisey C, Paisey NJ (2004) Student attendance in an accounting module–reasons for non-attendance and the effect on academic performance at a Scottish University. Acc Educ 13(sup1):39–53
17. Gump SE (2005) The cost of cutting class: attendance as a predictor of success. Coll. Teach. 53(1):21–26
18. Kottasz R (2005) Reasons for student non-attendance at lectures and tutorials: an analysis. Invest Univ Teach Learn 2(2):5–16
19. Braak D (2015) An exploratory investigation into factors affecting class attendance in a hospitality management module
20. Alzheimer Europe, Research methods. https://www.alzheimer-europe.org/Research/Understanding-dementia-research/Types-of-research/Research-methods
21. Crotty M (1998) The foundations of social research: meaning and perspective in the research process. Sage
22. Spata A (2003) Research methods: science and diversity. Wiley, Incorporated
23. Neuman WL (2014) Basics of social research. Pearson/Allyn and Bacon
24. Williams C (2007) Research methods. J Bus Econ Res (JBER) 5(3)
25. Taro Yamane (1967) Statistics an introductory analysis, 2nd edn. Harper and Row, New York
26. De Vaus DA (1996) Survey and social research. UCL Press Limited, London, England
27. Reynolds NL, Simintiras AC, Diamantopoulos A (2003) Theoretical justification of sampling choices in international marketing research: key issues and guidelines for researchers. J Int Bus Stud 34(1):80–89
28. Webometrics ranking web of Universities. https://www.webometrics.info/en/Africa/Ghana
29. uniRank top universities in Ghana. https://www.4icu.org/gh
30. Tremblay K, Lalancette D, Roseveare D (2012) Assessment of higher education learning outcomes: feasibility study report, volume 1 design and implementation. Organisation for Economic Co-operation and Development, Paris, France
31. Virvou M, Alepis E (2005) Mobile educational features in authoring tools for personalised tutoring. Comput Educ 44(1):53–68

Author Index

V. Singh et al. (eds.), *Computational Methods and Data Engineering*,
Advances in Intelligent Systems and Computing 1227,
https://doi.org/10.1007/978-981-15-6876-3